Smithsonian Field Guide to the Birds of North America

Ted Floyd

Edited by Paul Hess and George Scott

Designed by Charles Nix

Maps by Paul Lehman

PHOTOGRAPHS BY

Brian E. Small with Mike Danzenbaker, Brian Wheeler, Jim Zipp, Kevin T. Karlson, Bill Schmoker, Brian Patteson, Garth McElroy, Robert Royse, Gary Nuechterlein, Jeff Poklen, Les Chibana, and others represented by VIREO, The Academy of Natural Sciences

Collins

An Imprint of HarperCollins Publishers

For Kei, Hannah, and Andrew

Produced by
Scott & Nix, Inc.
150 West 28th Street, Suite 1103
New York, NY 10001-6103

www.scottandnix.com

Published by
HarperCollins Publishers
195 Broadway
New York, NY 10007

FIRST EDITION

The name of the "Smithsonian," "Smithsonian Institution,"
and the sunburst logo are registered trademarks of the
Smithsonian Institution.

Library of Congress Cataloging-in-Publication Data

Floyd, Ted, 1968–
        Smithsonian field guide to the birds of North America /
        Ted Floyd; edited by Paul Hess and George Scott;
        designed by Charles Nix; maps by Paul Lehman;
        photographs by Brian E. Small... [et al.].
            p.   cm.
        Includes bibliographical references and index.
        ISBN 978-0-06-112040-4
        1. Birds—North America.   I. Hess, Paul.
            II. Scott, George.   III. Title.
QL681.F56 2008
598.097—dc22

ISBN 978-0-06-112040-4

15   SCP   10   9   8   7   6

Manufactured in China

HarperCollins books may be purchased for educational, business,
or sales promotional use. For information, please e-mail:

Special Markets Department at
SPsales@harpercollins.com

Frontispiece: American Kestrel by Jim Zipp

Plumage, Molt, and Age Terminolgy

Plumage. The set of feathers that a bird wears at a particular time.

Molt. The process by which one plumage is replaced by another.

Juvenile. The plumage in which a bird fledges.

Immature. Any plumage following juvenile plumage and preceding adult plumage.

Subadult. Any pre-adult plumage; includes juvenile and immature.

Adult. A bird's mature plumage or plumages.

Cycle. Typically annual pattern of plumages and molts.

Wear (Worn). Abrasion or bleaching of feathers; distinct from molt.

ABA Codes

The American Birding Association (ABA) maintains a checklist of all birds recorded in the ABA Area, defined as the region of North America north of Mexico, which includes the continental United States, Canada, the French Islands of St. Pierre et Miquelon (off the coast of Newfoundland), and offshore waters to 200 miles. The list includes both naturally occurring species and established exotic birds. The status of each species is summarized in the ABA Checklist, as follows:

Code 1. Occurs widely and regularly in North America.

Code 2. Regular within a restricted region of North America.

Code 3. Rare but regular; occurs annually, usually in small numbers.

Code 4. Casual; well-defined pattern of occurrence, but not annual.

Code 5. Accidental; one or a few records with no defined pattern.

Code 6. Extinct or otherwise impossible to observe in the wild.

The current assigned ABA Code is included for each species in the main accounts throughout the *Smithsonian Guide* and in the species checklist.

Smithsonian Field Guide
to the Birds of North America

Contents

Introduction

Birds are among the most conspicuous forms of life on earth. American Robins and Northern Mockingbirds flourish in many of the largest cities in North America. Song Sparrows inhabit suburban yards, Great Blue Herons prowl at the edges of farm ponds, and Red-tailed Hawks scan from utility poles along Interstate highways. Birds seem to be everywhere, in the hottest deserts, atop the highest mountains, on distant islands, and far out at sea.

Birds are all around, but they never seem quite ordinary to us— they capture the human imagination perhaps more than any other creature. Most birds fly, and their flight is a symbol of freedom. Birds sing richly and gloriously, and some birds make astonishing annual migrations. The Sooty Shearwater, a common visitor to North American waters, may travel forty thousand miles in the course of a single year— nearly twice the circumference of the planet.

Every bird has its own story. The graceful Bonaparte's Gull builds its nest high in a spruce tree in the boreal forest. The American Dipper, fantastically adapted for swimming and diving in cold mountain streams, has its own story. So does the Red Crossbill, a complex of different populations that wander widely in search of pinecones. So does the Burrowing Owl, which stands guard over prairie dog colonies.

A Great Egret courts a mate with a magnificent display of specialized feathers called aigrettes—one of the countless sophisticated behaviors that make birds such a pleasure to observe. LOUISIANA, MAY

We are drawn to birds and their stories because they provide us with immediate access to the workings of nature. It takes only a few minutes outdoors to discover that birds inhabit a realm of diversity and complexity, of fecundity and commotion. Birds do things. Their behaviors are sophisticated, their populations are dynamic. Whether in the company of Carolina Chickadees outside the kitchen window or in the midst of Murphy's Petrels on the ocean, we are attracted to birds because they draw us into the fascinating world of nature.

We are also drawn to an abiding question: What is the name of that bird? A name is a tool for organizing our thoughts, for making sense of the world around us. Knowing the name of something makes it more important. Giving a name to something immediately triggers a cavalcade of questions, of discovery, and of wonder.

Identifying Birds

Bird identification may seem simple to observers with experience, but to the novice, it often seems more like wizardry. Even in the case of straightforward identifications, the observer is sometimes called upon to discriminate among physical features that are much less than an inch long—briefly glimpsed features that more often than not are obscured by shadow, distance, and motion.

How do experienced observers do it? It has often been remarked that the fundamental problem of bird identification was brilliantly and convincingly solved by Roger Tory Peterson in his 1934 *A Field Guide to the Birds*. But that is not the whole story. Peterson did not merely solve the problem of field identification; he also *conceived* it. Amateur bird study in the early 20th century had blended biology and sentimentalism in a manner that more resembles what we now call nature writing than what we now recognize as identification guides. Peterson created a bare-bones manifesto with an audacious agenda: to empower the user to quickly and efficiently name every bird. His field guide used the simplest possible combination of characters to uniquely identify each species. The Yellow Warbler, a fabulously complex marvel of physiology and behavior, is described in Peterson's 1934 formulation simply as the only small bird that is entirely yellow. The Chipping Sparrow, a more complex identification problem, is described as having only three important "field marks"—rufous cap, white line above the eye, and black line through the eye. That was all one needed to name it. Applied to hundreds of birds in his guide, the "Peterson System," as it still is known today, instantly caught on and helped popularize birding as an outdoor activity across North America.

Peterson's system continues to enjoy acceptance. Over the 75 years since the publication of his first guide, there have been modifications to the method in subsequent editions of his own books and in other field guides. *The Golden Guide* (1966) covered all of North America, and the *National Geographic Field Guide* (1983) set a new standard of accuracy. *The Kaufman Field Guide* (2000) employed computer-enhanced photography, and *The Sibley Guide* (2000) was instantly praised for its combination of extreme thoroughness and user-friendliness.

The *Smithsonian Guide* has two special emphases that reflect emerging trends in the field identification of birds. The first is a focus on natural variation within and among species, and the second is a "holistic" view of the bird as the sum of its behavioral, ecological, and morphological parts. This field guide includes a suite of information for each species. Descriptions are keyed on variation in appearance, behavior, habitat, ecology, molt strategy, and voice. Multiple photographic images depict the birds in their natural surroundings. Detailed maps show birds' ranges across the continent. A DVD of birdsong is included to assist in identifying birds by the sounds they make throughout the year and from region to region. The goal of this "holistic" approach is to equip users with the best possible tools to identify birds simply and enjoyably in the field.

COVERAGE

The *Smithsonian Guide* includes information for all bird species that occur regularly within the "ABA Area," as defined by the American Birding Association. This area consists essentially of Canada, Alaska, and the lower 48 states, and includes ocean waters up to 200 miles away from the mainland, plus the French islands of Saint Pierre et Miquelon off the eastern coast of Canada. Hawaii, Puerto Rico, and Greenland are not considered part of the ABA Area, and therefore are not treated in the *Smithsonian Guide*.

The American Birding Association Checklist currently recognizes 730 species as being of regular occurrence within the ABA Area. Of this total, 492 species are recognized by the ABA as widespread and regular, 169 as range-restricted but regular, and 69 as rare but annual. These categories are designated in the ABA Checklist as Code 1, Code 2, and Code 3, respectively.

Species not covered in full in the *Smithsonian Guide* are those designated by the ABA Checklist as casual (Code 4), accidental (Code 5), or extinct or impossible to find in the wild (Code 6). The ABA recognizes 92 species as being of casual occurrence, 108 as accidental, and 9 as extinct or impossible to find in the wild. An example of a Code 4 species is the Eurasian Curlew; it does not occur annually in North America, but there are enough records (mainly from the California coast) to indicate a pattern of casual occurrence. An example of a Code 5 bird is the Collared Plover; there is a single North American record (a stray to Texas) of this tropical species. And an example of a Code 6 species is the Eskimo Curlew, formerly common in North America but now presumed extinct.

The exact placement of species into the six categories of the ABA Checklist is in some cases an uncertain business, but the system is widely admired for its clarity and rationality. In borderline instances, the *Smithsonian Guide* does include Code 4 species, giving them limited treatment. Examples include Red-tailed Tropicbird (treated within the Red-billed Tropicbird account) and the Blue-footed and Red-footed Boobies (both treated within the Brown Booby account).

Text regarding such species is set apart from full species accounts in the *Smithsonian Guide* by a yellow-shaded background. In addition to naturally occurring Code 4 species currently on the ABA Checklist, certain non-native species ("exotics") not on the ABA Checklist are also indicated by a yellow-shaded

CODE 1: REGULAR OCCURRENCE IN THE ABA AREA. *Greater Yellowlegs has an extensive range across North America.* CALIFORNIA, SEPTEMBER

CODE 2: RANGE-RESTRICTED BUT REGULAR IN THE ABA AREA. *Mountain Plover occupies a smaller geographic range and is sometimes difficult to find.* CALIFORNIA, NOVEMBER

CODE 3: RARE BUT ANNUAL IN THE ABA AREA. *Sharp-tailed Sandpiper is recorded in small numbers every year in North America.* CALIFORNIA, MARCH

CODE 4: CASUAL VISITOR TO THE ABA AREA. *Eurasian Curlew is not recorded annually in the ABA area.* EUROPE, JANUARY

FLORIDA, FEBRUARY

Birds, like all living things, are arranged by scientists into groups. Birds make up the large class Aves, and are subdivided into orders, families, genera, and species. For example, Reddish Egret is part of the class Aves, order Ciconiiformes, family Ardeidae, genus Egretta, and classified as the species rufescens. The scientific name of Reddish Egret is Egretta rufescens.

background. Examples include Purple Swamphen and Orange Bishop—exotic species with established populations in North America that are not on the ABA Checklist.

GROUP ACCOUNTS

Most of this book's pages are accounts of the 730 bird species that occur regularly in North America. Beginning with the Black-bellied Whistling-Duck and ending with the Eurasian Tree Sparrow, they are arranged in a fashion that closely—but not strictly—follows the linear sequence adopted by the American

TEXAS, FEBRUARY

TEXAS, APRIL

Wing length is an excellent and underutilized field mark for many species that do not usually soar. Orange-crowned (left) and Yellow Warblers (right) are identical in body length (both 5 inches from bill tip to tail) but the Yellow Warbler shows a longer wing length that may correspond to its more than ten percent greater wingspan. Wing length can translate into a significant distinction with birds on the ground or in the bush. Yellow Warblers look notably long-winged in comparison with Orange-crowned Warblers—an important point of separation on confusingly marked females and males in the fall.

CALIFORNIA, JANUARY

CALIFORNIA, JANUARY

Even though American Robin (left) and Surfbird (right) are both on average 10 inches from bill tip to tail tip, the robin weighs 2.7 ounces compared to the 7-ounce Surfbird, more than twice the mass. This is consistent with most observers' impressions of significant size differences between the two species. The difference in wingspan between the two is also significant—the Surfbird's is 26 inches compared to the 17-inch wingspan of American Robin.

Ornithologists' Union (AOU). Embedded within the linear sequence are clusters of related species that are similar in many ways. Accordingly, the text describing these clusters begins with a group account that provides an overview of habitat, behavior, diet, population status, and conservation status. Group accounts also include statistics on the number of species per group in North America, broken down by their ABA codes. Several group accounts contain mini-essays on various topics concerning identification, conservation, and ecology to supplement information about particular groups. For example, the group account for Larks (Alaudidae) includes a brief essay about subspecies. These topics are included in the table of contents.

SPECIES ACCOUNTS

Group accounts are followed by individual species accounts that provide information in standardized locations on the pages of the book for ease of use.

Names

Each species has a standard English name, assigned by the AOU, endorsed by the ABA, and also employed in this field guide. Each species also has a scientific name, the "official" designation recognized by ornithologists worldwide.

Measurements

Measurements for each species include body length (L), wingspan (WS), and body weight (WT). Body length is measured in inches from bill tip to tail tip. Wingspan (measured from wing tip to wing tip) is useful not only for identifying birds in flight; it can also be used to gauge the relative lengths of the folded wing on perched birds. Weight can be an excellent indication of a bird's volume or bulk. It is given in pounds and ounces;

for birds that weigh on average a single ounce or less, equivalent weights are given in grams. This is useful for comparing the mass of smaller birds. In addition, when males average visibly larger than females or vice versa, symbols indicate this: ♂ > ♀ or ♀ > ♂.

The three measurements reported in the *Smithsonian Guide* are best used in concert. Together, they can help to create important impressions such as "sleek" or "squat," "top-heavy" or "pot-bellied." The measurements reported in this field guide are averages. They vary among populations, among individuals within a population, and even within a single individual over the course of a year.

Natural Variation

Three summaries of variation for each species are indicated by bullet points in the following order.

Molt. The number of adult molts per year is given, followed by molt strategy, if known. Molt is one of the major sources of variation in birds.

Sex-related, age-related, and seasonal differences. In general, differences are acknowledged in the text if they are strong or moderate; if no difference is mentioned, it can be assumed to be weak or nonexistent.

Other Variation. Geographic variation across the range of a species and individual variation within a region are described here, if useful for identification.

Habits and Ecology

Habitat, behaviors, and habits through the seasons, including feeding, mating, nesting, and other key facts, are described for each species.

TEXAS, MAY

TEXAS, NOVEMBER

The Laughing Gull offers a good example of age-related variation: black, white, and gray as an adult (left) and grayish with a subdued pattern in its first winter of life (right).

"Thick-billed" Fox Sparrow. CALIFORNIA, OCTOBER

"Slate-colored" Fox Sparrow. OREGON, APRIL

"Sooty" Fox Sparrow. OREGON, OCTOBER

"Red" Fox Sparrow. NEW JERSEY, DECEMBER

The Fox Sparrow is among the most dramatically variable of North American passerines (perching birds), with four distinct groups as well as intergrades between populations where they overlap. Some scientists consider the four major groups to be separate species.

MAINE, JUNE

NEW JERSEY, NOVEMBER

The White-throated Sparrow is a classic example of variation within a species. It has a bright, white-striped morph (left) and a much duller, tan-striped morph (right) that do not represent separate subspecies. The two morphs are found together throughout their range.

Vocalizations

The accounts describe major vocalizations and occasionally a bird's non-vocal sounds. Three primary classes of vocalizations are routinely referred to.

> *Song.* Long-distance vocalization that is sometimes complex and generally given in connection with courtship and territory defense.

> *Call note.* Typically a simple vocalization and often given during ordinary daytime behaviors, such as foraging, and aggressive interactions.

> *Flight call.* Usually given in flight and often quite distinct from the call note. Flight calls can be a tremendous aid in identification.

Not all vocalizations fall neatly into these three categories, and the treatment of bird sound in the *Smithsonian Guide* is accordingly flexible. It is important to keep in mind that songs in particular differ greatly among individuals, and so can call notes and flight calls. But the preceding caveat is often taken to an unreasonable extreme—namely to the conclusion that it is impossible to learn to identify birds by their songs and calls. Not at all! There are consistent patterns and diagnostic acoustic signatures for almost all species, and bird sound is a vital component of the identification process. In many situations, the birds are more readily identified by sound than by sight.

For about one in every five main species accounts, a symbol ◀ appears before the vocalization description text. The symbol indicates that MP3 audio files for this species are available on the DVD that accompanies the *Smithsonian Guide*. (See page 491 for more information about the birdsong and vocalization recordings on the DVD.)

Range Maps

A range map accompanies each main species account in the *Smithsonian Guide*. Breeding range is indicated in green ■; winter range is indicated in blue ■. Where the breeding and wintering ranges overlap, they are indicated in purple ■. Migratory routes are indicated in orange ■, and regions of rare occurrence are indicated in yellow ■.

Even the most versatile of species do not occupy every single available habitat within their indicated breeding, wintering, and migratory ranges. Birds are sometimes found away from their regular ranges in areas not indicated on the maps. The map information for each species' distribution is based on the knowledge of Paul Lehman, an expert on the distribution of North American birds, and regional experts across the continent.

Photographs and Caption Text

Most species in the *Smithsonian Guide* are illustrated by two or more photographs, the first most often showing the bird in its "easiest" plumage or pose. The main text of each photo caption describes key features—conventional field marks and behavioral and structural features—that are in most cases clearly present in the image. Each caption concludes with the location and month the image was taken. Bird populations are naturally variable, and this information establishes a context to the time of year and where the bird might be observed.

Natural History of Birds

Understanding where birds are, what they do, and what they look like are key elements in becoming a better birder. Matching up a bird to an image in a book is an excellent beginning, but learning more about each element in the natural history of birds will make birding a more enriching and enjoyable experience.

The Chestnut-sided Warbler is a characteristic species of second-growth deciduous woodlands, primarily in the northeastern U.S. and across much of southern Canada. TEXAS, APRIL

WHERE TO LOOK FOR BIRDS

Birds are influenced more than anything else by habitat type, both within landscapes and across landscapes. An old hayfield in Virginia is full of Eastern Meadowlarks and Grasshopper Sparrows, but a woodlot just a quarter mile distant is home to Wood Thrushes and Red-eyed Vireos. Birds choose coarse-level habitat types, but they also recognize distinctions *within* habitat types. Training oneself to recognize these fine-scale differences can greatly assist in identification. The overview that follows is necessarily selective, highlighting only a handful of the hundreds of fine-scale habitat differences that influence bird community structure.

Forests and Woodlands

Small vs. Large Tracts of Eastern Deciduous Forest. Even though two tracts of woodland may have the same tree species composition, bird communities may vary substantially with tract size and distance to the edge of the tract. Species such as Worm-eating and Kentucky Warblers are usually restricted as breeders to the interior of forest tracts that are at least 100 acres; they are generally absent from smaller tracts, as well as from the edges of large tracts. In contrast, "edge-tolerant" species, such

as Least Flycatchers and Black-capped Chickadees, routinely nest along woodland edges, as well as in the forest interior.

Western Montane Forests. As you drive along a Forest Service road in the western mountains, it looks and smells like "pine trees" all the way up. Actually, the composition of these conifer forests changes on the way up, with dramatic effects on birdlife. The drier, lower slopes in much of the interior West are dominated by pinyon pines and junipers, home to Western Scrub-Jays, Bushtits, and Black-throated Gray Warblers. At middle elevations, ponderosa pines take over, and so do Plumbeous Vireos, Western Tanagers, and Dark-eyed Juncos. Higher up, spruces and firs dominate, and so do Golden-crowned Kinglets, Hammond's Flycatchers, and Evening Grosbeaks. In traversing less than a vertical mile of conifer forest, you have visited at least three highly distinct bird communities.

Forest Health and Forest Bird Communities. Fires, drought, and insect outbreaks are important determinants of avian diversity and density, especially over the course of a few years. Black-backed and American Three-toed Woodpeckers are classic examples of species that quickly colonize recently burned forests. Yellow-billed Cuckoo numbers fluctuate annually in response to population surges of hairy caterpillars that attack broadleaf trees, and numbers of Cape May Warblers fluctuate as a function of outbreaks of the spruce budworm. Variation in cone crops, often influenced by precipitation, causes Red Crossbills to disperse great distances in their pursuit of conifer seeds.

Whether in its breeding range in the north or on its wintering grounds in the south, the Sprague's Pipit is best sought in its typical microhabitat within grasslands. NORTH DAKOTA, JUNE

Prairies and Meadows

Regional Differences. From the vantage point of a traveler along an Interstate highway, central North Dakota and eastern Colorado present similar vistas: grass, and a lot of it; and sparrows. But not the same sorts of sparrows. The tallgrass prairie outside Bismarck is home to breeding Clay-colored Sparrows, Nelson's Sharp-tailed Sparrows, and Chestnut-collared Long-spurs. In contrast, the shortgrass prairie just north of Denver supports breeding Cassin's Sparrows, Brewer's Sparrows, and McCown's Longspurs. Across large spatial scales, grassland bird communities exhibit major differences.

Microhabitat Differences. During the summer months, birders visit central South Dakota for encounters with two specialty species: Sprague's Pipit and Baird's Sparrow. And during the winter months, birders visit southeastern Arizona for the exact same reason: Sprague's Pipit and Baird's Sparrow. In both venues, the two species segregate into different microhabitats: Baird's Sparrows flush from dense clusters of tall grasses, whereas Sprague's Pipits walk quietly in short grasses along the edges of patches of open dirt. The birds

may well occur within a few hundred feet of each other, but they nonetheless observe microhabitat differences—an essential piece of knowledge for the seeker of these two hard-to-find species.

Variation in Rainfall. Grasslands are naturally susceptible to variation in rainfall, and this variation is reflected in the natural histories of several grassland bird species. A good case study is the Cassin's Sparrow, whose numbers vary greatly on both annual and monthly time scales, both within landscapes and across major regional boundaries. Sometimes, the relationship between precipitation and Cassin's Sparrow activity is clear, as when the species commences breeding during the monsoon season in the Desert Southwest. In other instances, the relationship is less clear, and somewhat conjectural, as when local density is thought to be influenced by precipitation months earlier.

Deserts and Shrublands

Sonoran Desert Cactus Gardens. In many
people's eyes, the columnar cactus stands
of south-central Arizona are the epitome
of the desert. Such stands are just about the
most productive of North American deserts, home to Harris's Hawks and Elf Owls,
Gila Woodpeckers and Gilded Flickers,
Brown-crested Flycatchers and Verdins,
Phainopeplas and Varied Buntings, and
many others. The Sonoran Desert receives
rain in both the winter and the summer,
resulting in a plant community that is
diverse both in terms of species composition and in vertical structure. The Sonoran
Desert may be the epitome of the desert,
but it is not at all austere and lifeless.

Chihuahuan Desert Creosote Bush Flats. The
Chihuahuan Desert, extending from eastern Arizona to central Texas, is immense.
And it is expanding, apparently in response
to overgrazing and climate change. Much
of this desert is characterized by near-
monocultures of creosote bush, home to
Black-throated Sparrows and few others.
Away from playas and bajadas dominated
by creosote bush, however, the Chihuahuan Desert comes to life. Dry washes and
even roadsides support breeding Black-
chinned Hummingbirds, Black-tailed
Gnatcatchers, Crissal Thrashers, Pyrrhu-
loxias, Canyon Towhees, Rufous-crowned
Sparrows, Scott's Orioles, and many others.

Great Basin Sagelands. The deserts of the
Great Basin are among the most austere
habitats in North America. Very little
precipitation falls, most of it in the form of
light snow in the winter months. Sagebrush is the defining component of the
Great Basin, a fact reflected in the names
of some of the most characteristic birds

The Sage Thrasher is an ecological specialist of Great Basin sagelands,
a favored habitat that is rapidly disappearing. CALIFORNIA, JUNE

of this region: Greater Sage-Grouse, Sage
Thrasher, and Sage Sparrow. Great Basin
sagelands have recently been devastated
by unnaturally immense wildfires, and the
entire ecosystem is threatened with conversion to unproductive alien cheatgrass.

Alpine and Arctic Tundra

High-elevation Alpine Tundra. Atop the high
peaks of the Rockies and the Sierra Nevada,
the naturally fragmented alpine tundra
biome is home to a small but distinctive assemblage of breeding bird species:
White-tailed Ptarmigan, American Pipit,
rosy-finches, and a few others. A little-
appreciated aspect of the alpine tundra
is that its avifauna swells in late summer,
with a regular incursion of post-breeding
dispersers from lower elevations: Sage
Thrashers, Canyon Wrens, and even Baird's
Sandpipers on migration to South America.
Historically, the alpine tundra has been
one of the least threatened habitat types
in North America, but there is recent and
worrisome evidence that global warming
might greatly diminish the extent of this
habitat type.

Some of the highest elevations on North American mountains are home to the Brown-capped Rosy-Finch. NEW MEXICO, DECEMBER

Wet Arctic Tundra. This is the tundra habitat of nature documentaries and coffee table books: an endless expanse of flat, low-lying lagoons and estuaries. And it is, indeed, a wonderfully diverse place in the summer months, home to breeding Long-tailed Ducks, Yellow-billed Loons, White-rumped Sandpipers, Red-necked Phalaropes, Sabine's Gulls, and many others. Where the sedges give way to alders and small birch trees, terrestrial species such as Bluethroat and Gray-cheeked Thrushes are found. The wet arctic tundra is justly famous for its frenetic pace of life: courtship, incubation, and care of young are literally round-the-clock affairs, all compressed into a period of less than three months.

Dry Arctic Tundra. Away from the low-lying coastal districts, the Arctic is drier and rockier where bird communities are distinctive in many ways. Depending on geography, Willow Ptarmigans give way to Rock Ptarmigans, Pacific Golden-Plovers yield to American Golden-Plovers, Black Turnstones are replaced by Surfbirds, and Pectoral Sandpipers switch places with Buff-breasted Sandpipers. Passerines include Northern Wheatears, American Tree Sparrows, Snow Buntings, and Hoary Redpolls. And in the most desolate of dry tundra habitats— the arctic cliffs and mountains—rare and declining Ivory Gulls are found.

Wetland and Aquatic Habitats

Riparian Woodlands. Throughout the West—and especially in the desert Southwest—seemingly insignificant streams and seeps provide sufficient moisture for cottonwoods, ashes, and other broadleaf trees. The attendant birdlife is remarkably diverse: Common Black-Hawks, Yellow-billed Cuckoos, Vermilion Flycatchers, Thick-billed Kingbirds, Bell's Vireos, Lucy's Warblers, Summer Tanagers, Blue Grosbeaks, and a host of others. Western riparian habitats are highly threatened by development and cattle grazing.

Swamp Forests. Extensive swamps, consisting of standing water amid tall trees are found primarily in the Southeast. Anhingas soar overhead, while Black-bellied Whistling-Ducks and White Ibises perch on snags. Prothonotary Warblers sing from thickets at the water's edge, and Acadian Flycatchers watch from overhanging branches. By night, the swamp comes alive with the clamor of Barred Owls, the wailing of Limpkins, and the squawking of night-herons. Despite their superficially pristine character, southeastern swamp forests have been ecologically ravaged; virtually all of the original timber was cleared, and essentially everything that remains is second growth.

Boreal Bogs. Across much of southern and central Canada and extending south into the Great Lakes and Appalachian regions, spruce and fir forests are dotted with waterlogged clearings that provide homes to some of the most characteristic bird species of the boreal forest biome: Yellow-bellied and Alder Flycatchers, Northern Waterthrushes and Connecticut Warblers, and others. The extent and quality of boreal bogs are influenced by everything from logging operations to beaver activity, with the result being that the attendant birdlife is dynamic and often unstable.

Rivers, Lakes, and Ponds. Only a few birds—for example, some grebe species—are able to build nests on open water away from emergent vegetation. But the open water of rivers, lakes, and ponds provides critical foraging grounds for a large array of species. The wintertime distribution of waterfowl and gulls is often governed by the availability of ice-free, fish-filled open waters including estuaries. In summer, Barn Swallows skim the surface for insects, American White Pelicans hunt for fish in the shallows, and American Coots pull out vegetation and crustaceans.

Freshwater Marshes. These highly productive habitats are characterized by emergent grassy vegetation—especially cattails. Passerines and nonpasserines alike are dependent upon freshwater marshes: Marsh Wrens, Swamp Sparrows, and Yellow-headed Blackbirds; Common Moorhens, Black Terns, and Franklin's Gulls. In much of the East, large marshland complexes are clearly preferred to small cattail stands along highways or on golf courses. In the West, however, this preference breaks down, and truly diminutive wetland patches frequently harbor such species as Pied-billed Grebes, Virginia Rails, and Soras.

Saltwater Marshes. Especially characteristic of the Atlantic and Gulf coasts, tidal marshes dominated by cordgrass (genus *Spartina*) are naturally susceptible to flooding, imperiled by development, and

The Common Moorhen flourishes in freshwater marsh habitats, where its food is plentiful. TEXAS, MAY

home to a diverse breeding avifauna. Typical breeding species in East Coast saltwater marshes include the Glossy Ibis, Clapper Rail, American Oystercatcher, Willet, Laughing Gull, Black Skimmer, and Seaside and Saltmarsh Sharp-tailed Sparrows. Coastal saltwater marshes also occur along the Pacific coast but are smaller overall, largely replaced by rocky headlands that quickly give way to terrestrial habitats.

Ocean Beaches. No habitat is more unstable and dynamic than the interface between ocean and land. Few species breed right on the ocean beaches, and the few that do—for example, Piping Plovers and Least Terns—face dire threats from development, disturbance, and projected rises in sea level. Nonbreeding visitors, however, flourish along beaches. Sanderlings chase after the waves, gulls poke around for detritus, and herons fish in the surf. Along the West Coast, a specialized guild of "rockpipers"—among them Black Oystercatchers, Surfbirds, Black Turnstones, Wandering Tattlers, and Rock Sandpipers—gather in the most violent of coastal microhabitats: surf-spattered, seaweed-slick rock outcroppings at the edge of the beaches.

The Black Oystercatcher is one of the most spectacular birds on rocky West Coast shores. CALIFORNIA, OCTOBER

Open Ocean. Beyond the breakers and beyond the jetty tips lies the vastest habitat on earth: the open ocean, or pelagic zone. It is tempting to say that it all looks the same, but the open ocean consists of varied habitats. Water depth, water temperature, salinity, and seabed topography are key determinants of the availability of marine foodstuffs, with corresponding influences on the occurrence and distribution of pelagic bird species. Off the East Coast, for example, the numbers of warm-water Audubon's Shearwaters vary predictably with sea surface temperature. Off the West Coast, the northward dispersal of Black-vented Shearwaters is reliably governed by annual variation in ocean temperature. After a few trips to sea, the birder realizes that the ocean is as complex and dynamic as forests, grasslands, and deserts.

Human Habitats

Certain places, such as large cities and suburban tracts, are more obviously influenced by humans than others. But the influence of humans is ubiquitous. The early twenty-first-century avifauna of every single major habitat type in North America is substantially different from what it was just a few human generations ago. There are familiar examples: the establishment of Cattle Egrets, Eurasian Collared-Doves, and House Finches; the recovery of Wood Ducks, Snowy Egrets, and Bald Eagles; the ascendancy of Canada Geese, Anna's Hummingbirds , and Great-tailed Grackles; the decline of Mountain Plovers, and Henslow's Sparrows; and the extinction of Eskimo Curlews, Ivory-billed Woodpeckers, and Bachman's Warblers. All of the preceding are widely acknowledged to be linked to human activity. And there are many more examples—entailing hundreds of species—of habitat shifts, range expansions and contractions, behavioral modifications, and other changes that are hypothesized to be related, directly or indirectly, to human activities. The verdict is not quite in, but everything from expanding populations of Northern Fulmars to declining numbers of Band-tailed Pigeons are plausible candidates for influence by humans. The challenge, really, is to identify a bird species or avian habitat that has not been influenced by humans.

BIRD BEHAVIOR

Bird behavior is directly relevant to the mastery of bird identification. In observing how birds fly, how they find food and mates, and how they behave in many other ways, birders become more skilled at observation. Many behaviors are themselves field marks, as essential to the identification process as plumage colors and patterns.

Food and Foraging

The diversity of birds is equaled by the almost endless variety of their diets and feeding behaviors. Some birds are strictly carnivorous, others vegetarian, and some are opportunistic omni-

Wilson's Warblers typically forage for food in mid-level vegetation rather than on the ground or in the tree canopy as many other species do.
CALIFORNIA, APRIL

vores, feeding on what is most readily available. When birds are hungry, they may let their guard down and be less vigilant, thus allowing observers to get a longer and closer look at them.

When and how they are feeding are also excellent clues to identification. Different species of warblers tend to feed at different levels of trees—some of them in the canopy, some of them on the forest floor, and still others in the midstory. A few warblers favor aquatic habitats along streams and ponds. All eat insects, but all tend to favor particular areas.

Some birds, such as kinglets and waxwings, hover and chase insects on the wing. Woodpeckers and nuthatches probe tree bark to look for insects and larvae. Wading birds use their specialized bills to poke into the mud in search of crustaceans and other small animals. Some birds of prey perch quietly at a vantage point to scan open areas looking for smaller birds.

It is not usually the case that foraging behavior is sufficient alone to clinch an identification. But it is very often the case that foraging behavior provides a strong first hint as to the identity of a poorly glimpsed bird.

Bird Vocalizations

Nearly all birds make some sort of vocalization as a means of communication. They vocalize to mark a territory, to attract a mate, to alarm other birds to potential predators, and sometimes simply because they are in flight. How and why birds vocalize is amazingly complex. For birders, however, learning to recognize the specific songs, call notes, and flight calls of birds can open a door to a new level of experience in the field. It is often the only way to identify a bird that is hidden or too far away to be observed by sight. For example, the swamp forests of the southeastern United States are dense with vegetation. The birds in this habitat seem forever out of view. But they can be gratifyingly vocal. The Carolina Wren, Swainson's Warbler, Summer Tanager, and many others are readily detected as soon as their vocalizations are learned.

In swamp forests of the Southeast and elsewhere, the "dawn chorus" of warbles and trills, of chirps and tweets, can seem overwhelming, but patterns start to emerge. The piping, pounding, repetitive song of the Carolina Wren provides a good basis for comparison with the similar songs of other species, for example, the Tufted Titmouse, Kentucky Warbler, and Northern Cardinal. The herky-jerky utterances of the Summer Tanager resemble those of other "robin-like" songsters: Warbling Vireo, Rose-breasted Grosbeak, and American Robin. And the pure-toned proclamations of an "invisible" Swainson's Warbler invite comparison with the ringing songs of Louisiana Waterthrushes and Yellow-throated Warblers.

Bird songs can be limited to certain times of the day and year, but most birds give call notes all day long and all year long. Moreover,

The Carolina Wren sings loudly and frequently, and its diverse vocalizations provide a good basis for comparison with other species. TEXAS, APRIL

call notes are less variable than songs. Many of the songs of Carolina Wrens are easily confused with those of other species, but the twangy alarm call of the species is unique. A Swainson's Warbler in a nearby bramble may refuse to sing, but the bird can usually be depended upon to give its loud, rich call note. And the disjointed singing of a distant Summer Tanager is easy to lose in the bedlam of other treetop singers, but the explosive, machine-gun-like call note of the species rarely escapes notice.

Recognizing the different vocalizations of birds is one of the most challenging skills to acquire. It takes patience and concentration, but it can be one of the most rewarding and pleasurable parts of birding.

Courtship and Breeding

Closely related to bird vocalization is the topic of courtship and breeding. Songs are an integral part of courtship, but courtship often involves elements other than song.

In March and April in the agricultural districts of the Intermountain West, nearly every bird species is engaged in some sort of courtship display. Sandhill Cranes pair up and perform synchronized dances. High overhead, Wilson's Snipes spread their tail feathers far apart, creating an eerie winnowing sound. A pair of Cinnamon Teals lands in an irrigation ditch, greeting each other with exaggerated head-bobbing. On a farm pond, a drake Common Goldeneye points its bill straight up and utters a harsh, nasal call; even though this bird and its mate do not breed here, they have already established a pair bond. Also at the pond, a male American Avocet lowers his head and charges at a receptive female. And everywhere, there are displaying male Red-winged Blackbirds, their flaming red epaulettes puffed out to maximum extent.

After courtship comes breeding, an activity that is time-consuming but rewarding to observe. Most birds hide their nests from view, and they tend to interact with their eggs and young in a way so as to avoid the notice of potential predators. But breeding bird behavior can often be observed indirectly: a female Red-winged Blackbird carrying a mouthful of food back to a concealed nest in a patch of cattails; a Wilson's Snipe feigning injury to lure an intruder away from a nest; and an American Avocet opting for the more dramatic method of flying straight at intruders, yelping shrilly all the while. Please note that breeding birds are highly susceptible to disturbance by

Red-winged Blackbirds flare their dazzling shoulder patches in both territorial and courtship displays. TEXAS, MARCH

humans, and that observations during the breeding season should be conducted as unobtrusively as possible.

Flight and Migration

Billions of birds make annual migrations in North America from southern wintering areas to summer habitats in the north. Some travel relatively short distances of only a few hundred miles or less, while others disperse thousands of miles to find suitable places for feeding and nesting. Many warblers spend the winter in Central and South America and on islands of the Caribbean and fly north to central Canada to mate and nest throughout the summer months. Not all birds migrate. Some remain in areas that provide all they need to survive year-round.

Certain locations across the continent allow observers to witness thousands of birds in a single day moving together in groups. In the fall, Monterey Bay off California offers birders an opportunity to see hundreds of thousands of migratory shearwaters. High Island, Texas, is an ideal spot in the spring and fall for warblers and other songbirds, and even urban parks, such as Central Park in New York City, can give birders a chance to see dozens of different species in a single area as

The Common Tern is a master of the air, both swift and maneuverable. CONNECTICUT, AUGUST

the birds move north or south. All such areas offer a condensed view of migration activity that occurs throughout the fall and spring seasons. Many birds migrate at night and can be experienced by birders at that time only by listening to their flight calls overhead. These birds all land in terrestrial habitats or settle onto aquatic habitats at some point during migration to rest and feed. During these periods, birders can be treated to unusually large numbers of birds in a single location.

Observing birds during migration can be challenging for the simple reason that they are in flight. They can be too far away to notice individual feathers and the exact color of the bare parts. Instead, we can focus on details of behavior and structure that are readily discerned on birds in flight. Common Terns appear longer-winged than Forster's Terns. Distant Surf Scoters appear lankier than Black Scoters. Long-tailed Ducks appear pointy,

Common Loons look hunchbacked, and Buffleheads in flight show an odd rocking motion.

Just as important as noting behavior and structure is paying attention to the timing of migration. For example, in the eastern Great Lakes, rare Pomarine Jaegers move late in the season, in October and November; a jaeger in September is more likely to be a Parasitic or maybe even a rare Long-tailed. Timing is an especially important component of understanding and systematically searching for vagrants—birds far off the beaten path. In the eastern Great Lakes, rarities like Hudsonian Godwits and Red Phalaropes appear on the heels of rough weather in early November, whereas Laughing Gulls and Buff-breasted Sandpipers are more likely in August and September. Every once in a while, a rarity shows up as if out of nowhere. More often than not, though, rarities are found by birders with detailed knowledge of the timing and patterns of vagrancy.

PARTS OF A BIRD

Knowing the basic parts of a bird, or its "topography," is an important step in becoming a better birder. Most of a bird's body is covered in feathers, so becoming familiar with the major feather groups (called "tracts") and how they are arranged on different types of birds is key to identifying many species.

The images of different types of birds in this section (duck, songbird, raptor, gull, and shorebird,) point out all the major visible body parts referred to in the *Smithsonian Guide*. Of course, all birds are individuals, and differences in appearance are to be expected as a result of age, health, wear, molt, posture, and other factors.

Many birders use multiple terms for some particular parts of a bird that are generally interchangeable, such as "moustache" for the malar region and "ear patch" for the auriculars. The use of these terms is standardized throughout this guide.

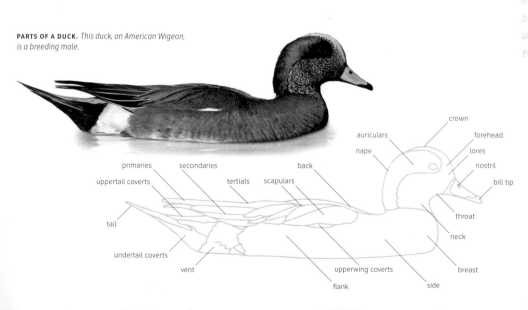

PARTS OF A DUCK. *This duck, an American Wigeon, is a breeding male.*

crown
auriculars
forehead
nape
lores
nostril
primaries
secondaries
back
bill tip
uppertail coverts
tertials
scapulars
throat
tail
neck
undertail coverts
vent
upperwing coverts
breast
flank
side

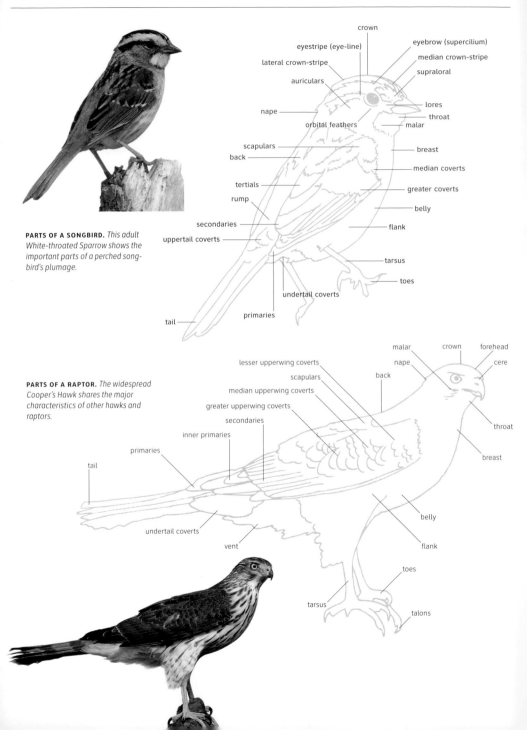

PARTS OF A SONGBIRD. *This adult White-throated Sparrow shows the important parts of a perched songbird's plumage.*

crown
eyestripe (eye-line)
eyebrow (supercilium)
lateral crown-stripe
median crown-stripe
supraloral
auriculars
lores
nape
throat
orbital feathers
malar
scapulars
breast
back
median coverts
tertials
greater coverts
rump
belly
secondaries
flank
uppertail coverts
tarsus
toes
undertail coverts
primaries
tail

PARTS OF A RAPTOR. *The widespread Cooper's Hawk shares the major characteristics of other hawks and raptors.*

malar
crown
forehead
lesser upperwing coverts
nape
cere
scapulars
back
median upperwing coverts
greater upperwing coverts
throat
secondaries
inner primaries
breast
primaries
tail
belly
undertail coverts
flank
vent
toes
tarsus
talons

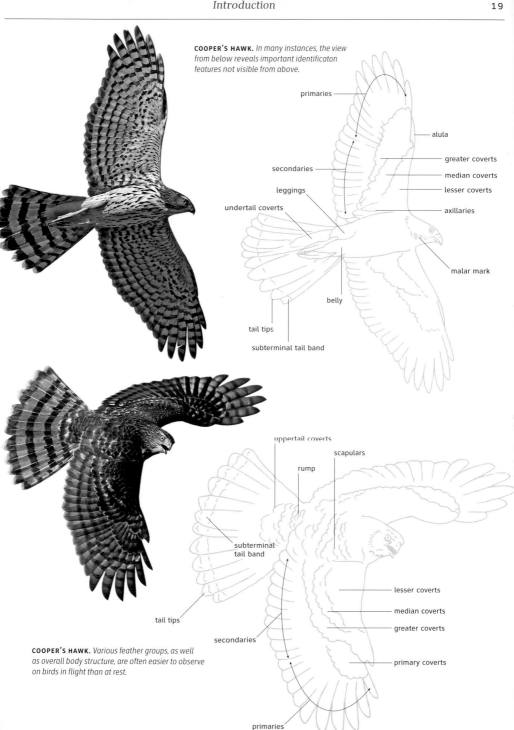

COOPER'S HAWK. *In many instances, the view from below reveals important identificaton features not visible from above.*

primaries

alula

greater coverts

median coverts

secondaries

lesser coverts

leggings

axillaries

undertail coverts

malar mark

belly

tail tips

subterminal tail band

uppertail coverts

scapulars

rump

subterminal tail band

lesser coverts

median coverts

greater coverts

tail tips

secondaries

primary coverts

COOPER'S HAWK. *Various feather groups, as well as overall body structure, are often easier to observe on birds in flight than at rest.*

primaries

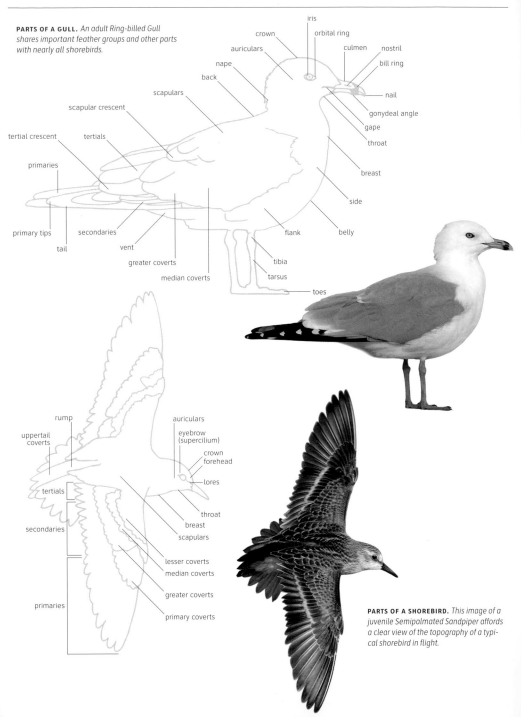

PARTS OF A GULL. *An adult Ring-billed Gull shares important feather groups and other parts with nearly all shorebirds.*

iris
crown
orbital ring
auriculars
culmen
nostril
nape
bill ring
back
scapulars
nail
scapular crescent
gonydeal angle
gape
tertial crescent
tertials
throat
primaries
breast
side
primary tips
secondaries
flank
belly
tail
vent
greater coverts
tibia
median coverts
tarsus
toes

rump
auriculars
uppertail coverts
eyebrow (supercilium)
crown
forehead
lores
tertials
throat
breast
secondaries
scapulars
lesser coverts
median coverts
greater coverts
primaries
primary coverts

PARTS OF A SHOREBIRD. *This image of a juvenile Semipalmated Sandpiper affords a clear view of the topography of a typical shorebird in flight.*

PLUMAGE AND MOLT

All birds have feathers, and feathers, with growth, age, season, and other factors, wear out and are replaced in a process called molt. Most adult birds molt at least once per year, many molt twice per year, a few molt three or more times per year, and a small number molt less than once per year.

A key point is that plumages and molts represent distinct stages in a bird's life. In any given period, a bird is either in one particular plumage (for example, juvenile or adult; breeding or nonbreeding), or it is undergoing a particular change in its plumage (for example, molting from juvenile to adult plumage, or from breeding to nonbreeding plumage).

It is useful to think about molt and plumage in three categories: sex-related plumage differences, seasonal differences, and age-related differences.

Sex-related plumage differences

Male and female plumages may be very different, somewhat different, or exactly the same. Where there are differences, it is usually—but not always—the case that the male plumage is brighter, sharper, or more colorful than the female plumage. Frequently, the sex that takes the lead with courtship is more boldly marked than the sex that is responsible for raising young. The male Painted Bunting and female Red Phalarope are spectacular, whereas the female bunting and male phalarope are comparatively dull—differences that correspond perfectly to sex roles in the two species.

Age-related plumage differences

In most (but not all) instances, this guide uses the term "juvenile" to refer to individuals in their first plumage that can be used for flight.

TEXAS, APRIL

TEXAS, APRIL

The Painted Bunting is an extreme example of sex-related plumage differences in birds. Throughout the year, adult males are brilliant red, green, and blue. Adult females are a lovely pale green but appear drab next to males of the species.

This juvenile plumage is the plumage in which a bird fledges. Birds that have not achieved adult plumage but that have molted out of their juvenile plumage are referred to as "immature." Age-related terminology is not consistently employed in field guides and other bird books. The strategy of the *Smithsonian Guide* is to recognize age-related distinctions that are both precise and easy to understand.

Seasonal plumage differences

Names associated with seasonal differences in appearance are famously non-standardized. In referring to the same bird, one book might say that it is in "breeding" plumage, another might refer to its "summer" plumage, and yet another might state that it is in "alternate" plumage. All three terms are correct, and there are instances in which any of them might be preferred. The usual practice of the *Smithsonian Guide* is to

JUVENILE. RHODE ISLAND, JULY

IMMATURE. TEXAS, MARCH

ADULT. FLORIDA, FEBRUARY

The Herring Gull, like a few other gull species, undergoes four complete annual cycles of molt and plumage as it ages from a juvenile early in its first year of life to a full adult. These age-related differences in plumage are important pointers to the identification of the species. For convenience in this book, the complex aging process of gulls is simplified into three categories: A juvenile (top) is dark brown all over, has a completely dark bill, and a dark eye. An immature (center) is paler overall; during the course of several years, and in the process of multiple molts, the immature becomes progressively more adult-like in appearance. An adult (bottom) has attained exquisite white, gray, and black plumage, a bright yellow bill with red gonydeal spot, and a bright red orbital ring.

label seasonally distinct plumages as either "breeding" or "nonbreeding." The reason for this decision is that in the large majority of North American bird species with two adult molts per year, there is a close correspondence between plumage and breeding condition.

Molt strategies

Birds employ one of four fundamental molt strategies during the span of their lives. This way of making sense of a bird's appearance is used in the bullet points and caption text throughout the *Smithsonian Guide* to assist users in understanding the larger picture of molt and its relationship to a bird's life cycle.

> *Simple Basic Strategy* (one plumage per year)
>
> *Complex Basic Strategy* (one adult plumage per year, plus a second plumage in the first year of life)
>
> *Simple Alternate Strategy* (two plumages per year)
>
> *Complex Alternate Strategy* (two adult plumages per year, plus a third plumage in the first year of life)

The Simple Basic Strategy and the Simple Alternate Strategy are exhibited only by non-passerines (non-perching birds).

Examples of species exhibiting the Simple Basic Strategy include the Black-capped Petrel, Turkey Vulture, and Barn Owl. Examples of species exhibiting the Simple Alternate Strategy include the Pelagic Cormorant, Pacific Loon, and Whimbrel. The House Sparrow and the Wrentit employ the Complex Basic Strategy, whereas the Yellow-rumped Warbler and the White-crowned Sparrow employ the Complex Alternate Strategy.

Molt can be an intimidating topic for novice and even experienced birders, but it

MAINE, JUNE

BRITISH COLUMBIA, JUNE

TEXAS, APRIL

NEW JERSEY, JANUARY

Yellow-rumped Warblers in the breeding season (top) are brightly patterned with streaks and splashes of color, but are much more drab in their nonbreeding (fall and winter) plumage (bottom).

Common Loons differ so greatly in breeding plumage (top) and nonbreeding plumage (bottom) that they might be mistaken by a beginner for two different species.

doesn't have to be. Molt, perhaps more than any other aspect of bird biology, exposes us to the foundational truth of natural variation in bird populations. It all boils down to just two fundamental parameters: first, whether or not a bird has an alternate plumage as an adult; second, whether or not a bird has an additional, or formative, plumage in its first year of life. There are, needless to say, countless nuances to delight and distract the aficionado, but there is also an underlying simplicity in the matter of molt. With experience, one learns that a basic understanding of molt aids the identification process. And regardless of your skill level or experience, paying attention to molt adds immensely to the enjoyment of bird study.

Other Differences in Appearance

Variation in the appearance of birds can be due to factors that are not directly related to molt. Feather tips may become worn down, bleached, or abraded—so much so that a bird looks as though it has molted from one plumage to another. A bleached Western Gull in early summer is strikingly different from how it appeared a few months earlier, but no molt has occurred. Another source of variation has to do with age-related changes that do not correspond to distinct plumages. For example, older adult Black-footed Albatrosses average whiter on parts of the wing than younger adults—but they are all in the same adult plumage.

Some birds have unusual appearances from birth as a result of an aberration. These atypical birds may exhibit leucism (pale to white feathers), albinism (completely white feathers, red eyes, and red legs), or melanism (darker than normal feathers). Aberrant plumages are uncommon in wild bird populations and are not encountered with regularity by birders in the field.

Lastly, there is the fundamental biological reality of individual variation: Even after we have accounted for plumages and molts, for wear and aging, there is the simple fact that all populations of organisms are variable.

How to Identify Birds

Bird identification can be a rousing challenge or a simple joy, an absorbing pastime or just quiet contemplation. Bird identification can, like singing and painting and playing chess, give fullness and richness to our lives. And at its core, the art of bird identification is personal and expressive.

One observer's formula—start young, study hard—might not be feasible for another observer. A different approach—make field sketches, record flight calls—could well be a turnoff for certain observers. Nearly all of us were first attracted to birds by a powerful personal experience: a Pileated Woodpecker at summer camp or a Roseate Spoonbill on a business trip; goldfinches and buntings in the backyard garden or ravens and raptors on a family vacation. The objective of bird identification is perhaps best thought of as recovering that initial delight, of reliving that revelatory thrill—one bird at a time, for the rest of one's life.

Bird identification is as much a way of thinking as it is a suite of skills or a body of knowledge. In this vein, the tips that follow have been conceived more with a broad thought process in mind than with specific techniques and particular methods. The actual mechanics of correct field identification can be mastered with sufficient time spent in the field. Certain tools can assist birders, such as a good pair of binoculars or spotting scope, a field guide, and, of course, the company of an experienced birder.

But how does one acquire the "correct" outlook or the "right" mindset? Could it have something to do with letting go and trusting your instincts?

Put Down Your Binoculars

Make an initial guess about an indistinct speck on a distant rooftop: a House Sparrow perhaps, or maybe a House Finch. A Townsend's Solitaire in another setting perhaps, or maybe an Olive-sided Flycatcher. Next, instead of looking at the bird through high-quality optics, start to walk toward the bird. Keep looking at the bird as you close in on it. Notice how various details start to come into view: the overall structure of the bird, a gross read on pattern, a basic impression of color. Keep looking. Notice how the bird behaves. Does it just sit there? Does it shift about? Or does it do something dramatic, like sally out from its perch and then return?

Some optical tools are so good that it is easy, in so many instances, to bypass the process of observation. But there is something more subtle going on here. Today's optics are so good that it is hard to ignore the fine details. Equipped with high-quality optics, we are naturally drawn toward close inspection of feather wear and bill aberrations and so forth. We need to step back, put down our binoculars,

The Northern Parula is a truly striking and memorable bird and has a distinctive song, especially during the breeding season. It is always worthwhile to linger with an "easy" species, such as this colorful warbler; lessons learned from such encounters are beneficial when applied to subsequent observations of much more difficult species. CONNECTICUT, JUNE

In some cases, critical identification points are better observed on birds in flight than while they are at rest. The white secondaries contrasting with darker primaries are excellent field marks on a Yellow Rail in flight. OKLAHOMA, OCTOBER

and look at those features that are obscured or distorted as a result of high magnification; a bird's movements, for example, are better assessed by the naked eye than through binoculars. Thirty seconds of observation without binoculars may reveal considerably more than thirty seconds with binoculars.

Close Your Eyes

In many settings, birds are more conspicuous aurally than visually. The eastern deciduous forest at dawn is a classic example, but there are many others: western pine forests, prairie grasslands, coastal salt marshes, even the southwestern deserts with their wide-open vistas. In the early morning hours in one of those habitats, it may be possible to identify more species by voice in five minutes than you might track down visually in the course of several hours.

Birds sing and produce other vocalizations for the single purpose of communicating

information. Although that information is not intended for human observers, we can make sense of the information in birdsong. In fact, some humans can recognize the songs of many more bird species than the birds themselves can! Yet it is an undeniable fact that many of us grossly underutilize—and that is an understatement—bird vocalizations in the identification process.

Open Your Eyes

The biggest impediment to learning birdsong—to learning identification more broadly—is sensory and conceptual disconnect. We tend to compartmentalize the learning process: learn plumages from field guides, learn songs from recordings; look at wing bars with our eyes, listen to flight calls with our ears. In the specific case of seeing and hearing, we have a tendency to think of sight and sound as discrete and unrelated aspects of in the identification process.

An excellent exercise—one that can never be overdone—is to watch a bird vocalizing. A song or call without a bird is just that: a disembodied voice, an abstraction. It is devoid of context, and it offers us no compulsion to learn

The Saltmarsh Sharp-tailed Sparrow's bill is relatively long for a sparrow, an adaptation for finding food in the mud. MAINE, JUNE

it. Go outside and listen to a song. Any song will do. Track down the songster and listen to it with your eyes. The experience of simultaneously seeing and hearing a bird is greater than the sum of its parts.

Keep Your Eyes Open

Sooner or later, it is going to happen: the bird flies away. Now is the time to start paying attention! Now is the time to start using your binoculars. So much about a bird's overall build is most readily ascertained on the bird in flight. Notice the relative proportions of the tail, wings, and head—all relative to the overall heft of the bird. Notice the manner of flight, and listen for flight calls. And above all, look at field marks. This is where color and pattern— and binoculars—come into play. Look at plumage patterns on the upper and lower surfaces of the wing and tail, and look for overall contrast across the body.

Watching birds in flight, like listening to birds sing, is of tremendous importance in the identification process. But both endeavors are worthwhile simply in and of themselves. Appreciating a Henslow's Sparrow in flight as ornate and intricate, or the utterances of Northern Fulmars around a trawler as arrestingly beautiful, are among the greatest rewards of birding.

Location and Timing

"Birds have wings," it is often said—shorthand for dismissing the importance of habitat and landscape in the identification process. It is helpful to think of the problem of location on three spatial scales. In the middle is the most familiar: habitats defined by plant communities. A bird such as the Gray Vireo is restricted in the breeding season to dry pinyon-juniper woods, whereas the Plumbeous Vireo breeds amid ponderosa pines and at higher elevations.

At a finer scale, birds often sort out within plant communities according to the physical structure of the vegetation. In the boreal forest of New England, for example, four or more species of warblers may be found in the same tree, but they minimize competition partly by foraging in different parts of the tree. And at a coarser scale, avian distribution and abundance may vary greatly across whole landscapes. Some eastern warbler species migrate more heavily through the Ohio River valley than east of the Appalachians, some are more likely along the coast in fall migration than along inland ridge tops, and some are as rare in Gulf coast fallouts as others are plentiful.

Just as there is a spatial component of bird identification, so there is a temporal component. Orange-crowned Warblers in the mid-Atlantic region are rare before late September, whereas Louisiana Waterthrushes in the same region are practically gone by that time. That sort of information can be very useful when one is confronted with, say, a drab first-fall female Yellow Warbler in early September or an unforthcoming Northern Waterthrush in early October.

Another aspect of timing involves the weather. Inland Red Phalaropes in the East in late fall tend to be associated with stormy weather, and it is good to think of "late fall" and "stormy weather" as points of identification on a par with "thick bill" and "gray back." In the West, credible records of Black Swifts far from the nest colonies often involve the timing of the weather, as foragers may disperse great distances and to low altitudes during bouts of inclement weather.

Location and timing, taken together, are essentially the same thing as what is often referred to more formally as "status and distribution." The key to mastery of status and distribution is integrative thinking. Stop for a moment: look around for habitat cues big and small; remind yourself of the time of year; and think about the weather. You've already made significant progress toward correct identification.

The Biological Context

Picture in your mind's eye the drabbest Pine Warbler, the dullest Yellow Warbler, the blandest Orange crowned Warbler, and the plainest Common Yellowthroat. We already know to look for them in specific habitats and within particular temporal windows. We also know from our exercises with and without the use of our binoculars that those four species differ in overall build and proportions. But how do we keep all that information straight? How do we memorize which warblers—or thrushes or shorebirds or whatever—have long vs. short wings, early vs. late migrations, and so forth?

The trick is to realize that such information is handily encoded in what we already know about the biology of the species. Pine Warblers are relatively omnivorous short-distance migrants; hence their relatively thick bills and short wings. Common Yellowthroats are relatively terrestrial; hence their long legs. Most Yellow Warblers have farther to migrate in the fall than most Orange-crowned Warblers; hence, the longer wings and earlier departure date of the former species.

In other words, it is advisable not to approach the problem as one of memorization. Bill and wing structure, body size and coloration, songs and calls, even molt—all those things make sense in light of biology. It is not necessary to remember that Saltmarsh Sharp-tailed Sparrows have long bills or that Bobolinks undergo two complete molts per year. Instead, these things simply make sense. The Saltmarsh Sharp-tailed Sparrow feeds in the mud, and its long bill is a logical adaptation to its intriguing shorebird-like ecology. The Bobolink is one of the longest-distance passerine migrants that breeds in North America, and it "needs"—in an evolutionary sense—two complete sets of feathers to complete its arduous travels.

Go Out on a Limb

Bird identification is inherently imperfect in two major ways. The first has to do with the observer. Our eyes and ears are imperfect, and our minds even more so. Our judgments are not foolproof, and our logic is not infallible. The second problem has to do with the thing observed. Birds do not always fit into tidy categories, regardless of the expertise or inclinations of the observer. Nature is full of variants and aberrations, hybrids and intergrades, surprises and delights.

Recognition of this aspect of bird identification should be liberating, not demoralizing. Each identification is conjectural and provisional. Think of an identification as a hypothesis—and remember that hypothesis testing is the touchstone of the scientific method. The upshot of this mode of thinking is paradoxical: It empowers us to try to identify every bird.

Don't worry about what the experts—whoever they are—might think. Make a guess, and say it out loud. Be keenly alert to the reality that your judgments might be wrong and your eyes and ears faulty, but be equally aware of the possibility that your insights might be valid and your intuition well-founded. Put a name on every bird. Then go back and assess a level of probability— "possibly" or "probably," "likely" or "very likely." Two labels, though, are to be avoided. The first is, "I am absolutely certain." The second is, "I have no idea." In between those unimaginative extremes lies the fascinating and fertile territory of bird identification.

CONSERVATION AND ETHICS

> Birds undeniably contribute to our pleasure and standard of living. But they are also sensitive indicators of the environment, a sort of "ecological litmus paper," and hence more meaningful than just chickadees and cardinals to brighten the suburban garden, grouse and ducks to fill the sportsman's bag, or rare warblers and shorebirds to be ticked off on the birder's checklist. The observation of birds leads inevitably to environmental awareness.
>
> — ROGER TORY PETERSON

In the spirit of Peterson's revelation, the *Smithsonian Guide* contains 45 group accounts, organized at the level of avian families or orders, each of which concludes with conservation information. Population trends, if known, are cited, and current threats are described. Conservation successes are discussed, too—pesticide bans, habitat protection, and other important efforts. Species accounts frequently contain conservation information, typically in the "Habits and Ecology" section. Such information is provided to the extent that it is relevant to the identification process. And the fact that such information is so routinely provided bears witness to the inextricable link between identifying and conserving birds.

Serious bird conservation, involving population monitoring and on-the-ground implementation, is obviously beyond the scope of the *Smithsonian Guide*. But it is emphatically not beyond the reach of ordinary birders. There are many conservation organizations and initiatives, but one of them deserves special mention here: Partners in Flight (www.partnersinflight.org) is a decentralized but highly effective cooperative partnership of hundreds of agencies, organizations, and institutions charged with or dedicated to bird conservation. A hallmark of the Partners in Flight approach is the meaningful engagement of amateurs.

Finally, there is the highly personal but vital matter of ethical conduct in the field. Birders, acting individually and corporately, have impacts on birds and the habitats that support them. The American Birding Association's "Code of Birding Ethics" is a widely admired and widely adhered to manual for ethical behavior in the field. Interested users of the *Smithsonian Guide* are encouraged to visit www.aba.org/abaethics.htm to learn more about this important and interesting topic.

Waterfowl
Whistling-Ducks, Geese & Swans, and Ducks
ORDER ANSERIFORMES

62 species: 41 ABA CODE 1   3 CODE 2   10 CODE 3   4 CODE 4   3 CODE 5   1 CODE 6

Waterfowl—the familiar ducks, geese, and swans—are widespread and well represented in North America. All species are affiliated with aquatic habitats during some or all of their life cycle, and most are legally hunted across large swaths of the continent. Waterfowl classification is in a state of flux, with three subfamilies currently recognized as occurring in North America: the goose-like whistling-ducks (Dendrocygninae) of the southern states; the large-bodied geese and swans (Anserinae) of chiefly northern climes; and the widespread and remarkably diverse ducks (Anatinae).

Favored habitats for waterfowl are sprawling wetland complexes both in coastal districts and inland. Every high-quality marsh in the northern Midwest, it seems, harbors a dozen or more species of breeding ducks. Twenty or more species of waterfowl are easily found in a day afield in fall or winter along the Pacific and Atlantic coasts. But it would not be accurate to say that waterfowl are habitat generalists, indiscriminately accepting any aquatic habitat. On the contrary, knowledge of microhabitat preferences plays an important role in identifying waterfowl.

The breeding biology of waterfowl is more easily observed than in most other birds. Courting is conspicuous, and the precocial young are frequently seen paddling furiously behind dutiful parents—hens only, in most species. Nesting always takes place in the general vicinity of water, but actual nest placement is often in surprisingly non-aquatic microhabitats: Gadwalls out in sagebrush flats, Canada Geese atop tall office buildings, Common Mergansers in caves on sheer cliffs. Most waterfowl species are intermediate-distance migrants, and a few can legitimately be classed as long-distance migrants. Migration is typically by day, and passages along the coast can be spectacular. Daily movements of species wintering along the coast are likewise impressive.

Waterfowl population health has been well studied for two reasons: first, wildlife agencies carefully monitor populations that are legally hunted; second, waterfowl are high-fidelity indicators of wetland quality, and their numbers provide an important baseline for conservation action. Some species—particularly in the subfamily Anserinae (geese and swans)—are enjoying sustained population growth, but many duck species are worrisomely declining, notably the American Black Duck and King Eider. The overarching threat to waterfowl populations is habitat loss. Bioaccumulation of toxins such as selenium is a local threat, and climate change may eventually prove to be a serious challenge for species with substantial arctic-breeding populations.

Black-bellied Whistling-Duck
Dendrocygna autumnalis CODE 1

ADULT. *Neck, breast, and back chestnut, contrasting strongly with black belly and gray face. Juvenile much duller: lacks red highlights of adult; plumage contrast lower overall.* TEXAS, MAY

ADULT. *Gooselike, with long neck and long legs. Bill and feet coral or reddish. Broad white wing stripe visible on standing bird, striking in flight. Often perches on branches and stumps.* TEXAS, MAY

L 21"   ws 30"   wt 1.8 lb

▸ one adult molt per year; complex basic strategy

▸ strong differences between juvenile and adult

▸ a few individuals have pale band at base of breast

Forages in open marshes and agricultural fields; roosts either in dense vegetation at ground level or up in trees. Gregarious; roosts and feeds in small to large flocks. North American range appears to be expanding; vagrants stray north of core range.

Flying birds, especially in flocks, give series of short, squeaky, whistled notes, with elements varying in emphasis: *pi pi pee pee pew.*

Fulvous Whistling-Duck
Dendrocygna bicolor CODE 1

ADULT. *Structurally similar to Black-bellied Whistling-Duck: gooselike, often stands erect. Smaller body evident in direct comparison with Black-bellied. Bill and feet blue-gray, wings dark.* TEXAS, MAY

IN FLIGHT. *At all ages, buff body contrasts with dark wings, strikingly so in flight. Compared to adult, juvenile averages more buff overall, with less contrast.* LOUISIANA, APRIL

L 19"   ws 26"   wt 1.5 lb

▸ one adult molt per year; complex basic strategy

▸ weak differences between juvenile and adult

▸ individual variation in extent of buff on tail

Like Black-bellied Whistling-Duck, occurs in rice fields and marshes; Fulvous prefers more-open habitat, but much overlap. Feeds from water's surface; Black-bellied is more of a grazer. Numbers fluctuate locally; sometimes wanders north of Gulf coast region.

Calls short and squeaky, like Black-bellied Whistling-Duck; in flight, frequently gives *pi pee!* and *pi dee! dee* calls, suggesting Great Kiskadee.

Greater White-fronted Goose
Anser albifrons CODE 1

ADULT. *All have white at base of bill, black on belly. "Greenland" (A. a. flavirostris), shown here, has orange bill. Widespread "Tundra" (A. a. frontalis) has pink bill. "Tule" (A. a. elgasi) larger, darker overall.* OHIO, MARCH

JUVENILE. *Lacks white at base of bill and black speckling on belly shown by adult. Bill dull yellow at first, but gradually assumes orangish or pinkish tones of adult.* CALIFORNIA, OCTOBER

SWAN GOOSE (Anser cygnoides). *Exotic waterfowl are common in the wild, and sometimes invite confusion with native species; note that exotics are often tame and disproportioned.* COLORADO, SEPTEMBER

IN FLIGHT. *Smaller than Canada Goose. White on vent and rump contrasts with mainly gray-brown body; from above, wing coverts dark blue-gray.* NEW MEXICO, NOVEMBER

GRAYLAG GOOSE (Anser anser). *Eurasian species; widely domesticated and frequently seen in feral situations. Variants resemble Greater White-fronted Goose.* EUROPE, JANUARY

| L 28" | WS 53" | WT 4.8 lb | ♂ > ♀ |

▸ one adult molt per year; complex basic strategy

▸ moderate differences between juvenile and adult

▸ geographic variation in size, bare parts, plumage

Breeds on open tundra. On migration and in winter, prefers wetland and agricultural settings. Except in core staging areas, often occurs in small numbers with much larger flocks of Canada Geese. Beneficiary of changing land use practices; numbers increasing.

Call higher pitched and less robust than Canada Goose, typically two- or three-noted and strongly upslurred: *hee-link?* or *hee-lee-wink?*

BARNACLE GOOSE (Branta leucopsis). *A few in North America are vagrants from Europe, but many are escapes from captivity. Smaller overall than Canada Goose, with more white on face.* EUROPE, JULY

Snow Goose

Chen caerulescens   CODE 1

ADULT "LESSER." (C. c. caerulescens) *All adults (Blue and Snow morphs, "Lesser" and "Greater" subspecies) have pink bills with black "grinning patch" on sides. "Greater" larger-bodied, longer-billed.* NEW MEXICO, NOVEMBER

L 28–31"   WS 53–56"   WT 5.3–7.5 lb   ♂ > ♀

▸ one adult molt per year; complex basic strategy

▸ strong differences between juvenile and adult

▸ complex variation in color and body size

Breeds on high-arctic tundra; winters in marshes, along lakeshores, in stubble fields. Forms medium to very large flocks. Outside main staging areas, singles or a few individuals may consort with other goose species. Most populations increasing rapidly.

Call a strident, high-pitched *whenk* or *h'wenk.* "Greater" average lower-pitched than "Lesser"; differences between Blue and Snow Geese negligible.

Variation in the Snow Goose is complicated by the fact that the species has two morphs and two subspecies. By convention, different subspecies are indicated by quotation marks ("Lesser" and "Greater"), but different morphs are not (dark-morph individuals are called Blue, light-morph individuals are often just called Snow).

JUVENILE. *Variable, but always duskier than adult; this individual (a Blue Goose) is at the dark end of the normal range of variation. Juvenile Ross's is smaller-bodied, smaller-billed, whiter.* NEW MEXICO, NOVEMBER

ADULT BLUE. *Typical Blue Goose is dark-bodied and white-headed, but variation is extensive. Some may resemble Emperor Goose in plumage, but bill structure distinctive.* NEW MEXICO, DECEMBER

IN FLIGHT. *All have black primaries, but amount of black in secondaries and extent of white on wing coverts variable. Blue Goose, with all flight feathers black, shown here.* NEW MEXICO, NOVEMBER

Ross's Goose

Chen rossii CODE 1

ADULT. *Key distinction from Snow Goose is the bill: shorter, more triangular; border with cheek vertical and straight (curved on Snow Goose); "grinning patch" reduced or absent.* CALIFORNIA, MARCH

JUVENILE. *Bare parts darker and plumage duskier than adult. Juvenile Snow Goose is considerably duskier. Like adult, build is delicate overall.* COLORADO, NOVEMBER

L 23"	ws 45"	wt 2.7 lb

▸ one adult molt per year; complex basic strategy

▸ moderate differences between juvenile and adult

▸ dark morph is rare but possibly increasing

Habits and habitats similar to Snow Goose. Often found in marshes and agricultural fields with Snow Goose, but Ross's tends to cluster together within larger, mixed-species flocks. Like Snow Goose, Ross's is increasing in number and expanding its range.

HYBRID. *Hybrids with Snow Goose occur. This individual has petite body structure of Ross's; bill is long like Snow, but with gray at base and no "grinning patch."* COLORADO, APRIL

Corresponding to Ross's smaller size, vocalizations are higher-pitched than those of Snow Goose; individuals and flocks usually not as noisy as Snow Goose.

Emperor Goose

Chen canagica CODE 2

L 26"	ws 47"	wt 6.1 lb

▸ one adult molt per year; complex basic strategy

▸ strong differences between juvenile and adult

▸ head and hindneck sometimes stained orangish

A marine species, both on the breeding grounds and in winter. As with many other geese, population is increasing. Away from core range in Bering Sea region, most observations are of lone birds wintering in western coastal regions, especially the Pacific Northwest.

Infrequently heard. Honk is a two-noted *la-la*, lacking downward inflection of Snow Goose; in agitation, gives a deep, hoarse *uh-rung*.

IMMATURE. *Superficially similar to Blue Goose, but note all-white crown and hindneck. Pale pink bill is small. Adult is cleaner white on head; juvenile has all-gray head.* CALIFORNIA, MARCH

Canada Goose
Branta canadensis   CODE 1

IMMATURE. *All subspecies of Canada and Cackling Geese have black necks and white chinstraps. Adult similar, but plumage of juvenile is softer and more loosely textured than adult.* CALIFORNIA, FEBRUARY

ADULT. *Canada is larger, longer, and deeper-voiced than Cackling, but variation is extensive; "Lesser" Canada (B. c. parvipes), shown here, is relatively small-bodied, short-necked, short-billed.* COLORADO, SEPTEMBER

L 36–45"   WS 53–60"   WT 6–12 lb   ♂ > ♀

▸ one adult molt per year; complex basic strategy

▸ weak differences between juvenile and adult

▸ extensive structural and moderate plumage variation

One of the most abundant, widespread, and familiar species in North America. In addition to historical haunts of marshes and farmland, increasingly occurs in urban districts. Range expanding and numbers increasing; rapidly adapting to human habitats.

Variable, but all subspecies give basic *honk* or *ha-ronk*; larger birds average lower-pitched. In agitation, thrusts head back, hisses, then lunges forward.

Cackling Goose
Branta hutchinsii   CODE 1

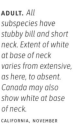

ADULT. *All subspecies have stubby bill and short neck. Extent of white at base of neck varies from extensive, as here, to absent. Canada may also show white at base of neck.* CALIFORNIA, NOVEMBER

ADULT. *Body structure, a key distinction between Cackling and Canada, is easily ascertained on birds in flight. This individual is dark overall, but not exceptionally so.* CALIFORNIA, NOVEMBER

IN FLIGHT. *Cackling Goose (front), Canada Goose (rear). In direct comparison, structural differences between Cackling and Canada Goose are obvious, but the two species are alike in plumage.* COLORADO, NOVEMBER

L 25–38"   WS 43–48"   WT 3–5 lb

▸ one adult molt per year; complex basic strategy

▸ weak differences between juvenile and adult

▸ weak structural and moderate plumage variation

Recently split from Canada Goose; status and distribution require further study. Most breed at higher latitudes and migrate greater distances than Canada Goose. Outside core staging areas, singles and small flocks occur in the company of larger flocks of Canada Geese.

Honk averages higher-pitched and squeakier, sometimes markedly so, than Canada Goose: *hink* or *ha-rink*. Both species, especially Canada, variable.

Brant

Branta bernicla CODE 1

L 25"	WS 42"	WT 3.1 lb	♂ > ♀

- one adult molt per year; complex basic strategy
- moderate differences between juvenile and adult
- three North American populations differ in plumage

Breeds in arctic salt marshes along ocean or in estuary systems. Winters widely on coasts, where large flocks feed on marine vegetation. Occurs in diverse settings: beyond the surf, in the shelter of inlets and jetties, or in back bays and cordgrass meadows.

Compared to other North American geese, call is mellower and lower-pitched. The sound of a large flock in chorus is rolling, murmuring: *rruh-luck.*

ADULT, "BLACK." (B. b. nigricans) *Dark overall, with variable but extensive black on belly. White necklace variable, but always conspicuous. "Black" is the western Brant, common in winter on Pacific coast.* ALASKA, JUNE

ADULT, "ATLANTIC." (B. b. hrota) *Body paler than "Black"; necklace less prominent. Intermediate "Gray-bellied" Brant, not shown, winters in Seattle area. Juveniles of all populations lack necklace.* NEW JERSEY, MAY

Tundra Swan

Cygnus columbianus CODE 1

ADULT. *Black bill usually shows tiny yellow patch near base. Border between bill and facial feathering is not perfectly straight; either curved or jagged.* COLORADO, NOVEMBER

L 52"	WS 66"	WT 14.4 lb	♂ > ♀

- one adult molt per year; complex basic strategy
- moderate differences between juvenile and adult
- bill yellower on "Bewick's" subspecies (*C. c. bewickii*), Asian stray

Breeds on arctic tundra; large flocks winter on bays, especially along coasts. Overland migration is conspicuous: long, thin lines migrate day or night, visible and audible at great distances. Populations are increasing and apparently changing their migration routes.

Formerly known as Whistling Swan, in reference to sonorous chorusing. Call is more hooting than whistled: a mellow, pure-tone *klew* or *kuh-lew.*

ADULTS. *Amount of yellow on bill is variable: individual at front is near one extreme; some show even less yellow than individual at rear.* COLORADO, NOVEMBER

JUVENILE. *Bill is extensively pink or pinkish orange; bill of juvenile Trumpeter less so. Plumage duskier than adult; some are quite dingy, others show just some dusky highlights.* CALIFORNIA, SEPTEMBER

Trumpeter Swan
Cygnus buccinator CODE 1

L 60"	WS 80"	WT 23 lb	♂ > ♀

▸ one adult molt per year; complex basic strategy

▸ strong differences between juvenile and adult

▸ prominence of crest variable

Breeds in and around large wetlands. Some populations move to agricultural complexes during the winter; others are sedentary. Endangered in the early 20th century, but now expanding rapidly. Introductions in Midwest and East are controversial.

Its namesake call is distinct from Tundra Swan: a hoarse honking sound, often doubled, which resembles the horns of cars from the mid 20th century.

ADULT. *When seen head on, Trumpeter Swan's bill pinches together in a deep V shape extending down from eyes; Tundra's bill extends from eyes in shallow U shape.* COLORADO, DECEMBER

ADULT. *Larger overall than Tundra. Bill usually all-black; long and sloping, with straight or smoothly contoured base. Juvenile much duskier than adult; bill pinkish toward tip.* MINNESOTA, JANUARY

Whooper Swan
Cygnus cygnus CODE 3

L 55–63"	WS 81–93"	WT 18–24 lb	♂ > ♀

▸ one adult molt per year; complex basic strategy

▸ moderate differences between juvenile and adult

▸ head and neck of adult sometimes smudged yellowish

Ecologically the Old World counterpart of Trumpeter Swan; habits similar. Regularly reaches North America only in the Aleutian Islands, where it has bred. A few records away from Alaska refer to genuine wild strays, but many are presumed escapes from captivity.

Honk a nasal, throaty *rang*. Often doubled or tripled: *rang-go-rang*. Small flocks whip themselves up into a frenzy, but most strays to Alaska are silent.

ADULT. *Bill has extensive yellow at base, extending farther out toward tip than on either subspecies of Tundra Swan. Juvenile duskier than adult.* JAPAN, NOVEMBER

ADULT. *In flight, adults of all North American swans are recognized at a great distance by their long necks held straight out and white wings with no black in the flight feathers.* JAPAN, NOVEMBER

Mute Swan
Cygnus olor CODE 1

L 60"	WS 75"	WT 22 lb	♂ > ♀

▸ one adult molt per year; complex basic strategy

▸ strong differences between juvenile and adult

▸ variation in plumage color and bill morphology

Introduced from Eurasia. Established and spreading in East, with especially strong toehold on Long Island; also established in lower Great Lakes region. This invader menaces native waterfowl and rips up vegetation, and is increasingly subject to control by wildlife managers.

Most commonly heard sound is a muffled but far-carrying whoosh of wings of birds in flight. In aggression toward humans and dogs, gives a loud hissing sound.

ADULT. *Bright orange bill; variable black knob at base. Swimming bird often holds wings away from body, holds neck in "graceful" S-curve that is actually an aggressive posture.* NEW JERSEY, MAY

IN FLIGHT. *Mute's tail is longer than other swans; note that feet fall short of tail tip. Adult may show some rusty or gray in plumage, as here; juvenile is more extensively gray.* NEW JERSEY, MAY

Wood Duck
Aix sponsa   CODE 1

BREEDING MALE. *Unique; plumage colors and pattern stunning. Eclipse males are much less colorful, but retain basic facial pattern, along with red bill and eye.* CALIFORNIA, JANUARY

ADULT FEMALE. *Distinctive white eye patch, white chin. Separated from male in eclipse plumage by brown eye and bill. On both sexes, note head structure: large with droopy crest.* CALIFORNIA, JANUARY

L 18.5"	WS 30"	WT 1.3 lb	♂ > ♀

▸ two adult molts per year; complex alternate strategy

▸ strong sex-related and seasonal differences

▸ prominence of crest variable

Reclusive and skittish; often flushes when detected. Arboreal nests in tree cavities; roosts and feeds in wooded swamps; often seen perched high up on horizontal bough. After breeding season, gathers in small flocks that feed and roost in quiet ponds and swamps.

◀ All calls wheezy. Flushing female gives rising *whoo-eek*, often repeated. Male call is thinner, more drawn-out than female: *schwuh-weeeeuh-weep.*

IN FLIGHT. *Both sexes appear dark above in flight. Large head, often thrust forward, stands out. Flight direct, often just above treetops; birds in pairs fly very close to one another.* MONTANA, MARCH

Muscovy Duck
Cairina moschata   CODE 3

L 25–31"	WS 38–48"	WT 3.3–6.6 lb	♂ > ♀

▸ Molt strategy and number of adult molts unknown

▸ moderate differences between juvenile and adult

▸ feral birds vary greatly in plumage and structure

Native range barely reaches southern Texas, where singles and pairs occur along wooded waterways. Wild birds are reclusive: hide at water's edge and fly low over treetops. Feral birds, including established and expanding population in Florida, favor parks and subdivisions.

Wild birds are seldom heard; quack does not carry far. Feral birds at "duck ponds" are tame and docile; quack and murmur when milling about feeding stations.

FERAL ADULT. *This individual has the black plumage and white wing patch of wild birds, but the extensive red facial skin is not usually seen in wild populations.* CALIFORNIA, DECEMBER

American Wigeon

Anas americana   CODE 1

BREEDING MALE. *In breeding plumage, held throughout much of year, broad green crescent on head and white or orangish crown are distinctive. Body color always has warm tones.*
CALIFORNIA, MARCH

ADULT FEMALE. *Head cold gray with fine mottling; female Eurasian has brownish or rufous tones on head. Blue-gray bill is short and thin.*
CALIFORNIA, MARCH

IN FLIGHT. *From below, all American Wigeons (adult male here) show pale wing linings that contrast with dark flight feathers; underwing of Eurasian Wigeon more uniformly gray.* ARIZONA, FEBRUARY

ʟ 20"	ws 32"	wт 1.6 lb	♂ > ♀

- two adult molts per year; complex alternate strategy
- strong sex-related and moderate seasonal differences
- adult male crown color varies from white to orangish

Occurs in varied wetland habitats throughout most of year. Gregarious; small to medium flocks gather at quiet ponds and marshes, as well as in creeks, rivers, and ocean inlets. In winter, forages on golf courses and lawns, often with American Coots.

◀ Wheezy, three-note call of male distinctive: *whee-whew!-whew*, with accent on second syllable. Female call, heard less frequently, is a low, guttural *rrup*.

Eurasian Wigeon

Anas penelope   CODE 3

IMMATURE MALE. *Genetically pure male has reddish head with peach forecrown. Breast with pink or orange tones, body gray. In flight, all plumages show uniformly gray underwings.* CALIFORNIA, MARCH

ADULT FEMALE. *Body averages colder and grayer than American, but head has warmer tones. Overall effect is one of especially low contrast. Check for gray underwings on bird in flight.* EUROPE, OCTOBER

ʟ 20"	ws 32"	wт 1.5 lb	♂ > ♀

- two adult molts per year; complex alternate strategy
- strong sex-related and moderate seasonal differences
- frequently hybridizes with American Wigeon

Old World counterpart of American Wigeon. Regular winter visitor to North America; most records from coasts, but strays can show up anywhere. Lone birds join large flocks of American Wigeon but tend to stay apart from the main flock.

Wheezy call of male similar to American, but usually shorter, consisting of just one or two notes: *wheeer* or *hweee-oo*. Female call similar to American.

American Black Duck

Anas rubripes CODE 1

L 23"	WS 35"	WT 2.6 lb	♂ > ♀

▸ two adult molts per year; complex alternate strategy
▸ weak differences between adult male and female
▸ hybrids with Mallard frequently encountered

ADULT MALE. *In all plumages, white wing linings contrast sharply with dark body. Note purple speculum lacking white borders.* NEWFOUNDLAND, NOVEMBER

Habits similar to closely related Mallard, but not as tame. Avoids especially small wetlands that are acceptable to Mallards; in winter, large flocks occur in estuaries and along breakwaters. Many populations are threatened with genetic invasion by Mallards.

ADULT FEMALE. *Plumage similar to male. Bill duller and darker than male; compare with female Mallard.* NOVA SCOTIA, JUNE

Overall less vocal than Mallard. Male and female vocalizations correspond to those of Mallard, but average lower-pitched and not as far-carrying.

The American Black Duck, Mallard, and Mottled Duck are members of a "superspecies complex"—a taxonomic concept referring to groups of distinct species that have evolved from a common ancestor.

HYBRID. *Dark body contrasts with paler head; bill duller, greener than Mallard. Many carry Mallard genes; note touch of green on crown, indicating introgression.* NEW JERSEY, FEBRUARY

Mallard

Anas platyrhynchos CODE 1

L 23"	WS 35"	WT 2.4 lb	♂ > ♀

▸ two adult molts per year; complex alternate strategy
▸ strong sex-related and seasonal differences
▸ hybrids and domestic variants frequently seen

BREEDING MALE. *All-green head separated from chestnut breast by white stripe. Male in eclipse plumage resembles adult female, but retains yellow in bill and chestnut on breast.* CALIFORNIA, JANUARY

In many regions, the familiar wild duck. Occurs in any aquatic habitat, from open ocean to decorative ponds; also frequents terrestrial habitats such as lawns and stubble fields. Often visits feeding stations, where domestic variants are most likely to be encountered.

ADULT FEMALE. *Body ruddy-brown with coarse vermiculation; head gray-brown with fine mottling. Bill is good mark: usually bright orange with irregular dark splotch at center.* CALIFORNIA, MARCH

◀ Honest-to-goodness *quack* given only by female, singly or in series. Male call softer, more sibilant, often given in series: *quishp quishp quishp.*

DOMESTIC. *Birds of domestic lineages, possible anywhere, may be oversized and awkwardly proportioned; compared to wild types, often less finely colored, more coarsely patterned.* COLORADO, OCTOBER

IN FLIGHT. *Light blue speculum broadly bordered by white. "Mexican" Mallard (A. p. diazi) of south-central borderlands, possibly a distinct species, shows white borders of speculum much reduced. Male "Mexican" has female-like plumage.* COLORADO, FEBRUARY

Mottled Duck
Anas fulvigula   CODE 1

| L 22" | ws 30" | wt 2.2 lb | ♂ > ♀ |

- two adult molts per year; complex alternate strategy
- weak sex-related and age-related differences
- hybrids with Mallard more common than previously known

Gulf coast representative of the Mallard superspecies complex. Unobtrusive; typically solitary or in pairs. Among the least gregarious of North American waterfowl. Destruction of coastal wetlands, grasslands and hybridization with Mallard pose a threat to the species.

ADULT MALE. *Bill yellow, like Mallard. Pale head contrasts with darker body, like American Black Duck. Overall coloration warmer, buffier than American Black Duck.* FLORIDA, FEBRUARY

Male and female vocalizations correspond to those of Mallard and American Black Duck; both sexes tend to be quiet for long periods of time.

IN FLIGHT. *Speculum, if fresh and seen well, is bluish (not purple as in American Black Duck), and has thin white edges or none at all (broader white edges in Mallard).* FLORIDA, MARCH

Gadwall
Anas strepera   CODE 1

| L 20" | ws 33" | wt 2 lb | ♂ > ♀ |

- two molts per year; complex alternate strategy
- strong sex-related and moderate seasonal differences
- individual variation in amount of white on upperwing

A prairie breeder whose population declined severely in the 20th century but is now on the rebound. Breeds in large, shallow, grassy marsh systems. Found in winter in diverse aquatic settings, including deep water; usually in small to medium flocks.

BREEDING MALE. *Gray body with fine vermiculation all over, black rump, orange-tinged scapulars. In flight, white inner secondaries contrast sharply with rest of upperwing.* CALIFORNIA, MARCH

Female quack an abrupt, sneezy *kvunk*; usually given singly, often heard at night. Male usually silent; sometimes gives quiet *rurp* or *rrep* notes.

ADULT FEMALE. *Like male, thin-billed and round-headed. If seen well, bill distinctive: mainly dark with orange sides. White patch on secondaries often visible on swimming female.* CALIFORNIA, JANUARY

IN FLIGHT. *White underwing coverts contrast brightly with gray primaries and secondaries. Dark gray breast and sides of male contrast with brownish head and neck.* BRITISH COLUMBIA, FEBRUARY

Northern Shoveler
Anas clypeata CODE 1

BREEDING MALE. *Head green, breast white, flanks chestnut. In eclipse plumage, during late summer and fall, head is dark and dusky with variable pale crescent at base of bill.* CALIFORNIA, JANUARY

IMMATURE FEMALE. *All shovelers, including drab female, show distinctive bill: very large, spatulate. Adult female bill orange, breeding male gray.* CALIFORNIA, DECEMBER

L 19"	ws 30"	wt 1.3 lb	♂ > ♀

▸ two adult molts per year; complex alternate strategy
▸ strong seasonal and sex-related differences
▸ markings on face of eclipse male variable

Breeds in shallow wetlands with mix of open water, dense cover. May be found in deeper water on migration and in winter; often forms large, dense feeding flocks. Foraging method distinctive: with body slung low, rapidly swipes bill back and forth across water's surface.

Female gives gruff, muffled quack and harsh, cackling *pap*. Male gives two-note muffled *woof-woof*. When feeding, both sexes murmur softly *fshew, fshew*.

Cinnamon Teal
Anas cyanoptera CODE 1

BREEDING MALE. *Brick-red all over. Male in eclipse plumage (late summer) approximates adult female in color, but retains red eye and reddish hue of plumage.* CALIFORNIA, FEBRUARY

ADULT FEMALE. *Plain face with no line through eye imparts blank expression. Bill slightly longer and wider than Blue-winged Teal, and overall size larger.* CALIFORNIA, JANUARY

L 16"	ws 22"	wt 14 oz	♂ > ♀

▸ two adult molts per year; complex alternate strategy
▸ strong sex-related and moderate seasonal differences
▸ hybrids with Blue-winged Teal sometimes seen

Much of breeding range overlaps with Blue-winged, but generally favors hotter, drier landscapes and microhabitats than Blue-winged. Arrives early in spring, opposite of Blue-winged. Not a hardy species; rarely noted in cold months north of core winter range.

Female quack similar to Blue-winged Teal, averaging lower-pitched. Male call rougher, reedier than male Blue-winged: *grep* or *grerp*. Both sexes fairly quiet.

Blue-winged Teal

Anas discors   CODE 1

BREEDING MALE. *Lead-gray head with white crescent. Brown body stippled with black; white patches on lower flanks.* CALIFORNIA, DECEMBER

ADULT FEMALE. *Similar to other female teals, but note thin eye stripe, white eye arcs, and whitish wash at base of bill. Bill smaller than Cinnamon, larger than Green-winged.* CALIFORNIA, JANUARY

IN FLIGHT. *Upperwing pattern of both Blue-winged (here) and Cinnamon Teal similar: both sexes have powder-blue secondary coverts; secondaries vary from green to nearly black.* TEXAS, APRIL

L 15.5"	WS 23"	WT 13 oz	♂ > ♀

▸ two adult molts per year; complex alternate strategy

▸ strong sex-related and seasonal differences

▸ plumage varies in brightness; East Coast birds darkest

Associated with muddy, grassy wetlands. Quiet but conspicuous: loafs on mudflats or banks; feeds along marsh edges or in open shallow water. Migrates farther south than most other North American ducks. Spring migration late, fall migration fairly early.

Female quack is harsh; not as far-carrying as Mallard or Northern Shoveler. Male call, given infrequently, is a shrill, whistled *wheer.*

Green-winged Teal

Anas crecca   CODE 1

BREEDING MALE. *Gray body; red head with green crescent. American subspecies (shown here) has vertical white bar on side. Old World and Bering Sea subspecies have horizontal white bar.* CALIFORNIA, JANUARY

ADULT FEMALE. *Dark overall; lacks pale face and blank stare of Cinnamon and whitish wash at bill base of Blue-winged. Small size sometimes obvious. Green wing patch may be hidden.* CALIFORNIA, JANUARY

L 14"	WS 23"	WT 12 oz	♂ > ♀

▸ two adult molts per year; complex alternate strategy

▸ strong sex-related and seasonal differences

▸ Asian subspecies, "Common Teal," is distinctive

In its extensive breeding range, favors diverse types of wetlands. Consorts with other teal species on migration and in winter; singles or small flocks often join Mallards at winter feeding stations. One of our most common ducks; populations apparently increasing.

More vocal than other teals. Most commonly heard call is given by male: sonorous and reedy *rreep,* singly or in series. Female quack is short, scratchy.

Northern Pintail

Anas acuta   CODE 1

| L 21–25" | ws 34" | wt 1.8 lb | ♂ > ♀ |

- two adult molts per year, complex alternate strategy
- strong seasonal and sex-related differences
- tail long, but length highly variable

Nests in grassy wetlands north to Arctic Ocean. Migrates early in spring; returns south early. Often seen singly or in small flocks, but large flocks are possible anywhere. Numbers fell in 20th century due to habitat destruction and hunting; habitat loss still a threat.

Female quack intermediate between Mallard and Gadwall; closer in pitch to Mallard, but nasal like Gadwall *quenk*. Male call a short, pure whistle: *kweer*.

BREEDING MALE. *White on breast extends as thin vertical stripe onto all-brown head. Body gray. Eclipse male duller; retains grayish body and brownish head.* CALIFORNIA, JANUARY

ADULT MALE IN FLIGHT. *Overall structure is obvious in the air. Tail of female and nonbreeding male lack long tail streamers, but appear pointy.* COLORADO, FEBRUARY

ADULT FEMALE. *Drab, but body structure distinctive: "pointy" in all respects. Wings, tail, and steel-gray bill are long. Neck often appears quite long.* COLORADO, FEBRUARY

Garganey

Anas querquedula   CODE 3

| L 15.5" | ws 24" | wt 13 oz | ♂ > ♀ |

- two adult molts per year; complex alternate strategy
- strong sex-related and seasonal differences
- purplish gloss on head of breeding male variable

Part of the "blue-winged" teal complex, which includes Northern Shoveler, Cinnamon and Blue-winged Teals. Garganey is the Old World ecological counterpart of Blue-winged. Highly migratory; regular in western Aleutians. Casual elsewhere; many records well inland.

Usually silent in North America. Female quack teal-like, most closely resembling Green-winged. Male display call a short, stuttering rattle; vaguely wren-like.

BREEDING MALE. *Prominent white eyebrow extends far behind eye. Crown nearly black; face and neck have wine-colored tint. Extensive gray on sides visible from afar.* EUROPE, JUNE

ADULT FEMALE. *Has large bill (like Cinnamon) compared to other teals and a more angular head profile. Head pattern of female Garganey stronger than other female teals.* ASIA, NOVEMBER

Redhead
Aythya americana CODE 1

BREEDING MALE. *Red on head variable; sometimes orange-tinted, sometimes brown-tinged. Breast black; body gray and vermiculated. Eclipse male is similar but dusky overall.* ARIZONA, DECEMBER

ADULT FEMALE. *Soft brown all over, paler on face. Bill usually distinctly tricolored: black tip, white spot, blue-gray base. Flight feathers, not visible on swimming bird, are gray.* ARIZONA, DECEMBER

| L 19" | WS 29" | WT 2.3 lb | ♂ > ♀ |

▸ two adult molts per year; complex alternate strategy

▸ strong sex-related and moderate seasonal differences

▸ individual variation in intensity of red on male head

In many ways intermediate between dabbling (genus *Anas*) and diving (genus *Aythya*) ducks. Plumages, molts, and structure like *Aythya*, but occurs in shallow marshes with dabblers. Common in West. Local and absent from parts of East, with recent population losses there.

Male courtship call a series of breathy, far-carrying hoots: *hooh hoo hoooh! hooh.* Female quack a low-pitched *grurp*, guttural but not harsh.

Canvasback
Aythya valisineria CODE 1

BREEDING MALE. *Resembles Redhead, but body is white (Redhead gray) without vermiculation and head usually not as bright. Eclipse male similar to breeding male, but duskier overall.* ARIZONA, JANUARY

ADULT FEMALE. *Paler than Redhead. Best distinction of Canvasback, visible at great distance, is structure: long sloping bill and forehead, coupled with long neck.* ARIZONA, DECEMBER

| L 21" | WS 29" | WT 2.7 lb | ♂ > ♀ |

▸ two adult molts per year; complex alternate strategy

▸ strong sex-related and moderate seasonal variation

▸ hybrids with Redhead are occasionally noted

Tied to deep water throughout its life cycle. In winter, forms conspicuous rafts on bays and large lakes, often well offshore. Occurs in both mixed-species and single-species flocks. Reclusive on breeding grounds, where sometimes a victim of nest parasitism by Redheads.

Male courtship call recalls Redhead, but with two-syllable notes not uttered in series: *hooloooo.* Female quack similar to Redhead; averages deeper, rougher.

Common Pochard
Aythya ferina CODE 3

ADULT MALE. *Resembles Canvasback and Redhead. Has sloping forehead of Canvasback, but bill pattern more like Redhead. Female forehead structure and bill pattern similar to male.* JAPAN, MARCH

ADULT FEMALE. *Similar to female Redhead and Canvasback. Pale brown head usually shows lighter "bridle" behind eye. Dark bill broken by white ring in center.* JAPAN, MARCH

L 16–19"	WS 26–30"	WT 2 lb	♂ > ♀

- ▸ two adult molts per year; complex alternate strategy
- ▸ strong sex-related and moderate seasonal differences
- ▸ hybrids with Redhead and Canvasback possible

A typical diving duck, likely to be seen on open water in sheltered bays or lagoons. Uncommon but regular in migration in Alaska, especially outer Aleutians. Birds seen away from Alaska may be either escapes from captivity or rare vagrants.

Male courtship call resembles Redhead, Canvasback: throaty hoots, *whip wheep whip whip*. Female quack a deep *grrrap*. Usually silent in North America.

Tufted Duck
Aythya fuligula CODE 3

ADULT MALE. *Structure, plumage similar to scaup, Ring-necked Duck. On swimming bird, white sides contrast with otherwise black plumage. Actual and apparent extent of tuft variable.* CALIFORNIA, MARCH

ADULT FEMALE. *Resembles Ring-necked Duck and scaup. Most show hint of a tuft. Face feathering variable, usually more uniformly brown than Ring-necked or scaup. Bill nail prominent.* CALIFORNIA, MARCH

L 17"	WS 26"	WT 1.6 lb	♂ > ♀

- ▸ two adult molts per year; complex alternate strategy
- ▸ moderate age-related, sex-related, seasonal differences
- ▸ tuft on head varies greatly in size and shape

Regular in western Alaska in small numbers. Vagrants have strayed throughout North America, but escapes from captivity occur. Habits and behavior similar to scaup and Ring-necked Duck; true vagrants and escapes often associate with those species.

Female quack a low growl, *grerp*, like other *Aythya* females. Male display, not likely to be observed in North America, a short series: *eepeepee*.

Ring-necked Duck
Aythya collaris   CODE 1

BREEDING MALE. *Boldly marked bill and plumage; black overall with pale gray sides; gleaming white breast band appears as vertical wedge on swimming bird.* ARIZONA, JANUARY

ADULT FEMALE. *Peaked crown like male. Facial feathering variable; usually shows whitish patch at base of bill, gray cheeks, brown cap. Dark bill usually has pale patch in middle.* CALIFORNIA, JANUARY

L 17"	WS 25"	WT 1.5 lb	♂ > ♀

- ▸ two adult molts per year; complex alternate strategy
- ▸ strong age-related, sex-related, and seasonal differences
- ▸ plumage variable, but peaked crown usually evident

Common migrant and winter resident in diverse aquatic settings, including large bays, freshwater ponds, and shallow marshes. Occurs in mixed-species or single-species flocks, often packed densely. Populations apparently increasing.

Female quack is like other *Aythya* females: a low, rolling *grrurp*, with muffled quality. Male in display gives a weak, gasping hoot, *hheeee*.

Greater Scaup
Aythya marila   CODE 1

BREEDING MALE. *Best mark is head structure: crown smoothly rounded, with high point near or in front of eye. Bill larger than Lesser Scaup, bill nail usually more extensive.* CALIFORNIA, FEBRUARY

ADULT FEMALE. *Head and bill structure like male. Variable white patch on face usually more extensive than on Lesser. In flight, both sexes show more white on wing than Lesser.* CALIFORNIA, FEBRUARY

IN FLIGHT. *On both sexes (female shown here) primaries and secondaries are extensively white, creating a long bar visible from afar.* BRITISH COLUMBIA, FEBRUARY

L 18"	WS 28"	WT 2.3 lb	♂ > ♀

- ▸ two adult molts per year; complex alternate strategy
- ▸ strong age-related, sex-related, seasonal differences
- ▸ individual plumage variation, especially adult females

Compared to similar Lesser Scaup, averages more northerly distribution, more marine orientation; often favors larger bodies of water. Generally scarce inland, except in Great Lakes region. Populations declining, in part because of heavy-metal contamination.

Rarely heard except in courtship. Call most likely to be heard is female quack, a grating *grawrp*. Male display calls are a variety of soft hoots and whistles.

Lesser Scaup
Aythya affinis CODE 1

L 16.5"   WS 25"   WT 1.8 lb   ♂ > ♀

▸ two adult molts per year; complex alternate strategy

▸ moderate age-related, sex-related, seasonal differences

▸ head gloss of adult male depends on lighting

Closely related and similar in appearance to Greater Scaup. Inland, Lesser is much more common than Greater. Coastally, Lesser prefers freshwater lagoons; Greater prefers brackish waters. Migrates late in spring, late in fall. Occurs in small to very large rafts. Numbers declining.

BREEDING MALE. *Distinguished from Greater by peaked crown with highest point behind eye, narrow bill nail, variable purplish gloss on head. Smaller size usually not apparent.* CALIFORNIA, DECEMBER

Rarely heard away from breeding grounds. Female call is a growling *grarp, grarp*, especially when flushing. Male in display gives quiet whistle, *heewp*.

ADULT FEMALE. *Like male, shows peaked crown, narrow bill nail. Variable white at base of bill averages less extensive than on Greater. All characters of both sexes are variable.* CALIFORNIA, DECEMBER

IN FLIGHT. *With good view, wing pattern distinct from Greater Scaup: Lesser (male here) flashes bright white only in secondaries, Greater flashes white in both primaries and secondaries.* NEW JERSEY, FEBRUARY

Common Eider
Somateria mollissima CODE 1

L 24"   WS 38"   WT 4.7 lb   ♂ > ♀

▸ two adult molts per year; complex alternate strategy

▸ strong age-related, sex-related, seasonal differences

▸ western and eastern populations distinctive

Highly marine; typically found just offshore, where it dives for crustaceans. In the East, aggregations of thousands are seen regularly south to New York. In the West, large numbers are rarely noted south of Alaska.

BREEDING MALE. *White face with black cap; black underparts and white back. Bill distinctive in all plumages, even on oddly marked nonbreeding, subadult, and transitional males.* MANITOBA, JUNE

ADULT FEMALE. *Color varies from warm rufescent (eastern birds) to cold gray (western birds); usually shows strong barring, especially on sides. Long, sloping bill always a good mark.* EUROPE, MARCH

Generally quiet. All calls are low, groaning; hard to hear over the pounding surf. Female gives grouse-like clucks; courting male gives melancholy hoots.

King Eider

Somateria spectabilis   CODE 1

L 22"	WS 35"	WT 3.7 lb	♂ > ♀

▸ two adult molts per year; complex alternate strategy
▸ strong age-related, sex-related, seasonal differences
▸ individual plumage variation, especially adult males

This spectacular but generally unfamiliar duck is the less common counterpart of the Common Eider, averaging farther north on the breeding grounds and wintering grounds; only small numbers travel as far south as New England. Populations apparently declining.

Calls similar to Common Eider: female, a series of muffled grunts, *guff guff guff*; male, a soft hooting, *woo hooo! hoo.*

ADULT FEMALE. *Bill and head shape combine to give a less sloping profile than Common Eider; bill of King Eider shorter, forehead steeper. Flanks less strongly barred than Common.* NORWAY, MARCH

IMMATURE MALE. *Yellow bill easily discerned at great distance; close up, note characteristic bill structure. Overall chocolate-brown coloration contrasts with splotchy pale breast.* CALIFORNIA, SEPTEMBER

IN FLIGHT. *All eiders appear dark below in flight. Flash patterns (for example, contrasting white axillaries on this nonbreeding male) can be useful in the field.* NEW JERSEY, DECEMBER

BREEDING MALE. *Ornate: head blue, green, yellow; bill red. Some individuals less ornate; on these, note small bill, steep forehead, and at least hints of ostentatious coloration.* ALASKA, JUNE

Spectacled Eider

Somateria fischeri   CODE 2

L 21"	WS 33"	WT 3.4 lb	♂ > ♀

▸ two adult molts per year; complex alternate strategy
▸ strong age-related, sex-related, seasonal differences
▸ individual plumage variation, especially adult male

Like King and Common Eider, Spectacled is an arctic breeder that winters in marine habitats. Breeders depart for offshore molting grounds by late May, lingering until early fall; entire species then winters in central Bering Sea. Listed in U.S. as Threatened.

Female gives grouse-like series of clucks, grunts; male in display gives soft, syncopated hooting.

BREEDING MALE. *Green head with white "goggles"; bill orange. Female, immature male, and eclipse male duller, but all show goggles, short bill, sloping forehead.* ALASKA, JUNE

Steller's Eider
Polysticta stelleri CODE 2

L 17"	WS 27"	WT 1.9 lb	♂ > ♀

- two adult molts per year; complex alternate strategy
- strong age-related, sex-related, and seasonal differences
- adult female head varies from cold brown to warm buff

Steller's stands apart in many ways from the three other eider species. Away from the breeding grounds, it is less likely to be found beyond the surf or out at sea. Population is declining, and species is listed by U.S. as Threatened.

Calls varied, mainly rolling and guttural. Feeding flocks may be talkative; otherwise, rarely heard. Listen for female *hooey* amid clucking of both sexes.

BREEDING MALE. *Like other eiders, boldly patterned black-and-white with suffusions of other colors. Bill lacks the odd shape and bright colors shown by other eiders.* EUROPE, APRIL

ADULT FEMALE. *Easily overlooked; plumage brown overall, face blank and staring. Best mark is inner wing: dark in the middle, edged with white, creating a two-barred appearance.* EUROPE, MARCH

Harlequin Duck
Histrionicus histrionicus CODE 1

L 16.5"	WS 26"	WT 1.3 lb	♂ > ♀

- two adult molts per year; complex alternate strategy
- strong age-related, sex-related, seasonal differences
- Pacific male averages more orange on head than Atlantic

A habitat specialist. On wintering grounds, where most often found, typically seen perched on or swimming just off jetties or other marine rock formations. Breeds along turbulent streams, often in mountains and well inland. Numbers declining, especially in East.

Difficult to hear along rushing streams or pounding surf, but actually fairly noisy. Flocks at sea give a whistled *puh-wee, puh-wee*, often in rapid series.

BREEDING MALE. *In all plumages, tail is fairly long and bill is fairly small.*
NEW JERSEY, FEBRUARY

BREEDING MALES. *Exquisite; note crisp blue-and-white markings with variable orangish and chestnut highlights.* NEW JERSEY, FEBRUARY

ADULT FEMALE. *Lacks intense hues and striking contrast of male. Large white spot on cheek is usually conspicuous; note also variable whitish splotching around eyes and base of bill.* ALASKA, MARCH

IN FLIGHT. *Dark overall; wings blackish above and below. Usually flies in small, tightly packed flocks.*
NEW JERSEY, FEBRUARY

Surf Scoter

Melanitta perspicillata CODE 1

ADULT MALE. *Black overall with bright white patches on nape and forecrown give "skunk head" appearance. Bill distinctive: long and sloping, large overall, colorful.* CALIFORNIA, JANUARY

ADULT FEMALE. *Bill size and structure similar to male, but gray overall. Head usually shows dark crown and paler cheeks, variable white patches behind eye and at base of bill.*
CALIFORNIA, JANUARY

IN FLIGHT. *Like Black Scoter, a "dark-winged" scoter. Appears lankier in flight than Black. Two lead birds are females; other five are adult males.* NEW JERSEY, FEBRUARY

L 20"   WS 30"   WT 2.1 lb   ♂ > ♀

▸ two adult molts per year; complex alternate strategy

▸ strong age-related and sex-related differences

▸ extensive variation in plumage and bill color

Breeds on shallow inland lakes; winters primarily along both coasts. Seasonal and daily movements, involving thousands of birds, are impressive. Widely noted in smaller numbers on passage inland, especially in Great Lakes and Northeast regions. Populations declining.

Fairly quiet at sea, but listen on breeding grounds for a low *krruk* (female) or whistled *biddle ip* (male). All scoters produce loud wing whistle in flight.

Black Scoter

Melanitta nigra CODE 1

PAIR. *Female sooty brown all over, except for contrastingly paler face. Male body entirely black; dark bill has variable yellow protuberance at base, often visible at a distance.* NEW JERSEY, MARCH

FEMALE. *Can appear similar to female or nonbreeding male Ruddy Duck. Similarities include dark plumage overall with paler cheeks, and tail posture.*
CALIFORNIA, NOVEMBER

IN FLIGHT. *Black Scoters (here three males and a female) appear stockier in flight than the other scoter species with all-dark wings, Surf Scoter, which is lankier.* NEW JERSEY, FEBRUARY

L 19"   WS 28"   WT 2.1 lb   ♂ > ♀

▸ two adult molts per year; complex alternate strategy

▸ strong age-related and sex-related differences

▸ some individual variation in bill color and structure

Like other scoters, breeds at high latitudes and winters along coasts; widely noted inland, usually in small numbers. Black Scoter may be found accompanying other scoters and sea ducks, but is somewhat more clannish and more likely to be found in pure flocks.

Flocks at sea vocal. Male has two calls, both with begging quality: shrill peep notes; thin, high *eeeee*. Female quack is throaty, similar to other scoters.

White-winged Scoter
Melanitta fusca CODE 1

L 21"	WS 34"	WT 3.7 lb	♂ > ♀

- two adult molts per year; complex alternate strategy
- strong age-related and sex-related differences
- extent of whitish markings on face variable

Best known from coastal wintering grounds, where large flocks migrate by day; often joins Black and Surf Scoters and other "sea ducks" in communal roosting and feeding aggregations. South of breeding grounds, migrants may be noted in small numbers anywhere.

Usually quiet at sea; distant birds hard to hear over waves. Male display call distinctive: melancholy *eeeee-urr*, both notes sweet, wavering.

ADULT MALE. *Intense black contrasts with eye patch. White wing patch, striking in flight, often visible on swimming bird. Peculiar bill, colorful and tapered.* CALIFORNIA, FEBRUARY

FEMALE. *Individuals other than adult males can be difficult to age and sex. On all White-winged, note combination of bill structure, white wing patch, and large overall size.* CALIFORNIA, MARCH

FEMALE. *Bill dark; colorful on male. Sloping forehead and long bill evident in all plumages. White patches on face extensive but variable on young birds; reduced on older birds.* CALIFORNIA, JANUARY

Bufflehead
Bucephala albeola CODE 1

L 13.5"	WS 21"	WT 13 oz	♂ > ♀

- two adult molts per year; complex alternate strategy
- strong age-related, sex-related, and seasonal differences
- extent and color of head gloss on male variable

Generally common and conspicuous. Outside the breeding season, occurs in many habitats, from quiet ponds to large bays; large flocks are seen inland as well as coastally. In the breeding season, prefers wooded waterways; requires tree cavities for nesting.

Both sexes give soft, low-pitched calls. Male *grrrk*, often doubled or trebled, calls to mind a distant Common Raven. In general, a quiet duck.

IN FLIGHT. *Both sexes show peculiar rocking motion in flight. Breeding male (here) in flight shows high-contrast black-and-white all over; female darker overall.* COLORADO, FEBRUARY

BREEDING MALE. *Large white patch on black head; the black head feathers iridescent in good light. White body contrasts with black back. Nonbreeding male shows much less white overall.* CALIFORNIA, JANUARY

ADULT FEMALE. *Dusky overall. Dark head has contrasting white patch below and behind eye. Both sexes are small-bodied. Juvenile similar to adult female, but white patch reduced.* CALIFORNIA, JANUARY

Long-tailed Duck
Clangula hyemalis   CODE 1

BREEDING MALE. *Adult males are strikingly black-and-white all year, with very long tails. Plumages variable; darker overall in summer than in winter.* MANITOBA, JULY

BREEDING FEMALE. *Females are variably dark-and-white throughout the year, almost always showing white on the belly and around the eye.* MANITOBA, JULY

IN FLIGHT. *All plumages (adult female and two nonbreeding adult males here) are dark-winged above and below, with extensive white on belly.* NEW JERSEY, FEBRUARY

L 16.5–21"   WS 28"   WT 1.6 lb   ♂ > ♀

▸ up to three plumages per year; complex alternate strategy

▸ strong age-related, sex-related, seasonal differences

▸ strong individual plumage variation in both sexes

Breeds north of tree line; winters widely along coasts, both out at sea and in protected inshore lagoons. Coastal migration is usually near shore. Large flights are noted inland in the Northeast and Great Lakes region, smaller numbers inland elsewhere.

Highly vocal on offshore wintering grounds. Familiar call, a far-carrying *owl omelet*, accented on last syllable. Distant flocks in chorus mellow, evocative.

IMMATURE MALE. *All males usually show some pink on bill; females usually lack pink. Note dark-and-white plumage overall, with white on belly and around eye.* CALIFORNIA, JANUARY

Common Goldeneye

Bucephala clangula CODE 1

L 18.5"	WS 26"	WT 1.9 lb	♂ > ♀

- two adult molts per year; complex alternate strategy
- strong sex-related and moderate age-related differences
- female bill color variable; subadult plumage variable

Breeds on ponds in wooded districts in the north; disperses in broad southerly front to winter on bays and lakes. Adults form pairs in winter, far in advance of breeding season and often far from breeding grounds. Winter birds active, often chasing and splashing.

Most familiar sound is wing whistle: steady hum, surprisingly loud, with strange mechanical quality. Male display call, *p'peeent*, suggests American Woodcock.

IN FLIGHT. *Both sexes (breeding male here) show more white on upperwing than Barrow's Goldeneye.* COLORADO, JANUARY

BREEDING MALE. *Black head, iridescent in good light, has large white spot at base of bill. Scapulars more extensively white than Barrow's; effect conspicuous on swimming bird.* CALIFORNIA, JANUARY

ADULT FEMALE. *Typically has mainly dark bill, sloping forehead. Some individuals approach female Barrow's in head structure, especially when diving, and in bill color.* CALIFORNIA, JANUARY

IMMATURE MALE. *Plumage variable; some individuals show extensive white spot at base of bill, others do not. Sloping forehead usually a good field mark on all but youngest birds.* CALIFORNIA, JANUARY

Barrow's Goldeneye

Bucephala islandica CODE 1

L 18"	WS 28"	WT 2.1 lb	♂ > ♀

- two adult molts per year; complex alternate strategy
- strong sex-related and moderate age-related differences
- hybrids with Common Goldeneye sometimes noted

Breeds along pond edges in semi-wooded habitats. Most winter along coasts, but also at scattered locations throughout the West. Usually outnumbered by Common on wintering grounds. Two discrete populations: One breeds in the Rocky Mountain region, the other in Québec.

Wing whistle not as shrill and loud as Common. Male display call very different from Common: a short, sputtering *pack-ack*, accented on second syllable.

IN FLIGHT. *Both sexes (adult female here) have less white on upperwing than Common Goldeneye.* ALASKA, FEBRUARY

BREEDING MALE. *Vertical white patch at base of bill is crescent-shaped. Forehead very steep. Swimming bird typically shows more dark on sides than Common.* CALIFORNIA, JANUARY

ADULT FEMALE. *Bill extensively yellow. Forehead steep, but effect depends on bird's behavior and observer's angle. Some variants and hybrids are indistinguishable from Common.* CALIFORNIA, JANUARY

IMMATURE MALE. *Immatures, especially females, can be difficult or impossible to distinguish from Common. Differences in head structure are usually evident by early winter.* CALIFORNIA, JANUARY

Common Merganser

Mergus merganser CODE 1

BREEDING MALE. *Black head, iridescent green in good light, and black back contrast with white body. Bill red or red-orange. Some show rose or peach suffusion on breast.* CALIFORNIA, JANUARY

ADULT FEMALE. *Chestnut head and throat are sharply demarcated from gray breast. White chin patch prominent. Forehead slopes smoothly into bill base. Nonbreeding male similar.* CALIFORNIA, JANUARY

| L 25" | WS 34" | WT 3.4 lb | ♂ > ♀ |

▸ two adult molts per year; complex alternate strategy

▸ strong seasonal and sex-related differences

▸ variable wash on breast of male related to diet

Breeds along rivers and lakes in wooded regions. Widespread in winter; more common inland in West than East. In coastal areas, pre- fers sheltered freshwater lagoons; Red-breasted usually favors more saline environments. Migrates early in spring, fairly late in fall.

Most commonly heard call, apparently given by both sexes, *mruff*, often in series, with muffled quality. Flushing birds often noisy; otherwise, usually silent.

Red-breasted Merganser

Mergus serrator CODE 1

BREEDING MALE. *Black head, glossed with green, has variable shaggy crest. Ruddy breast is separated from dark head by white neck. Nonbreeding male resembles female.* CALIFORNIA, MARCH

ADULT FEMALE. *Female and nonbreeding male Red-breasted (here) and Common show white secondaries contrasting with dark upperparts. Male of both species shows more white in flight.* NEW JERSEY, MARCH

IMMATURE. *Pale overall; orangish head blends into grayish body. Compared to Common, forehead steeper and bill base narrower, but structural distinction muted on diving birds.* CALIFORNIA, JANUARY

| L 23" | WS 30" | WT 2.3 lb | ♂ > ♀ |

▸ two adult molts per year; complex alternate strategy

▸ strong seasonal and sex-related differences

▸ head structure affected by feeding behavior

Disperses widely from northern breeding grounds; most birds winter in saltwater habitats along the coasts. Large inland passage over eastern Great Lakes region. Significant population fluctua- tions are apparently normal in this species.

Calls are varied, but infrequently heard: *mreck mreck* of female is not nearly as gruff and muffled as corresponding call of Common Merganser.

Hooded Merganser
Lophodytes cucullatus CODE 1

BREEDING MALE. *Whether compressed or erect, black-and-white crest is striking. Sides orange, rest of bird black-and-white. Nonbreeding males resemble females.* CALIFORNIA, NOVEMBER

ADULT FEMALE. *Warm reddish and gray all over; warmest tones on crest. Bill straw-yellow. Overall structure distinct in all plumages: small and dumpy, large-headed, and small-billed.* TEXAS, NOVEMBER

L 18"	ws 24"	wt 1.4 lb	♂ > ♀

▸ two adult molts per year; complex alternate strategy

▸ strong seasonal and sex-related differences

▸ male crest can be quickly lowered or raised

During most of year associated with quiet freshwater habitats, especially favoring deep water with trees. Both breeding and wintering ranges are expanding into new regions. Winters in small flocks scattered across a landscape, not in large rafts.

Calls varied, but soft and infrequently heard; vocalizations of both sexes are grating, croaking, sometimes froglike. Wing whistle shrill, modulated.

Smew
Mergellus albellus CODE 3

BREEDING MALE. *Exquisite; mainly white, with black facial mask and sparse, thin black lines on body. In flight, largely black wings contrast with white body.* CALIFORNIA, JANUARY

ADULT FEMALE. *Relationship with mergansers evident on female: bill long, pointed; head brownish; throat white; body gray. Nonbreeding male and immature similar to adult female.* CALIFORNIA, JANUARY

L 15–17"	ws 22–27"	wt 1.3–1.6 lb	♂ > ♀

▸ two adult molts per year; complex alternate strategy

▸ strong seasonal and sex-related differences

▸ extent and hue of brown on females variable

Breeds along wooded waterways in northern regions of Old World; regularly reaches western Alaska. Records elsewhere hard to assess; some are escapes from captivity, but true vagrants are possible anywhere. Wild birds are active and flighty; escapes tame and confiding.

Varied vocalizations have froglike quality similar to Hooded Merganser, but lower, more growling, more complex. Vagrants to North America usually silent.

Ruddy Duck
Oxyura jamaicensis CODE 1

BREEDING MALE. *Body rich chestnut, head black with large white cheek patch; bill bright blue. Spiky tail may be held erect or carried flat on water's surface.* CALIFORNIA, FEBRUARY

ADULT FEMALE. *Variable dusky line across pale cheek. All Ruddy Ducks are oddly proportioned: dumpy and large-headed, like a toy rubber duck.* CALIFORNIA, JANUARY

NONBREEDING MALE. *Pattern similar to breeding male, but bill duller and chestnut replaced by grayish brown. Some resemble Black Scoter, but note differences in overall structure.* CALIFORNIA, FEBRUARY

| L 15" | ws 18.5" | wt 1.2 lb | ♂ > ♀ |

- two adult molts per year; complex alternate strategy
- strong seasonal and sex-related differences
- rare morph with all-black head suggests Masked Duck

Breeds mainly in West, but opportunistic and may nest far from main range. In winter, congregates in large rafts. During most of year, especially on migration and in winter, it is seemingly lethargic. On the breeding grounds, it is highly animated and aggressive.

During display, male gives rapid sputtering series suggesting Sedge Wren. Female note one syllable, abrupt and sharp. Generally silent away from breeding grounds.

Masked Duck
Nomonyx dominicus CODE 3

BREEDING MALE. *Chestnut overall like Ruddy Duck, but usually with extensive black splotching. Note also black cheek and chestnut nape. Dumpy, like Ruddy, but smaller.* TEXAS, APRIL

ADULT FEMALE. *Recalls Ruddy Duck, but buff face is crossed by two horizontal black stripes; on Ruddy, whitish face crossed by one horizontal black stripe.* TEXAS, APRIL.

| L 13.5" | ws 17" | wt 13 oz | ♂ > ♀ |

- two adult molts per year; molt strategy unknown
- strong seasonal and sex-related differences
- amount of black on body of breeding male variable

Reaches range limit in south Texas, where status poorly known. Population may be sizeable, but the species is secretive; sightings, usually of single birds, are uncommon. Inhabits small, quiet ponds with dense vegetation; generally less forthcoming than Ruddy Duck.

Rarely heard, but vocal repertoire diverse. Male calls have rolling, cooing quality: *krroo-krroo-krroo* or *droo-d'loo*, often given in series.

Upland Game Birds
Chachalacas, Grouse & Allies, and New World Quail
ORDER GALLIFORMES

---

24 species: 10 ABA CODE 1   14 CODE 2

The birds of this order are the upland game birds, inextricably linked in many people's minds with America's rich hunting culture. Some are among our best-known species—for example, the Ring-necked Pheasant, Wild Turkey, and Northern Bobwhite. Three families occur in our area: the tropical Cracidae, represented in the ABA Area only by the Plain Chachalaca of southern Texas; the diverse Phasianidae, including grouse, pheasants, turkeys, and others; and the New World quail, in the family Odontophoridae.

Most species of upland game birds are closely tied to terrestrial habitats: grasslands and agricultural fields, shady forest floors, and harsh alpine tundra. In such habitats, coveys forage unobtrusively for plant matter, especially seeds. For much of their annual cycle, game birds often sit tight, not flushing until practically stepped on. During the courtship season, however, many species (especially in the family Phasianidae) give spectacular displays. For example, the communal courtship rituals, known as leks, of male Greater Prairie-Chick-ens are among the most stirring natural phenomena in North America.

Upland game birds are largely nonmigratory. Locomotion is often by foot—for example, quietly ambling amid undergrowth, quickly scurrying across clearings. But all of our game birds are capable of flight, sometimes impressively so, and the wing-whirring of a flushing game bird is often startlingly loud. Many species cluck and cackle softly while foraging. Vocalizations are highly varied, but generally unmusical. Examples include the frenzied chorusing of Plain Chachalacas, the eerie gulping and clicking and foot-stomping of lekking Sharp-tailed Grouse, and the ringing proclamations of most quail species.

Because most upland game birds are hunted, conservation is a complicated matter. Four species (Chukar, Himalayan Snowcock, Gray Partridge, and Ring-necked Pheasant) are not native; they were introduced from the Old World. Others, although native, have been released outside their natural range; these include the White-tailed Ptarmigan, Wild Turkey, and Northern Bobwhite. Dependence on high-quality grassland and rangeland in the interior threatens both sage-grouse species, the two prairie-chickens, and Sharp-tailed Grouse. Hunting *per se* is not usually what imperils game bird populations. Rather, the prime culprit is habitat conversion, and two major threats have recently emerged: loss of sagebrush to huge wildfires, and oil-and-gas exploration in the western Great Plains.

Plain Chachalaca

Ortalis vetula   CODE 2

| L 22" | WS 26" | WT 1.2 lb | ♂ > ♀ |

▸ one adult molt per year; complex basic strategy

▸ moderate differences between juvenile and adult

▸ red throat patch variable, often hard to see

ADULT. *Frequently seen in trees, both in dense vegetation and on open perches. On birds in flight or clambering about, note white tips on tail feathers.* TEXAS, NOVEMBER

The only member of the tropical family Cracidae to reach North America. Like other cracids, a large forest bird that makes a lot of noise. In our area, occurs in a variety of wooded haunts: dry scrublands, wet riparian forests, and even shady residential neighborhoods.

Very loud. Small flocks call antiphonally. Basic element is repeated *choch-ah-lock* or *choch-ah-lock-ah.* Male and female "choruses" can be distinguished.

Gray Partridge

Perdix perdix   CODE 2

| L 12.5" | WS 19" | WT 14 oz |

▸ two adult molts per year; complex alternate strategy

▸ moderate differences between adult male and female

▸ extent of belly patch and overall ground color variable

Introduced; most stock of eastern European origin. Widely but thinly distributed in North America. Where established, does best in farming regions; usually encountered in small coveys. Populations fluctuate greatly, but long-term trend is a decline.

Most common call is a grating *keeeeer-ah,* rather tern-like. Feeding birds cluck quietly; flushing coveys create loud wing-whirring and agitated cackles..

ADULT MALE. *All adults are rotund and portly. Extensively gray body has rusty bars on flanks and black patch on belly. Head dull rusty.* EUROPE, APRIL

Chukar

Alectoris chukar   CODE 2

| L 14" | WS 20" | WT 1.3 lb | ♂ > ♀ |

▸ one adult molt per year; complex basic strategy

▸ moderate differences between juvenile and adult

▸ bare parts of adult vary greatly in color

Introduced from South Asia. Well established in arid, rocky foot-hills and grasslands of Intermountain West; favors degraded grasslands and shrublands, especially those invaded by exotic cheatgrass. Elsewhere in North America widely released for hunting.

Calls simple, often given as series, for example, *luck luck luck* or *daluck daluck daluck.* Feeding birds cluck softly, flushing birds excitedly.

ADULT. *Pale overall; juvenile duller, duskier. Cream-colored face has distinctive black border, absent in juvenile; creamy-white flanks have bold black barring. Bill red.* CALIFORNIA, NOVEMBER

Himalayan Snowcock
Tetraogallus himalayensis CODE 2

L 28"	ws 36"	wt 5 lb

▸ one adult molt per year; complex basic strategy

▸ moderate differences between juvenile and adult

▸ ground color varies from cold gray to warmer buff

Introduced from Pakistan in mid 20th century. In North America, established only in the Ruby and Humboldt Mountains of northeastern Nevada. Favors alpine meadows with boulder-strewn slopes. Daily movements extensive; wary and difficult to glimpse, even in prime habitat.

Like many game birds, gives various cackling notes. One call is highly distinctive: a curlew-like wailing that rises sharply, *wheee-eeee-eep!*

ADULT MALE. *Large; proportioned like an oversized Chukar. Gray head and "bib" contrast with dark body. Juvenile and adult female paler overall; white of adult male replaced by dull gray.* NEVADA, JANUARY

Ring-necked Pheasant
Phasianus colchicus CODE 1

L 21-35"	ws 31"	wt 2.5 lb	♂ > ♀

▸ one adult molt per year; complex basic strategy

▸ strong differences between adult male and female

▸ hybrids and domestic variants sometimes encountered

Introduced long ago to North America; still widely released. Occurs in various semi-open habitats; especially favors agricultural districts. Sometimes wanders out into pastures, but usually sticks close to cover. Runs swiftly, flies strongly over short distances.

◀ Male display call is an explosive double honk, *arnk! awk*, followed by wing-whirring. Honking carries well, but wing-whirring inaudible at a distance.

GREEN PHEASANT (Phasianus versicolor). *Our pheasants derive from diverse lineages. Intergrades within species are frequent; hybrids with other species such as Green Pheasant also occur.* ASIA, JUNE

INDIAN PEAFOWL (Pavo cristatus). *Unmistakable: huge, with unique parasol-shaped crest. Male attire is famously outlandish. Established in California.* ASIA, JANUARY

ADULT MALE. *Large-bodied and slender, with long tail. Green head has large red facial wattles. Body bronzy all over, with intricate variegation; white neck ring often conspicuous.* CALIFORNIA, APRIL

ADULT FEMALE. *Confusion with female Sharp-tailed Grouse possible; female pheasant shows warmer tones overall; tail on flushing bird warm brown on pheasant, whitish on grouse.* CALIFORNIA, APRIL

Sharp-tailed Grouse

Tympanuchus phasianellus CODE 2

ADULT MALE. *In display, lavender sacs on sides of neck are inflated, yellowish comb above eye is exposed. Note spotting on breast and belly, white tail.* MONTANA, JUNE

ADULT. *Female and nondisplaying male are similar. Whitish, pointed tail good mark of separation from prairie-chickens. Body is speckled and spotted; prairie-chickens are barred.*

MONTANA, JUNE

L 17"	WS 25"	WT 1.9 lb	♂ > ♀

- one adult molt per year; complex basic strategy
- moderate differences between female and displaying male
- paler toward southern limit of range than elsewhere

A bird of steppe habitats, but sometimes ranges into dry woodlands. In much of range, especially toward south, occurs on rangeland where grazing is not too intense. Southern population of "Columbian" (*T. p. columbianus*) declining; extirpated from some areas.

In early spring, adults assemble at communal courtship grounds called leks. Male gives hooting and swallowing sounds interspersed with cackling and clapping.

Greater Prairie-Chicken

Tympanuchus cupido CODE 2

ADULT MALE. *Inflates air sacs on sides of neck and orange-yellow eye combs; oversized "ears" (pinnae) held erect above head. Except in display, male resembles female.* MINNESOTA, JUNE

ADULT FEMALE. *Nearly identical to Lesser Prairie-Chicken, but ranges are disjunct. Sharp-tailed Grouse also similar, but prairie-chickens' tail is darker, more rounded, more barred.* MINNESOTA, JUNE

L 17"	WS 28"	WT 2 lb	♂ > ♀

- one adult molt per year; complex basic strategy
- strong differences between female and displaying male
- "Attwater's" (*T. c. attwateri*) more brightly colored

Range greatly reduced because of hunting and habitat loss. Local, but common in some places; seen in agricultural districts and along roadsides. Mainly restricted to tallgrass prairie in the Midwest; "Attwater's" Prairie-Chicken of coastal Texas is endangered.

Lekking males strut slowly and prance wildly, giving long, low-pitched hoots: *hoom badoooom*. Hoots interspersed with higher-pitched whistling and squealing.

Lesser Prairie-Chicken

Tympanuchus pallidicinctus CODE 2

ADULT MALE. *Posture of courting male similar to Greater Prairie-Chicken, but note reddish air sacs of Lesser (Greater yellow). Adult male similar to female except in display.* NEW MEXICO, APRIL

ADULT FEMALE. *Nearly identical to female Greater, but current ranges do not overlap. Breast paler and barring on breast thinner than Greater; slightly smaller overall than Greater.* NEW MEXICO, APRIL

L 16"   WS 25"   WT 1.6 lb   ♂ > ♀

▸ one adult molt per year; complex basic strategy

▸ strong differences between female and displaying male

▸ air sac color variable, but always with reddish hue

Rare and declining. Currently restricted to southern Great Plains; individual populations increasingly isolated from one another. Occurs in arid shortgrass prairie with sandy soils and widely scattered yuccas and shrubs.

Males display at communal leks. Compared to Greater Prairie-Chicken, male display calls higher-pitched, more bubbly and musical; reminiscent of Sandhill Crane.

Greater Sage-Grouse

Centrocercus urophasianus CODE 1

ADULT MALE. *In display, stands in place; does not prance and run about. As if to accentuate its size, head and tail feathers are held erect, breast is puffed out.* CALIFORNIA, MARCH

ADULT FEMALE. *All show some amount of contrast on underparts: splotchy black belly separated from gray-brown breast by variable white patch. Large size usually apparent.* CALIFORNIA, MARCH

L 22–28"   WS 33–38"   WT 3.3–6.3 lb   ♂ > ♀

▸ two adult molts per year; complex alternate strategy

▸ strong sex-related differences in size and plumage

▸ individual variation in overall plumage contrast

Dependent on high-quality sagebrush habitat. Populations are declining everywhere, sharply in some places. Large-scale movements do not occur, but individuals and family groups may wander widely within appropriate habitat.

Lekking males give complex series of gulping, popping, and whooshing sounds. Most call elements are soft, but pops and gulps carry over great distances.

Gunnison Sage-Grouse
Centrocercus minimus　CODE 2

DISPLAYING MALE. *Filoplumes (specialized head feathers) are longer than in Greater Sage-Grouse and are waved about more wildly than Greater.* COLORADO, APRIL

ADULT FEMALE. *Nearly identical in plumage to female Greater, but substantially smaller; ranges of the two species do not overlap. Tail of Gunnison is paler than Greater.* COLORADO, AUGUST

| L 18–22" | WS 26–30" | WT 2.4–4.6 lb | ♂ > ♀ |

- two adult molts per year; complex alternate strategy
- strong sex-related differences in size and plumage
- individual variation in overall paleness may occur

Resident in the cold high desert of Colorado's Gunnison Basin. Unknown until the late 20th century; not "officially" described to science until 2000. Although the species was overlooked, it actually differs in many respects from Greater Sage-Grouse.

Lekking males give series of low *plup* sounds, interrupted in midseries by faint wing-whooshing. Compared to Greater, Gunnison sounds are simple.

Spruce Grouse
Falcipennis canadensis　CODE 2

ADULT MALE. *A dark bird; black overall with blood-red eye combs. In most of range, tail is buff-tipped, as here; "Franklin's" Grouse (F. c. franklinii) of Pacific Northwest has all-black tail feathers.* MICHIGAN, MAY

ADULT FEMALE. *Similar to other forest grouse, but barring is more extensive and overall build is especially rotund. Ground color varies from fairly cold, as here, to more rufescent.* MICHIGAN, MAY

| L 16" | WS 22" | WT 1 lb | ♂ > ♀ |

- one adult molt per year; complex basic strategy
- strong differences between adult male and female
- "Franklin's" of Pacific Northwest distinctive

Famously tame; known as the "fool hen" for its approachability. Intimately tied to the boreal zone; feeds on pine and spruce needles in dense conifer forests, often up in trees; also forages on or near the ground. Populations fluctuate due to natural causes.

Displaying male gives series of low, muffled hoots: *brr brr brr brr.* "Franklin's" adds wing-clapping at end of display. Both sexes cackle while feeding.

Sooty Grouse
Dendragapus fuliginosus CODE 2

ADULT MALE. *Both Sooty and Dusky (once classified as a single species) average darker in humid north, paler in arid south. At corresponding latitudes, tail band of Sooty is paler and more contrasting.* CALIFORNIA, APRIL

ADULT MALE. *In display, air sacs on sides of neck are yellow or yellow-orange (raspberry on Dusky) Note that bird is displaying from tree (Dusky on ground or fallen log).* CALIFORNIA, APRIL

L 20"	WS 26"	WT 2.3 lb	♂ > ♀

▸ one adult molt per year; complex basic strategy
▸ moderate differences between adult male and female
▸ hybridizes with Dusky Grouse in zone of overlap

Occurs farther west than Dusky Grouse, in generally wetter forests; extent of overlap in Columbia River valley imperfectly known. Like Dusky, widespread in mountain forests, but also ranges down to sea level, as well as on forested islands offshore.

Male display distinct from Dusky. From high in trees, Sooty delivers a series of muffled and well-spaced hoots audible over great distances.

Dusky Grouse
Dendragapus obscurus CODE 2

ADULT MALE. *In display, inflates raspberry neck sac, puffs up red-orange eye comb. Most have all-gray tail; others show pale tail tip, but not as contrastingly as Sooty.* WYOMING, JULY

ADULT FEMALE. *Dark, with white spotting. Lacks extensive thin barring of female Spruce Grouse, thick barring of Ruffed. Females feeding in sagebrush sometimes mistaken for sage-grouse.* COLORADO, JULY

L 20"	WS 26"	WT 2.3 lb	♂ > ♀

▸ one adult molt per year; complex basic strategy
▸ strong differences between adult male and female
▸ generally darker in northern part of range, paler south

Rocky Mountain counterpart of Sooty Grouse. Occurs widely in high-elevation conifer forests, but also wanders into adjacent grasslands; foragers are sometimes seen more than one mile out in sagebrush scrublands, where confusion with female sage-grouse possible.

Male displays from fallen log or on ground; gives series of quiet hoots, inaudible except at close range. Both sexes cluck softly while feeding; wingbeats noisy when flushing.

Ruffed Grouse

Bonasa umbellus   CODE 1

ADULT GRAY MORPH. *Adults of both morphs are splotchy in variable earth tones; all show black bars on flanks. Note crest; usually erect, sometimes not. Pale tail has black subterminal band.* MINNESOTA, JANUARY

ADULT RED MORPH. *Structure and plumage pattern identical to gray morph, but note reddish highlights. Juvenile similar, but "ruff" (erectable feathers on hindneck) not fully developed.* QUÉBEC, JULY

L 17"	WS 22"	WT 1.3 lb	♂ > ♀

▸ one adult molt per year; complex basic strategy

▸ moderate differences between juvenile and adult

▸ two color morphs; gray mainly north, red mainly south

A fairly large grouse of hardwood and mixed forests. Highest densities occur in early successional forests recovering from logging or agriculture. Almost never wanders away from woods. Heavily hunted and preyed upon, but populations are generally stable.

Displaying male produces low-frequency drumming with wings. Begins slowly, then speeds up, like a motor starting. At a distance, first part of drum inaudible.

Willow Ptarmigan

Lagopus lagopus   CODE 1

BREEDING MALE. *Large ptarmigan with relatively large bill. Breeding male has dark ruddy head and neck; rest of body variably mottled with black and white.* ALASKA, JUNE

BREEDING FEMALE. *Cryptically colored like other ptarmigan species. Overall tone is warmer than White-tailed. Very similar to female Rock. Willow has larger bill.* ALASKA, JUNE

L 15"	WS 24"	WT 1.2 lb	♂ > ♀

▸ three adult molts per year; complex alternate strategy

▸ strong seasonal and sex-related differences

▸ subtle variation in structure, size, ground color

Widespread and numerous in its northern haunts. Inhabits wet and shrubby habitats in subarctic and subalpine zones; generally favors less extreme arctic conditions than similar Rock Ptarmigan. Some populations migratory, others mainly sedentary.

Male gives comical chuckles, strangely similar to male human voice: *go back, go back!* and *tobacco, tobacco, tobacco!* Both sexes cluck while feeding.

White-tailed Ptarmigan

Lagopus leucurus CODE 2

BREEDING FEMALE. *Intricate variegations of brown, black, and white blend perfectly with tundra ground cover.* COLORADO, JULY

BREEDING MALE. *Splotchy black-and-white all over. Red eye comb is hard to see except on displaying males.* COLORADO, JUNE

L 12.5"   WS 22"   WT 13 oz   ♂ > ♀

▸ three adult molts per year; complex alternate strategy

▸ strong seasonal and sex-related differences

▸ northern birds average smaller and grayer

Like all ptarmigans, well adapted to tundra life; prefers areas with rocky outcroppings. White-tailed is the southernmost ptarmigan, reaching south at high elevations to New Mexico. Populations stable, but warming climate may be a threat.

Calls of displaying male range from low and growling to strident and cackling; for example, *kree kree! krick.* Both sexes cluck continually when feeding.

WINTER ADULT. *Plumage completely white, including tail; other ptarmigans have black on tail. A small-billed, small-bodied ptarmigan.* COLORADO, APRIL

SUMMER–AUTUMN ADULT. *After breeding, becomes mainly gray with reddish highlights and variable white splotching. Note rocky habitat.* COLORADO, JUNE

IN FLIGHT. *Flight is noisy and usually brief. White in tail is prominent even in otherwise dark breeding plumage.* COLORADO, JULY

Rock Ptarmigan
Lagopus mutus   CODE 1

BREEDING MALE. *Breeding-season males are variable, especially in Bering Sea region. Most acquire extensive brown or black in summer.*
ALASKA, JUNE

ADULT MALE. *Like Willow, white overall with black on tail. Note black line through eye, variable red comb above eye. Unlike other ptarmigan, all-white plumage may be worn by breeding and nonbreeding males.*
ALASKA, JUNE

BREEDING MALE. *In populations on the outer Aleutians, formerly treated as a separate species, breeding males are mainly black on upperparts.*
ALASKA, MAY

L 14"	WS 23"	WT 15 OZ	♂ > ♀

▸ three adult molts per year; complex alternate strategy

▸ strong seasonal and sex-related differences

▸ much geographic variation in color and contrast

Ranges farther north than other ptarmigan species. Occurs in rockier and more barren habitats than Willow Ptarmigan, but some overlap. A short-distance migrant, sometimes dispersing in impressive flocks of several hundred birds. Very tame; can be easily approached.

Display includes aerial components; male gives low-pitched musical hoots, bizarre toneless rattles, and soft sibilant calls. Vocalizations lack "human" qualities of Willow Ptarmigan.

BREEDING FEMALE. *Similar to other female ptarmigans; bill is smaller than Willow; ground color of plumage not as cold as White-tailed.*
ALASKA, JULY

Wild Turkey
Meleagris gallopavo CODE 1

ADULT FEMALE. *Long tail; long neck; small head with little if any reddish color and protuberances. Southwestern birds barred with white, eastern birds darker.*
NEW MEXICO, NOVEMBER

L 37–46"	WS 50–64"	WT 9.2–16.2 lb	♂ > ♀

▸ one adult molt per year; complex basic strategy

▸ moderate differences between male and female

▸ geographic and individual variation in size and color

This adaptable native game bird is widely domesticated and frequently released outside its original range. In many places, it is unclear whether local populations are natural, introduced, or a mixture. Populations generally increasing, and colonizing urban areas.

◀ Male in display gives familiar gobble: begins startlingly loud, then tapers off into bubbly, mumbled notes. Both sexes cluck, cackle.

ADULT MALE. *In display, erects feathers and looks massive. Note peculiar tuft of feathers jutting out of breast. Note also extensive naked red on head, with variable protuberances.* CONNECTICUT, JUNE

NONBREEDING MALE. *Similar to female, but most show some reddish on head, bare part protuberances, breast tuft. Pale wing panel usually more conspicuous on male.*
CONNECTICUT, JUNE

Montezuma Quail
Cyrtonyx montezumae CODE 2

L 8.75"	WS 15"	WT 6 oz	♂ > ♀

▸ one adult molt per year; complex basic strategy

▸ strong differences between adult male and female

▸ in both sexes, prominence of crest variable

A shy bird that stays in dense cover; can be difficult to find, even in high-quality habitat. Prefers high desert grasslands with shrubs and scattered oaks. Compared to other quail, less apt to congregate in coveys; instead, usually seen singly or in pairs.

Male display call a quavering, downslurred whistle, hoarse and gasping, that trails off at end, *hee-e-e-e-e-oo-oo*; calls to mind Eastern Screech-Owl.

ADULT MALE. *Oddly proportioned, with round body and round head. Often called Harlequin Quail for its bizarre facial pattern; note also white spots all over dark body.*
TEXAS, APRIL

ADULT FEMALE. *Duller than male, but shares basic head pattern, including variable crest. Plumage details can be hard to see on flushing or hiding birds; note body structure.*
ARIZONA, JUNE

Northern Bobwhite
Colinus virginianus   CODE 1

ADULT MALE. *Most adult males have a bright white throat (puffed up in display) and broad white eyebrow. Ground color rufous; paler in Midwest, darker in some Florida populations.* FLORIDA, FEBRUARY

L 9.75"	WS 13"	WT 6 OZ

- ▸ two adult molts per year; complex alternate strategy
- ▸ moderate differences between adult male and female
- ▸ geographic and individual variation in size and color

Only native quail in eastern North America. Favors woodland edges, thickets. Population declining, but being restocked for hunting in many areas. Sometimes released well outside native range. Endangered "Masked Bobwhite" in southeastern Arizona is probably extirpated.

◀ Male *erp whoyt!* ("Bobwhite!") is well known in East and Midwest. Call sometimes preceded by soft *pwur* note, or shortened to a one-syllable *whoyt*.

ADULT FEMALE, "MASKED." (C. v. ridgwayi) *A few may still exist in southeastern Arizona. Both sexes darker and more rufous than other populations of Northern Bobwhite, especially males.* ARIZONA, JANUARY

ADULT MALE. *This individual is at the pale end of the normal range of variation in Northern Bobwhite.* TEXAS, NOVEMBER

ADULT FEMALE. *Head pattern similar to adult male, but more muted: white of male replaced with low-contrast buff on female.* FLORIDA, NOVEMBER

Gambel's Quail
Callipepla gambelii CODE 1

ADULT MALE. *Unmistakable in most of range, which overlaps slightly with California Quail. On Gambel's, note buff belly with large black splotch. Male topknot may be broken or missing.* ARIZONA, APRIL

ADULT FEMALE. *Duller overall than male. Topknot shorter; belly usually lacks black splotch. Female often accompanied by chicks.* ARIZONA, APRIL

L 10"	WS 14"	WT 6 OZ

- one adult molt per year; complex basic strategy
- strong differences between adult male and female
- hybridizes with California and Scaled Quail

A characteristic bird of the desert, but ventures into suburbs and even into major cities. Gathers in tight coveys. Greatest densities in washes, along bosques, and in residential areas with plantings. Wary; dashes quickly through clearings and across streets.

◀ Male call a crowing, musical *chew wee! buh* ("Chicago!"), often repeated. "Chicago" call frequently preceded by agitated clucking. Flushing coveys are noisy.

California Quail
Callipepla californica CODE 1

ADULT MALE. *Similar to Gambel's, with which California sometimes hybridizes. Browner overall than Gambel's. Best distinction is California's scaly belly.* CALIFORNIA, MARCH

ADULT FEMALE. *Note topknot. Similar to female Gambel's, but note scaly belly. Most females are grayish, as here; birds of wet coastal mountains more brownish.* CALIFORNIA, MARCH

L 10"	WS 14"	WT 6 OZ

- one adult molt per year; complex basic strategy
- strong differences between adult male and female
- averages paler in driest regions of range

Northern and western counterpart of Gambel's Quail. Occurs in cold Great Basin deserts and in warmer settings west of the Sierra. A familiar sight in agricultural regions; also ranges well up into mountains. Popularly known as Valley Quail.

Male "Chicago!" call similar to Gambel's, but lower-pitched, less musical. Male also has *quark* call that can be confused with Mountain Quail.

Mountain Quail
Oreortyx pictus   CODE 1

L 11"   WS 16"   WT 8 oz

▸ one adult molt per year; complex basic strategy

▸ moderate differences between adult male and female

▸ darker overall in northern portions of range

A quail of far western mountains. Favors slopes with extensive shrub cover; occurs to near treeline, rarely penetrates below foothills. Makes significant altitudinal migrations; some birds move upslope in winter, but most retreat to lower elevations.

Male courtship call a loud, crowing *quee-weerk*; far-carrying. Heard nearby, note has wavering, tremulous quality, but effect inaudible at greater distances.

ADULT MALE. *Has general body plan of* Callipepla *quails. Rusty and gray tones contrast colorfully. Long double plumes on head typically held straight. Female is similar but duller.*
CALIFORNIA, JUNE

Scaled Quail
Callipepla squamata   CODE 1

ADULT. *Popularly known as "Cottontop" for variable tuft of white feathers atop head. Male sandy gray-brown; female a bit duller. Underparts scaled; scaling extends to back and nape.* TEXAS, NOVEMBER

ADULT. *At close range, note intricate structure of the contour feathers of the body: oblate gray "scales" interspersed with smaller, lanceolate, burnt-sienna down feathers.* TEXAS, APRIL

L 10"   WS 14"   WT 6 oz

▸ two adult molts per year, complex alternate strategy

▸ weak differences between adult male and female

▸ some eastern birds have chestnut on belly

Occurs in both shortgrass prairie and harsh Chihuahuan Desert scrublands. Individuals and coveys often seen running across washes, darting into cover. Populations are declining in and withdrawing from some regions, apparently a result of overgrazing.

Two male call types: a rasping *kweesh!*, lacking the musical quality of most other quails; a steady *churp churrr, churp churrr* ("Pe-cos, Pe-cos"), repeated.

HYBRID. *All quail in the genus* Callipepla *are susceptible to hybridization. Scaled Quail sometimes hybridizes with Gambel's, as here, and more rarely with Northern Bobwhite.*
NEW MEXICO, APRIL

Loons

ORDER GAVIIFORMES

5 species: 3 ABA CODE 1   2 CODE 2

There are only five species in the order Gaviiformes, and all five can be found in our area. They are called loons in North America, and divers in Great Britain. Both names are apt: The wailing of loons, especially the Common Loon, is mirthful, even lunatic; and all species are expert divers, submerging hundreds of feet in the pursuit of large-bodied fish. The Common and Yellow-billed Loons form a close taxonomic pair, as do the Pacific and Arctic Loons. The Red-throated Loon, meanwhile, is something of an outlier within the loon clan.

Loons are aquatic, and all five species follow the same basic annual cycle: Breeding takes place on lakes and large ponds in the north; migration is chiefly coastal, although often with a substantial overland component; and the wintering grounds are also mainly coastal, with many individuals occurring just offshore or in large inshore bays and inlets. Except when diving, loons often appear clumsy. Most species are incapable of launching from land, and must shuffle down to the water to take off. Also, loons in flight strike a labored and humpbacked pose, even though they are strong fliers.

Loons undergo a striking "personality change" when they leave the breeding grounds. In summer, they are devoted parents, and it is not uncommon to see adults carrying young on their backs. They are defensive parents, too, frequently warding off intruders, even attacking and killing them. Most of all, loons on the breeding grounds are legendary for their "laughter" and haunting "yodeling." On the wintering grounds, however, loons are aloof, occurring singly or in small assemblages loosely spread over thousands of square feet of water. They are generally quiet during the winter, and usually forage and roost in deep water.

Loon populations have had their ups and downs over the years, but all five species are considered to be in good shape, if somewhat tenuously, at the present time. As all five species are obligate fish-eaters, they face two major threats: depletion of fisheries and bioaccumulation of environmental toxins, such as mercury. Several species have been much more frequently noted inland on migration in the West in recent years, for reasons that are unclear. Is it just a matter of better detection by observers in the field? Or is a large-scale range shift underway? There is not yet a satisfactory answer.

Common Loon
Gavia immer CODE 1

BREEDING ADULT. *Large dark bill, checkered back, and vertical black-and-white stripes on neck. Steep forehead, peaked crown, and large size distinctive. Often rides low in water.* MONTANA, JUNE

NONBREEDING. *Plain, dull grayish and white. Similar to juvenile, but upperparts entirely gray without bright feather edges.* NEW JERSEY, JANUARY

JUVENILE. *Jagged edge between dark and light areas on head. Dark ridge (culmen) along upper mandible. Upperparts brightly speckled by white fringes of fresh juvenile feathers.* CALIFORNIA, DECEMBER

INCUBATING ADULT. *Common Loon pairs build nests of wet plant material collected by both the male and the female nearby. Both members of pair share incubation duties.* WISCONSIN, JULY

L 32"	WS 46"	WT 9 lb	♂ > ♀

- ▸ two adult molts per year; simple alternate strategy
- ▸ strong seasonal and moderate age-related differences
- ▸ weak geographic variation in wing and bill length

At all times of year, closely tied to large bodies of water with big fish. Nests along lakes in northern woods. Winters mainly coastally, especially offshore, singly or in loose flocks. Diurnal migration along coasts and inland is easily detected.

◀ Two songs: a short, maniacal laugh, *a hee a hiya heea yay*, heard year-round; evocative wailing ("yodeling"), *eeeee waaaaay ooooh*, given on breeding grounds.

IN FLIGHT. *All loons appear humpbacked in flight, with large paddle-shaped feet. On this nonbreeding adult, note solidly dark upperparts and long gray bill.* CALIFORNIA, OCTOBER

Yellow-billed Loon

Gavia adamsii CODE 2

BREEDING ADULT. *Similar to Common but larger, with upturned yellow bill and more extensive white on back; overall impression, especially at a distance, is of a paler bird.* ALASKA, JUNE

JUVENILE. *Similar to Common, but note generally paler tones and especially pale yellow bill. Nonbreeding adult darker, not as speckled. In all plumages, shape of crown is angular.* COLORADO, NOVEMBER

L 35"	ws 49"	wt 11.8 lb	♂ > ♀

- two adult molts per year; simple alternate strategy
- strong age-related and seasonal differences
- white on face variable in nonbreeding plumages

Habits and ecology similar to Common Loon, but breeding grounds largely non-overlapping; Yellow-billed breeds on tundra lakes near the Arctic Ocean. Most winter in coastal waters off Canada and Alaska; increasingly sighted inland in West, especially in late fall.

Both song types similar to Common Loon. Cackle ("tremolo") is slightly lower-pitched. Wail ("yodel") is deeper and more growling, often lacking third syllable.

Red-throated Loon

Gavia stellata CODE 1

BREEDING ADULT. *Small size and petite proportions distinctive. Adult has plain gray face, neck with thin red patch in front, nape with long black-and-white stripes, and unmarked black back.* ALASKA, JUNE

JUVENILE. *Face blank and extensively white. Back heavily spotted with white. May show white along flanks. Bill thin and upturned. Nonbreeding adult similar but has darker cap.* CALIFORNIA, DECEMBER

L 25"	ws 36"	wt 3.1 lb	♂ > ♀

- two adult molts per year; simple alternate strategy
- strong age-related and seasonal differences
- on swimming birds, white along flanks varies with angle

Breeds in northern coastal districts. Away from breeding grounds is more marine than other loons, although wanderers can show up anywhere. Often seen feeding in surf or otherwise close to shore. Sometimes found on shore; unlike other loons, can launch from land.

Yodeling is a simpler and more gull-like than other loons: a descending *waaaay* or *waaaay-oh*, with mournful quality.

Pacific Loon
Gavia pacifica CODE 1

ADULT BREEDING. *Smoothly rounded, pale gray head contrasts with dark back. Sides of neck have long black-and-white lines; front of neck is velvety purple in good light.* MANITOBA, JUNE

JUVENILE. *Head and neck smoothly rounded; bill relatively small and straight. Dark back lightly scalloped. Nonbreeding adult similar, but back all dark. Flanks dark in all plumages.* CALIFORNIA, JANUARY

L 25"	WS 36"	WT 3.7 lb	♂ > ♀

▸ one adult molt per year; simple alternate strategy

▸ strong age-related and seasonal differences

▸ thin, dark chinstrap of nonbreeders is variable

Breeds along edges of northern lakes; favors less wooded landscapes than Common, but much overlap. Winters in near-shore habitats along Pacific coast. Rare inland in winter, but inland dispersal, especially in fall migration, is more extensive than previously thought.

Yodeling is shorter and shriller than Common: *er-weeee* or *er-ya-weeee*, often with rising inflection. Usually silent away from breeding grounds.

Arctic Loon
Gavia arctica CODE 2

BREEDING ADULT. *Paler overall than Pacific. Best mark—visible on sitting birds at great distance—is variable white patch on flanks.* ALASKA, MAY

SUBADULT. *All plumages distinguished from Pacific by white on flanks; otherwise nearly identical to Pacific. Structure similar to Pacific, but Arctic is larger, with proportionately larger bill.* EUROPE, APRIL

L 27"	WS 40"	WT 5.7 lb	♂ > ♀

▸ two adult molts per year; simple alternate strategy

▸ strong age-related and seasonal differences

▸ prominence of white flank patch highly variable

Old World counterpart of Pacific Loon. Breeding population extends east across Siberia to Alaska's Seward Peninsula, where Pacific Loon also breeds. Most winter in Old World, but a few wander southeast along West Coast, where separation from Pacific Loon is difficult.

Yodeling is deeper than smaller-bodied Pacific; rarely heard in North America away from restricted breeding grounds in Alaska

Grebes

ORDER PODICIPEDIFORMES

---

7 species   6 ABA CODE 1   1 CODE 2

Grebes are lumped in many people's minds with loons, and the two orders are indeed similar in various ways. Grebes, like loons, breed mainly inland on ponds and lakes, and many migrate to coasts in fall. Grebes and loons are also similar in nestling care, food requirements, and foraging behavior. But the two groups are only distantly related. Among the North American grebes, relationships are rather diffuse. Clark's and Western Grebes are close cousins and ecologically very similar to each other; Horned, Eared, and Red-necked Grebes are in the same genus, but are distinct in many ways from one another; and Pied-billed and Least Grebes, in separate genera, round out this diverse group.

Compared to loons, grebes in North America breed in wetlands that average less northerly and more marshy. Winter habitats are more convergent, with both grebes and loons co-occurring on large bays and near-shore ocean waters. (The tropical Least Grebe deviates broadly from this pattern, favoring densely vegetated marshes year-round.) Grebes are prone to vagrancy, and wayward individuals sometimes stray hundreds or even thousands of miles off course, especially in fall. Even the ordinarily sedentary Least Grebe has wandered widely from its core range in southern Texas and points south.

All grebes are divers, and all undergo the same "personality change" as loons when they leave the breeding grounds. Like loons, they are animated and aggressive around their nest sites, but quiet and somewhat antisocial in winter. An important difference is that grebes migrate mainly by night, whereas loon migration is often conspicuous by day. A similarity is that migrants of both groups are helpless on land, and stranded grebes are sometimes rescued from parking lots and similar habitats.

Grebe populations are generally healthy, but most face the threats associated with dispersal, often by night, among diverse and often degraded habitats. The Eared Grebe is both extraordinary and exemplary in this regard. It is the most abundant grebe in the world, but large numbers die each year during migration. Sudden changes in the weather create large-scale mortality, and there is the constant threat of habitat deterioration on special molting sites (for example, Mono Lake in eastern California) that lie between the breeding and wintering grounds. The Eared Grebe has always faced challenges on migration, but it faces additional perils from human changes to the landscape.

Pied-billed Grebe

Podilymbus podiceps CODE 1

BREEDING ADULT. *Bulbous bill has thick black ring, creating "pied" effect. Muddy brown all over with black chin. Dumpy; when resting, looks like a brown clump on the water.* TEXAS, APRIL

NONBREEDING ADULT. *Black on bill and chin disappears on nonbreeding birds. In all plumages, puffy white stern often exposed. Juvenile similar but has diffuse streaks on head.* CALIFORNIA, JANUARY

| L 13" | ws 16" | wt 1 lb | ♂ > ♀ |

- two adult molts per year; complex alternate strategy
- weak age-related and moderate seasonal differences
- extent of black on throat of adult is variable

The most widely distributed grebe in our area, also occurring south to southern South America. Breeds in various freshwater settings; winters in both salt water and fresh water. Rarely seen in flight; dives often. Swims low to water; sometimes only head is visible.

◀ Deep, loud hooting, seemingly mismatched to so small and unobtrusive a bird; a gulping, cuckoo-like *kee kee coo! coo! coo coo cow cow cowlp cowlp.*

Horned Grebe

Podiceps auritus CODE 1

BREEDING ADULT. *Plumage recalls Eared Grebe, but neck chestnut, golden "horns" more extensive than "ears" of Eared. Larger; often looks low-slung, flat-headed. Bill thicker than Eared.* MONTANA, JUNE

NONBREEDING ADULT. *Black cap usually set off sharply from white face. Molting birds look messy, can closely resemble Eared. At close range, note white spot at tip of bill.* CALIFORNIA, JANUARY

| L 14" | ws 18" | wt 1 lb |

- two adult molts per year; complex alternate strategy
- weak age-related and strong seasonal differences
- individual variation in amount of white on upperwing

Builds floating nest in densely vegetated freshwater ponds; highly territorial during breeding season. Migrates by night mainly to coasts; smaller numbers winter inland on large bodies of water. Coastal winterers prefer brackish waters to fresh water.

In elaborate pair dance, loud, slow trilling preceded by short introductory note: *buh-dee!-dee-dee-dee-dee.* Individual notes in trill are sharp, staccato.

Least Grebe

Tachybaptus dominicus CODE 2

L 9.5"	WS 11"	WT 4 oz

▸ two adult molts per year; complex alternate strategy
▸ moderate seasonal and age-related differences
▸ amount of white in primaries and secondaries variable

Widespread tropical species that ranges north into southern Texas. Although nonmigratory, vagrants have occurred at widely scattered locations in the southern states. Favors densely vegetated ponds, including man-made bodies of water; easily overlooked.

Call a strange, thin, dry high-pitched rattle that keeps going and going; usually preceded by a coot-like *shweenk*. Also gives single shrill *weenk!* notes.

BREEDING ADULT. *Dumpy and small like, but distinctive. Slate gray all over, with black chin; chin pale in other plumages. Bill thin and pointed, unlike Pied-billed.* ARIZONA, APRIL

Eared Grebe

Podiceps nigricollis CODE 1

NONBREEDING ADULT. *Patterned like Horned, but without sharp demarcation between black and white on head. Note body structure: petite overall, pointy-headed, and thin-billed.* CALIFORNIA, JANUARY

13"	WS 16"	WT 11 oz

▸ two adult molts per year; complex alternate strategy
▸ strong age-related and seasonal differences
▸ individual variation in amount of white in wing

Physiologically, one of the most remarkable birds in North America; undergoes frequent extreme changes in weight, and is flightless for 9–10 months of the year; huge numbers congregate in fall at salt lakes, then complete their coastward "fall" migration in winter.

BREEDING ADULT. *Note golden "ear" of fine plumes behind red eye. Body mainly black; flanks chestnut. Dainty compared to Horned Grebe, with a thin upturned bill.* MONTANA, JUNE

Like all grebes, both sexes perform elaborate dances. Courtship call a squeaky, high-pitched *poo-wee!-ik* or *po-week*, often repeated.

IN FLIGHT. *In all plumages (nonbreeding adult here), white secondaries contrast with otherwise dark wing. Horned usually shows a white leading edge along with white secondaries.* COLORADO, NOVEMBER

Red-necked Grebe

Podiceps grisegena   CODE 1

L 18"	ws 24"	wt 2.2 lb

- two adult molts per year; complex alternate strategy
- moderate age-related and strong seasonal differences
- extent and intensity of yellow on bill variable

Like other grebes, breeds inland and migrates by night to coastal wintering grounds. Generally, the most coastal wintering grebe; midwinter sightings inland are uncommon. For breeding, usually selects semi-wooded landscapes.

Courtship call begins with a growling two-syllable note, then goes into a trill *wah-ow, d'd'd'd'*. Outside breeding season, utters occasional croaks.

BREEDING ADULT. *Long chestnut neck, white face, black cap, and long yellow bill create strong contrast. Note young riding on back, a behavior exhibited by all grebe species.* MONTANA, JUNE

DOWNY YOUNG. *Chicks of Red-necked and other grebes are precocial, often seen swimming. Striking black-and-white plumage and colorful bare parts quickly diminish as bird ages.* MONTANA, JUNE

NONBREEDING ADULT. *High-contrast breeding plumage becomes variably muted; confusion with Horned Grebe possible, but Red-necked has long bill, always retaining some yellow.* CALIFORNIA, DECEMBER

Clark's Grebe

Aechmophorus clarkii   CODE 1

L 25"	ws 24"	wt 3.1 lb

- two adult molts per year; complex alternate strategy
- moderate differences between summer and winter adult
- variants can resemble Western; hybrids uncommon

Range overlaps extensively with similar Western Grebe; generally less numerous, especially toward eastern and northern limits of range. Migration routes and timing are not well understood. In Pacific coast region, less likely than Western to occur offshore.

Display nearly identical to Western; the only difference involves first step in elaborate courtship ceremony. Call *rrreek-ick*, more slurred than Western.

BREEDING ADULT. *Bill averages brighter yellow than Western; sometimes tinged with orange. Black cap does not extend to eyes and lores. Shows more white on flanks and wings.* UTAH, MAY

NONBREEDING ADULT. *Variable. Head pattern sometimes similar to breeding, as here; often becomes dusky around eyes and lores, suggesting Western. Bill usually remains yellow.* CALIFORNIA, DECEMBER

Western Grebe
Aechmophorus occidentalis CODE 1

BREEDING ADULT. *Large, long-necked, strikingly black-and-white; contrast subtly stronger than on Clark's. Black on cap extends to eyes, lores. Long bill is dull yellow-green.* UTAH, MAY

NONBREEDING ADULT. *Black-and-white pattern of head muted, sometimes approximating Clark's. Bill usually remains yellow-green; Clark's is yellow or orange-yellow.* CALIFORNIA, JANUARY

RUSHING CEREMONY. *After meeting, male and female scoot across the water's surface in tandem formation called the "Rushing Ceremony."* MINNESOTA, APRIL

L 25"   WS 24"   WT 3.3 lb

▶ two adult molts per year; complex alternate strategy

▶ moderate differences between summer and winter adult

▶ averages larger-bodied in northern portions of range

Breeds in extensive inland marsh complexes. In winter, most move by night toward Pacific coast region but small numbers are widespread inland. Range broadly overlaps with similar Clark's Grebe, but the two species often segregate at locations where both occur.

◀ Famous for its elaborate synchronized courtship displays. Outside breeding season, listen for far-carrying, grating *rrick-reeeeek!* Often calls at night.

WEED CEREMONY. *The second half of the Western Grebe's display consists of the "Weed Ceremony," highlighted by "Weed Dancing," shown here.* MINNESOTA, MAY

Tubenoses
Albatrosses, Shearwaters & Petrels, and Storm-Petrels
ORDER PROCELLARIIFORMES

---

47 species: 9 ABA CODE 1   12 CODE 2   6 CODE 3   7 CODE 4   13 CODE 5

The "tubenoses"—so named for tubular protuberances on the upper mandible—are the quintessential birds of the ocean. Even the most wide-ranging and abundant species, such as the Sooty Shearwater and Wilson's Storm-Petrel, are rarely seen over land. Three families are distinguished. The gigantic albatrosses (Diomedeidae) are notable as large-bodied and very long-winged. The shearwaters and petrels (Procellariidae) are considerably smaller, but similar in body plan: powerful, with long, swept-back wings. The storm-petrels (Hydrobatidae) are smaller still, and very different in flight; many appear dainty and fluttery.

All species are oceanic, but habitat differences are significant, and often important in the identification process. Off the West Coast, for example, gadfly petrels occur far offshore where they zoom about in high, bounding arcs; shearwaters are found closer to shore, in upwelling zones, and they fly close to the surface, alternating powerful flapping with effortless glides. Other differences are imperceptible to the human eye, but nonetheless significant. Off the East Coast, for example, the occurrence of Manx, Greater, Cory's, and Audubon's Shearwaters is predictable by seemingly minor differences in sea surface temperature. Feeding behavior and flight style differ markedly among species, and veteran seabirders rely more heavily on these behavioral distinctions than on comparatively minor differences in physical appearance. Tubenoses at sea are silent for long periods of time, but they can become surprisingly vocal, especially in feeding flocks around boats. Most species are highly migratory, and several undergo annual migrations greater in distance than any other organisms on earth. A key aspect in the identification process is simply knowing what to expect and when to expect it. The converse is also true, perhaps more thrillingly so than for any other group of birds: Heart-stoppingly rare tubenoses have a penchant for showing up thousands of miles out of range, sometimes in the completely "wrong" ocean.

Many tubenoses are species of conservation concern. The breeding grounds—often thousands of miles from North America—are vulnerable to devastatingly efficient introduced predators, and human persecution remains a threat at some sites. At sea, entanglement in drift nets and hooking and drowning on longline fishing gear are well-documented hazards. Less clear, but nonetheless worrisome, are threats at a regional and global level: pollution, fisheries depletion, and potentially human-influenced changes in sea surface temperature.

Black-footed Albatross

Phoebastria nigripes CODE 1

ABOVE. *Huge and dark; at first glance, comes across as an oversized Sooty Shearwater. White uppertail coverts conspicuous on adult, absent or faint on juvenile.*

PACIFIC OCEAN, MARCH

BELOW. *Mainly sooty brown. Vent and undertail coverts white, more extensively on older adults and worn birds, as here.*

CALIFORNIA, OCTOBER

L 32"	WS 84"	WT 7 lb	♂ > ♀

▸ one adult molt per year; simple basic strategy

▸ moderate differences between juvenile and adult

▸ extent of white on face of adult increases with wear

Fairly common in late summer off Pacific coast. Visits boats a few miles offshore. Formerly bred more widely and was more numerous. Populations may be declining due to the use of drift nets and longlining by commercial fishing operations.

Usually silent at sea, but may become noisy in feeding flocks; listen for grunts interspersed with squeaks, squeals, and bill-clapping.

Laysan Albatross

Phoebastria immutabilis CODE 2

FRESH ADULT. *Similar in size and structure to Black-footed Albatross. Even at great distances, contrast between dark wings and white body is striking.*

PACIFIC OCEAN, MARCH

BELOW. *Underparts mainly white; dark on Black-footed. Underwing linings usually show much white, but the extent is variable.* PACIFIC OCEAN, MARCH

L 32"	WS 78"	WT 6.6 lb	♂ > ♀

▸ one adult molt per year; simple basic strategy

▸ moderate differences between fresh and worn plumages

▸ occasionally hybridizes with Black-footed Albatross

A scarce visitor to North America; most numerous in southern Bering Sea region. Range expanding: recently established as breeder off Mexico and could eventually colonize islands off California. Populations may be increasing, but drift nets and longlining pose a threat.

Generally silent at sea, but elements of courtship display are sometimes observed: long mooing sounds, descending whinnies, woodpecker-like rattles.

Short-tailed Albatross
Phoebastria albatrus CODE 3

ADULT. *Requires 10–20 years to mature from all-brown juvenile plumage to black-and-white adult plumage with extensive yellow on head and nape. Huge pink bill acquired at early age.* MIDWAY, MAY

IMMATURE. *Massive; appears stockier in flight than Black-footed. Huge pink bill visible from afar.* CALIFORNIA, APRIL

L 35"   ws 88"   wt 9.5 lb

▸ one adult molt per year; simple basic strategy

▸ strong differences among numerous age classes

▸ individual plumage variation within each age class

An endangered species, slowly recovering. Rare off West Coast, but being seen more frequently. Formerly abundant. With protection on its breeding grounds, sightings may continue to increase, although long-line and drift net fishing are a threat.

Usually silent at sea in North American waters. Courting birds give long growls that might be heard in busy feeding flocks.

Northern Fulmar
Fulmarus glacialis CODE 1

LIGHT INDIVIDUAL. *On all Northern Fulmars, note stocky build overall, stubby "tubenose" bill, and shearwater-like flight consisting of long glides interspersed with rapid flapping.* CALIFORNIA, AUGUST

DARK INDIVIDUAL. *Dark sooty overall; some are even darker than this bird. Note that plumage restoration in Northern Fulmar is continuous, such that discrete color morphs do not exist.* CALIFORNIA, SEPTEMBER

INTERMEDIATE INDIVIDUAL. *On many Northern Fulmars, the inner primaries are contrastingly paler than the rest of the wing.* CALIFORNIA, OCTOBER

L 18"   ws 42"   wt 1.3 lb   ♂ > ♀

▸ one adult molt per year; simple basic strategy

▸ adult males are larger-billed than females

▸ ground color varies from dark gray to nearly white

Abundant; undergoing a remarkable long-term population increase and range expansion, especially in the Atlantic Ocean region. Nests are placed on steep cliffs, usually by the sea. Aggressive; often follows ships.

Birds in feeding flocks give repeated rasping notes, *uhrnt, uhrnt, uhrnt.* Also listen for low moaning growls, rising or dropping in pitch.

Black-capped Petrel
Pterodroma hasitata CODE 2

FRESH ADULT. *Black cap contrasts with white head, some extensive white on nape. Long black tail separated from dark mantle by white uppertail coverts.*
NORTH CAROLINA, AUGUST

WORN ADULT. *Contrasting black-and-white plumage evident at distance. Note especially the contrast between the dark wings and back and the white collar and uppertail coverts.*
NORTH CAROLINA, MAY

L 16" WS 37" WT 10 oz

▸ one adult molt per year; simple basic strategy
▸ molting birds have pale splotches on wings
▸ individual variation in extent of white on collar

Most common gadfly petrel off the East Coast. Formerly thought to be rare, but now known to disperse in sizeable numbers north into the Gulf Stream during warm months. Birds entrained in hurricanes sometimes show up well inland, especially in mid-Atlantic region.

BELOW. *Mainly white. Striking underwing pattern, highlighted by diagonal black stripe across secondary coverts.*
NORTH CAROLINA, JUNE

Birds at breeding sites in West Indies give strange wailing sounds.

Herald Petrel
Pterodroma arminjoniana CODE 3

L 15" WS 35" WT 12 oz

▸ one adult molt per year; simple basic strategy
▸ weak differences between fresh and worn plumages
▸ polymorphic, with much individual variation

Rare but regular visitor to the Gulf Stream; numerous records off North Carolina. Single birds usually seen in our area; dark morph birds more often than light morphs.

Not likely to be heard in North American waters. Birds on breeding grounds utter long growls, characteristic of genus *Pterodroma*.

INTERMEDIATE MORPH. *All morphs generally brownish; other East Coast gadfly petrels are more grayish. Confusion with jaegers and Sooty Shearwater possible, but note high arcing flight. Atlantic form called "Trinidade" Petrel.* EUROPE, APRIL

ABOVE. *All morphs are mainly brown above, with little contrast. Wings are proportionately long for a gadfly petrel.* NORTH CAROLINA, AUGUST

Fea's Petrel
Pterodroma feae　CODE 3

BELOW. *Underwing mainly dark, contrasting with largely white breast and belly. Gray brown upperparts extend to edges of breast.*
NORTH CAROLINA, JUNE

ABOVE. *Sleek and extensively gray-brown; darker brown bar cuts across secondaries. Nape pale gray-brown. Pale tail is fairly long and narrow.*
NORTH CAROLINA, MAY

L 15"　　WS 34"　　WT 11 oz

▸ one adult molt per year; simple basic strategy

▸ weak differences between fresh and worn plumages

▸ extent of light and dark on underparts variable

Only recently discovered to be regular in the Gulf Stream; now dozens of records off North Carolina and several from Nova Scotia. Status in North American waters still being worked out.

Not likely to be heard off North America. On nesting islands in the eastern Atlantic, gives eerie wails and soft chattering notes at night.

Murphy's Petrel
Pterodroma ultima　CODE 3

ADULT. *The dark brown gadfly petrel off Pacific coast; worn birds slightly duller. Suggests jaegers and Sooty Shearwater; note diffuse pale patch at base of primaries.*
CALIFORNIA, NOVEMBER

ADULT. *High arcing flight distinctive even at considerable distances. Note fairly extensive white on face of this individual.*
CALIFORNIA, NOVEMBER

L 15"　　WS 35"　　WT 13 oz

▸ one adult molt per year; simple basic strategy

▸ weak differences between fresh and worn plumages

▸ extent of white on face varies among individuals

Disperses widely from Southern Hemisphere nesting sites, some reaching North American waters off Pacific coast, especially in spring. Usually seen only far offshore; recent explorations off southern California have turned up several dozen per day in April and early May.

Vocalizes rarely, if at all, in North American waters. Around nesting colonies, gives various cries and moans, especially at night.

Mottled Petrel

Pterodroma inexpectata CODE 3

ABOVE. *Extensive gray above contrasts with pale coloration on long tail. Gray of adult, shown here, nearly uniform; juvenile more scaly.* CALIFORNIA, JUNE

BELOW. *Distinctively marked: broad black bar flashes across white underwings; extensive dark on belly contrasts with white tail and breast.* ASIA, DECEMBER

L 14"   WS 32"   WT 11 oz

▸ one adult molt per year; simple basic strategy

▸ weak differences between juvenile and adult

▸ extent of gray and black varies among individuals

Breeds on islands in Southern Hemisphere; disperses far north during austral winter, reaching into North Pacific during boreal summer. Status in North American waters unclear; may be most numerous off Aleutian Islands, but few at-sea observations are made there.

Unlikely to be heard in North American waters. On breeding grounds gives a complex series of cackles and groans, varying in pitch and volume.

Cook's Petrel

Pterodroma cookii CODE 2

ABOVE. *Delicate: small-bodied with a thin bill. Gray overall, with wings darker than back and scapulars; juvenile has paler feather tips. All plumages extensively white below.* SOUTH PACIFIC, FEBRUARY

BELOW. *Extensively white below. Underwing pattern suggests Black-capped Petrel, but black reduced. Thin bill is a good field mark, if discernible.* SOUTH PACIFIC, JANUARY

L 13"   WS 30"   WT 7 oz

▸ one adult molt per year; simple basic strategy

▸ weak differences between juvenile and adult

▸ individual variation in darkness of upperparts

Probably the most numerous and most regular West Coast gadfly petrel. Look for it in small numbers far offshore during the warm months. Engages in high arcing flight typical of the genus *Pterodroma*, but also makes erratic, nighthawk-like movements.

Like other gadfly petrels, not likely to be heard in North America. Birds at sea may give a soft chattering, sometimes frenzied: *ch'ch'ch!ch!ch!ch'ch'.*

Cory's Shearwater
Calonectris diomedea CODE 1

L 18"   WS 44"   WT 1.8 lb   ♂ > ♀

- one adult molt per year; simple basic strategy
- weak differences between fresh and worn plumages
- subspecific variation in plumage color and pattern

Fairly common off East Coast, but numbers vary from year to year. Seen from land during fine weather on light winds and while feeding on baitfish. Appears relaxed in flight, with wings held loose and wingbeats appearing "easy."

Usually silent at sea. In flocks, sometimes gives low-pitched hoots singly or in series; hooting not strongly in two syllables, as in Greater.

BELOW. *Appears large-bodied and powerful. Body entirely white below; wings extensively white. Large yellow bill prominent and often visible at considerable distance.*
NORTH CAROLINA, MAY

ABOVE. *Brown in all plumages; worn adults, as here, show irregular white splotching. Accidental Cape Verde Shearwater (C. edwardsii), not shown, is darker overall and has darker bill.*
NORTH CAROLINA, JULY

ADULT. *Up close, Cory's massive, deeply hooked, prominently "tubed" bill is even more impressive than at a distance. Thick neck adds to the imposing appearance.* NORTH CAROLINA, JULY

Greater Shearwater
Puffinus gravis CODE 1

L 18"   WS 42"   WT 1.8 lb

- one adult molt per year; complex basic strategy
- moderate differences between juvenile and adult
- color of upperparts, smudge on belly variable

This Atlantic Ocean species prefers slightly cooler waters than Cory's, but much overlap. Compared to Cory's, wings are held straighter, stiffer, with less "flex." Greater is readily attracted to boats, where it can be observed plunge-diving and surface-diving.

Feeding flocks near boats are sometimes noisy. Listen for syncopated hooting, like distant sea ducks: *hoo-hoo! hoo-hooa!*

JUVENILE. *Dark above; pale below with dark smudge on belly. Dark cap contrasts with white chin and neck; dark rectrices contrast with white uppertail coverts. Adult has more white on hindneck.* MAINE, JUNE

Audubon's Shearwater
Puffinus lherminieri CODE 1

L 12"	ws 27"	wt 6 oz

- one adult molt per year; simple basic strategy
- moderate differences between fresh and worn plumages
- extent of dark, especially on underwings, variable

Breeds mainly on tropical and subtropical islands; wanders north into the Gulf Stream during warm months. Favors warmer, more southerly waters than Manx Shearwater, but much overlap. Flight style distinctive: choppy and fluttery, interspersed with short glides.

Not usually heard at sea. Rarely utters a long, drawn-out moan with dry, scratchy quality: *errrrrch*; or *errrrrch'ch'ch* with terminal rattle or trill.

ADULT. *Small-bodied and long-tailed. Dark undertail coverts contrast with white belly. Upperparts dark, but slightly paler in fresh plumage.* NORTH CAROLINA, AUGUST

Sooty Shearwater
Puffinus griseus CODE 1

ABOVE. *Uniform dark brown above. In flight, alternates stiff-winged glides with deep, powerful flapping.* MARYLAND, JUNE

BELOW. *Pale wing linings contrast with dark body, often creating flashing effect on bird in banking flight. Worn bird shows small white flecks and splotches on wings.* MASSACHUSETTS, JUNE

L 17"	ws 40"	wt 1.7 lb

- one adult molt per year; complex basic strategy
- weak differences between fresh and worn plumages
- extent of white on underwing variable, light-dependent

A masterful flier. Sometimes the most common tubenose in North American waters. Off East Coast, widespread in small numbers. Off West Coast, huge flocks are routine. A common sight from land on Pacific coast; infrequent but regular from land on East Coast during spring.

In feeding assemblages, utters two-syllable hoots, with accent on second syllable: *hoo hoo!* or *hoo haw!* or *hee hee!* Hooting varies greatly in pitch.

FLOCK AT SEA. *Gargantuan flocks, often numbering hundreds of thousands, gather over deep water off the central Pacific coast.* CALIFORNIA, NOVEMBER

Short-tailed Shearwater
Puffinus tenuirostris　CODE 2

L 16"	WS 38"	WT 1.2 lb

▸ one adult molt per year; complex basic strategy
▸ weak differences between fresh and worn plumages
▸ extent of gray patch on underwing variable

Smaller counterpart of Sooty Shearwater. Although uncommon off much of the West Coast, Short-tailed is abundant in the Bering Sea region. Late-summer flights in the millions are seen from coastal promontories off western Alaska. Numbers apparently declining.

Rarely heard in our waters. Calls variable. Similar to Sooty, but not as strongly two syllables; gruffer, lower-pitched overall.

ADULT. *Similar to Sooty Shearwater, but wing linings usually not as contrastingly pale. On many birds, hood is darker than breast, especially in worn plumage. Occurs later in fall off West Coast than Sooty.* CALIFORNIA, NOVEMBER

Manx Shearwater
Puffinus puffinus　CODE 2

L 13.5"	WS 33"	WT 1 lb

▸ one adult molt per year; simple basic strategy
▸ weak differences between fresh and worn plumages
▸ individual variation in extent of dark on underwing

Wingbeats fast. Status in North America in flux; generally increasing. Began breeding off eastern Canada in late 20th century. Seen in small numbers in warm months off Atlantic coast, sometimes in winter. Has recently become rare but regular off Pacific coast.

Around nest sites, listen for scratchy, grunting calls interspersed with high-pitched squeals: *grrrt, grrrt, peee!*; usually silent at sea.

ADULT. *White underparts contrast with dark upperparts; upperparts darkest in fresh plumage, duller in worn. At close range, note post-auricular crescent or "ear surround."* EUROPE, JULY

Black-vented Shearwater
Puffinus opisthomelas　CODE 2

L 14"	WS 34"	WT 9 oz

▸ one adult molt per year; complex basic strategy
▸ moderate differences between fresh and worn plumages
▸ much individual variation, sometimes approaching Manx

Flight fluttery. Breeds off Baja California, disperses north into waters off California. Occurrence dictated by sea surface temperature; population possibly declining. Favors waters fairly close to shore, where it plunge-dives and surface-dives for food.

Usually heard only at night at or near breeding grounds. Birds at sea occasionally give hissing and growling sounds.

TYPICAL ADULT. *Not as starkly black-and-white as Manx. Most have variable black on undertail coverts and indistinct face pattern, but a few can be nearly as pale as Manx.* CALIFORNIA, SEPTEMBER

Pink-footed Shearwater
Puffinus creatopus   CODE 1

DARK ADULT. *Sizeable minority shows extensive dusky on throat, underwing, and belly, as here. A few are even darker. Molting birds can show extensive white mottling above.* CALIFORNIA, NOVEMBER

L 19"	ws 43"	wt 1.6 lb

▸ one adult molt per year; complex basic strategy

▸ moderate differences between fresh and worn plumages

▸ plumage variable, sometimes approaching Flesh-footed

The more common counterpart of Flesh-footed Shearwater. Frequently seen off West Coast in warm months, singly or in large mixed-species assemblages. Tends to prefer shallow waters, fairly close to land. Flight style like Flesh-footed: relaxed, but powerful.

Like Flesh-footed and other shearwaters, rarely heard at sea. Occasionally gives hoots like Flesh-footed, interspersed with agitated, wheezy whinnies.

JUVENILE. *Pinkish bill and generally "dirty" plumage typify nearly all birds; most are dusky brown above, off-white below. Bulky in build, but a strong flier.* CALIFORNIA, NOVEMBER

Flesh-footed Shearwater
Puffinus carneipes   CODE 3

L 17"	ws 40"	wt 1.4 lb

▸ one adult molt per year; simple basic strategy

▸ weak differences between fresh and worn plumages

▸ underwing can appear to flash white in bright light

Breeds on islands in the Southern Hemisphere. Disperses widely during austral winter; some reach eastern North Pacific during northern summer. Uncommon off West Coast, usually seen singly. In flight, appears powerful, with long glides and slow flaps.

Not usually heard at sea, but individuals in large feeding flocks may utter variable hoots and cackles, *hoohoo!* or *hoohoo!ah,* accented on second syllable.

TYPICAL ADULT. *Hefty build and strong flight like Pink-footed. The two are sometimes considered conspecific. Dark all over, becoming slightly paler in worn plumage.* CALIFORNIA, NOVEMBER

Buller's Shearwater
Puffinus bulleri   CODE 2

ABOVE. *Upperwing pattern prominent: lovely gray-and-white in fresh plumage; more contrasting brownish-and-white when worn.* CALIFORNIA, SEPTEMBER

BELOW. *Extensively white underparts contrast sharply with black cap. With good view, note long, thin, grayish bill, distinct from gadfly petrels.* CALIFORNIA, NOVEMBER

L 16"   WS 40"   WT 13 oz

▸ one adult molt per year; complex basic strategy

▸ moderate differences between fresh and worn plumages

▸ upperwing pattern depends on light, but always striking

Fairly uncommon off West Coast by midsummer, but numbers vary annually according to water temperature and food availability. Population generally increasing. Feeds mainly on the surface; flight is arcing and effortless, reminiscent of gadfly petrels.

Rarely if ever heard in North American waters; even among noisy other species in feeding flocks, tends to be silent but may give occasional bleats or whines.

Wilson's Storm-Petrel
Oceanites oceanicus   CODE 1

WORN ADULT. *In all plumages and in all poses, shows extensive white at base of tail. Tail short and square; feet dangle behind.* NORTH CAROLINA, JULY

ADULT. *Pale upperwing bar reduced in fresh plumage. Wilson's "walks on water" while feeding. White at base of tail visible from every angle.* MASSACHUSETTS, AUGUST

L 7.25"   WS 18"   WT 1.2 oz

▸ one adult molt per year; complex basic strategy

▸ moderate differences between fresh and worn plumages

▸ individual variation in paleness of underwing

One of the most common tubenoses off East Coast. Sometimes forms large flocks, but usually in loose assemblages of a few individuals. Most sightings are fairly close to land. Flight swallow-like; feeds by pattering over surface. Readily attracted to boats, chum slicks.

Not often heard in North American waters, but birds in large feeding flocks sometimes give a scratchy *chff* note, singly or in long series.

Band-rumped Storm-Petrel

Oceanodroma castro CODE 2

ADULT. *Similar to Wilson's, but larger-bodied and shorter-legged; less white on tail than Wilson's. Dark in all plumages; upperwing of juvenile averages paler.*
NORTH CAROLINA, OCTOBER

L 9"   WS 19"   WT 1.5 oz

▸ one adult molt per year; simple basic strategy
▸ weak differences between juvenile and adult
▸ individual variation in uniformity of dark coloration

Flight buoyant, steady. Status in North American waters unclear; possibly in flux. Formerly thought to be rare here; with better coverage at sea, now known to occur regularly in warm water off East Coast. Population may be reduced from previous levels.

Not likely to be heard in North American waters, but birds in feeding flocks may occasionally give squeaky notes interspersed with lower-pitched growls.

WHITE-FACED STORM-PETREL (Pelagodroma marina). *Casual visitor off mid-Atlantic coast. Black-and-white plumage vaguely phalarope-like; bouncy feeding motions distinctive.*
NORTH CAROLINA, OCTOBER

Leach's Storm-Petrel

Oceanodroma leucorhoa CODE 1

ABOVE. *White-rumped individuals, as here, occur off East Coast and off West Coast south to central California. Note forked tail and pale upperwing patch (carpal bar).*
ATLANTIC OCEAN, APRIL

BELOW. *White uppertail coverts, when present, are less prominent than Wilson's; black feathers fade in worn plumage. On all birds, contortionist manner of flying distinctive.*
ATLANTIC OCEAN, APRIL

L 8"   WS 20"   WT 1.4 oz

▸ one adult molt per year; simple basic strategy
▸ weak differences between fresh and worn plumages
▸ dark-rumped birds occur off southern Pacific coast

Flight style distinctive: erratic, with abrupt changes in speed and direction, recalling Common Nighthawk. Population large and stable, but Leach's does not usually form large flocks and is not usually attracted to ships; therefore, it can be difficult to detect.

Most readily heard at night around nesting colonies, where it gives excited squeaky notes: *pee pee pee!* or *pip pip!*

Black Storm-Petrel
Oceanodroma melania   CODE 2

JUVENILE. *Large; appears powerful in flight. Extensively black in fresh plumage, less so when worn. Wings long; tail long and forked. Pale carpal bar contrasts with dark wings.* CALIFORNIA, AUGUST

MIXED-SPECIES FLOCK. *With experience, Pacific Ocean storm-petrels can be separated by flight style and wing shape. This flock contains mostly Black and Ashy Storm-Petrels.* CALIFORNIA, SEPTEMBER

L 9"	WS 22"	WT 2.1 OZ

- one adult molt per year; simple basic strategy
- weak differences between fresh and worn plumages
- prominence of pale bar on upperwing depends on lighting

Nests on arid islands off California and Baja California. Like all storm-petrels and many other tubenoses, is active by night at nest sites. Outside breeding season, disperses both north and south from colonies; occurrence apparently linked to presence of warm water.

In flight, usually around colonies at night, gives a harsh *pchuck-a-rroo*. Typically silent at sea, but may call occasionally when in large feeding flocks.

Ashy Storm-Petrel
Oceanodroma homochroa   CODE 2

WORN ADULT. *Paler above, especially in worn plumage, than other dark storm-petrels. Tail long and forked, often swept about.* CALIFORNIA, SEPTEMBER

BELOW. *Paler overall than other dark Pacific storm-petrels; pale ashy wing linings contrast with relatively darker flight feathers.* CALIFORNIA, SEPTEMBER

L 8"	WS 18"	WT 1.3 OZ

- one adult molt per year; simple basic strategy
- weak differences between fresh and worn plumages
- appearance of ashy coloration depends on lighting

The most northern dark storm-petrel of the eastern Pacific. Dispersal from breeding colonies on offshore islands is generally less extensive than in other storm-petrel species. Total population low. Wingbeats shallow, flight fluttery.

Usually silent away from breeding colonies. On nesting grounds by night, birds in flight utter strident *krick-a-reeu*, similar to Black Storm-Petrel.

Least Storm-Petrel

Oceanodroma microsoma CODE 2

ABOVE. *Superficially like a small Black Storm-Petrel, but build is different: Least is short-tailed with rounded wings. Upperwing bar usually indistinct.*
CALIFORNIA, AUGUST

BELOW. *Small size and swift flight convey an unusual and distinctive impression. Tail wedge-shaped. Worn individuals may be somewhat splotchy or paler overall.*
CALIFORNIA, AUGUST

L 5.75"   WS 15"   WT 0.74 oz

▸ one adult molt per year; complex basic strategy
▸ weak differences between fresh and worn plumages
▸ prominence of upperwing bar depends on light

Breeds off western Mexico. Disperses irregularly into North American waters, probably in response to variation in sea surface temperature; rare in some years, fairly common in others. Despite small size, flight is swift and direct, and wingbeats deep.

Usually silent in North American waters. By night at colonies, gives chuckling call similar to Black and Ashy, but softer, squeakier, more twittering.

Fork-tailed Storm-Petrel

Oceanodroma furcata CODE 2

ADULT. *Soft blue-gray plumage, if seen in good light, is distinct from all other storm-petrels, including Ashy. Tail forked, more strongly in adult than juvenile.*
CALIFORNIA, APRIL

ADULT. *Compare with photo above: Lighting and viewing angle are critical in at-sea identification of all tubenoses, greatly affecting impressions of plumage contrast.*
CALIFORNIA, APRIL

L 8.5"   WS 19"   WT 1.9 oz

▸ one adult molt per year; simple basic strategy
▸ weak differences between juvenile and adult
▸ northern populations average larger, paler

Numerous, especially north of California; some populations apparently increasing. Active by night; also frequently by day. Attracted to lights and often follows boats; sometimes seen from headlands and promontories. Often alights on water. Wingbeats shallow, steady.

Noisy around breeding grounds; gives a variety of squeaks and dry chatters, often on approach to nest site. Usually silent away from colonies.

Pelicans & Allies Tropicbirds, Boobies & Gannets, Pelicans, Cormorants, Darters, and Frigatebirds

ORDER PELECANIFORMES

20 species: 10 ABA CODE 1   1 CODE 2   4 CODE 3   4 CODE 4   1 CODE 5

Pelicans, cormorants, tropicbirds, frigate-birds, and others—the North American representatives of this order—are all large aquatic species clad mainly in browns, blacks, and whites. Additionally, all are fish-eaters to some extent, although very different foraging methods are employed by the various species within the order. A further point of similarity is how conspicuous they are. They gather in large flocks, they roost and nest in prominent places, and their movements are typically diurnal and impressive. In our area, it is convenient to think of the six families of Pelecaniformes in coarse landscape-level terms: tropicbirds (Phaethontidae) over warm tropical seas; boobies (Sulidae) over both offshore and near-shore oceanic waters; pelicans (Pelecanidae) both inland and along the coast; cormorants (Phalacrocoracidae), like pelicans, inland and coastal; Anhingas (Anhingidae) in inland swamps; and frigatebirds (Fregatidae), like tropicbirds, over warm oceans.

If there is a way to catch fish, species in the order Pelecaniformes have figured out how to do it. Consider the differences between American White and Brown Pelicans. The latter engages in magnificent plunge-dives just beyond the ocean breakers; the former assembles in shoulder-to-shoulder "foraging canopies" that devour fish just beneath the surface. Also compelling is the Magnificent Frigatebird, master of ocean skies yet ironically terrified of the water; it gingerly plucks food from the surface, snags flying fish in midair, or swoops after prey dropped by other seabirds. In some cultures, Great Cormorants are actu-ally trained in underwater "falconry," chasing down fish and retrieving them for humans.

Many species in this order are supremely graceful in flight, but they are singularly ungainly in other respects. The calls of most species are some combination of grunts, growls, and groans. Colonies are foul-smelling and copiously spattered with guano. And individuals often assume hulking and haggard poses when roosting in trees or sitting on sandbars.

Several species of Pelecaniformes have benefited greatly from the environmental reforms of the late 20th century. The Double-crested Cormorant and especially the Brown Pelican suffered major population declines during the DDT era, but populations of both have rebounded sharply—to the point that cormorants are reckoned locally as pests. Most species in this order nest colonially, and face the hazards inevitably associated with group living: introduced predators, habitat destruc-tion, and occasional human persecution.

Red-billed Tropicbird

Phaethon aethereus CODE 3

JUVENILE. *Lacks tail streamers; bill is yellow or orange-yellow, unlike adult. Separation from juvenile White-tailed difficult; Red-billed shows more dark above than White-tailed.* NORTH CAROLINA, MAY

ADULT. *Mainly white; long tail streamers, if present, are striking. Combination of blood-red bill and prominent black eyebrow separates Red-billed from White-tailed.* CARIBBEAN, OCTOBER

L 18-36"   WS 44"   WT 1.6 lb

▸ one adult molt per year; complex basic strategy

▸ moderate differences between juvenile and adult

▸ bill color variable; tail streamers can break off

A bird of warm tropical oceans, generally south of North America. Widely scattered records along the East Coast, where rare. Off the West Coast, records restricted to California; apparently regular far off southern California. Often flies high above water.

RED-TAILED TROPICBIRD (P. rubricauda). *Casual visitor to warm waters well off North America. Adult has red bill and red tail streamers, is mainly white above.* PACIFIC, MARCH

Birds at sea usually silent. Birds in flocks sometimes give a clucking series, *ch'ch'chick'chick'chick!chick!ch'ch'*, accelerating and decelerating.

White-tailed Tropicbird

Phaethon lepturus CODE 3

L 15-29"   WS 37"   WT 11 oz

▸ one adult molt per year; complex basic strategy

▸ moderate differences between juvenile and adult

▸ extent of black in primaries varies geographically

Regular in summer off the Southeast coast. Most sightings are at sea, but has been seen coastally or even inland after powerful storms. Often attracted to boats, where it may be seen flying directly overhead.

ADULT. *Plumage mainly white, but note bold black patches on upperwing; juvenile is barred black and white above. Bill is orange-yellow in adult, yellow in juvenile.* FLORIDA, APRIL

ADULT. *Tail streamers extremely long, but sometimes missing or hard to see against bright sky; juvenile lacks streamers, but tail is distinctively long and tapered.* PACIFIC, MARCH

Birds at sea usually silent, but sometimes give high, thin notes interspersed with lower calls; for example, *wink wink wink wurp wink wurp wink wink.*

Northern Gannet
Morus bassanus   CODE 1

ADULT. *Like a flying cross: tail and neck are conspicuously long. Extensive black on wing visible at great distance; golden wash on head evident at closer range.*
NEW YORK, SEPTEMBER

IMMATURE. *Acquisition of mainly white adult plumage requires several years. Subadults are easily told by large size and distinctive shape; often associate closely with adults.*
VIRGINIA, JANUARY

L 37"   WS 72"   WT 6.6 lb

▸ one adult molt per year; simple basic strategy

▸ strong differences among multiple age classes

▸ extensive variation within subadult age classes

At all seasons, Northern Gannets plunge like missiles into the sea to catch fish. During breeding season, they gather at noisy colonies on sea cliffs. On migration, they are seen moving along the coasts just beyond the shoreline; strays are occasionally noted inland.

Feeding flocks are highly vocal: *aak...aak...aak.* Near colonies, gives a muffled but excited *ooff oof oof! oof! oof,* repeated often.

Masked Booby
Sula dactylatra   CODE 3

ADULT. *Black and white like Northern Gannet, but differs in having black tail and secondaries, plus black at base of bill.* SOUTH AMERICA, NOVEMBER

JUVENILE. *Can be confused with adult Brown Booby, but brown on head does not extend to breast; feet are brown (yellow-tan on adult Brown Booby).*
ASIA, JANUARY

L 32"   WS 62"   WT 3.3 lb   ♀ > ♂

▸ one adult molt per year; simple basic strategy

▸ strong differences between juvenile and adult

▸ adults average smaller in Atlantic than Pacific

Widespread in tropical oceans, occasionally reaching North American waters. More frequent in warm Gulf of Mexico and Gulf Stream than in cold waters off West Coast. Engages in spectacular plunge-dives. Local populations often unstable, subject to extirpation.

Not usually heard in North America. Around nesting colonies, male gives a whistled *ch'weeeee!oh;* both sexes give a muttering, unsteady *wek wek wek wek.*

Brown Booby
Sula leucogaster CODE 3

ADULT FEMALE. *In all adults, dark breast contrasts sharply with white belly. Pacific female has dark head, as here; male has pale head. In Atlantic/Caribbean, head dark in both sexes.* PACIFIC OCEAN, MARCH

JUVENILE. *Brown overall, but contrast between dark brown hood and paler brown belly usually evident. Bill gray; feet yellow-tan.* HAWAII, NOVEMBER

| L 30" | WS 57" | WT 2.4 lb | ♀ > ♂ |

▸ one adult molt per year; simple basic strategy

▸ strong age-related and weak sex-related differences

▸ Atlantic and Pacific populations differ in plumage

Widespread but generally rare off Atlantic and Caribbean coasts, southern Pacific coast. Not as rare off southern Florida, and may have bred in small numbers on the Florida Keys. This pantropical seabird can be seen both near and far from shore.

Most calls similar to Masked Booby. Male gives "wolf-whistle," female a honk or quack; both sexes give low chattering, clucking. Rarely heard in North America.

Brown Boobies are expert fishers, but they sometimes acquire a meal by stealing food from other species. This behavior is called kleptoparasitism. It has evolved many times in birds, and it is especially common in fish-eating species. Perhaps the best-known kleptoparasites are the jaegers, which habitually chase down gulls and force them to relinquish their quarry. Gulls and terns also engage in kleptoparasitism, with Brown Pelicans being a frequent target. The preferred victim of the Brown Booby seems to be the Magnificent Frigatebird.

BLUE-FOOTED BOOBY (S. nebouxii). *Feet bright baby-blue on adult, duller on juvenile. Adult shows lower contrast overall than Brown Booby, with dull gray bill and dusky head.* CALIFORNIA, NOVEMBER

RED-FOOTED BOOBY (S. sula). *Plumage of adult resembles Masked Booby, but most have white tail and gray bill; Masked has black tail and yellow bill.* HAWAII, MARCH

Brown Pelican
Pelecanus occidentalis   CODE 1

NONBREEDING ADULT. *White head and neck contrast with brown body. Pattern on juvenile reversed: head, neck, and breast are dusky brown, but belly is white.*
CALIFORNIA, SEPTEMBER

BREEDING ADULT. *Atlantic and Gulf coast breeders show rich chocolate brown where Pacific coast breeders show red.* FLORIDA, FEBRUARY

IN FLIGHT. *All plumages (juvenile shown here) appear dark overall in flight. Brown Pelican flies with neck tucked in; American White often extends neck.* CALIFORNIA, SEPTEMBER

L 51"	WS 79"	WT 8.2 lb	♂ > ♀

- ▸ three adult molts per year; simple alternate strategy
- ▸ moderate seasonal and strong age-related differences
- ▸ Pacific birds larger and more colorful when breeding

Coastal counterpart of American White Pelican. Awkward on land, but a masterful flier and expert diver. Engages in amazing plunge-dives, often just offshore. Flocks fly single-file just above the ocean's surface. Populations have increased rapidly in recent years.

Various noises when courting and at nest sites, including a distinctive couplet *herk! ... herk!*, repeated slowly. Rarely heard outside breeding season.

BREEDING ADULT. *On West Coast, breeders show intense red on throat and bill. Brown Pelicans on all coasts are very large, although not as strikingly huge as American White.* CALIFORNIA, SEPTEMBER

American White Pelican

Pelecanus erythrorhynchos CODE 1

BREEDING ADULT. *Vertical bill plate distinguishes breeders from nonbreeders. Structure of bill plate variable; apparently plays a role in both courtship display and ritualized combat.* OREGON, APRIL

L 62"	WS 108"	WT 16.4 lb	♂ > ♀

▸ three adult molts per year; simple alternate strategy

▸ complex but easily overlooked seasonal variation

▸ rare variants with reduced black sometimes seen

Nests and feeds in marshes well inland. Awkward on land, but spectacular in flight, often soaring on outstretched wings at great altitudes. American White does not dive, as Brown Pelican does; instead, large groups cooperatively scoop up fish just below the surface.

Silent away from colonies. Adults and especially nestlings are noisy at nest sites, where various grunts and squawks are heard; for example, *uhrnk uhrnk.*

NONBREEDING ADULT. *Huge-bodied and huge-billed. Mainly white in all plumages; adult becomes dusky-headed in early summer; juvenile is duskier overall.* CALIFORNIA, JANUARY

IN FLIGHT. *White overall with contrasting black flight feathers of wings. Flocks twist and turn as if synchronized. Soaring flight effortless; takeoff flight steady and powerful.* FLORIDA, JANUARY

Great Cormorant

Phalacrocorax carbo CODE 1

L 36"	WS 63"	WT 7.2 lb

▸ two adult molts per year; simple alternate strategy

▸ strong age-related and moderate seasonal differences

▸ extent of white on breeding individuals variable

Globally widespread, but in North America restricted to East Coast. Population increasing here, expanding south. Mainly marine, but flocks sometimes roost in fresh water a few miles inland. Wintering birds may disperse significant distances during a single day.

Silent most of the time, but noisy in colonies and at roost sites, where various calls, often three-parted, are heard; for example, a low-pitched *urt urt urt.*

ADULTS. *White at base of bill always present; extensive in breeding season, reduced but still prominent at other times. Breeding adults have white patch on flanks, often concealed, but when visible in flight a good field mark.* QUÉBEC, AUGUST

JUVENILE. *Paler on belly than breast, the opposite of Double-crested. All Great Cormorants are heavy-built, large-billed; side by side, Great appears larger than Double-crested.* NEW JERSEY, DECEMBER

Neotropic Cormorant
Phalacrocorax brasilianus CODE 1

BREEDING ADULT. Throat patch of adult less extensive, less orangish than Double-crested; outlined in white. Breeders acquire wispy plumes on sides of head, held briefly in spring. TEXAS, APRIL

JUVENILE. Paler overall than adult, but rarely as pale underneath as juvenile Double-crested. Long-tailed and small-billed compared to Double-crested. TEXAS, JANUARY

IN FLIGHT. Dark and rangy like Double-crested, but note differences in shape: Neotropic is small-billed and long-tailed, causing it to look less front-heavy than Double-crested. TEXAS, APRIL

L 25"   WS 40"   WT 2.6 lb   ♂ > ♀

▸ two adult molts per year; simple alternate strategy

▸ strong age-related and moderate seasonal differences

▸ some variation in plumage of breeding birds

Widespread in Neotropics. Occurs regularly north to New Mexico, Texas, Louisiana; vagrant farther north in Great Plains. Apparently increasing and expanding to new sites. Adaptable; occupies diverse wetland habitats, from desert oases to coastal situations.

More likely than other cormorants to be heard away from colonies. Calls varied, but most are pig-like: grunting *glronk*; also *grlunk glrunk glrunk* in series.

Double-crested Cormorant
Phalacrocorax auritus CODE 1

BREEDING ADULT. Wispy plumes (the "double crest") are held briefly in spring; more conspicuous in West. At all times of year, adult shows orange throat patch and all-black body. CALIFORNIA, MARCH

JUVENILE. Bill and throat patch extensively orange or orange-yellow; has thinner bill than Great Cormorant. Paler below than above; palest (sometimes nearly white) on breast. CALIFORNIA, MAY

IN FLIGHT. Kink in neck visible at great distance; appears large-bodied. Immature, especially when molting, as here, is variable; adult strikingly dark in flight. CALIFORNIA, DECEMBER

L 33"   WS 52"   WT 3.7 lb   ♂ > ♀

▸ two adult molts per year; simple alternate strategy

▸ strong age-related and moderate seasonal differences

▸ geographic variation in size, crest, and bare parts

An adaptable species, rapidly expanding its range; increasingly in conflict with humans. Found in a variety of aquatic habitats from inland to out at sea. Frequently seen swimming and flying; roosts on ground and in trees. Catches fish by surface-diving.

At colonies or roosts, a variety of grunts and hisses, all with grating quality. Note especially a descending series of belches, *grrp, grrp, grr, g', g'.*

Brandt's Cormorant
Phalacrocorax penicillatus CODE 1

NONBREEDING ADULT. *Dark overall with a variable pale throat patch; in late spring, acquires blue facial skin and wispy white crests on sides of face.* CALIFORNIA, MARCH

MOLTING ADULT. *In flight, some show kink in neck like Double-crested. Large West Coast cormorant; short tail imparts front-heavy look, especially noticeable in flight.* CALIFORNIA, JANUARY

L 34"	WS 48"	WT 4.6 lb

▸ two adult molts per year; simple alternate strategy

▸ moderate age-related and seasonal differences

▸ breeding birds vary in plumage and bare parts

A West Coast cormorant, usually found in marine settings. Sizeable flocks roost on rock outcroppings just offshore, and disperse in large numbers along the coast. Inland sightings are rare. Populations vary annually, perhaps in response to sea surface temperature.

Silent during most of the year. Even around nest site, a quiet cormorant. Calls are soft and gulping; for example, a muted *hrunk*, typically repeated.

Pelagic Cormorant
Phalacrocorax pelagicus CODE 1

BREEDING ADULT. *Small-bodied; small-proportioned. Bill slender; neck thin. Breeders acquire red facial skin, variable yellow on bill, and wispy dark plumes on greenish head.* CALIFORNIA, MARCH

NONBREEDING ADULT. *Separation from Red-faced Cormorant frequently difficult; characters such as bill color (typically pale in Red-faced, dark in Pelagic) are notoriously variable.* CALIFORNIA, FEBRUARY

L 28"	WS 39"	WT 3.9 lb

▸ two adult molts per year; simple alternate strategy

▸ weak age-related and moderate seasonal differences

▸ northern birds average larger, but effect is variable

An uncommon West Coast species. Global population low, and sightings are usually of singles or small flocks, often amid larger numbers of Brandt's Cormorants. Despite name, usually not seen in the true pelagic zone; instead, favors rocky habitats close to shore.

Reclusive and quiet away from nesting and roosting areas. At aggregation sites, listen for various croaking and hissing sounds, fairly soft.

Red-faced Cormorant
Phalacrocorax urile CODE 2

L 29"   WS 46"   WT 4.6 lb

▸ two adult molts per year; simple alternate strategy
▸ moderate age-related and seasonal differences
▸ variant adults with yellow faces sometimes seen

In North America, largely restricted to seacoasts of southern Alaska. Rarely seen inland or out at sea; instead, prefers coastal islands and rocky shorelines. Usually occurs in small, loose flocks. Population status unclear; may be undergoing a range shift.

Rarely heard away from nesting and roosting areas. At aggregation sites, gives a growling, tremulous *br'r'r'r'k*, plus various groans and hisses.

BREEDING ADULT. *Most show extensive red on face and conspicuous yellow on bill, more so than on Pelagic. Juvenile lacks bright colors, but pale bill usually stands out.* ALASKA, JULY

IN FLIGHT. *Neck not as long and thin as Pelagic, but this trait can be subjective. White flank patch on breeding adult can be absent, as here, on this and other cormorant species.* ALASKA, JULY

Anhinga
Anhinga anhinga CODE 1

IN FLIGHT. *When soaring, tail appears oversized. Long neck, head, and bill create javelin-like shape. Overall impression is of rear-heavy bird; compare cormorants.* TEXAS, JUNE

L 35"   WS 45"   WT 2.7 lb

▸ two adult molts per year; simple alternate strategy
▸ strong sex-related and age-related differences
▸ facial plumes and head shape of breeding male variable

A distinctive resident of freshwater wetlands in the Southeast; vagrants seen far north and west of core range. Often swims with only its head above the water; perches in trees, where it dries its feathers. A strong flier, often seen soaring.

Generally quiet, but sometimes highly vocal. Most calls have chattering and buzzy qualities, for example, *zurr zurr zurr zurrdd*.

BREEDING MALE. *On both sexes, white on wing contrasts with black body; bill dagger-like and tail long. Breeding male acquires rusty plumes on sides of face.* FLORIDA, MARCH

ADULT FEMALE. *Head and breast pale brown, contrasting with black belly and back; juvenile similar but duller. Male head all black. All Anhingas often assume "spread eagle" pose.* FLORIDA, MARCH

Magnificent Frigatebird
Fregata magnificens CODE 1

ADULT MALE. *Body small relative to rest of bird: bill long and thin, tail long and forked, wings very long. Male shows red throat, inflated preposterously in display.* FLORIDA, APRIL

ADULT FEMALE. *Distinguished from adult male by white breast; plumage of male is wholly black. Except around roosts and nests, frigatebirds are likely to be seen only in flight.* FLORIDA, APRIL

| L 40" | ws 90" | wt 3.3 lb | ♀ > ♂ |

▸ one adult molt per year; simple basic strategy

▸ strong sex-related and age-related differences

▸ complex variation among and within subadult plumages

Flies effortlessly and endlessly, almost never alighting on the water. Picks food from the surface or in midair; occasionally steals from other birds. Wanders north of usual range along coasts, and has been recorded far inland.

Usually silent except during courtship and at nest. At colonies, gives various whirring, chattering, and clicking calls: *rick rick rick* and *rack rack.*

JUVENILE. *Head and breast white on juveniles; as bird matures, becomes more adult-like. Progression toward adult plumage poorly understood; requires several years.* FLORIDA, APRIL

ADULT MALE. *At breeding colony, red throat pouch of male hangs loosely; in a display to attract passing females, the bird inflates throat pouch with air like a large balloon.* FLORIDA, JULY.

ADULT FEMALE. *The breeding season is lengthy, beginning with egg-laying in the fall and extending into early summer with the care of dependent young.* FLORIDA, JULY

Herons, Ibises & Spoonbills, and Storks

ORDER CICONIIFORMES

26 species: 17 ABA CODE 1 1 CODE 4 8 CODE 5

The herons and their relatives in the order Ciconiiformes are long-legged waders that are often conspicuous in the marshes, swamps, and mudflats they call home. Three families are represented in North America. The herons and bitterns (Ardeidae) are in many ways the flagship family—familiar sights at wooded swamps and pond edges continent-wide. The sickle-billed ibises and spatulate-billed spoonbills (Threskiornithidae) are best represented in warmer regions of the continent. The bare-headed storks (Ciconiidae) are even more strongly affiliated with southerly climes. At various points in the past, the New World vultures have been linked to the order Ciconiiformes, but a close relationship is currently thought to be unlikely.

The birds in this order are aquatic during parts or all of their life cycle. A few, like the bitterns, are highly secretive, but many occupy wide-open habitats, such as rice fields. They are much given to loafing, and it is common to see one, a few, or even hundreds resting communally in tall trees. Many are skilled aerialists, sometimes soaring at great heights. Their diet consists of diverse animal matter: spoonbills filter invertebrates from muck-filled waters, ibises eat earthworms, Green Herons sometimes fish using insects for bait, and Great Blue Herons eat anything and everything.

Most species are gregarious, and most are migratory. Nests are typically placed in dense treetop colonies, sometimes numbering thousands of birds. These aggregations may consist of just one species, but nesting colonies with ten or more species are not unheard of. With their strong powers of flight, vagrants are frequently seen—sometimes spectacularly out of range. The order Ciconiiformes is, in general, not known for vocal ability; most species give unmusical grunts, squawks, and cackles. The mellow cooing of the Least Bittern is a notable exception.

Egrets most famously, but also many other wading birds, were brought to the brink of extinction in the early 20th century by the fashion industry. (Their elaborate nuptial plumes, or "aigrettes," were all the rage a century ago.) But strict protections were enacted just in time, and populations rebounded quickly. These birds also benefited from the DDT ban. The current concern, as with so many other birds, is habitat loss; building development in coastal wetlands is an especially severe threat.

Great Blue Heron
Ardea herodias CODE 1

NONBREEDING ADULT. *Body mainly blue-gray, with variable black at bend of wing. Pale head topped by black cap. Long neck is often tucked in, especially in flight.* TEXAS, NOVEMBER

BREEDING ADULT. *Legs become brighter red and plumes are more prominent at height of courtship season.* FLORIDA, MARCH

46"	ws 72"	wt 5.3 lb	♂ > ♀

- one adult molt per year; complex basic strategy
- moderate age-related and seasonal differences
- "Great White" and "Würdemann's" forms distinctive

Widespread and adaptable; often seen around piers and degraded wetlands, or standing motionless at edge of a pond or river. Nests colonially in tall trees, but typically solitary away from nest. Colonies susceptible to disturbance, but overall population doing well.

When flushing, a deep and harsh *rrrup* or *rrr-uh-unk*. Call is far-carrying; startlingly loud up close. Heard any time of day or night.

JUVENILE. *Similar to adult, but plumage tends to show lower contrast overall. Long foreneck streaked dusky; adult foreneck usually "cleaner."* CALIFORNIA, SEPTEMBER

"GREAT WHITE" HERON. (A.h. occidentalis) *Regular in southern Florida; vagrants sometimes noted farther north. Substantially larger than Great Egret; legs and feet yellowish (black on Great Egret).* FLORIDA, JANUARY

"WÜRDEMANN'S" HERON. *The name given to birds that are intermediate in appearance between "Great White" Heron and typical Great Blue Heron. Pale overall, especially on head.* FLORIDA, MAY

Great Egret

Ardea alba   CODE 1

ADULT. *Large; plumage completely white. Bill long and yellow; feet long and black. Juvenile similar. Similar to "Great White" Heron, but smaller overall, bill less massive, legs and feet blacker.*
FLORIDA, MAY

COURTING ADULT. *In courtship, ornate white plumes (aigrettes) are broadly fanned. Lores bright green in courtship, dusky yellow otherwise.*
FLORIDA, APRIL

L 39"	WS 51"	WT 1.9 lb	♂ > ♀

- one adult molt per year; complex basic strategy
- moderate differences between breeders and nonbreeders
- plumes and bare parts of breeders variable

Accepts diverse wetland settings, from secluded farm ponds to open tidal flats. Occurs singly or in large groups. Nests in mixed-species colonies, but tends to form single-species flocks elsewhere. Has largely recovered from severe population losses a century ago.

Vocalizations varied, but most are rough and grating, descending in pitch; for example, *glurp* and *grr-lurp*. Usually silent away from roosts or nests.

Snowy Egret

Egretta thula   CODE 1

NONBREEDING ADULT. *Smaller than Great Egret; plumage completely white. Thin, black bill; legs black and feet yellow. Juvenile has darker legs and paler bill, suggesting nonbreeding Cattle Egret.*
FLORIDA, FEBRUARY

COURTING ADULT. *At height of courtship season, acquires and displays fine white plumes; feet and especially facial skin intensify, often becoming reddish.* FLORIDA, APRIL

L 24"	WS 41"	WT 13 oz

- one adult molt per year; complex basic strategy
- moderate seasonal and age-related differences
- substantial variation in color of bare parts

Occurs in many wetland types; favors large wetlands with dense vegetation. Especially common along coasts, but also found far inland. Feeds frenetically, prancing about. Population devastated by plume hunters a century ago, but numbers have rebounded.

Gives a variety of rough, unmusical grunts and croaks, like Great Egret, but higher-pitched and more drawn-out: for example, *rrrrip* or *rrrup* or *rick-rick*.

Cattle Egret
Bubulcus ibis CODE 1

BREEDING ADULT. *Acquires Creamsicle-orange highlights; bill and legs become bright orange-red or red. Terrestrial at all times of year, perching in fields and trees.* TEXAS, MAY

NONBREEDING ADULT. *Superficially like Great Egret and especially juvenile Snowy Egret, but bill stout, legs short, stouter overall. Juvenile bare parts darker than juvenile Snowy.* TEXAS, NOVEMBER

L 20"   WS 36"   WT 12 oz

▸ one adult molt per year; complex basic strategy

▸ strong seasonal and moderate age-related differences

▸ plumage and bare parts of breeders variable

Among herons, a landlubber; frequents roadsides and pastures, often in the company of cows or horses. Arrived in North America in 20th century; quickly became extensively established. Still expanding in West, but Northeastern population in decline.

Vocalizations diverse. Most are harsh, many have *k* sounds: for example, *rick rick rick* and *urk urk urk*, frequently repeated in series.

Little Blue Heron
Egretta caerulea CODE 1

ADULT. *Similar in build to Snowy Egret, a close relative. Body and wings slate-gray, head and neck dull purple. Birds in flight, often high up, are strikingly dark.* FLORIDA, FEBRUARY

JUVENILE. *Distinguished from other all-white herons by bare parts: two-tone bill with pale gray on inner half, dark on outer half; dull greenish-yellow legs; wing tips usually dark.* FLORIDA, FEBRUARY

IMMATURE. *Splotchy blue-and-white plumage, progressing from mainly white to mainly dark. Bare parts as on other age-classes.* FLORIDA, FEBRUARY

L 26"   WS 36"   WT 13 oz

▸ one adult molt per year; complex basic strategy

▸ strong differences between adult and juvenile

▸ older immatures highly variable

A southeastern heron, most abundant along the Gulf coast; vagrants are widely noted far out of range. Breeding range is slowly expanding. Feeds in large marshes, solitarily or in small numbers. Roosts and nests in mixed-species assemblages.

Gives varied grunts and growls. Many calls have drawn-out, wavering quality; for example, *uuuurrrk* and *uuurrrp*. Quiet away from nesting and roosting areas.

Tricolored Heron

Egretta tricolor　CODE 1

NONBREEDING ADULT. *Lanky: long-necked and long-billed; often assumes droopy poses. White belly contrasts with dark breast. Breeders acquire golden plumes on throat and back.* FLORIDA, FEBRUARY

JUVENILE. *Plumage similar in pattern to adult, but more extensively russet, especially on neck and upperwing coverts. Note extremely long neck; bill lengthens as juvenile ages.* TEXAS, JUNE

IN FLIGHT. *Long, lanky, and droopy profile often prominent on flying birds. Under most lighting conditions, appears paler in flight than adult Little Blue Heron.* FLORIDA, MAY

L 26"	ws 36"	wt 13 oz

▸ one adult molt per year; complex basic strategy

▸ strong age-related and moderate seasonal differences

▸ on adults, shades of blue, purple, and gold variable

Occurs mainly in coastal habitats such as salt marshes and estuaries, but frequently wanders far inland, where it may be found in a variety of wetlands. Usually hunts by itself or in loose, very small assemblages. Population apparently declining.

Calls average less harsh than Little Blue Heron and other herons. Listen for somewhat mellow groans; for example, *ummmm* and *rrummm*.

Reddish Egret

Egretta rufescens　CODE 1

DARK MORPH ADULT, BREEDING. *Like an oversized, brightly colored Little Blue Heron; head and breast russet, body slate-blue. High-contrast bill: pink base, dark tip. Note animated feeding behavior.* FLORIDA, FEBRUARY

WHITE MORPH ADULT, BREEDING. *Distinguished from other all-white herons by bill: thick-based and straight, with high-contrast pattern visible from afar.* FLORIDA, MAY

WHITE MORPH JUVENILE. *Similar to Little Blue Heron, but bill and facial skin are dark on both white and dark juvenile and nonbreeding adult Reddish Egrets. (Little Blue has pale facial skin and bill base.)* TEXAS, OCTOBER

L 30"	ws 46"	wt 1 lb

▸ one adult molt per year; complex basic strategy

▸ moderate differences between adult and subadult

▸ highly dimorphic; white morph rarer than dark morph

Among our herons, the most intimately associated with coastal habitats. Reddish Egret is uncommon in North America, but it is conspicuous by its berserk feeding behavior: lunging, galloping, and dashing about in open saltwater lagoons.

Calls are varied. Suggest Tricolored Heron, but most are softer and shorter, not as drawn-out; for example, *urpa* and *urr-uh*. Generally silent.

Yellow-crowned Night-Heron

Nyctanassa violacea CODE 1

ADULT. *More svelte than Black-crowned, with longer neck and legs. Elegant black-and-white head pattern is distinctive; thin white plumes on head are often reduced or absent.* TEXAS, MAY

JUVENILE. *Body structure as adult. Color similar to Black-crowned, but streaking and spotting are finer; further separated from Black-crowned by all-dark bill.* FLORIDA, MARCH

IN FLIGHT. *Yellow-crowned Night-Heron is longer-legged than Black-crowned. On flying birds, the legs dangle well beyond tip of the tail.* TEXAS, APRIL

L 24" ws 42" wt 1.5 lb

▸ one adult molt per year; complex basic strategy
▸ moderate differences between adult and subadult
▸ individual variation in bill size of adults

Occupies various aquatic habitats in core range in Southeast, but favors wooded wetlands. Recovering from population losses; numbers rebounded dramatically in 20th century. Range expansion apparently underway; vagrants often noted far from current range.

Alarm call similar to Black-crowned Night-Heron, but higher-pitched and less jarring: *kwep!* and *kweerk!* Usually heard at dusk or at night.

Black-crowned Night-Heron

Nycticorax nycticorax CODE 1

ADULT. *Stocky and rotund. Mainly pale gray, but with contrasting black cap and back. Bill dark, eye intense ruby. Yellow legs shorter than Yellow-crowned Night-Heron.* TEXAS, APRIL

JUVENILE. *Chocolate brown all over with coarse spotting and barring; pale yellow on bill is extensive. All night-herons strike a hunched, dopey pose when perched.* CALIFORNIA, MARCH

L 25" ws 44" wt 1.9 lb

▸ one adult molt per year; complex basic strategy
▸ strong differences between adult and subadult
▸ subadult plumages show modest variation

Widespread and fairly common, but often overlooked because of its nocturnal habits; hides in dense vegetation by day, emerges at dusk to hunt. Occurs in both freshwater and saltwater habitats, usually with woody vegetation nearby. Populations increasing.

◀ Powerful *quock!*—usually given at night by birds in flight—is abrupt and jarring. Also gives a higher-pitched version of call: *querp!*

Green Heron
Butorides virescens CODE 1

ADULT. *Dark and rotund; flushing birds show rounded wings and trail yellow legs. Head and breast chestnut, with variable pale streak down center of breast; upperparts dull green.* TEXAS, JUNE

JUVENILE. *Breast and foreneck with splotchy dark streaks on pale background. Green-backed like adult, but with more spotting and splotching. Dark cap often prominent.* CALIFORNIA, OCTOBER

L 18"   WS 26"   WT 7 oz

▸ one adult molt per year; complex basic strategy

▸ moderate differences between adult and subadult

▸ western birds grayer above and larger

Wary. Throughout range, often the most widely distributed heron; occurs in wetlands too small for other herons. Usually solitary; sometimes forms loose flocks. Prowls marshes, swamps, and stream edges. Has the unusual habit of baiting fish with invertebrates.

Piercing *skyow!*—sometimes doubled or trebled—is a characteristic sound of summer in wetlands. Also clattering *skuck* and *skup* notes, often repeated.

American Bittern
Botaurus lentiginosus CODE 1

ADULT. *Dirty brown all over, with coarse streaking below; juvenile slightly duller. Bill longer, more yellow than night-herons; streaking below is bolder and better defined.* TEXAS, MAY

IN FLIGHT. *Similar to juvenile night-herons and Green Heron, but upperwing of American Bittern distinctly two-toned; dark flight feathers contrast with pale wing coverts.* LOUISIANA, APRIL

L 28"   WS 42"   WT 1.5 lb

▸ one adult molt per year; complex basic strategy

▸ weak differences between juvenile and adult

▸ courting birds sometimes erect obscure white plumes

A highly solitary inhabitant of dense marshes, where it stands motion-less for long periods. Flushes infrequently; flying birds most often seen at dawn or dusk. Prefers freshwater marshes, but also occurs in saltwater habitats. Populations are in large-scale decline.

Song a strange three-syllable gulping: *ooong-uh-roonk*, repeated. In flight, an abrupt *kwark!*—most often heard at night—recalls Black-crowned Night-Heron.

Least Bittern
Ixobrychus exilis CODE 1

ADULT MALE. *A tiny heron; much smaller than even Green Heron. Tawny below, extensively dark above; greater plumage contrast on male than female.*
TEXAS, APRIL

ADULT FEMALE. *Not as boldly patterned as adult male. Black cap, back, and wings of male replaced by buff-brown on female. Juvenile similar but duller.*
NEW JERSEY, JULY

L 13"   ws 17"   wt 2.8 oz

▸ one adult molt per year; complex basic strategy
▸ moderate age-related and sex-related differences
▸ rare "Cory's" morph shows extensive dark chestnut

Skulks in marshes filled with dense vegetation, sometimes wandering into view. Common in some areas, inhabiting surprisingly small bodies of water. Occurs in both freshwater and brackish wetlands. Populations generally declining, probably due to habitat loss.

Song completely unlike American Bittern: a wooden *coo coo coo*, repeated slowly, reminiscent of Black-billed Cuckoo; soft and ventriloqual, but carries well.

White Ibis
Eudocimus albus CODE 1

ADULT. *Plumage all-white except for black tips of outer primaries, mainly concealed except in flight. Bare-part colors variable, usually carmine or coral; bill strongly decurved.*
FLORIDA, FEBRUARY

SUBADULT. *Juveniles extensively brown at first; plumage becomes whiter and bare parts redder as bird nears adulthood. Always whiter than corresponding age-classes of other ibis species.* TEXAS, MAY

L 25"   ws 38"   wt 2 lb   ♂ > ♀

▸ two adult molts per year; simple alternate strategy
▸ strong differences between adult and subadult
▸ bill length and overall structure are variable

In North America, occurs in lowlands near Southeast coast; vagrants wander far north and west. Forms large flocks, large colonies. Occurs in various aquatic habitats, often with trees. Accepts both fresh water and salt water, but fresh water essential for rearing young.

Usually silent. Most frequent call, typically given in flight, is a series of gruff honks: *runk runk runk*. Also gives higher-pitched squeals.

White-faced Ibis
Plegadis chihi   CODE 1

BREEDING ADULT. *Red facial skin and eye, broadly bordered in white; breeding Glossy can show white border, but narrower. Bill gray; plumage averages brighter and glossier than Glossy.* UTAH, JUNE

IMMATURE. *Very similar to Glossy, but eye and some facial skin are red; juvenile similar, but duller. Extent of interbreeding between the two species unknown; possibly extensive.* TEXAS, MAY

| L 23" | ws 36" | wt 1.3 lb | ♂ > ♀ |

▸ two adult molts per year; simple alternate strategy

▸ moderate age-related and seasonal differences

▸ variants resemble Glossy; hybrids sometimes reported

Western counterpart of Glossy Ibis, sharing many ecological and physiological similarities. Mainly a freshwater species, but accepts highly saline conditions; often ranges into flooded pastures to look for earthworms and other invertebrates. Vagrants widely noted.

Gives soft grunting and chuckling notes, especially when in feeding flocks or when in flight. All vocalizations similar to Glossy, but average lower-pitched.

Glossy Ibis
Plegadis falcinellus   CODE 1

BREEDING ADULT. *Like a dull White-faced Ibis. Typical breeding adult has brown eyes, dark facial skin, horn-colored bill, and dull legs. Thin powder-blue lines on face often present.* NEW JERSEY, JUNE

JUVENILE. *Many cannot be distinguished from juvenile White-faced. As young birds age, they slowly acquire the bare part colors characteristic of adults of their respective species.* NEW JERSEY, AUGUST

IN FLIGHT. *All plumages appear quite dark. Lone birds or single-file flocks flying at a distance may call to mind Double-crested Cormorant.* NEW JERSEY, JUNE

| L 23" | ws 36" | wt 1.2 lb | ♂ > ♀ |

▸ two adult molts per year; simple alternate strategy

▸ moderate age-related and seasonal differences

▸ variants and hybrids approach White-faced in appearance

Mainly a bird of the Southeast coastal region; range expanding north, vagrants often noted west and north of core range. Large flocks feed and roost in shallow fresh water or salt water. Day and night, probes for invertebrates with tactile sensors in its bill.

Various croaks and crows, many with soft and subdued quality; for example, *hoommm*. Also higher-pitched duck-like cackling: *hack-ack-ack*. Often vocal at night.

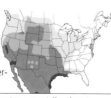

Roseate Spoonbill

Platalea ajaja CODE 1

ADULT. *Snow white and hot pink; pink intensifies to blazing red at bend of wing. Flat bill, broadening greatly at tip, is distinctive.* TEXAS, JUNE

JUVENILE. *Much paler than adult, but most show some pink. Distinctive bill structure as adult. Pale individuals—sometimes nearly white—occur in all age-classes.* TEXAS, JUNE

L 32"	WS 50"	WT 3.3 lb

▸ two adult molts per year; simple alternate strategy

▸ strong differences between adult and subadult

▸ intensity of pink variable, related to diet

A mainly tropical species that occurs throughout the Gulf coast region; sometimes wanders far from normal range. Seen in flocks in freshwater and saltwater habitats. Feeds by sweeping its spatulate bill through shallow water. Recovering from earlier population losses.

Infrequently heard. Gives a variety of low grunts and muffled quacks: *wuck* or *ruck*. Often gives calls in series: *wuck wuck wuck.*

Wood Stork

Mycteria americana CODE 1

ADULT. *Large overall; huge bill decurved. Black flight feathers of wings and tail prominent in flight, mainly concealed at rest. Naked head and neck not fully acquired until adult.* FLORIDA, MARCH

JUVENILE. *Similar in build and overall plumage to adult, but note differences on head: juvenile has variable dusky feathering on neck and pale yellow bill.* FLORIDA, JANUARY

L 40"	WS 61"	WT 5.3 lb	♂ > ♀

▸ one adult molt per year; complex basic strategy

▸ moderate differences between adult and subadult

▸ bare parts on adult head vary in extent and color

A lethargic resident of swamps in the Southeast; often roosts and feeds in the open, where it is conspicuous. Feeds in medium-sized flocks, roosts singly or in flocks; individuals or small flocks often seen in soaring flight. Vagrants sometimes noted far out of range.

Generally silent, but gives hissing and groaning calls, variable in duration, when at nest or interacting with other adults: *uhhhhh* and *uhh-uhhhhhh.*

Flamingos

ORDER PHOENICOPTERIFORMES

---

1 species: 1 ABA CODE 3

With their long legs, long necks, and aquatic habits and habitats, it is easy to imagine that the flamingos are closely allied to the herons and their relatives in the order Ciconiiformes. Yet the flamingos are placed in an entirely different order.

Besides their spectacular pink plumages, some distinctive characteristics of the flamingos include gregariousness, webbed feet, and a complex and unique method of filter feeding in mud and saltwater. Flamingos feed their young with a pink milk-like secretion called "crop milk" that is high in fat and protein. Young birds do not attain the pink feather coloration until adulthood. This color develops variably over time from their diet of algae, fish, and aquatic invertebrates.

ESCAPES FROM CAPTIVITY OR NATURAL VAGRANTS?

It is always a thrill to see a flock of flamingos—perhaps working the edge of a tidal pool, or flying in formation above a distant sandbar. And then the inevitable question a birder asks: Do they "count?" Did the birds arrive on their own wing power from Mexico or the Caribbean? Or did they escape from Disney World? Sometimes the answer is obvious. A drake Mandarin Duck with a metal band around its leg surely originated in a zoo or a private collection. And a Spotted Redshank (a species rarely kept in captivity) swept landward in a Nor'easter presumably made the trans-Atlantic crossing unaided.

But what of all the "gray-area" cases—a male Painted Bunting or Northern Cardinal well out of range? If the birds do not have obvious signs of prior captivity, such as leg bands on passerines, jesses (tethers) on raptors, and pinioned wings on large waterbirds, how can you tell? Sometimes escapes exhibit excessive tameness, unusual (often reduced) feather wear, and abnormal molt timing. But sometimes they do not. In some instances, we simply cannot know whether an out-of-range bird is a true vagrant or an escape from captivity.

Even if we do know (or are fairly certain) that an individual is an escape from captivity, it is worthwhile to document the bird's presence. Perhaps it is a "pioneer"—one of the first individuals of a population that is soon to become established. And even if it is not, its story and circumstances are surely of interest. A presumably escaped Chilean Flamingo (christened "Pink Floyd" by local birders) persisted for years at the Great Salt Lake, providing observers with detailed study of a morphologically unique and behaviorally fascinating species in a wild setting.

Greater Flamingo
Phoenicopterus ruber   CODE 3

CHILEAN FLAMINGO (P. chilensis). *Exotic flamingos such as this bird sometimes escape from captivity; many persist for years, especially in coastal and inland saltwater habitats.* SOUTH CAROLINA, NOVEMBER

ADULT. *All flamingos are instantly recognizable by their extensive pink, gangly shape, and "upside down" bill. Greater is the only flamingo that strays naturally to North America.* FLORIDA, NOVEMBER

L 46"	ws 60"	wt 5.6 lb	♂ > ♀

- one adult molt per year; complex basic strategy
- strong differences between adult and subadult
- extent of pink highly variable on adult

Widespread, but barely reaches North America. Some birds are vagrants from the Caribbean or Mexico; many others are escapes from captivity. Separation of vagrants from escapes is often difficult. Feeds in shallow saltwater lagoons; seen singly or in small flocks.

Flocks in chorus can be noisy: *rut rut rut* and higher-pitched *eeerp eeerp eeerp,* typically repeated; most vocalizations have barking or bugling qualities.

IN FLIGHT. *Extensive deep pink feathering of adult contrasts with black primaries and secondaries. Juvenile gray and black; deep pink acquired over several years.* CARIBBEAN, MARCH

FLOCK. *Usually seen singly or in very small flocks in our area, flamingos are colonial nesters in the Bahamas and elsewhere in the Caribbean region, where they gather in dense groups.* CARIBBEAN, MARCH

New World Vultures, Hawks, and Falcons

ORDER FALCONIFORMES

42 species:	24 ABA CODE 1	7 CODE 2	1 CODE 3	5 CODE 4	4 CODE 5	1 CODE 6

Diurnal raptors are found throughout the continent, and most species are conspicuous. This order includes familiar birds like the Bald Eagle, Red-tailed Hawk, and Peregrine Falcon, as well as little-known species that occur only at the periphery of our area—especially in the southern states. There are three families of diurnal raptors: the vultures and condors in the family Cathartidae; the kites, eagles, Osprey, and various hawks in the family Accipitridae; and the falcons and caracaras in the family Falconidae. A close relationship between the families Accipitridae and Falconidae is well established, but placement of the family Cathartidae is considered by many authorities to be uncertain.

Not surprisingly, many diurnal raptors are seen in the open, where they can best scan for prey. Utility poles and wires are the favorite perches of several species, and the open sky (especially during midday) is a good place to look for many species. For most of the year, raptors are seen singly or in small numbers. During migration, however, impressive congregations move by day along the coast and along ridge tops.

Except for the carrion-eating vultures, caracaras, and condors, the species in this order are predatory. Although a few are generalists, most are fairly specialized. For example, Common Black-Hawks eat mainly frogs and fish, Northern Harriers prey on small mammals, and Mississippi Kites catch large insects. An extreme specialist is the Snail Kite, which eats only one or two species of apple snail. Raptors also differ in their methods of capturing prey. Accipiters and falcons tend to snag birds in flight, whereas hawks in the genus *Buteo* often pounce on medium-sized mammals on the ground.

In the past, raptor populations suffered major losses because of human activities. Accipiters in particular were mercilessly persecuted in the mistaken belief that predators are bad for natural communities and because some species prey on farmers' fowl. Numbers of Bald Eagles and Ospreys plummeted during the DDT era. And a diverse suite of raptor species was affected by egg-collecting, falconry, and other disturbances around the nesting grounds. Fortunately, public attitudes toward raptors are much more enlightened than they used to be, and most species are on the rebound. The endangered California Condor has recently been reintroduced into the wild, following more than a decade when all were kept in captivity. Habitat conversion is a looming threat, however, and many species—from the American Kestrel to the Ferruginous Hawk—are at risk again.

Turkey Vulture
Cathartes aura CODE 1

L 26" WS 67" WT 4 lb

▸ one adult molt per year; simple basic strategy

▸ moderate differences between juvenile and adult

▸ smaller overall and glossier in Southwest

A familiar sight, but very few birders have observed the bird on its well-concealed nest. Usually seen in rocking, soaring flight; less often seen roosting in trees or dining at carcasses. An obligate scavenger; never kills its own food. Range expanding north.

When disturbed, gives an ominous hiss, sometimes up to several seconds long: *kssssssssshhh-uhhhh.* Wing flaps are audible, especially on molting birds.

ADULT. *Naked red head is small and shriveled; some show blue around eye. Juvenile similar, but head gray-brown and feathers duller.* FLORIDA, MARCH

CALIFORNIA CONDOR (Gymnogyps californianus). *Huge. All were captured in late 20th century; they and their descendants are being reintroduced to the wild in California and Arizona.* CALIFORNIA, AUGUST

IN FLIGHT. *Soars in easy circles; holds wings at a slight angle above the horizontal. From below, pale flight feathers contrast with dark wing coverts. Head small; tail long.* CALIFORNIA, MARCH

Black Vulture
Coragyps atratus CODE 1

L 25" WS 59" WT 4.4 lb

▸ one adult molt per year; simple basic strategy

▸ moderate differences between juvenile and adult

▸ body size increases slightly to the north

In North America restricted mainly to the Southeast; populations increasing, apparently expanding north. Often seen in flight; flaps more than Turkey Vulture. Frequents garbage dumps, carcasses. More aggressive than Turkey Vulture. Generally nonmigratory.

Gives a deep and foreboding hiss at intruders or when aggressive: *whoooooooooshhhh-ah.* Wing flaps are often audible.

IN FLIGHT. *Dark overall like Turkey Vulture, but stockier and somewhat smaller. Wings held level; tail short. Six outer primaries pale.* TEXAS, MAY

ADULT. *Legs long and sturdy-looking; bill longer than Turkey Vulture. Extensive unfeathered skin on head is ashy gray. Juvenile shows less contrast between head and body.* FLORIDA, FEBRUARY

Osprey

Pandion haliaetus   CODE 1

ADULTS. *Massive stick nest, placed on large dead tree or artificial structure, is sometimes visible more than a mile away. Adults are dark brown above and white below; dark eye-stripe, white crown, and gull-like flight are good marks.* NEW JERSEY, JULY

L 23"	WS 63"	WT 3.5 lb

▸ one adult molt per year; complex basic strategy

▸ weak sex-related and moderate age-related differences

▸ individual variation in extent of dark and pale markings

Diet consists almost exclusively of live fish, which it catches by diving and hitting the water feet first. Breeds wherever there are fish, from lakes high in the Rockies to coastal inlets. Numbers have rebounded from severe losses during the DDT era.

◀ Most frequently heard call a series of rapid, piping whistles, often increasing in speed and volume, then trailing off: *pee pee pee! peep! peep! pee pee.*

JUVENILES. *Size, structure, and pattern similar to adult, but note variable spotting on upperparts. Some show browns and buffs where adults are darker.* CONNECTICUT, JULY

ADULT FEMALE. *Some females show a heavy "necklace" of dark streaks, which males usually do not. All show a sharp kink or bend in the wing. Contrasting head pattern visible from afar.* FLORIDA, MARCH

Hook-billed Kite

Chondrohierax uncinatus   CODE 3

L 18"	WS 36"	WT 10 oz

▸ one adult molt per year; complex basic strategy

▸ strong age-related and sex-related differences

▸ has pale and dark morphs; dark morph rare

A tropical species that barely reaches south Texas, where it is uncommon. Spends much time foraging deliberately and roosting in trees in dense woodlots, where it is difficult to detect. Sometimes seen flying across clearings or soaring in the open.

Most commonly heard vocalization is a rapid *ee-dee-dee-dee-dee-dee-dee*, with anxious quality. Individual notes are short and staccato.

ADULT MALE. *Flight profile distinctive in all plumages: wings broad and rounded, tail long and narrow; impression is of an exaggerated Accipiter shape.* TEXAS, MAY

ADULT MALE. *Mainly gray. Belly finely barred gray and white; tail with much broader gray and white bands. Female upper parts gray-brown, underparts barred orange and white. Oversized bill with prominent downward hook is distinctive in all plumages.* TEXAS, MAY

White-tailed Kite
Elanus leucurus CODE 1

ADULT. *High-contrast plumage noticeable especially from above: black-and-gray wings; white underparts and tail.* TEXAS, OCTOBER

ADULT. *Selects conspicuous perches. General impression is of a small raptor, pale overall with contrasting darker areas.* TEXAS, OCTOBER

JUVENILE. *Plumage pattern generally as adult, with mainly white tail and contrasting black-and-gray wings; but more mottled overall, often with buff and brown highlights.* TEXAS, MAY

L 15"   WS 39"   WT 12 oz

- one adult molt per year; complex basic strategy
- moderate differences between juvenile and adult
- body size increases northward, but trend variable

A conspicuous inhabitant of semi-open country in the lowlands. Occurs in grasslands and marshes, as well as in intensively farmed regions. Has recovered from earlier population losses, and is expanding into new areas. A skilled aerialist, but often seen perched.

Vocalizations include a descending whistle, *tyeerp* or *tyeert*, and a two-syllable *whee-ichk*, first note whistled, second note harsh and grating.

Mississippi Kite
Ictinia mississippiensis CODE 1

ADULT MALE. *Both sexes are gray and black overall, but male shows paler head and secondaries, creating higher-contrast plumage than female.* COLORADO, JUNE

JUVENILE. *Barred and splotchy all over. Underparts with strong rusty tones; upperparts colder gray. Long wings and languid flight style as adult.* COLORADO, AUGUST

ADULT MALE. *In flight, all Mississippi Kites appear dark overall and long-winged. Proportions falcon-like, but manner of flying more languid and erratic.* COLORADO, JUNE

L 14"   WS 31"   WT 10 oz

- one adult molt per year; complex basic strategy
- strong age-related and moderate sex-related differences
- extensive plumage variation recently recognized

Catches dragonflies and other large insects on the wing; also eats birds and mammals. Sometimes forages and roosts in flocks. Eastern birds nest in dense forests; Western birds prefer towns and golf courses. Range expanding in West. Highly migratory.

Call a melancholy, two-syllable *fee-feeeeeee*, remarkably pure in tone; similar in timbre to Eastern Wood-Pewee. Also gives three- and four-syllable calls.

Swallow-tailed Kite

Elanoides forficatus   CODE 1

ADULT. *Perched birds of all ages are dark above and pale below; tail longer on adult than juvenile. Adult strikingly black-and-white; juvenile has variable buff wash on paler areas.*
FLORIDA, APRIL

ADULT ABOVE. *Lower contrast above than below (black-and-gray above, black-and-white below).*
CONNECTICUT, APRIL

ADULT BELOW. *Long wings and tail streamers accentuated by seemingly effortless flight; abrupt but graceful twists and turns are distinctive even at great distances.* FLORIDA, MAY

L 22"   WS 51"   WT 15 OZ

▸ one adult molt per year; complex basic strategy

▸ moderate differences between juvenile and adult

▸ some individuals have broken or missing tail feathers

Much admired for its gracefulness and elegance; flies low, searching for aerial insects or for food in the leaves of tall trees. Small flocks forage together, and sometimes hundreds roost together. Numbers low overall; range has contracted greatly.

Gives squeaky, high-pitched calls: *eeep eeep eeep, wee wee wee.* Notes pure-toned, usually given in rapid succession; overall effect is one of agitation.

Snail Kite

Rostrhamus sociabilis   CODE 2

ADULT MALE. *Only the adult male is extensively slaty; both sexes show white vent and base of tail. In all plumages bill is long, decurved; bare parts orange-yellow to blood red.* FLORIDA, JUNE

ADULT FEMALE. *Splotchy dark and pale. Juvenile similar but browner. In flight, broad-winged and short-tailed.*
FLORIDA, MARCH

L 17"   WS 42"   WT 15 OZ

▸ one adult molt per year; complex basic strategy

▸ strong age-related and sex-related differences

▸ individual variation in color of bare parts

Unusual among raptors, a dietary specialist that feeds primarily on only a few species of snail. Found in large, fairly open wetlands that support its prey. Endangered in U.S., where range is reduced from its historical extent. Gregarious, roosts communally.

Calls not as pure and whistled as other kites; gives wooden, cackling notes with "*k*" sounds: *eh-keh-keh-keh-keh* and *ka-ka-ka.*

Sharp-shinned Hawk
Accipiter striatus CODE 1

ADULT. *Blue-gray above, sometimes with white splotches; barred orange and white below. Cap of male slightly more contrasting than female. Note small body and short, squared tail.*
ARIZONA, DECEMBER

JUVENILE. *Body profile distinctive: short, squared-off tail; small head held close to body; rounded wings held forward. Wear and angle of view affect appearance of tail*
NEW JERSEY, SEPTEMBER

JUVENILE. *Dull overall: upperparts fairly solid brown; pale underparts streaked with brown. Body structure usually distinctive: small head, plump body, squared-off tail.* MINNESOTA, SEPTEMBER

| 11" | ws 23" | wt 5 oz | ♀ > ♂ |

one adult molt per year; complex basic strategy

strong age-related and weak sex-related differences

much variation in body mass and tail shape

feisty, bird-eating inhabitant f forests. Widespread but thinly istributed; usually occurs in small umbers, except on fall migration the East when hundreds may be een from ridge tops or along the past. Generally wary; rarely allows lose approach.

all is a frantic series, *kek kek kek...* or *kip kip kip....* otes usually given rapidly, but sometimes more slowly: k... ick... ick....

Cooper's Hawk
Accipiter cooperii CODE 1

ADULT MALE. *Both sexes like Sharp-shinned: blue above, barred orange and white below. Long, rounded tail often has pale terminal band. Black cap of male more contrasting than female.* ARIZONA, JANUARY

JUVENILE. *In leisurely flight, tail is longer and more rounded than Sharp-shinned, head more prominent, wingbeats less jerky; beware effects of angle of view and feather wear.* NEW JERSEY, OCTOBER

JUVENILE. *Patterned like Sharp-shinned: brown above, streaked below. Ground color averages warmer than Sharp-shinned, but variable. Long, rounded tail has white terminal band.*
NEW JERSEY, OCTOBER

| L 16.5" | ws 31" | wt 1 lb | ♀ > ♂ |

▸ one adult molt per year; complex basic strategy

▸ strong age-related and moderate sex-related differences

▸ birds average larger in East than West, but variable

Like Sharp-shinned, an inconspicu-ous denizen of forests, except on fall migration in the East, when some-times numerous and conspicuous. Hunts birds by quick chases through dense woodlands; prey are often surprisingly large. Formerly perse-cuted, but numbers recovering.

◀ Calls reminiscent of Sharp-shinned Hawk and Northern Goshawk, but distinctly lower-pitched and more nasal: *cack cack cack...* and *kuck kuck kuck...*

Northern Goshawk

Accipiter gentilis CODE 1

ADULT MALE. *Brighter blue above than Cooper's or Sharp-shinned. Fine blue-gray bars below; barring coarser on female. Black-and-white face pattern and large size usually obvious.* MINNESOTA, OCTOBER

JUVENILE. *Brown overall. Pale eyebrow usually prominent; tail bands wavy. In flight, all goshawks show characteristic Accipiter shape: thin tail and broad wings. Proportionately shorter-tailed and longer-winged than Cooper's.* MINNESOTA, OCTOBER

L 21"	WS 41"	WT 2.1 lb	♀ > ♂

▸ one adult molt per year; complex basic strategy

▸ strong age-related and weak sex-related differences

▸ larger and slightly paler in northern portions of range

A powerful and maneuverable woodland raptor. Nests in extensive forests, often in mountains; some birds spread out into more open habitat in winter. Even in core habitat, a low-density and seldom-seen species. Highly aggressive around nests, attacking human intruders.

Call is wild and urgent-sounding, closer in timbre to Sharp-shinned than to Cooper's Hawk: for example, *ki ki ki...* or *kak kak kak...* or *wick wick wick...*

Northern Harrier

Circus cyaneus CODE 1

ADULT MALE. *Ghostly; blue-gray above, largely pale below. In flight, black primary tips and trailing edge of wing contrast with otherwise pale underparts.* NEW YORK, MARCH

ADULT FEMALE. *Gray-brown above, streaked brown below. All plumages show conspicuous white rump, visible at a great distance, and owl-like facial disk.* COLORADO, MARCH

JUVENILE. *Extensively tawny below; not as streaky as adult female In typical flight, as here, all harriers appear slender-bodied, long-tailed, and long-winge* CALIFORNIA, DECEMBER

L 18"	WS 43"	WT 15 oz	♀ > ♂

▸ one adult molt per year; complex basic strategy

▸ strong age-related and sex-related differences

▸ extensive individual variation in plumage

Occurs broadly in open habitats such as marshes and grasslands. Somewhat colonial, especially in winter, when large roosts sometimes are formed. Usually flies low on wings held in a shallow V, but migrants can achieve significant altitude.

In courtship, gives series of piping notes with anxious quality: *peep! peep! peep pee pee pee.* Otherwise, produces occasion *kep* or *cup* notes in flight: *cack...* and *kuck kuck kuck...*

Bald Eagle

Haliaeetus leucocephalus CODE 1

ADULT. *All-white tail and head contrast with dark body. In all plumages, head and bill proportionately larger, but tail shorter, than Golden Eagle.* ALASKA, MARCH

IMMATURE, THIRD YEAR. *Transition from juvenile to adult plumage requires at least three years. By third year, head extensively white but with variable dark patch behind eye.* ALASKA, MARCH

ADULT. *Unmistakable overhead; horizontal plane of wings, with no dihedral, helps to identify an eagle at great distance. Spread "fingers" of outer primaries enhance aerodynamic lift.* ALASKA, MARCH

31"	ws 80"	wt 9.5 lb	♀ > ♂

- one adult molt per year; complex basic strategy
- much year-to-year variation in immature plumages
- body mass variable; generally increases northward

A conservation success story; listed as Endangered following the DDT era, but numbers rebounded greatly in late 20th century. Population still increasing. Occurs in various habitats, but largest aggregations usually along coasts and rivers, especially in winter.

IMMATURE, SECOND YEAR. *Variable, but always extensively dark and dusky; large head and oversized bill stand out. In flight, underwing pattern more diffuse than Golden Eagle.* ALASKA, MARCH

IMMATURE, SECOND YEAR. *Despite its reputation for being an opportunistic scavenger, many Bald Eagles may hunt live prey (such as coots and ducks) in swift sideways attacks near or at the water's surface.* ALASKA, MARCH

Most frequently heard call is a chatter, variable in length but usually short. Often descends and slows, changing pitch midway: *k'k'k'kip'kip'kip.* Most vocalizations musical.

Golden Eagle
Aquila chrysaetos   CODE 1

ADULT. *Dark overall with buff ("golden") highlights, most prominent on nape; buff on wing coverts and tarsal feathers can also be conspicuous.* CALIFORNIA, DECEMBER

JUVENILE. *Large like Bald Eagle, but smaller-headed and longer-tailed. White patches in wing and tail are sharply defined, unlike juvenile Bald Eagle. Has golden nape like adult.* COLORADO, APRIL

L 30"	WS 79"	WT 10 lb	♀ > ♂

- one adult molt per year; complex basic strategy
- moderate differences between adult and subadult
- in all populations, plumage varies among individuals

A bird of rugged, lonely, often arid country: mountains, deserts, and cliffs, often arid. At present, mainly a northern and western species, but increased sightings and reintroduction efforts inspire hope that the species may reoccupy former breeding grounds in the eastern U.S.

Sometimes vocalizes in flight or in interactions with other birds. Most calls short and rough: *unk* or *unk unk; swurpah* or *suh-werp.*

Despite its name, the Golden Eagle is not closely related to the Bald Eagle. Instead, the Golden Eagle may be a close ally of the "buzzards" in the genus Buteo *(for example, Red-tailed and Swainson's Hawks). The Golden Eagle may also be a close relative of some of the most spectacular raptors of the New World tropics, among them the Harpy Eagle and Ornate Hawk-Eagle.*

ADULT. *At distance, suggests Turkey Vulture: large and dark; small-headed and long-tailed. Golden wash on nape extensive, but prominence affected by lighting and angle.* WYOMING, JUNE

Ferruginous Hawk
Buteo regalis CODE 1

LIGHT-MORPH ADULT. *"Ferruginous" means rust-colored; note rusty highlights above and below on extensively pale body. Wings long with white flashes at bases of primaries.* WYOMING, JUNE

DARK-MORPH ADULT. *White flight feathers contrast strongly with dark-brown body. Large bill, often with extensive yellow, stands out at a distance.* FEBRUARY, COLORADO

LIGHT-MORPH ADULT. Extensive rufous on upperparts, bright rufous-and-white tail, and long dark-tipped primaries are a distinctive combination. Boldly barred leggings extend to toes. WYOMING, JUNE.

23"	ws 56"	wt 3.5 lb	♀ > ♂

- one adult molt per year; complex basic strategy
- moderate differences between juvenile and adult
- has light and dark morphs; dark morph uncommon

A raptor of open, fairly arid country, mainly in the Intermountain West. Short-distance migrant, but local movements complex and not well understood. Prey is chiefly mammals. Numbers declining in sagebrush country and perhaps elsewhere.

LIGHT-MORPH JUVENILE. *Note long wings; paler overall below than other juveniles in genus Buteo. Juvenile dark morph resembles adult, but generally lacks ferruginous highlights.* FEBRUARY, COLORADO

LIGHT-MORPH ADULT. *Variable, but most are fairly pale with scattered rufescent highlights, seen here on tail and scapulars. A large, long-winged, lanky hawk.* COLORADO, MARCH

Call is a screech like Red-tailed Hawk, but usually shorter in duration, less raspy, not as powerful. Vocalizes throughout the year, but infrequently.

Rough-legged Hawk
Buteo lagopus CODE 1

LIGHT-MORPH ADULT MALE. *Paler overall than adult female; in particular, lacks extensive black on belly. In all plumages, note small bill.* COLORADO, DECEMBER

LIGHT-MORPH ADULT FEMALE. *Variable, but most have extensive black on belly, lacking on male. All plumages are small-billed for a* Buteo. COLORADO, JANUARY

LIGHT-MORPH ADULT FEMALE. *Boldly patterned: belly extensively black, wing linings buff-and-black; tail white with contrasting black band near tip. Juvenile tail less contrasting.* COLORADO, JANUARY

L 21"	WS 53"	WT 2.2 lb	♀ > ♂

▸ one adult molt per year; complex basic strategy

▸ moderate age-related and sex-related differences

▸ has dark and light morphs; light morph more common

An Arctic breeder that descends in winter to much of lower 48; winters in extensive marshes, prairies. Migrates late in fall, fairly early in spring. Like many arctic predators, population levels are tied to cyclic prey populations. Often perches on branch tips.

Call is not usually heard away from breeding grounds: like Red-tailed Hawk, but more musical, more whistled, a shrill, descending *pweeeeeerv.*

DARK-MORPH ADULT MALE. *Black overall, with prominent white bands on black tail. Dark-morph females are brown rather than black (although some males appear brownish).* MISSOURI, DECEMBER

DARK-MORPH ADULT MALE. *From below, tail similar to light morph, but body and wing linings nearly black. From above, nearly black; light-morph adult male gray-and-white above.* COLORADO, MARCH

DARK-MORPH ADULT JUVENILE. *Buff-and-black underwing pattern like light-morph, but rest of underparts largely dark.* MISSOURI, DECEMBER

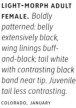

Swainson's Hawk

Buteo swainsoni CODE 1

LIGHT-MORPH ADULT. *Dark flight feuthers of wing contrast with white body and wing linings. Chin and face white; head gray-brown; breast extensively chestnut.* COLORADO, SEPTEMBER

L 19"	WS 51"	WT 1.9 lb	♀ > ♂

▸ one adult molt per year; complex basic strategy

▸ strong age-related differences in light morph

▸ has multiple morphs; much plumage variation overall

A long-distance migrant that breeds in the American West. Occurs in diverse open-country settings, including agricultural areas. Severe mortality caused by ingestion of pesticide on Argentine wintering grounds recently documented, but the species remains fairly common.

◀ Call is a high-pitched, pure-tone, descending squeal: *eeeeeee.* Vocalizations sometimes shorter, given in series; the effect is gull-like.

LIGHT-MORPH ADULT. *In most areas, the most common morph of Swainson's. Most show dark bib contrasting with white chin. Long wings impart a sleek look, even at considerable distance.* ARIZONA, APRIL

LIGHT-MORPH JUVENILE. *Below, pale wing linings contrast with dark flight feathers of wing, like adult. Head markings highly variable; range from extensively dark to nearly white.* COLORADO, SEPTEMBER

INTERMEDIATE-MORPH ADULT. *Has white chin and face like light morph, hut belly and wing linings darker. All Swainson's Hawks have long wings, extending past the tail on perched birds.* COLORADO, AUGUST

DARK-MORPH ADULT. *Body dark; wing linings suffused with dark chestnut. Flight feathers of wing and tail about as dark as in other morphs. Dark-morph juvenile similar to adult.* COLORADO, SEPTEMBER

Red-shouldered Hawk

Buteo lineatus CODE 1

EASTERN ADULT. *In flight, surprisingly reminiscent of genus Accipiter. Adult tail boldly barred black-and-white, juvenile less so. White crescent toward wing tip often prominent.* TEXAS, DECEMBER

CALIFORNIA ADULT. *More intensely rufous than Eastern Red-shouldered. All juveniles less brightly colored than adults, but usually show some rufous on wings and underparts.* CALIFORNIA, NOVEMBER

FLORIDA ADULT. *Not as bright as Eastern Red-shouldered, but a few Easterns can appear this pale. Note red "shoulder," present on all adults and most juveniles of all populations.* FLORIDA, MARCH

L 17"	WS 40"	WT 1.4 lb	♀ > ♂

- ▸ one adult molt per year; complex basic strategy
- ▸ strong differences between juvenile and adult
- ▸ eastern and California populations differ moderately

A woodland *Buteo.* Eastern birds favor swampy groves and river bottom forests. Western birds occur in a variety of native and exotic woodlots. Flight style is *Accipiter*-like: swift and agile, with gliding and flapping interspersed. Eastern population has declined.

Vocal, especially on breeding grounds. Gives short descending squeals, *whee-ee-ah* or *whee-ah*, reminiscent of sapsuckers. Call often imitated by Blue Jay.

Broad-winged Hawk

Buteo platypterus CODE 1

LIGHT-MORPH ADULT. *Smallest* Buteo. *Solidly brown above. Mottled underparts variable: from coarsely barred brown and white, as here, to finely barred rufous and white.* COLORADO, SEPTEMBER

LIGHT-MORPH ADULT. *Brown wings largely unmarked above; in flight, often appears pointy-winged (name "Broad-winged" misleading). Black-and-white tail bands prominent.* MINNESOTA, SEPTEMBER

LIGHT-MORPH JUVENILE. *Thin white edgings on fresh dark scapulars and wing coverts are distinctive. Brownish tail has narrow dark bands and wider subterminal band.* MINNESOTA, SEPTEMBER

JUVENILE. *Extensively brown above, with variable paler highlights. In most of range, no other* Buteo *is as extensively brown above.* MINNESOTA, SEPTEMBER

L 15"	WS 34"	WT 14 oz	♀ > ♂

- ▸ one molt per year; complex basic strategy
- ▸ strong differences between juvenile and adult
- ▸ has light and dark morphs; dark morph rare

A hawk of forests in eastern North America. Loosely distributed during the breeding season, but forms impressive concentrations, known as "kettles," on fall passage through Appalachian and Texas coastal regions. Seen with increasing frequency west of main range.

◀ Call is a long, thin, steady, pure-tone whistle, *w'weeeeeeee,* easily imitated. Often given by birds in soaring flight above summer forests and woodlands.

Short-tailed Hawk

Buteo brachyurus CODE 2

DARK-MORPH JUVENILE. *All black, except for variable white splotching below; adult lacks white splotching. When perched, all Short-tailed Hawks hide in treetops and are hard to find.* FLORIDA, NOVEMBER

LIGHT-MORPH JUVENILE. *Underparts mainly white. Note broad dark malar. Adult similar, but shows variable, broken rufous collar.* FLORIDA, NOVEMBER

DARK-MORPH ADULT. *Pattern similar to Turkey Vulture, but body shape typical of genus Buteo: broad-winged and fairly short-tailed. Pale tail usually shows darker terminal band.* MEXICO, FEBRUARY

LIGHT-MORPH ADULT. *Extensively pale beneath. From below, dark tips of primaries and secondaries frame otherwise white wings.* MEXICO, FEBRUARY

| 16" | ws 37" | wt 15 oz | ♀ > ♂ |

▸ one adult molt per year; complex basic strategy

▸ moderate differences between juvenile and adult

▸ has light and dark morphs; dark morph more common

A tropical species that ranges into North America in Florida, and is apparently becoming established in the Desert Southwest. Florida population is sparse. Nests in a variety of woodland habitats; away from nest, is rarely seen except in high, soaring flight.

Most frequently heard is a short, descending, whistled *eeeeer*, richer and fuller than equivalent calls of Swainson's and Broad-winged Hawks.

White-tailed Hawk

Buteo albicaudatus CODE 2

ADULT MALE. *The palest individuals—typically older males—have largely white chins and smooth gray upperparts. Rufous patch on upperwing is shown by all but the youngest birds.* TEXAS, FEBRUARY

ADULT FEMALE. *All adults show distinctly short tail with high-contrast black subterminal band. Amount of dark on face and throat variable; females and younger adults generally darker.* TEXAS, FEBRUARY

IMMATURE. *Some are surprisingly similar to darker Red-tailed Hawks, but most have extensive white on breast and head.* TEXAS, FEBRUARY

| L 20" | ws 51" | wt 2.3 lb |

▸ one adult molt per year; complex basic strategy

▸ strong age-related and weak sex-related differences

▸ older adults average more extensively white on chin

Widespread in tropics; reaches North America in the Texas coastal plain, where it occurs in open country. Usually seen soaring above agricultural fields, chaparral, and marshes. Sometimes seen walking on ground. North American population possibly increasing.

Vocal repertoire varied. Gives squealing *er-weeeee*, characteristic of genus *Buteo*; also harsher sequences, *ick... ick... ick...* and *ick-ick ick-ick...*

Red-tailed Hawk

Buteo jamaicensis   CODE 1

ADULT, EASTERN. *Red tail striking from above, and usually evident from below. Eastern birds are extensively pale on throat; usually have broad dark band across belly.* CONNECTICUT, OCTOBER

LIGHT-MORPH ADULT, WESTERN. *Similar to Eastern, but throat usually all-dark. Black leading edge of wing usually present on light-morph individuals.* CALIFORNIA, FEBRUARY

JUVENILE. *Dirty-brown tail finely speckled and barred with white. Many show distinctive upperwing pattern: from above, outer half pale, inner half darker.*
MINNESOTA, OCTOBER

L 19"	WS 49"	WT 2.4 lb	♀ > ♂

- one adult molt per year; complex basic strategy
- strong differences between juvenile and adult
- plumage variation extreme; few species are as variable

Accepts a huge diversity of habitats, from remote western wilderness to midtown Manhattan. Most populations are healthy, and several appear to be increasing. Has generally benefited from deforestation, fire suppression, and other ecologically dubious practices.

◀ Well-known call a descending harsh scream, *ksheeeeer*. Call frequently imitated by songbirds, especially jays and European Starlings. Juveniles beg with repeated *kree... kree... kree.*

ADULT, "HARLAN'S." (B. j. harlani) *This distinctive subspecies has a dusky white tail and black body with white splotching on breast; compare with Rough-legged Hawk. Breeds Alaska, winters Great Plains.* COLORADO, NOVEMBER

INTERMEDIATE-MORPH ADULT, WESTERN. *Especially in the West, Red-tailed shows great variation. Note chestnut breast and dark wing linings on this individual.*
CALIFORNIA, NOVEMBER

ADULT, EASTERN. *Perched, Red-tailed Hawk is stocky and relatively short-tailed. Eastern adults often show coarse white splotches on upperparts.*
CONNECTICUT, JANUARY

ADULT, "KRIDER'S." (B. j kriderii) *Several mid-continent populations are pale, with uncommon "Krider's" of Great Plains strikingly so. Wings fall short of tail tip, as on most Red-tailed Hawks.*
TEXAS, FEBRUARY

ADULT, "HARLAN'S." (B. j. harlani) *Long wings extend nearly to tail tip. Longer wings typically relate aerodynamically to longer-distance migration; "Harlan's" travels farther than other subspecies.*
COLORADO, NOVEMBER

INTERMEDIATE-MORPH ADULT, WESTERN. *Red-tailed morphs are not discrete; they vary continuously from pale to dark.*
COLORADO, DECEMBER

DARK-MORPH, ADULT WESTERN. *Appears uniformly dark at distance; close up, note extensive rufescent highlights. "Harlan's" Hawk colder, having black-and-white plumage with almost no rufous.* ARIZONA, JANUARY

Harris's Hawk

Parabuteo unicinctus CODE 1

ADULT. *All Harris's Hawks perch conspicuously, often assuming upright pose. Body mainly gray-brown with contrasting rufous on wings and "trousers."* TEXAS, FEBRUARY

JUVENILE. *In flight, comes across as an oversized Accipiter. All ages show rufous on wing linings and trousers, combined with boldly patterned tail: white at base, black toward tip.* ARIZONA, DECEMBER

JUVENILE. *Some juveniles are heavily barred and streaked all over, but all show some rufous on wings and trousers. Note also long tail with conspicuous white at base.* TEXAS, FEBRUARY

| L 20" | WS 42" | WT 2 lb | ♀ > ♂ |

▸ one adult molt per year; complex basic strategy

▸ moderate differences between juvenile and adult

▸ Arizona birds slightly darker than Texas birds

An inhabitant of our southern deserts, found in open country with suitable perches: trees, saguaros, utility poles; also in towns and even cities. A communal species, hunting cooperatively and nesting in small groups. Breeds year-round.

Call is a strained, squeaky *eeee-ah* or *eeee-ur*, typically repeated. In interactions when hunting also gives shorter, raspier notes.

Zone-tailed Hawk

Buteo albonotatus CODE 2

ADULT. *Tail bands more contrasting than on juvenile. Female typically has three white tail bands; male has two. Unlike other hawks in genus Buteo, Zone-tailed lacks light morph.* ARIZONA, APRIL

JUVENILE. *Tail finely barred. Similar to Turkey Vulture, but head feathered in black, feet and cere yellow. Most juveniles show a little white on otherwise black underparts.* NEW MEXICO, JULY

| L 20" | WS 51" | WT 1.8 lb | ♀ > ♂ |

▸ one adult molt per year; complex basic strategy

▸ moderate age-related and weak sex-related differences

▸ pale spotting and splotching on juveniles variable

Widespread but generally uncommon in Southwest. Occurs in diverse habitats, from lowland riparian woods to arid mountain canyons. In flight, appears similar to unrelated Turkey Vulture; even in areas of core abundance, typically outnumbered by Turkey Vulture.

Call a melancholy whistle, rich in tone and long in duration, somewhat wavering: *eeeee-eeee-urr*. Gives shorter, harsher notes around nest.

Gray Hawk
Buteo nitidus CODE 2

ADULT. *Underparts extensively and distinctly barred gray and white all over; upperparts solid blue-gray. Tail has prominent black-and-white bands. Female slightly darker than male.* MEXICO, FEBRUARY

JUVENILE. *Superficially similar to Prairie Falcon: note dark malar and eye stripe on largely pale face; cere and feet yellow; coarsely streaked below. Averages longer-tailed than Broad-winged.* MEXICO, JUNE

JUVENILE. *In flight, gives Accipiter-like impression: long-tailed and fairly short-winged, with similar flight style. Juvenile has at least five, often more, low-contrast pale tail bands.* ARIZONA, MAY

Common Black-Hawk
Buteogallus anthracinus CODE 2

ADULT. *Rotund. Prominent white band at base of short, black tail; some have white on face. With brief glimpse, can be confused with adult Zone-tailed Hawk, but structure very different.* TEXAS, APRIL

JUVENILE. *Proportions less "extreme" than adult, but still appears short-tailed and broad-winged. Tail barred black-and-white; wings show light-dependent pale "window" toward tip.* MEXICO, FEBRUARY

L 17"   ws 34"   wt 1.2 lb   ♀ > ♂

▸ one adult molt per year; complex basic strategy

▸ strong age-related and weak sex-related differences

▸ individual variation in tail markings of juveniles

Inhabits riparian woods in Gila and lower Rio Grande drainages. Scarce but apparently increasing; for example, population increased in response to habitat improvement in late 20th century along San Pedro River, Arizona. Flight style *Accipiter*-like: flap, flap, glide.

Vocal in woodlands. Calls similar to other hawks in genus *Buteo*: long, thin *pweeeee-eeee.* Also gives short notes, in series: *ee-whee, ee-whee, ee-whee.*

L 21"   ws 46"   wt 2.1 lb

▸ one adult molt per year; complex basic strategy

▸ strong differences between juvenile and adult

▸ extent of small white patch on face variable

A frog- and fish-eating raptor of lowland riparian habitats. Usually requires extensive old-growth woodlands. Can be hard to see: spends much time in dense tree-tops, flushes reluctantly. Apparently expanding northwest into upper Colorado River drainage.

Call a falcon-like series, chattering and rising in volume: *whee whee whee! whee! wheee!* Around nest, gives various chatters, usually with excited quality.

American Kestrel
Falco sparverius CODE 1

ADULT MALE. *A small, unusually colorful raptor: male has blue wings; female has rufous. Both sexes have bright rufous back and tail, plus high-contrast head pattern.* NEW JERSEY, JUNE

L 9"	WS 22"	WT 4.1 oz	♀ > ♂

▸ one adult molt per year; complex basic strategy

▸ strong differences between male and female

▸ extensive variation in size and color

In many places, the commonest falcon. A bird of diverse parkland settings, including heavily urbanized areas. Nests in cavities excavated by flickers and other birds. Often hovers over prey before pouncing. Numbers declining in some regions.

◀ A vocal raptor; calls often, usually in flight, throughout the year. Call a series of 3–6 anxious *klee* notes. In interactions, gives short chattering notes.

Raptors are among the most prominent of diurnal (daytime) migrants. During favorable conditions, hundreds and even thousands of migrating raptors may be recorded daily at professionally staffed "hawk watches."

ADULT MALE. *All kestrels hover on fluttering wings as they scan for prey. Note prominent black tail tip; black at tip of female is much reduced.* COLORADO, MARCH

ADULT FEMALE. *Broadly and diffusely streaked below; male sparingly spotted. In normal flight, body shape of kestrels characteristic of falcons: long-tailed and pointy-winged.* MINNESOTA, SEPTEMBER

APLOMADO FALCON. *(Falco femoralis) Formerly bred in arid grasslands of Southwest; has been reintroduced in Texas and New Mexico. Underparts tricolored: white, black, and rufous.* TEXAS, MAY

Merlin
Falco columbarius CODE 1

IN FLIGHT. *Merlins in flight (adult female "Northern" shown here) sometimes behave aggressively toward other species.*
NEW JERSEY, SEPTEMBER

ADULT MALE, NORTHERN ("TAIGA"). (F. c. columbarius) *Widespread. Dark blue above; faint rufous below with dark streaking; female browner. All Merlins show fainter malar than American Kestrel or Peregrine.*
NEW JERSEY, OCTOBER

IN FLIGHT. *All Merlins (adult male "Prairie" shown here) appear pointy-winged and band-tailed in flight.*
COLORADO, MAY

ADULT MALE, "PRAIRIE." (F. c. richardsonii) *Occurs mainly in midcontinent. Male pale blue-gray above; female sandy brown. Both sexes look more washed-out below than "Taiga" and "Black."*
COLORADO, FEBRUARY

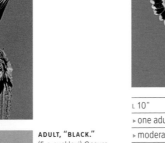

L 10"	WS 24"	WT 6.5 oz	♀ > ♂

▸ one adult molt per year; complex basic strategy

▸ moderate sex-related and weak age-related differences

▸ three populations differ in color and other characters

ADULT, "BLACK." (F. c. suckleyi) *Occurs mainly in Pacific Northwest. Shows extensive chocolate brown; adult male darker below than adult female. Always note faint malar.*
ALASKA, SEPTEMBER

Fearsome and fearless; stands its ground against intruders much larger than itself. Hunts in semi-open parkland habitats. Breeding range expanding south, especially into cities. Variably migratory; for example, "Black" Merlin of Pacific Northwest is largely sedentary.

Call similar to American Kestrel, but averages faster, more hurried and anxious, wilder sounding, with more notes in series: *keee kee kee kee kee...*

Prairie Falcon
Falco mexicanus CODE 1

ADULT. *Appears large and powerful in flight; black axillaries are distinctive. Pointy-winged like other falcons, but not as long-winged as Peregrine.* WYOMING, JUNE

JUVENILE. *All Prairie Falcons are sandy brown above with brown malar. Adults are pale below with sparse spotting; juveniles also pale below, but with coarse streaks.* WYOMING, JULY

L 16"   WS 40"   WT 1.6 lb   ♀ > ♂

▸ one adult molt per year; complex basic strategy

▸ moderate differences between juvenile and adult

▸ extent of faint barring on tail of adult variable

Characteristic species of lonely country in arid Intermountain West. Ranges widely in hilly, rocky habitats, from low deserts to alpine tundra. In winter disperses east to central Great Plains; some seen farther east are falconers' birds; others may be wild vagrants.

Call a throaty, raspy *reeer reer reer reer...* Often ten or more notes in series. Carries far; can be surprisingly ventriloqual amid cliffs and canyons.

Peregrine Falcon
Falco peregrinus CODE 1

ADULT. *In flight, all plumages show long, pointed wings. Underparts finely barred all over; thick black malar conspicuous. Male slightly darker, more contrasting than female.* CONNECTICUT, NOVEMBER

ADULT. *Black on malar often expands to fill much of face. West Coast birds average darker, but much variation. Wings comparatively longer than Prairie Falcon or Gyrfalcon.* WASHINGTON, MARCH

JUVENILE. *Extensive steel-blue above of adult replaced by muddy brown on juvenile. Underparts of juvenile streaked; underparts of adult barred.* WASHINGTON, MARCH

L 16"   WS 41"   WT 1.6 lb   ♀ > ♂

▸ one adult molt per year; complex basic strategy

▸ moderate age-related and weak sex-related differences

▸ plumage variation complicated by introduction efforts

Regarded by many as the ultimate symbol of power and splendor. One of the most widely distributed birds in the world; occurs throughout North America, but uncommon everywhere. Habitats include high mountains, remote seacoasts, and, increasingly, major urban centers.

Call variable, but generally rougher than Prairie Falcon: *ur-chur-chur-chur-chur...* Effect is sometimes muted and cackling, sometimes wild and frenzied.

Gyrfalcon
Falco rusticolus CODE 2

GRAY-MORPH ADULT. *All plumages distinguished from Peregrine by larger size, longer tail, and lower-contrast plumage. On bird at rest, tail falls short of wing.* ALASKA, JUNE

GRAY-MORPH ADULT. *Pointy-winged with a long, broad tail. On powerful wingbeats, quickly achieves great speed.* ALASKA, JUNE

WHITE-MORPH ADULT. *All Gyrfalcons are pointy-winged and hefty-tailed. White morph extensively pale below, barred black-and-white below; rarely seen as far south as lower 48.* ALASKA, JUNE

L 22"	WS 47"	WT 3.1 lb	♀ > ♂

- one adult molt per year; complex basic strategy
- moderate differences between juvenile and adult
- white, gray, and dark morphs, with much variation

Breeds in the Arctic, where it is not rare. Only a few winter south of Canada; wintering birds intersperse long bouts of inactivity with brief, spectacular sorties for prey. Hunts large prey, mainly birds: ptarmigan in summer, pheasants in winter.

Call similar to Prairie Falcon and especially Peregrine Falcon: a series of harsh notes, *ur urt urt urt ur urt...* Wintering birds are rarely heard.

Crested Caracara
Caracara cheriway CODE 1

23"	WS 49"	WT 2.2 lb

- one adult molt per year; complex basic strategy
- moderate differences between juvenile and adult
- substantial plumage variation among subadults

Tropical species that reaches Florida, Texas, and Arizona; vagrants noted elsewhere across southern U.S. Conspicuous both in flight and walking on ground. Gregarious, especially around kills. Florida population declining, but expanding; Texas population possibly increasing.

Call strange, like a 20th-century typewriter: *tick tick tk'tk'tk'tk,* starts slow, speeds up rapidly; exceedingly dry and wooden, with no musical quality.

ADULT. *Body mainly dark. White head topped by black cap; variable facial skin brightly colored and extensive. Long legs and tail noticeable as bird struts on ground.* TEXAS, DECEMBER

SUBADULT. *Pale areas variably more buffy than on adult. All caracaras strike a distinctive flight profile; wings and tail long; head especially long, as if stretched out.* FLORIDA, FEBRUARY

Rails, Limpkin, and Cranes

ORDER GRUIFORMES

17 species: 8 ABA CODE 1   4 CODE 2   1 CODE 4   4 CODE 5

Seemingly a taxonomic grab-bag, the order Gruiformes is represented in North America by three families. The largest family (Rallidae) is itself something of a grab-bag, consisting of both the furtive rails and the often conspicuous coots and gallinules. The Limpkin, suggesting a cross between a crane and an ibis, is in its own family (Aramidae). And the cranes are in yet another family (Gruidae), which may be fairly closely related to the Limpkin. Outside North America, there are additional families, many of them consisting of just a few species of uncertain taxonomic affinities. Despite their diverse appearances, the order Gruiformes is generally thought to be a valid taxon of related species, held together by various anatomical and biochemical similarities.

North American members of the order Gruiformes are generally aquatic. Multiple species of rails, coots, and gallinules may be found together in large marshland complexes, with rails sticking to dense vegetation along the edges, and coots and gallinules swimming out into the open. The Limpkin is also a bird of marshes, particularly those harboring apple snails. In North America, cranes are the least aquatic members of the order Gruiformes, with large congregations of migrants often staging in upland agricultural fields; at other times of the year, however, cranes forage and roost in wetlands. Most species live

a round-the-clock lifestyle, and several are voices of the night—especially the Black Rail and Limpkin.

Flightlessness has evolved on multiple occasions in the order Gruiformes, especially in the family Rallidae. All North American species can fly, but some are famously reluctant to do so—for example, Black and Yellow Rails. Yet all of our rails are capable of sustained migratory flight and are well known to stray far out of range. Limpkins, although largely resident, sometimes wander far north, and cranes are fabled for their migrations. Most impressive of all is the "Lesser" Sandhill Crane, which migrates in substantial numbers all the way to Siberia from the Great Plains.

Several species of Gruiformes are recovering from earlier population losses, with Whooping and Sandhill Cranes among the most notable. Black and especially Yellow Rails are probably in decline, but it is very difficult to monitor populations of this pair of notoriously secretive species. Most species in the order Gruiformes are legally hunted in North America, apparently with little overall impact on population health.

Clapper Rail
Rallus longirostris CODE 1

ADULT ATLANTIC COAST. *Variable, but nearly always grayer than any adult King Rail. Juvenile diffuse gray all over, with reduced facial markings and drab bill.*
SOUTH CAROLINA, APRIL

ADULT WESTERN. *Several western subspecies occur; all extensively orange. King Rail absent from West, so smaller and more boldly marked Virginia Rail is the only similar species.*
CALIFORNIA, JANUARY

ADULT GULF COAST. *More colorful than Atlantic coast adult. Gulf coast adult very similar to King Rail, but averages more extensively gray on cheeks and less crisply streaked above.* TEXAS, FEBRUARY

| 14.5" | ws 19" | wt 10 oz | ♂ > ♀ |

▸ one adult molt per year; complex basic strategy

▸ moderate differences between juvenile and adult

▸ geographic and individual variation extensive

Diverse and adaptable; most inhabit salt marshes, but some prefer freshwater. Look for it along ditches and tidal creeks in coastal cordgrass marshes or along major waterways in the Southwest. Most populations sedentary, but prone to vagrancy; range expanding in places.

Gives a series of dry *kek* notes, irregular in delivery, often slowing down and speeding up unpredictably. Delivery of notes is steadier in King Rail.

King Rail
Rallus elegans CODE 1

ADULT. *Very similar to Clapper Rail; flanks more boldly barred, upperparts more strongly patterned. More rufous than Clapper, especially male.* TEXAS, MAY

JUVENILE. *Darker and duskier than adult. Plumage averages brighter and higher-contrast than plumage of Clapper at same stage of development, but separation is difficult.* VIRGINIA, MAY

| L 15" | ws 20" | wt 13 oz | ♂ > ♀ |

▸ one adult molt per year; complex basic strategy

▸ strong age-related and weak sex-related differences

▸ variants and hybrids cannot be distinguished from Clapper Rail

Most occur in southeastern coastal plain and lower Mississippi drainage; widespread but very scarce inland in East. Inland breeders migrate coastward at night in fall. Hides in dense plant cover, occasionally swims into view. Population declining, especially inland.

Gives a series of very dry notes, slowly and at regular intervals, *chech chech chech chech chech...* Other vocalizations varied; most are dry and chattering.

Virginia Rail
Rallus limicola CODE 1

ADULT. *More brightly colored and substantially smaller than King Rail. Strong rusty tones, especially on breast; flanks barred black-and-white; face gray; bill long, reddish, and decurved.* TEXAS, MAY

OLDER JUVENILE. *Much darker than adult; dark lores often contrast with white supraloral stripe.* CALIFORNIA, JULY

L 9.5"	WS 13"	WT 3 oz	♂ > ♀

▸ one adult molt per year; complex basic strategy

▸ strong age-related and weak sex-related differences

▸ extensive plumage and structural variation in juveniles

A wetland species; secretive but fairly common. Usually glimpsed prowling about dense vegetation or scampering through clearings, less often flying and swimming. Breeds in freshwater habitats; some winter on saltwater. Becomes flightless during late-summer molt.

◀ Vocalizations varied, including a series of descending pig-like grunts, *woonga woonga woonk woonk woonk...* Also a harsh *ticket* or *ch'ticket*, repeated.

Sora
Porzana carolina CODE 1

BREEDING MALE. *Smaller and much shorter-billed than the Virginia Rails it often associates with. Black-and-gray contrast on face of breeders is muted in fall and winter.* CALIFORNIA, FEBRUARY

JUVENILE. *Low-contrast buff plumage and short yellowish bill invite confusion with rarer Yellow Rail. Sora prefers wetter habitats and is less reclusive.* BRITISH COLUMBIA, SEPTEMBER

L 8.75"	WS 14"	WT 2.6 oz	♂ > ♀

▸ one adult molt per year; complex basic strategy

▸ strong sex-related and moderate seasonal differences

▸ plumage of older immature birds variable

Fairly common and widespread; reaches high densities in some marshes, seemingly absent from others. Easily overlooked. Breeds mainly inland in freshwater marshes; retreats south in fall, with many wintering coastally in saltwater habitats. Migrates by night.

◀ Song an explosive *ser-wee!*—sometimes doubled; more froglike than birdlike. The upslurred *ser-wee!* song is often followed by a downward series, *deeeee-dee-dee-dee-dee...*

Black Rail

Laterallus jamaicensis CODE 2

L 6" WS 9" WT 1.1 oz

▸ one adult molt per year; complex basic strategy

▸ weak age-related and sex-related differences

▸ eastern adult less colorful; bill larger and stouter

Famously mysterious, a voice in the night. Formerly considered mainly coastal, but recently discovered in freshwater marshes in Colorado and Arizona. Favors shallow-water habitats. Presumably declined during 20th century, but some evidence of recent recovery.

Song loud and distinctive: a clearly enunciated *kicky-jeeer*, third note much lower and harsher. Sometimes lengthened to *kih-kicky-jeeer*.

ADULT. *Dark blue-gray below; black above with white speckling. Nape rusty, especially in West. Adults and males have stronger plumage contrast than juveniles and females.* CALIFORNIA, JANUARY

Yellow Rail

Coturnicops noveboracensis CODE 2

L 7.25" WS 11" WT 1.8 oz

▸ one adult molt per year; complex basic strategy

▸ moderate age-related and weak seasonal differences

▸ adults variable, possibly representing color phases

Even more elusive than other rails. Breeds around edges of northern freshwater marshes; winters mainly along coasts in weedy fields and wet meadows. Migrants rarely detected. Scarce across its breeding range, but sometimes forms aggregations in winter.

Gives series of dry clicks, often in couplets or trios, but variable: *tick, tick-tick, tick-tick, tick, tick-tick-tick...* Some calls of Virginia Rail similar.

Rails are famously hard to catch a glimpse of, but knowing which microhabitat to look in can significantly improve your chances of finding one. Yellow Rails, for example, are especially fond of wet, weedy meadows dominated by sedges and bulrushes. They are often the only rail species within their favored microhabitat. All rail species, however, have a propensity for getting lost, and the handful of Yellow Rails that are actually seen on migration are sometimes found in atypical habitats.

ADULT. *Small; bill stubby and yellow. Upperparts patterned in dusky golden and blackish brown. Juvenile finely speckled, bill duller. Adults somewhat brighter in breeding season.* MINNESOTA, JUNE

IN FLIGHT. *White secondaries contrasting with darker primaries and wing coverts are an excellent field mark, but beware tricks of lighting on briefly glimpsed flushing Soras.* OKLAHOMA, OCTOBER

Common Moorhen

Gallinula chloropus CODE 1

BREEDING ADULT. *Like an American Coot, but more colorful. Frontal shield and bill bright red; bill tipped with yellow. Sitting or swimming, shows variable white on flanks and tail.* TEXAS, MAY

JUVENILE. *Duller overall than breeding plumage: bill less colorful; throat sometimes paler. Nonbreeding adult similar but not as dull; has less white on throat than juvenile.* CALIFORNIA, OCTOBER

L 14"	WS 21"	WT 11 OZ	♂ > ♀

▸ one adult molt per year; complex basic strategy

▸ strong age-related and moderate seasonal differences

▸ extent of white on throat of adult is variable

Widespread in East, more local in West. Western populations less migratory. In most of range, prefers extensive marshlands, but also found in small ditches in urban areas, especially in Florida. Adaptable, but has generally declined due to wetland loss.

Some calls are cootlike, such as an excited *urp!* Others are louder, more complex, far-carrying: *wheee-a* and *wheee-urp* and *ur-purr-purr-purr.*

Purple Gallinule

Porphyrio martinica CODE 1

ADULT. *More lustrous than Common Moorhen; frontal shield pale. Prowls on long yellow legs. Nonbreeding plumage similar, but duller.* TEXAS, MAY

JUVENILE. *Duller and less colorful than adult; buff below, dull greenish above. Structure and behavior similar to adult; note especially the gangly legs and awkward foraging motions.* TEXAS, NOVEMBER

L 13"	WS 22"	WT 8 OZ	♂ > ♀

▸ one adult molt per year; complex basic strategy

▸ strong age-related and moderate seasonal differences

▸ color intensity varies with lighting and diet

Inhabits muck-filled marshes along the southern coastal plain; does not seem averse to degraded wetlands. Clambers among floating vegetation and flotsam. Susceptible to long-distance vagrancy, sometimes winding up in tiny "wetlands" such as swimming pools.

Gives sputtering cackles, averaging squeakier and sharper than moorhen: *sp't!t't't'* and *spurt!* and *g'g'g'g'geeek!* Also a cootlike *geeert!*

PURPLE SWAMPHEN (*P. porphyrio*). *Native to Eurasia; recently established and spreading in Florida. Similar to Purple Gallinule, but larger and bluer with red frontal shield.* FLORIDA, APRIL

American Coot

Fulica americana CODE 1

NONBREEDING ADULT. *Globular; rounded black head is darker than rest of body. Bill and frontal shield white; top of frontal shield is blood red in breeders, dull gray otherwise.* CALIFORNIA, JANUARY

JUVENILE. *Much paler than adult; ground color tends toward brownish gray (adult lacks brownish tones). Body structure and jerky movements when swimming are similar to adult.* CALIFORNIA, JULY

| 15.5" | ws 24" | wt 1.4 lb | ♂ > ♀ |

▶ one adult molt per year; complex basic strategy

▶ strong differences between juvenile and adult

▶ frontal shield averages larger in East than in West

Abundant, especially in West. Occurs in natural wetland complexes inland and coastally, but also flourishes at golf courses and other artificial habitats. Breeds mainly in fresh water; winters by the thousands in various habitats. Takes off noisily and splashingly.

Most calls are one syllable, and many flatulent: *fzzt* or *ffick* or *zht.* Also gives less-flatulent calls; rising *wurp?* or *wurpa?*

Whooping Crane

Grus americana CODE 2

ADULT. *Taller than even the tallest "Greater" Sandhill Crane; red on head more extensive. Feathers pure white, except for black primaries; juvenile mottled brown and white.* FLORIDA, FEBRUARY

IN FLIGHT. *Black primaries contrast with otherwise white plumage. Variant Sandhill Cranes with nearly white plumage are sometimes seen, but they lack all-black primaries of Whooping.* FLORIDA, FEBRUARY

| L 52" | ws 87" | wt 15 lb |

▶ one adult molt per year; complex basic strategy

▶ moderate differences between juvenile and adult

▶ red on face variable, can intensify suddenly

A conservation success story. Total population fell to 15 in 20th century; now in the hundreds and increasing. Wild birds breed in far-northern Alberta and winter at Aransas National Wildlife Refuge, Texas. Experimental populations are managed in Florida and elsewhere.

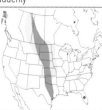

Call an ear-piercing *chu-looo!* or *cheer-ah-loo!* audible over tremendous distances. Performs exquisitely choreographed dances, like Sandhill Crane.

Sandhill Crane

Grus canadensis CODE 1

ADULT "GREATER." (G. c. tabida) *Long-necked and gray overall; not as large as Whooping Crane. Flies with neck straight out, unlike most herons.* NEW MEXICO, NOVEMBER

JUVENILE. *Quickly gains adult stature, but retains much brown in plumage. Juvenile further separated from adult by pinkish-brown bill and absence of red on crown.* WISCONSIN, JULY

L 41–46"   WS 73–77"   WT 7.3–10.6 lb   ♂ > ♀

▸ one adult molt per year; complex basic strategy

▸ moderate differences between juvenile and adult

▸ great variation in plumage and especially size

Highly social; forms tight-knit family groups in summer, huge aggregations in winter and on migration. Early-spring gathering along the Platte River, Nebraska, is one of the most stirring avian phenomena in North America. Many populations are increasing and expanding.

COMMON CRANE (G. grus). *Breeding range overlaps with Sandhill Crane in eastern Siberia, and a few migrate with Sandhills to North America. Bulk of records from central Great Plains.* EUROPE, AUGUST

◀ Call a hollow bugling, audible at great distance: *k'dr'dr'dr'dr'*, typically with rolling, clucking quality. Famous for elaborate "dances" by courting pairs.

Limpkin

Aramus guarauna CODE 2

L 26"   WS 40"   WT 2.4 lb

▸ one adult molt per year; complex basic strategy

▸ weak differences between juvenile and adult

▸ body size and amount of yellow on bill variable

Odd inhabitant of freshwater marshes and river edges in Florida and southeastern Georgia. Specializes on snails and clams, which it plucks from shallow water. Loosely colonial; moves locally and occasionally wanders far from normal range, but otherwise nonmigratory.

Voice a far-carrying *waaaaay-oh*, wailing and trumpeting; sounds "tropical" and "exotic." In courtship, male lunges suddenly at female.

ADULT. *Like a cross between a crane and an ibis. Brown body speckled with white; head and neck pale with dark streaking. Speckling and streaking slightly less prominent on juvenile.* FLORIDA, FEBRUARY

Plovers

---

17 species: 6 CODE 1   4 CODE 2   1 CODE 3   3 CODE 4   3 CODE 5

Powerful fliers and strong runners, the plovers are a relatively uniform family adapted for ecological opportunism in harsh habitats. Along with the diverse sandpiper family and several families with fewer species (for example, oystercatchers and jacanas), the plovers are considered to be shorebirds. The shorebirds are probably not a coherent evolutionary unit, but rather a loose assemblage that has converged upon similar life history strategies.

The regularly occurring North American plovers fall into two genera: *Charadrius*, most of which have prominent breast bands; and *Pluvialis*, all of which are strikingly black-bellied in breeding plumage.

North American plovers are found in open habitats—usually on mudflats and sandbars, but also in dry upland habitats. The Mountain Plover is notably not aquatic; its breeding and wintering ranges are concentrated in arid inland habitats. During the winter, Black-bellied Plovers and several other plover species roost on sea rocks along coasts, especially on the West Coast.

Most plovers exhibit the distinctive behavior of running quickly and then stopping, whether in flocks or as individuals. Flocks and individuals may be inactive for long periods of time, or they may fly about restlessly. All plovers feed in more or less the same manner: by plucking invertebrates on or just beneath the ground. Plovers do not usually form large and tight-knit assemblages as many sandpiper species do, although massive concentrations of American Golden-Plovers on spring migration in the Midwest are a notable exception.

Most plovers routinely disperse over large distances, and several species breed far north. Birds in flight—whether in short movements or in sustained migration—habitually give flight calls that can be very useful in identifying them. Around their nest sites, plovers are conspicuously defensive. And although most species are restricted as breeders to either the high Arctic or remote barrier beaches, the widespread Killdeer provides endless fascination for a student of breeding bird biology.

Like most shorebirds, plovers are threatened by habitat loss and conversion. Piping and Snowy Plovers have disappeared from sites that have been lost to development or subjected to disturbance. The Mountain Plover, an indicator species for the shortgrass prairie ecosystem, has declined sharply with the degradation of native grasslands. On a positive note, plover populations are actively monitored by various governmental agencies, and habitat protection and restoration projects have benefited some plover populations.

Killdeer
Charadrius vociferus CODE 1

ADULT. *Double breast bands always distinctive, but vary in width and sharpness. Juvenile similar, but upperparts finely scalloped, and breast bands often separated by buff (white on adult).* TEXAS, MAY

IN FLIGHT. *Flashes extensive orange at base of long tail and conspicuous white stripe on long wings. Usually noisy.* CALIFORNIA, SEPTEMBER

L 10.5"	WS 24"	WT 3.3 OZ	♀ > ♂

▸ one adult molt per year; complex basic strategy

▸ weak differences between juvenile and adult

▸ southern birds smaller overall, often redder

Has struck an uneasy truce with humans; flourishes at golf courses and construction sites, but constantly annoyed by joggers and dog-walkers. Often found far from shore, but habitat usually includes standing water. Migrates early in spring, departs late in fall.

◀ Continually agitated; day or night, gives "killdeer" call, *kill-deeer* or *kill-dee-ur* or *k'dee*. Also *dee-dee-dee*. From crouched position, gives nervous trill.

CHICK. *Has single breast band. Within a few hours of hatching, capable of running rapidly; can be confused with Semipalmated Plover, but proportions and bare parts differ.* CALIFORNIA, JUNE

NORTHERN LAPWING (Vanellus vanellus). *Ecological counterpart of Killdeer in Old World; casual visitor to East Coast. All plumages have wispy crest and high-contrast plumage.* EUROPE, JUNE

DEFENSIVE DISPLAY. *Crouched and flattened, fans long tail and gives agitated trill. If young are nearby, lures predators away with deception called "broken wing display."* NEW YORK, JUNE

Semimpalmated Plover
Charadrius semipalmatus CODE 1

BREEDING ADULT. *Upperparts muddy brown; white underparts with contrasting single black breast band. Bill orange with black tip; legs yellow or yellow-orange.* TEXAS, MAY

ADULT. *Long wings flash diffuse white stripe, which is less conspicuous than Common Ringed Plover. Rounded tail is dark toward tip, paler toward base. Often calls when flushed.* CALIFORNIA, AUGUST

L 7.25"　　WS 19"　　WT 1.6 OZ

▸ two adult molts per year; complex alternate strategy

▸ moderate age-related and seasonal differences

▸ eastern birds average larger-bodied, longer-winged

Breeds widely in Arctic, sparingly farther south. Widespread as migrant inland, usually stopping at mudflats along large lakes; migrants typically gather in small flocks. Winters widely on coasts. Generally approachable. Population apparently increasing.

JUVENILE. *Finely scalloped above; bill mainly dark and breast band gray-brown. Nonbreeding adult like juvenile, but lacks scalloping above.* NEW YORK, AUGUST

Common call, often given in flight, a rising *cher-wee?*—rather flat and rough. Gives various calls, most with scratchy quality, in feeding flocks.

Common Ringed Plover
Charadrius hiaticula CODE 2

L 7.5"　　WS 20"　　WT 2.1 OZ

▸ two adult molts per year; complex alternate strategy

▸ seasonal and age-related differences as Semipalmated

▸ extent and pattern of dark facial feathering variable

A mainly Eurasian species that also breeds in North America. Population in well-birded western Alaska is small, but farther east, in sparsely-birded northern Canada, numbers may be considerably larger. Most North American breeders migrate directly to the Old World.

BREEDING ADULT. *Even well-marked individuals, as here, are hard to separate from Semipalmated. Higher-contrast overall, with broader black breast band and broader white eyebrow.* EUROPE, MAY

Most commonly heard call, often given by birds in flight, is a mellow *dew-lup;* usually distinguishable from the scratchier call of Semipalmated.

Wilson's Plover
Charadrius wilsonia CODE 1

BREEDING ADULT. *Large bill conspicuous in all plumages; often strikes alert pose, as here. Breeders—especially females—acquire rufous tones on head and upperparts.* TEXAS, MAY

NONBREEDING. *Drabber overall and more uniformly colored than breeding adult; breast-band becomes browner, rufous highlights absent in nonbreeding plumage.* FLORIDA, MARCH

L 7.75"	WS 19"	WT 2.1 OZ

- two adult molts per year; complex alternate strategy
- moderate seasonal and age-related differences
- color and intensity of breast band are variable

Throughout the year usually found only along coasts; migrates coastally, although a few vagrants are seen inland. Found in areas with sparse vegetation and high salinity. Population faces large-scale threat of habitat loss and local threat of human disturbance.

Calls are varied, mainly short. Many with liquid qualities: *badeet* or *badeer* or *biddy-leet*. Also sharper notes when alarmed: *peet!* or *peet peet!*

Snowy Plover
Charadrius alexandrinus CODE 1

BREEDING ADULT. *Pale sandy above, white below; has broken black breast band. Separated from similar Piping Plover by thin black bill and feet and by black auriculars.* TEXAS, APRIL

NONBREEDING. *Juvenile and nonbreeding adult more weakly marked than breeder; black auriculars and broken breast band fade or disappear. Bill color and structure always a good mark.* FLORIDA, FEBRUARY

L 6.25"	WS 17"	WT 1.4 OZ

- two adult molts per year; complex alternate strategy
- moderate seasonal and age-related differences
- three separate populations vary slightly in coloration

Ecologically similar to Wilson's Plover. Two populations coastal; a third breeds far inland. Nest placed on bare ground, typically in harsh environments such as alkali flats. Numbers low overall; Pacific coast population listed as Threatened.

Birds in flight give rough *brrrt*, often doubled or trebled. Birds in feeding flocks or around the nest give quieter, sweeter *cheee-wee*, slurred and rising.

Piping Plover
Charadrius melodus CODE 2

BREEDING ADULT. *Very pale. Black collar often broken at center of breast. Feet orange; stubby orange bill has black tip. Inland breeders slightly darker than Atlantic coast breeders.* MAINE, JUNE

NONBREEDING. *Black on head and collar fade to same shade of gray as rest of upperparts; bill becomes black, feet remain orange. Always paler than corresponding plumage of Semipalmated.* FLORIDA, MARCH

JUVENILE. *Scalloping on back and scapulars separates juvenile from adult. In flight, all plumages show conspicuous white wing stripe.* TEXAS, SEPTEMBER

L 7.25"   WS 19"   WT 1.9 oz

▸ two adult molts per year; complex alternate strategy
▸ moderate seasonal and age-related differences
▸ two populations differ in bill and plumage details

Nests in dynamic habitats, such as gravel bars and sand dunes, that are susceptible to natural flooding, erosion, and predation; development and disturbance are additional threats. Widely separated breeding populations inland and along coast; both winter coastally.

Namesake call is a mournful, pure-tone *peep-low* or *pee-lup*. Around nest or in feeding flocks, gives harsher, shriller *weech* and *weech, weech.*

Mountain Plover
Charadrius montanus CODE 2

BREEDING ADULT. *Similar in overall heft to Killdeer, but not as boldly marked. Tan above, white below. Breeders have black patch on crown; nonbreeders more dusky, indistinct.* COLORADO, JULY

JUVENILE. *Scaly back marks it as a juvenile; pale sandy brown overall. Nonbreeding adult has uniformly dusky-tan back; overall appearance of plumage more dusky than sandy.* COLORADO, JULY

IN FLIGHT. *Body plan closer to small Charadrius plovers than to Killdeer: short tail has dark tip; inner wing brown; outer wing blackish with white flash.* COLORADO, JULY

L 9"   WS 23"   WT 3.7 oz   ♀ > ♂

▸ two adult molts per year; complex alternate strategy
▸ moderate age-related and seasonal differences
▸ contrast on head of older first-year birds variable

Nests in dry flatlands, often in ranching country, mainly in western Great Plains. Largest winter concentration is in Central Valley of California. Threatened by habitat conversion on breeding grounds, pesticide ingestion in winter. Infrequently detected on migration.

Usually shy and quiet, the "un-Killdeer." Flight call is a rough, tern-like *grrrt*, sometimes expanded to *grt'rt'rt'rt'rt'*. Also gives a sweeter *weerp?*

Lesser Sand-Plover

Charadrius mongolus   CODE 3

L 7.5"	WS 22"	WT 2.5 OZ

▸ two adult molts per year; complex alternate strategy

▸ strong seasonal and age-related differences

▸ leg color and extent of orange plumage variable

Formerly known as Mongolian Plover. A Eurasian species, rare but regular east to the Bering Sea region of Alaska, where it has bred. Casual farther south along the West Coast; accidental elsewhere. Usually found on mudflats and outer beaches, not in upland habitats.

Most commonly heard call, often given in flight, a hard *brrk* or *bik*, sometimes doubled. Also gives a trill, not as harsh: *drrrrr'r'.*

BREEDING ADULT. *Size and behavior like other small plovers in genus Charadrius, but extensive orange on underparts sets it apart. White throat contrasts with black mask.* ASIA, AUGUST

JUVENILE. *Suggests a small Pluvialis plover: gray-brown upperparts finely scalloped; bill thick and black. Unmarked white throat stands out. Nonbreeding adult similar.* ASIA, AUGUST

Black-bellied Plover

Pluvialis squatarola   CODE 1

BREEDING ADULT. *Adults in fresh alternate (breeding) plumage unmistakable, but birds molting into or out of alternate plumage—often seen on migration— can be oddly patterned and colored.* ALASKA, JUNE

L 11.5"	WS 29"	WT 8 OZ

▸ two adult molts per year; complex alternate strategy

▸ strong age-related and seasonal differences

▸ birds in spring and fall molt are highly variable

Breeds far north, winters farther north than most shorebirds. Widespread inland on migration, usually in small numbers. Migrants and winterers feed on expansive mudflats; winter flocks often roost on jetties, outcroppings. Dopey much of the time, just standing around.

◀ Flight call a lonely, evocative *pur-a-wee* or *tee-a-wee*; carries well. Gives hoarser, shorter calls, too; mainly on breeding grounds but also in feeding flocks.

NONBREEDING ADULT. *Brown-tan above, dirty white below; head shows lower contrast than other nonbreeding Pluvialis plovers. Large overall, thick bill.* CALIFORNIA, JANUARY

IN FLIGHT. *In all plumages, black axillaries prominent; visible from afar. Note spots on breast of this juvenile; nonbreeding adult has uniform gray-white breast.* CALIFORNIA, SEPTEMBER

JUVENILE. *In fresh juvenile plumage, brightly speckled all over; nonbreeding adult duller, more uniform. In all plumages, flashes white on wings and shows much white on tail.* CALIFORNIA, SEPTEMBER

American Golden-Plover
Pluvialis dominica CODE 1

BREEDING ADULT. *In fresh alternate (breeding) plumage, underparts are uniformly black; birds in molt often have some white below, inviting confusion with Pacific and European.* ALASKA, JUNE

JUVENILE. *Similar to Black-bellied Plover, but daintier, with smaller bill and stronger contrast on head. Pacific Golden-Plover similar; Pacific is brighter, shorter-winged.* CALIFORNIA, NOVEMBER

L 10.5" WS 26" WT 5 OZ
- two adult molts per year; complex alternate strategy
- strong age-related and seasonal differences
- birds in spring and fall molt are highly variable

Breeds in Alaska and northern Canada; migrates along very different trajectories spring and fall. In spring, moves north through middle of continent, mainly eastern Plains; huge flocks in distance resemble clouds of smoke. In fall, most migrate well off Atlantic coast.

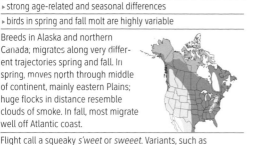

Flight call a squeaky *s'weet* or *sweeet*. Variants, such as *s'weet'ah* or *tee-a-sweet*, approximate Black-bellied, but sharper and squeakier.

IN FLIGHT. *Both nonbreeding adult (here) and juvenile show less contrast than Black-bellied. Weak contrast between tail and body; long brown wings flash little white.* NEW JERSEY, AUGUST

EUROPEAN GOLDEN-PLOVER (Pluvialis apricaria). *Casual, mainly in early spring, to eastern Canada; absent some years. Shorter-winged than American; paler overall, with more white below.* EUROPE, JUNE

Pacific Golden-Plover
Pluvialis fulva CODE 2

L 10.25" WS 24" WT 4.6 OZ
- two adult molts per year; complex alternate strategy
- strong age-related and seasonal differences
- appearance of birds during fall molt is variable

Mainly Siberian breeder whose range extends into western Alaska, where it does not interbreed with American. Compared to American, favors wetter, lower-elevation nest sites. Winters widely in Pacific Ocean region, including small numbers along California coast.

Flight call a weak *chur-week*, gull-like. In alarm, gives a similar, but richer *chur-wheel*. When feeding with other birds, sometimes chatters softly.

BREEDING ADULT. *Gold-flecked upperparts and black underparts separated by white band running from forecrown all the way to undertail coverts; molting American Golden-Plover similar.* ALASKA, JUNE

IMMATURE. *Like other Pluvialis plovers, juvenile is brightly speckled; nonbreeding adult similar, but more uniformly colored. Pacific has shorter wings and longer legs than American.* HAWAII, MARCH

Oystercatchers, Stilts & Avocets, and Jacanas

FAMILIES HAEMATOPODIDAE, RECURVIROSTRIDAE, AND JACANIDAE

7 species: 4 ABA CODE 1   1 CODE 4   2 CODE 5

These three families are among the most visually striking shorebirds, and all of them employ peculiar feeding strategies that are easily observed in the field. The hulking oystercatchers (family Haematopodidae) use their laterally compressed bills for extracting meat from clams and other bivalves. The thin-billed stilts and avocets (family Recurvirostridae) pluck tiny invertebrates from the surface of shallow, often saline, waters. The jacanas (family Jacanidae) clamber over lilies in freshwater marshes, where they search for insects on floating vegetation.

EVOLUTIONARY RELATIONSHIPS

One of the organizing principles of taxonomy is "common descent"—the idea that related groups of organisms evolved from a single ancestor. Often, the traits of that ancestral population are preserved in its present-day descendants; for example, woodpeckers share various morphological and life history traits that presumably trace back in geological time to a woodpecker-like common ancestor. A major challenge for biologists is to determine whether similarities among organisms are due to common descent or whether they are due independently to "convergent evolution" among unrelated species or groups of species.

The multiple families of shorebirds provide a good case study for understanding the challenges that scientists face when trying to determine evolutionary relationships among avian groups. The plover and sandpiper families (Charadriidae and Scolopacidae, respectively) share many ecological, behavioral, and physiological similarities, but they probably are not each other's closest relatives. Instead, plovers may be more closely related to gulls than they are to sandpipers. And what of the families that fall "in between" the plovers and sandpipers on the checklist of North American birds? The oystercatchers and stilts and avocets apparently group with the plovers, whereas the jacanas seem to be allied with the sandpipers.

Evidence for or against evolutionary affinities among birds has traditionally relied on studies of external (plumages and molts) and internal (anatomical) morphology. In recent years, behavioral, ecological, and especially biochemical analyses of DNA have been applied to studies of evolutionary relationships. These different methods are sometimes portrayed as being in conflict with one another, but integrating them provides the most successful approach to understanding avian evolution.

The evolutionary relationships among birds—especially at the family level and above—are currently a red-hot research topic, and major revisions to the avian "family tree" are widely anticipated.

American Oystercatcher

Haematopus palliatus   CODE 1

ADULT. *Bright orange bill is striking. Hood black, upperparts brown, underparts white. Bill of juvenile is duller; brown upperparts of juvenile variably speckled or scalloped.* FLORIDA, MARCH

IN FLIGHT. *Large size and high-contrast plumage visible at great distance: white underparts contrast with black hood; dark wing has broad white stripe.* NEW JERSEY, JUNE

| 17.5" | WS 32" | WT 1.4 lb | ♀ > ♂ |

- two adult molts per year; simple alternate strategy
- weak differences between juvenile and adult
- variants or hybrids with Black rare on Pacific coast

saltwater species; apparently expanding and increasing. In winter, forms loose flocks along shorelines; in summer, retreats to cordgrass salt marshes. Rare outside preferred habitat. Bill adapted for prying into bivalve mollusks, a major food item of the species.

flight, gives a piercing *peeeeer*, pure-tone and mournful; so an excited chatter, *whip-plp-plp-plp-plp*, as when birds are quarreling over food.

Black Oystercatcher

Haematopus bachmani   CODE 1

ADULT. *Body entirely black; compare American. Bare parts, including striking orange bill, identical to American. Like American, juvenile has duller bare parts, more scaly back.* CALIFORNIA, OCTOBER

WORN ADULT. *Some adults show dark sooty, rather than all-black, backs. On this individual, note extensive feather abrasion, indicating worn alternate (breeding) plumage.* CALIFORNIA, JULY

| L 17.5" | WS 32" | WT 1.4 lb | ♀ > ♂ |

- two adult molts per year; simple alternate strategy
- weak differences between juvenile and adult
- southern birds average browner, some with white flecks

Intimately tied to rocky intertidal zone of West Coast. Spottily distributed; absent from many areas of seemingly suitable habitat. Mainly resident, but some movement southward after breeding. Despite impressive size, is hard to detect amid wave-washed outcroppings.

Calls nearly identical to American. In flight, a clear *peeeeer*, descending slightly. Chattering averages richer than American: *peep peep peep-peep...*

American Avocet

Recurvirostra americana   CODE 1

NONBREEDING MALE. *Both sexes lose orange on head in basic (nonbreeding) plumage, but striking black-and-white pattern is retained. Male bill slightly upturned.* CALIFORNIA, JANUARY

BREEDING FEMALE. *Large, long-legged, and long-billed. Orange head and black-and-white wings distinctive. Female has more sharply upturned bill than male.* UTAH, JUNE

IN FLIGHT. *Black-and-white wings especially prominent in flight. Bill differences reflect feeding strategies: female (below) swishes bill from side to side; male (above) pecks in water.* UTAH, JUNE

L 18"	WS 31"	WT 11 oz

▸ two adult molts per year; complex alternate strategy

▸ strong seasonal and moderate sex-related differences

▸ extent and intensity of orange on head variable

Breeds mainly inland in West; favors highly saline microhabitats. Disperses widely after breeding, with many birds winding up along coasts. Highly aggressive at nest. Range has contracted from historical extent; habitat loss and environmental toxins probably responsible.

Especially around intruders, constantly utters piercing *weet!* or *d'weet!* Fairly quiet when unprovoked, but may give weaker versions of basic *weet!* call.

Black-necked Stilt

Himantopus mexicanus   CODE 1

ADULT MALE. *Gangly; long red legs give the impression of being unstable. Adult male crisply black and white; female slightly duller, juvenile slightly paler.* CALIFORNIA, APRIL

IN FLIGHT. *Long wings are solidly black above and below. High-contrast black-and-white plumage visible from afar. Long legs dangle far behind body; bill fine and needlelike.* CALIFORNIA, SEPTEMBER

L 14"	WS 29"	WT 6 oz

▸ one adult molt per year; complex basic strategy

▸ weak age-related and moderate sex-related differences

▸ individual variation in extent of black on head

Widely but sporadically distributed in naturally occurring shallow wetlands with emergent vegetation; also accepts "artificial" habitat associated with sewage treatment plants and rice fields. Range expanding in several regions. Often mingles with American Avocets.

Like American Avocet, becomes highly agitated at nest; gives excited *yep* notes, often in couplets, in long series. Call flatter, not as piercing, as American Avocet.

Sandpipers & Phalaropes

FAMILY SCOLOPACIDAE

55 species: 33 ABA CODE 1   4 CODE 2   12 CODE 3   9 CODE 4   5 CODE 5   2 CODE 6

These inhabitants of interior wetlands, coastal salt marshes, and ocean beaches are the quintessential shorebirds—mainly drab brown birds that spend much of their annual cycle engaged in long-distance migrations. The relationship of the family Scolopacidae to other shorebirds is not well resolved, and it may be the case that sandpipers and phalaropes are not closely allied with other shorebirds such as plovers, oystercatchers, and stilts and avocets. Within the family Scolopacidae, there is considerable taxonomic instability; for example, the genus *Tringa* was recently reorganized and expanded to include the Willet and tattlers.

Most sandpipers occur in open habitats with abundant water and little vegetation. Migratory stopover sites both inland and along the coasts may harbor more than twenty species at a time, with the total number of birds present reaching well into the thousands. Many sandpipers breed in wet arctic-tundra habitats, and their courtship and breeding behaviors are both little-known and spectacular. Favored wintering sites, like migratory stopover sites, tend to consist of expansive shallow wetlands.

Sandpipers, like other shorebirds, feed primarily on aquatic invertebrates. Several species employ tactile sensors in their bills that enable them to detect food buried in the mud. Others pick at bits of food on the surface of mudflats and sea rocks. And phalaropes spin rapidly on the water, creating a vortex that pulls invertebrates up to the surface from more than ten feet down. Shorebird feeding behavior, in addition to being fascinating in itself, can provide important cues in the identification process. For instance, an unusual "peep" (a term used generally for smaller *Calidris* shorebirds) may initially attract attention for its different feeding style, and a lone Stilt Sandpiper in a flock of dowitchers is often first detected by its feeding posture.

Although sandpipers are not songbirds, their vocalizations are among the loveliest in the bird world—and their flight calls are indispensable for identification of many species. On the breeding grounds, most sandpipers give elaborate songs that often have repetitive and pulsing qualities.

Sandpiper populations, on the whole, are undergoing serious declines. Population losses cut broadly across taxonomic and ecological lines, involving arctic-breeding species like the Red Knot and the Sanderling, as well as mid-latitude breeders such as Upland Sandpiper and Long-billed Curlew. Habitat loss and degradation—on the breeding and wintering grounds, and most of all along the migratory routes—appears to be the prime culprit in present-day population losses, and several species are at or near the brink of endangerment.

Spotted Sandpiper
Actitis macularius CODE 1

BREEDING ADULT. *Dark spots on white underparts variable, but always prominent; upperparts brown; bill and feet largely orange. All individuals are stout and pot-bellied.* COLORADO, JULY

L 7.5"	WS 15"	WT 1.4 OZ

▸ two adult molts per year; complex alternate strategy

▸ strong differences between summer and winter adults

▸ extent of spotting in breeding plumage variable

Nests throughout much of continent; seen singly or in small numbers. Occurs along streams and pond edges, less so on mudflats or coastlines with other shorebirds. Teeters constantly when foraging or standing. Flushes in broad arc, flying out over water and then back.

◀ Call a sharp *tweet* when flushed, usually uttered three or more times. Song, given in flight, consists of pulsing *tweet* notes superimposed on a soft trill.

IMMATURE. *Lacks spots on underparts, but white wedge at base of breast prominent. Bare parts less colorful than in breeding season. Juvenile similar.* CALIFORNIA, JANUARY

IN FLIGHT. *Flushes low over water, with wings bowed below plane of body. Wings notably short; dark with thin white stripe.* COLORADO, MAY

Common Sandpiper
Actitis hypoleucos CODE 3

L 8"	WS 16"	WT 1.6 OZ

▸ two adult molts per year; complex alternate strategy

▸ weak age-related and seasonal differences

▸ leg color of juvenile varies from straw yellow to gray

In many ways, the Old World counterpart of Spotted Sandpiper. Breeds well east in Siberia, with vagrants regularly reaching the Bering Sea Islands of Alaska. Most records are from spring. Like Spotted, tends to show up in freshwater settings such as pond edges.

Call similar to Spotted: a ringing *twee, twee, twee...* Display flight similar to Spotted, but not likely to be observed in North America.

BREEDING ADULT. *All plumages resemble nonbreeding Spotted Sandpiper, but ranges do not overlap. Note white wedge at base of breast compared to Spotted, bill is duller, tail is longer.* EUROPE, JULY

Solitary Sandpiper
Tringa solitaria CODE 1

BREEDING ADULT. *Dark brown above with extensive white speckling; white eye ring usually prominent. Suggests Lesser Yellowlegs, but bill shorter and thicker, and legs shorter and duller.* NEWFOUNDLAND, MAY

. 8.5" ws 22" wt 1.8 oz
- two adult molts per year; complex alternate strategy
- moderate age-related and seasonal differences
- eastern birds average slightly smaller and darker

NONBREEDING. *Juvenile and nonbreeding adult similar: juvenile less crisply marked than breeding adult; nonbreeding adult even duller. Most show prominent white eye ring.* TEXAS, JANUARY

Breeds in northern bogs. Migrates broadly across central and eastern North America; uncommon in West. True to name, migrants are seen singly or in small, loose flocks, in such habitats as flooded fields and pond edges; ventures out onto mudflats infrequently.

Flight call, often heard at night, a sharp *sweep-it*, not unlike Barn Swallow. In interactions with other birds, gives a weaker pee-pee or *pip pip pip pip*.

IN FLIGHT. *Tail of flushing bird, if seen well, is highly distinctive: black-and-white bars broken by brown streak down middle of tail. At a distance, wings appear completely dark.* MARYLAND, SEPTEMBER

Wood Sandpiper
Tringa glareola CODE 2

8" ws 21" wt 2.3 oz
- two adult molts per year; complex alternate strategy
- moderate age-related and seasonal differences
- aberrant Lesser Yellowlegs can resemble Wood Sandpiper

An Old World breeder that is in many respects halfway in size and shape between Solitary Sandpiper and Lesser Yellowlegs. Regular visitor to the Bering Sea region, mainly the western Aleutians in spring, where breeding has occurred. Like Lesser Yellowlegs, feeds at edges of ponds.

Vocalizations typical of genus *Tringa: tee teep!* or *tee tee teep*, mellower than Solitary Sandpiper; similar to Lesser Yellowlegs.

BREEDING ADULT. *Similar in build and plumage to Solitary Sandpiper, but note different facial pattern: long black lores and long white supraloral stripe; eye ring less prominent. Tail lacks brown streak down center.* EUROPE, MAY

JUVENILE. *Juvenile and nonbreeding adult less speckled than breeding adult. Facial pattern as on breeding adult; eye ring may be conspicuous.* EUROPE, AUGUST

Lesser Yellowlegs
Tringa flavipes   CODE 1

BREEDING ADULT. *Plumage similar to Greater Yellowlegs, but note structural differences: shorter, thinner bill; longer, often duller legs; slightly longer wings; smaller size overall.* TEXAS, APRIL

JUVENILE. *Finely patterned overall; note especially fine spotting on back and wings. Nonbreeding adult variable, but averages more coarsely and dully patterned.* NEW YORK, JULY

IN FLIGHT. *All plumages show mainly white tail and uppertail coverts contrasting with mainly dark plumage (finely speckled on this juvenile). Wings notably long and dark.* CALIFORNIA, AUGUST

L 10.5"   WS 24"   WT 2.8 oz

▸ two adult molts per year; complex alternate strategy

▸ moderate age-related and seasonal differences

▸ leg color variable; occasionally fairly dull

Nests in muskeg with scattered trees, in drier habitats than Greater Yellowlegs. Migrates broadly across continent, with passage more easterly in fall than in spring. Migrates singly or in small flocks; occasionally in large, but loosely scattered, flocks.

All calls resemble Greater, but most are distinctly flatter, less ringing and urgent: *tu* or *tu tu.* Birds in feeding flocks give quiet *tip* and *tup* notes.

Greater Yellowlegs
Tringa melanoleuca   CODE 1

BREEDING ADULT. *Like all adult Tringa sandpipers in spring, shows extensive speckling. Larger overall, longer-billed, and relatively shorter-legged and shorter-winged than Lesser.* CALIFORNIA, APRIL

NONBREEDING. *Juvenile and nonbreeding adult not as prominently speckled as breeding adult. Bill structure always helpful: longer than head, slightly upturned, and often two-toned.* CALIFORNIA, OCTOBER

L 14"   WS 28"   WT 6 oz

▸ two adult molts per year; complex alternate strategy

▸ moderate age-related and seasonal differences

▸ central Canadian breeders slightly longer-winged

Often thought of as the larger counterpart of Lesser Yellowlegs, but recent studies show that the two species are not each other's closest relative. Wary, flighty. Migrants occur singly or in small numbers. Returns early in spring, often lingers well into autumn.

◀ Calls loudly at the slightest provocation: a ringing, far-carrying *tee tee teeu!* or *tee tee tee teu,* typically with three or four syllables.

Willet
Tringa semipalmata CODE 1

BREEDING ADULT, "WESTERN." (T. s. inornata) *All Willets are large, with long, sturdy legs and a long, sturdy bill. Western Willet is larger and longer-billed than "Eastern," and paler in breeding plumage.* MONTANA, JULY

BREEDING ADULT, "EASTERN." (T. s. semipalmata) *Willets that breed in eastern salt marshes are dark and muddy with extensive pale speckles; shorter-billed and smaller-bodied than "Western."* NEW JERSEY, MAY

NONBREEDING. *Low-contrast plumage is dusky gray overall. Body structure and especially flash-pattern in flight separate Willet from other drab fall and winter shorebirds.* CALIFORNIA, OCTOBER

L 15"	WS 26"	WT 8 OZ

- two adult molts per year; complex alternate strategy
- moderate differences between spring and fall adults
- "Eastern" and "Western" subspecies differ substantially

Two disjunct populations: Eastern subspecies breeds in Atlantic coastal salt marshes, inland subspecies on prairies in West. Western subspecies winters on both coasts; Eastern vacates continent.

IN FLIGHT. *Although drab at rest, Willets in flight are striking, flashing large black-and-white patches on both wing surfaces. A hint of the wing pattern may be evident on birds at rest.* TEXAS, MAY

Both subspecies give basic *pill-will-willet* call, but Eastern consistently faster, higher-pitched; both give simpler *dyweep* notes, gruffer in Western.

Common Greenshank
Tringa nebularia CODE 3

L 14"	WS 28"	WT 6 OZ

- two adult molts per year; complex alternate strategy
- moderate age-related and seasonal differences
- intensity of greenish coloration on legs variable

In many respects, Old World counterpart of Greater Yellowlegs; similarity between the two species reflected in recent taxo-nomic revision of genus *Tringa*. Common Greenshank is a regular visitor to western Alaska, mainly in spring. Active: runs about, feeds animatedly.

Vocalizations are like most other members of genus: clear and ringing. Series of three or four *tee* notes is the call most likely to be heard in North America.

BREEDING ADULT. *Plumage and structure similar to Greater Yellowlegs, but legs dull green. Juvenile and nonbreeding adult less heavily speckled. All plumages show long white wedge up back.* EUROPE, MAY

Wandering Tattler
Tringa incana CODE 1

BREEDING ADULT. *At a distance, appears dark and oddly low-bodied; bill straight, sturdy. Close up, note that the undersides are extensively and finely barred with white and dark gray.* CALIFORNIA, MAY

L 11"   WS 26"   WT 4 OZ

▸ two adult molts per year; complex alternate strategy

▸ moderate age-related and seasonal differences

▸ coloration of bare parts variable; related to diet

NONBREEDING. *Juvenile and nonbreeding adult lack barring on underparts shown by breeding adults. Similar to Surfbird in plumage, but longer-billed and not as dumpy.* CALIFORNIA, SEPTEMBER

Breeds near rivers in arctic mountains and foothills; teeters and bobs, like other species that inhabit fast-flowing waters. Winters along West Coast, sticking exclusively to rocky intertidal habitats. A few forgo reproduction and remain along coast in summer.

IN FLIGHT. *Dark overall, lacking bright flash-pattern of most other "rockpipers." Flies low to water; because of low-contrast plumage, can get lost from view amid swells and surf.*

CALIFORNIA, AUGUST

Call a series of *weep* notes given repeatedly and rapidly. Nonbreeders mostly quiet, but sometimes become very vocal, highly audible above the din of the surf.

Gray-tailed Tattler
Tringa brevipes CODE 3

L 10"   WS 25"   WT 1.6 OZ

▸ two adult molts per year; complex alternate strategy

▸ moderate age-related and seasonal differences

▸ length of nasal groove variable in both tattlers

Old World equivalent of Wandering Tattler, breeding along stony streams in the hill country of Siberia. Regular in small numbers in Bering Sea region, especially in fall. More prone than Wandering to occur on mudflats and lagoons away from rocky habitats.

Call normally distinct from Wandering: a rising *poo-weet* or *poo-ah-weet*; mellow, rich, and reedy. Birds in feeding flocks may give harsher notes.

BREEDING ADULT. *Paler than similar Wandering Tattler: barring underneath less heavy; pale eyebrow more prominent. Juvenile and nonbreeding adult lack barring underneath.* JAPAN, MAY

Upland Sandpiper
Bartramia longicauda CODE 1

BREEDING ADULT. *Plumage Tringa-like. Note strange body shape: pot-bellied, with thin neck and oddly small head; erect posture and prominent eye impart an alert appearance.* NORTH DAKOTA, JUNE

IN FLIGHT. *Looks powerful; wings and especially tail are notably long. Fall adult and juvenile slightly paler than spring adult, shown here. Short yellow bill always distinctive.* NEBRASKA, MAY

L 12"   WS 26"   WT 6 OZ

▸ two adult molts per year; complex alternate strategy

▸ weak differences between spring and fall adults

▸ averages slightly larger-bodied toward west

Even among upland shorebird species, notable for its disdain of aquatic habitats. In core range (northern Great Plains) favors tallgrass prairie. Farther east, occurs in artificial habitats such as airports and blueberry farms. Populations declining, especially in East.

Song, usually given in flight, a spooky *b'b'b'b'b'b'b'bweee-eeeeee!-eeee*. Flight call is three or four *dyip* notes in rapid succession.

ADULT. *Long tail and wings impart a slender, attenuated appearance, especially when the bird is walking.* FLORIDA, MAY

Terek Sandpiper
Xenus cinereus CODE 3

L 9"   WS 22"   WT 3 OZ

▸ two adult molts per year; complex alternate strategy

▸ weak age-related and seasonal differences

▸ coloration of bare parts, especially legs, is variable

A mainly Siberian breeder that extends west to eastern Europe; regularly reaches the Bering Sea region of Alaska, mainly in coastal settings in the western Aleutians. Feeding behavior is distinctive: rushes about in pursuit of prey, with abrupt changes in direction.

Vocalizations varied, with most calls suggesting genus *Tringa*: a ringing *tweet* or slurred *t'weet*, both with a shrill quality.

NONBREEDING. *Structure distinctive: body is low-slung; long bill is upturned. Legs orange or orange-yellow. Juvenile and nonbreeding adult slightly duller than breeding adult.* CALIFORNIA, SEPTEMBER

Whimbrel

Numenius phaeopus CODE 1

ADULT. *Large-bodied; bill long and decurved; shows prominent stripes on head, lacking on plain-faced Long-billed Curlew.* CALIFORNIA, JANUARY

IN FLIGHT. *Colder brown or grayish brown than other curlews; flight is powerful and direct. Freshly molted adults more finely patterned than juveniles and worn adults.* CALIFORNIA, JULY

L 17.5"	WS 32"	WT 14 oz	♀ > ♂

- ▸ two adult molts per year; simple alternate strategy
- ▸ weak age-related and seasonal differences
- ▸ two disjunct breeding populations in North America

Breeds on tundra and taiga; one population around Hudson Bay, the other mainly in Alaska. Winters along southern coasts; migrates mainly coastally, but small numbers routinely noted inland. Winters on beaches and in salt marshes; on West Coast also gathers on sea rocks.

Most commonly heard call, at all times of year, a rapid chatter: *pip-pip-pip-pip...* At nest site, and uncommonly elsewhere, gives a clear, low whistle.

Long-billed Curlew

Numenius americanus CODE 1

ADULT. *Even compared to other large-bodied and long-billed shorebirds, Long-billed Curlew is exceptionally large-bodied and long-billed. Lacks prominent head stripes of Whimbrel.* TEXAS, SEPTEMBER

IN FLIGHT. *All plumages show some reddish, often most prominent on the wings in flight. In flight, resembles Marbled Godwit more than other curlews.* TEXAS, SEPTEMBER

L 23"	WS 35"	WT 1.3 lb	♀ > ♂

- ▸ two adult molts per year; simple alternate strategy
- ▸ weak differences between juvenile and adult
- ▸ much individual variation in bill length and body size

A grassland breeder; favors shortgrass and mixed-grass prairie. Breeding range has contracted; currently limited mainly to western Plains and Great Basin. Coastal wintering grounds also reduced from historical extent; currently absent from mid-Atlantic coast.

In aerial display gives repeated *weesh* notes, interspersed with machine-gun-like *b'r'b'r'brick* outbursts. All year, gives rising *curleee!* and falling *curlick*.

Bristle-thighed Curlew
Numenius tahitiensis CODE 2

BREEDING ADULT.
Large-bodied and
extensively rusty.
Long bill is straight or
nearly so; orangish at
base, darker toward
tip. Underparts barred
in summer, unmarked
pale rusty in winter.
MONTANA, JULY

Marbled Godwit
Limosa fedoa CODE 1

IMMATURE. *All plumages resemble Whimbrel, but Bristled-thighed is a paler bird; note especially the pale buff rump (best seen in flight) and pale flanks and belly of Bristle-thighed.* MIDWAY, MARCH

IN FLIGHT. *Buff rump and buff bars on tail contrast with darker, duskier upperparts; contrast is not quite as strong on worn adult and juvenile.* ALASKA, JUNE

IN FLIGHT. *Shows extensive rusty on wings, similar to Long-billed Curlew. Wings are proportionally shorter than other godwits, corresponding to shorter-distance migration.* TEXAS, APRIL

L 17"	WS 32"	WT 1 lb	♀ > ♂

▸ two adult molts per year; simple alternate strategy
▸ weak age-related and seasonal differences
▸ individual variation in plumage color and brightness

L 18"	WS 30"	WT 13 oz

▸ two adult molts per year; simple alternate strategy
▸ moderate differences between summer and winter adults
▸ isolated Alaskan population has shorter wings

Entire breeding population
restricted to two small clusters in
western Alaska. After breeding, all
birds fly nonstop to small islands
and atolls in the middle of the
Pacific Ocean. Population low, possibly threatened by exotic predators
on wintering grounds.

In spectacular aerial display, gives complex song: rhythmic and
pulsing, with most notes rich but pure-tone: *lurr deedee! lurr
deedee! lurrrl laurel...*

Most breed in northern Plains;
favors prairie pothole country with
tall grass and scattered ponds.
Avoids agricultural areas. One of
the earliest shorebirds to head
south; its "fall" migration is underway by mid-June. Winters mainly
on mudflats near coasts.

"Godwit" call a rising, gull-like *too-wit!* or more-drawn-out
tee-weeet. Often builds up to a chatter: *too-wit... too-wit... wyil...
whyeet whyeet whyuck.*

Hudsonian Godwit

Limosa haemastica CODE 1

BREEDING MALE. *Smaller and darker than Marbled; bill two-toned like Marbled, but slightly upturned. In breeding plumage, male is dark chestnut below; female averages paler.* MANITOBA, JUNE

IMMATURE. *Chestnut underparts of breeding plumage absent in fall; plumage variable, but overall tone is always cold. Juvenile similar, but a little more warmly colored.* TEXAS, SEPTEMBER

IN FLIGHT. *Striking above and below. Below, note black wing linings and paler flight feathers; above, note high-contrast white rump and white flash at base of flight feathers.* CALIFORNIA, SEPTEMBER

L 15.5"	WS 29"	WT 11 OZ	♀ > ♂

▸ two adult molts per year; complex alternate strategy

▸ strong seasonal and moderate sex-related differences

▸ much individual variation in plumage and bare parts

Breeds in discrete, far-flung clusters across Arctic; breeding concentrations associated with major wetland systems. Powerful flier: in fall, migrates across Atlantic Ocean from Canada to South America; in spring, migrates north in a narrow front through central Plains.

"Godwit" call harsher and higher-pitched than Marbled: *wit-wheet.* Also gives a series of scratchy notes, insistently so: *scrit scrit scrit...*

Bar-tailed Godwit

Limosa lapponica CODE 2

BREEDING MALE. *Long bill is slightly upturned. Breeding male shows unmarked rufous on underparts. Breeding female is much paler below; juvenile and nonbreeding adult even more SO.* ALASKA, JUNE

IN FLIGHT. *In all plumages, upper surface of wing is grayish overall with little contrast. Powerful flight aided by long wings. Tail is barred black-and-white.* EUROPE, JULY

L 16"	WS 30"	WT 12 OZ	♀ > ♂

▸ two adult molts per year; complex alternate strategy

▸ strong seasonal and moderate sex-related differences

▸ vagrants to East paler than Alaskan breeders

Widely distributed in Old World; range extends east to Alaska, where breeding occurs in coastal plain of Arctic Ocean and Bering Sea. Alaskan birds among the world's most amazing migrants; fly nonstop to New Zealand. Vagrants along East Coast come from western Eurasia.

Vocalizations generally lower and richer than other godwits. "Godwit" call: *tuloot,* repeated. Also frenzied, gull-like series: *kooa kooa kooa kooa.*

Black-tailed Godwit
Limosa limosa CODE 3

Ruddy Turnstone
Arenaria interpres CODE 1

ADULT MALE. *Deep orange on breast gives way to black-and-orange barring on belly; female much less colorful. Two-toned bill (orange toward base, black toward tip) is nearly straight.* ICELAND, JULY

JUVENILE. *Fairly uniform buff below; dark above with heavy buff speckles. Bill like adult: strongly two-toned and straight or nearly so.* EUROPE, AUGUST

IN FLIGHT. *Flash pattern above much as Hudsonian: black wings with white flash; black tail and white tail coverts. Below, wings are mainly white (black-and-gray on Hudsonian).* EUROPE, SEPTEMBER

L 16.5" WS 29" WT 11 oz ♀ > ♂

▸ two adult molts per year; complex alternate strategy

▸ strong age-related, sex-related, seasonal differences

▸ East Coast vagrants brighter than wanderers to Alaska

Old World species that wanders to North America: regular in small numbers in spring in western Aleutians; occurs less frequently along Atlantic coast. Typically found at major shorebird stopovers, sometimes with other godwit species. European population declining.

"Godwit" call variable but usually flat, for example, *wippa.* Often runs to three or more syllables: *paddy-wip* or *piddywippa* or *widdywippa.*

BREEDING MALE. *One of our most striking shorebirds, with extensive reddish above and elaborate black-and-white pattern on face and breast. Female averages browner above.* TEXAS, MAY

NONBREEDING ADULT. *Less colorful and contrasting than breeding plumage; breast is blotchy black, throat white, legs orange, bill sharp. Juvenile similar, but scalloped above.* CALIFORNIA, SEPTEMBER

IN FLIGHT. *All plumages show strong contrast in flight: tail is black at tip and white at base; dark wings flash extensive white.* TEXAS, APRIL

L 9.5" WS 21" WT 4 oz ♀ > ♂

▸ two adult molts per year; complex alternate strategy

▸ strong seasonal and weak sex-related differences

▸ extent of black on head and rufous on wings variable

High-arctic breeder. Winters along coasts; migratory movements complex and poorly studied. Opportunistic feeder: often on jetties and rocks, but also on beaches, where it pokes at horseshoe crabs, French fries, cigarette butts, anything. Tame, approachable.

Calls throaty: *chee'itchy'itchy'itchy* or *r'tit r'tit r'tic* or *p'p'p'pip.* Silent when foraging or roosting, but sometimes noisy in flight.

Black Turnstone

Arenaria melanocephala CODE 1

BREEDING ADULT. *Like a dark Ruddy Turnstone: mainly black above; white on head and breast much reduced compared to Ruddy. Legs vary from dull reddish-brown, as here, to off-yellow.*
CALIFORNIA, MAY

IMMATURE. *Browner than breeding adult; white belly contrasts with brown breast and upperparts. Juvenile similar to nonbreeding adult, but has shorter scapulars and wing coverts.* CALIFORNIA, JANUARY

IN FLIGHT. *All plumages show the same pattern of white and dark as Ruddy Turnstone, but Black lacks reddish highlights of Ruddy. Both species are squat, with sharp, upturned bills.*
CALIFORNIA, JULY

L 9.25"	WS 21"	WT 4 OZ	♀ > ♂

▸ two adult molts per year; complex alternate strategy

▸ moderate age-related and seasonal differences

▸ a few are bright-legged, approaching Ruddy Turnstone

Breeds in coastal meadows of western mainland Alaska. Winters on sea rocks along West Coast. Densities high on breeding grounds and at major migratory stopovers, but occurs singly or in small flocks during winter. Like Ruddy, feeds by turning over shells, detritus.

Vocal, especially around nest. Calls are trills and rattles; lower, faster, and richer than Ruddy, *k'r'r'r'r'r'r* or *ch't't't't't't*, often frenzied.

Surfbird

Aphriza virgata CODE 1

BREEDING ADULT. *Structure distinctive: rotund overall with stubby bill and sturdy-looking legs. Breeding adult coarsely patterned overall: belly has black spots; scapulars show rufous.*
CALIFORNIA, MARCH

NONBREEDING. *Plumage extensively smooth gray; white belly has sparse dark spots. Plumage and coloration recall Wandering Tattler, but bill and body structure different.*
CALIFORNIA, JANUARY

IN FLIGHT. *Like most "rockpipers," flashes extensive white on upperwing. White tail has broad black tip; flight feathers of the wing are dark gray with extensive white at bases.*
CALIFORNIA, JULY

L 10"	WS 26"	WT 7 OZ	♀ > ♂

▸ two adult molts per year; complex alternate strategy

▸ moderate differences between summer and winter adults

▸ spotting on belly and rufous on scapulars are variable

Breeds in dry, rocky, harsh, high-elevation sites in Alaska and western Canada; stages in huge numbers in protected coves and inlets of Prince William Sound, Alaska; then winters in small flocks on sea rocks along West Coast south all the way to Straits of Magellan.

Most calls rough and grating, usually doubled or trebled: *kuh-row! kuh-row! kuh-row!* or *kiddyrow kiddyrow.* Some notes purer: *reeer reeer.*

Rock Sandpiper
Calidris ptilocnemis CODE 2

NONBREEDING ADULT. *Similar to Purple Sandpiper, but winter ranges of the species do not overlap. Note dark plumage overall; decurved bill and short bill show variable dull orange.* ALASKA, MARCH

BREEDING ADULT. *Population breeding on the Pribilof Islands, frequently visited by birding groups, is Dunlin-like. Pale-headed, rufous above, and pale below with a black patch on belly.* ALASKA, JUNE

BREEDING ADULT. *Aleutian Island populations darker overall and more smudgy than Pribilof Islands breeders.* ALASKA, JUNE

Purple Sandpiper
Calidris maritima CODE 1

NONBREEDING ADULT. *Dull purple-gray above; belly is paler. Bill is fairly long and obviously decurved, usually with some orange. Feet orangish.* NEW JERSEY, JANUARY

BREEDING ADULT. *Browner than nonbreeding adult: white underparts spotted and streaked; upperparts with coarse scalloping. Lacks extensive rufous of adult Dunlin and Rock Sandpiper.* NEW JERSEY, MAY

IN FLIGHT. *Appears dark overall, with medium-contrast white wing stripe. Wintering birds (nonbreeding adult here) hole up around jetties and rarely fly far.* NEW JERSEY, JANUARY

L 9"	ws 17"	wt 2.5 oz	♀ > ♂

▸ two adult molts per year; complex alternate strategy
▸ strong seasonal and moderate age-related differences
▸ strong geographic variation in plumage and body size

West Coast counterpart of Purple Sandpiper. Breeds on lowland tundra, winters in intertidal zone. Uncommon south of northern California. Animated on breeding grounds, but a generally dopey representative of the West Coast wintertime "rockpiper" guild.

On breeding grounds, gives series of rich, pulsing notes, *rrruu rrruu rrruu...* and *rich rich rich...* Flight call is a short, scratchy, insect-like *chreet.*

L 9"	ws 17"	wt 2.5 oz	♀ > ♂

▸ two adult molts per year; complex alternate strategy
▸ moderate seasonal and weak age-related differences
▸ complex variation in bill and leg length, overall size

The quintessential East Coast "rock-piper." Winters on sea rocks north to Greenland; south of New England, occurs on jetties and seawalls. Expanding south in winter. Small flocks stay for days or all winter at favored sites, retreating only short distances when flushed.

In winter gives harsh *wick* and *wicka* calls. Flight song, heard on arctic breeding grounds, long, rhythmic series of rolling calls superimposed on trill.

Sanderling
Calidris alba   CODE 1

NONBREEDING ADULT. *Distinctively pale nonbreeding plumage is the one most frequently seen at mid-latitudes. Black forewing appears as a "shoulder" patch on bird at rest.* TEXAS, APRIL

BREEDING ADULT. *Extensive rufous on head and breast can invite confusion with other small "peeps" in genus Calidris; note Sanderling's larger size and distinctive feeding behavior.* TEXAS, MAY

L 8"	ws 17"	wt 2 oz	♀ > ♂

▸ two adult molts per year; complex alternate strategy

▸ strong age-related and seasonal differences

▸ individual variation in overall brightness and size

Well-known for its charming foraging method: small flocks chase after the retreating surf, looking for invertebrates left behind. Breeds far north, winters mainly south of North America. Common coastal migrant; more numerous in East. Widespread but uncommon inland.

Call note an abrupt, scratchy *chit*; feeding flocks give various *weet* and *weep* and *wick* notes. Flight song a series of *wick* and *wrrrrr* notes.

JUVENILE. *Like most sandpipers, juvenile is more intricately marked above than nonbreeding adult. This difference is strongly expressed in Sanderling.* CALIFORNIA, AUGUST

IN FLIGHT.
White wing stripe contrasts strongly with black forewing and black of outer flight feathers.
TEXAS, APRIL

IN FLIGHT. *Most Calidris sandpipers, including Sanderling, fly in compact flocks that twist and turn erratically.* NEW JERSEY, SEPTEMBER

Red Knot
Calidris canutus CODE 1

BREEDING ADULT. *Head and underparts extensively brick red, especially males; some reddish coloration on back. Molting birds are splotchy red and gray below.* NEW JERSEY, MAY

ADULTS. *Large flocks in late spring consist of adults in varying stages of transition between gray nonbreeding plumage and orange breeding plumage.* NEW JERSEY, JUNE

JUVENILE. *Large for Calidris, but body plan typical of genus: long-winged, short-tailed, short-billed. Juvenile shows fine scalloping above; nonbreeding adults plain gray.* CALIFORNIA, SEPTEMBER

IN FLIGHT. *All plumages are pale above in flight. Short tail and uppertail coverts are especially pale; white flash in wing does not contrast strongly with gray of long wings.* TEXAS, APRIL

10.5"	ws 23"	wt 5 oz	♀ > ♂

▸ two adult molts per year; complex alternate strategy

▸ strong seasonal and moderate age-related differences

▸ geographic variation in wing length, overall brightness

High-arctic breeder; winters from Washington and New England to southern South America. Large numbers feast on horseshoe crab eggs at Delaware Bay in late May. Most migrate coastally, favoring open beaches, sea rocks, salt marshes. Some populations declining swiftly.

NONBREEDING ADULT. *Drab gray-brown above, off-white below; upperparts lack scalloping of juvenile. Has Calidris body plan, but is notably large.* FLORIDA, FEBRUARY

Fairly quiet away from breeding grounds; listen for low *grrut*. Courtship song, given in flight, a succession of mournful whistles, *weep...* or *wheeep...*

Semipalmated Sandpiper
Calidris pusilla CODE 1

ADULT MALE. *Nondescript: muddy brown above, pale below. Little if any rufous in plumage; head subtly patterned with dark cap, white eyebrow, and dark eye line. Males are shorter-billed than females.* ALASKA, JUNE

JUVENILE FEMALE. *Scaly upperparts are brown and black; even fresh juveniles, as here, show little if any bright rufous (compare juvenile Western). Note bill: short, straight, tapered.* NEW JERSEY, AUGUST

IN FLIGHT. *Dark above, with little contrast; pale wing stripe usually evident, but rarely prominent. Worn adults, as here, are drab and virtually lack field marks.* NEW YORK, AUGUST

L 6.25"	ws 14"	wt 1 oz (28 g)	♀ > ♂

▸ two adult molts per year; complex alternate strategy

▸ moderate age-related and seasonal differences

▸ western birds average shorter-billed than eastern

Arctic breeder; nearly all vacate North America by winter. In fall, large numbers transit Atlantic Ocean from Canada to South America, but many inland, too. Passage mainly overland in spring. Flocks sometimes huge, numbering in six digits, but small flocks often occur.

Call a flat, mellow *churk* or *chur*, usually distinct from Western Sandpiper. In feeding flocks, both species give a variety of calls, with some overlap.

Western Sandpiper
Calidris mauri CODE 1

BREEDING ADULT. *Spotted and speckled all over, with three discrete areas of rufous: on the crown, behind the eye (auriculars), and on the scapulars that fall over the wing at rest.* ALASKA, JUNE

NONBREEDING. *Worn juvenile and nonbreeding adult lack fine patterning and rufous highlights of fresh juvenile and breeding adult. Eyebrow less prominent than Semipalmated.* CALIFORNIA, FEBRUARY

JUVENILE. *Fresh juvenile shows two thin rufous bars (scapulars) that Semipalmated Sandpiper does not. Bill is longer, drooping, and fine-tipped, unlike Semipalmated.* CALIFORNIA, SEPTEMBER

L 6.5"	ws 14"	wt 1 oz (28 g)	♀ > ♂

▸ two adult molts per year; complex alternate strategy

▸ moderate age-related and seasonal differences

▸ individual variation in bill length and structure

The generally western counterpart of Semipalmated Sandpiper, but much range overlap. Western tends to forage in deeper water than Semipalmated. Population large, but breeding range small; massive migratory flocks stage along West Coast and in western Great Basin.

Call harsher and higher-pitched than Semipalmated: a raspy *screet* or *jeeeeet*. On breeding grounds, gives elaborate series of slow, sweet trills.

Least Sandpiper
Calidris minutilla CODE 1

BREEDING ADULT. *A small brown sandpiper; in direct comparison with other small* Calidris *sandpipers, appears smaller still. Bill is short, thin, and droops downward.* TEXAS, APRIL

JUVENILE. *At all ages, dull yellow green legs are diagnostic, but the color may be hidden, as here, when covered with mud. All plumages are brownish across breast.* NEW JERSEY, AUGUST

6"	WS 13"	WT 0.7 oz (20 g)

- two adult molts per year; complex alternate strategy
- moderate age-related and seasonal differences
- legs can appear dark because of mud or bad lighting

Breeds farther south than other "peeps"; migrates broadly across continent both spring and fall. Flocks small to medium, typically disorganized and spread out. When in the company of other "peeps," tends to wander away from water's edge and out toward grassy margins.

IN FLIGHT. *Wings are proportionately shorter than other "peeps," corresponding to shorter-distance migration. Overall impression is of a dark bird with a low-contrast wing stripe.* CALIFORNIA, AUGUST

Flight call a rising *kreee?* Birds in feeding flocks give soft chattering notes. Song consists of high-pitched series alternating with low-pitched series.

Long-toed Stint
Calidris subminuta CODE 3

6.25"	WS 13.5"	WT 0.7 oz (20 g)

two adult molts per year; complex alternate strategy

moderate age-related and seasonal differences

leg color exhibits moderate individual variation

Breeds across southern Siberia; wanders to Bering Sea islands, mainly spring. Modest numbers sometimes on outer Aleutians. Usually in freshwater habitats, less so on tidal flats. In mixed-species assemblages, a stray Long-toed often stays away from the main flock.

Flight call a low, rolling *chrrr* or a flatter *churp*, both distinct from call of Least Sandpiper. In interactions with other birds, gives a sharper *chik*.

BREEDING ADULT. *Like Least Sandpiper, a small brownish "peep" with yellowish legs. Face shows more contrast than Least: dark cap and forecrown set off from pale eyebrow.* ASIA, APRIL

Red-necked Stint

Calidris ruficollis CODE 3

L 6.25"	WS 14"	WT 1 oz (28 g)	♀ > ♂

- ▸ two adult molts per year; complex alternate strategy
- ▸ strong age-related and seasonal differences
- ▸ intensity and extent of rufous-orange variable

Breeds mainly in eastern Siberia, less so in western and northern Alaska. Extent of Alaskan breeding population unknown; probably varies from year to year. Regular migrant throughout Bering Sea region; widely scattered records of vagrants elsewhere in North America.

Calls variable; gives both a hollow *chit* like Semipalmated and a higher *cheek* like Western. Song a series of *whoo-oo* notes interspersed with short trills.

BRIGHT ADULT. *Birds in fresh breeding plumage (often still shown by early fall migrants) show extensive bright rufous on face and neck. Wings mainly dark.* CALIFORNIA, SEPTEMBER

LITTLE STINT. (Calidris minuta) *Closely related to Red-necked Stint; casual stray to Alaska and coasts. Adult rusty in spring, becoming duller by late summer.* CALIFORNIA, JULY.

JUVENILE. *Drab and gray; easily mistaken for Semipalmated Sandpiper. Long wings and short legs impart a low-slung, tapered appearance.* ASIA, AUGUST

Temminck's Stint

Calidris temminckii CODE 3

L 6.25"	WS 13"	WT 0.8 oz (23 g)

- ▸ two adult molts per year; complex alternate strategy
- ▸ weak age-related and seasonal differences
- ▸ bill and leg color exhibit individual variation

Nests on wet tundra east to the Chukchi Peninsula, within sight of St. Lawrence Island, Alaska. Migrants regular in low numbers to Bering Sea islands. Habits and habitats recall Least Sandpiper: feeds slowly in freshwater pools, sometimes venturing into grassy margins.

Flight call is a dry, insect-like rattle, *d'dr'dr'dr'dr'*, often repeated. In most cases, clearly has many syllables, and is thus quite unlike other "peeps."

A number of sandpipers that are annual in small numbers on Bering Sea Islands are accidental or even unrecorded elsewhere in the ABA Area. Temminck's Stint, for example, has been recorded on many occasions on Bering Sea islands, but there are only two accepted records from outside Alaska (from British Columbia and Washington). "Weird" sandpipers in the Lower 48 are likely to be just that—weirdly or aberrantly marked individuals of mainland North American species, not rare vagrants from Eurasia.

BREEDING ADULT. *Extensively gray and low-contrast overall. Suggests a large-bodied Least Sandpiper; like Least, has yellow legs. Nonbreeding adult shows more solidly gray breast.* EUROPE, MAY

JUVENILE. *Similar to adult, but scalier. All plumages are long-tailed and short-winged for Calidris; all also show white outer tail feathers, often visible at rest, as here.* JAPAN, SEPTEMBER

Baird's Sandpiper

Calidris bairdii CODE 1

BREEDING ADULT. A large "peep," but smaller than most sandpipers. All plumages suggest a large Least Sandpiper (brown overall), but with different structure and bare parts. TEXAS, MAY

NONBREEDING. Dull gray-brown all over. Breast is usually extensively washed with gray-brown. Long wings and thin, straight bill are always good marks. COLORADO, AUGUST

JUVENILE. Like adult, but coarsely scaled above. Both juvenile and adult have long wings, extending well past the tail. Bill is all black, fairly long, thin, and nearly straight. CALIFORNIA, AUGUST

IN FLIGHT. Pale below and dark above with weak wing stripe, like other "peeps." Wings notably long and pointed. Note scaling above on this juvenile. CALIFORNIA, SEPTEMBER

7.5"	ws 17"	wt 1.3 oz	♀ > ♂

two adult molts per year; complex alternate strategy

moderate seasonal and age-related differences

paleness of throat variable, especially on nonbreeders

Breeds in arctic, winters in South America. Migration route narrow, especially in spring; main concentration in western Plains. Feeding habits diverse: shallow water and drier, grassier habitats. Some migrate at high elevations, stopping in tundra in southern Rockies.

Call low and gruff; a rolling *rrrrrt* or *r'r'r'r't*, often at night. Song rhythmic, with low and rolling elements, repeated in series: *lucluclu dididi...*

White-rumped Sandpiper

Calidris fuscicollis CODE 1

BREEDING ADULT. Basic body plan like Baird's: large for a peep and long-winged. Breeding adult is prominently streaked below, with crisp streaking often extending onto flanks. NEW JERSEY, MAY

NONBREEDING. In all plumages, slightly decurved bill shows a bit of yellow at base. Pale eyebrow has a curve or kink. White rump difficult to see except in flight. NEW YORK, SEPTEMBER

IN FLIGHT. All peeps show some white on tail, but only on White-rumped is the rump entirely white; effect is strongest on flying birds. Long wings obvious in flight. NEW JERSEY, MAY

L 7.5"	ws 1.5 oz	wt 1.5 oz	♀ > ♂

▸ two adult molts per year; complex alternate strategy

▸ moderate seasonal and age-related variation

▸ prominence of white rump subject to lighting and angle

Like various other shorebird species, migrates north from South America through Great Plains, south from high arctic mainly over Atlantic Ocean. Migratory flocks usually large; gather on extensive coastal or inland mudflats, typically in the presence of other species.

Call distinctive: a squeaky *squeeet*, thinner and higher-pitched than Western Sandpiper. Elaborate flight song consists of various squeaky elements.

Pectoral Sandpiper

Calidris melanotos CODE 1

BREEDING ADULT. *Breast finely speckled black and white; border between breast and all-white belly usually sharp. Mildly decurved bill usually has dull yellow or orange highlights.* ALASKA, JUNE

DISPLAYING MALE. *In courtship, male puffs out breast feathers; breast of breeding male has more black than breeding female. Nonbreeding adult (both sexes) has duller breast markings.* ALASKA, JUNE

L 8.75"	WS 18"	WT 2.6 oz	♂ > ♀

▸ two adult molts per year; complex alternate strategy

▸ moderate seasonal, age-related, sex-related differences

▸ males average larger than females; variable

Breeds on wet arctic tundra, winters to southern South America. Migrants return early in spring; fall passage is drawn out. Small to medium flocks of migrants spread out over mudflats, and into adjoining grassy margins; sometimes feed in pastures and flooded fields.

Call a low, growling *churk* or *churr-uk*. Striking courtship song consists of deep hoots; partial version of full song sometimes given by migrants on passage.

JUVENILE. *Like adult, finely patterned breast sharply set off from belly. Juvenile Sharp-tailed brighter buff overall, with more-prominent cap and less-prominent breast streaking.* CALIFORNIA, SEPTEMBER

IN FLIGHT. *Like a large peep in flight: wings dark and largely unmarked. Flocks of Pectorals do not usually bunch as tightly as flocks of smaller sandpipers.* CALIFORNIA, SEPTEMBER

Sharp-tailed Sandpiper

Calidris acuminata CODE 3

L 8.5"	WS 18"	WT 2.6 oz	♂ > ♀

▸ two adult molts per year; complex alternate strategy

▸ moderate seasonal and strong age-related differences

▸ extent of streaking on sides and breast variable

The Siberian counterpart of Pectoral Sandpiper. On journey to Australian wintering grounds, a fair number, mainly juveniles, stop off in western Alaska; less common in spring. Rare but regular in fall along Pacific coast; strays show up elsewhere in North America.

Call note *wheep* or *wheepa*, sweeter than Pectoral. Courtship song, like Pectoral, an impressive outpouring of deep hoots and rich trills.

ADULT. *Similar to Pectoral, but contrast between breast and belly not as strong; rufous cap more strongly separated from pale eyebrow than on Pectoral.* CALIFORNIA, MARCH

JUVENILE. *Breast has bright buff wash and diminished streaking; rufous cap prominent. Away from Alaska, juvenile Sharp-tailed is rarely seen before September.* CALIFORNIA, OCTOBER

Dunlin

Calidris alpina CODE 1

BREEDING ADULT. *A boldly marked sandpiper: extensive black below, diffuse reddish above; extent of black and reddish varies geographically. Bill long, thin, and decurved.* TEXAS, MAY

NONBREEDING. Has a "dirty" aspect overall, with low contrast: extensively gray-brown above, white with some dusky below. Bill structure always a good mark.

NEW JERSEY, JANUARY

JUVENILE. *Dark overall: upperparts are splotchy black and brown, underparts extensively dark. Juveniles molt before fall migration and are rarely seen in the lower 48.*

ALASKA, SEPTEMBER

| L 8.5" | WS 17" | WT 2 oz | ♀ > ♂ |

▸ two adult molts per year; complex alternate strategy

▸ strong seasonal and age-related differences

▸ complex geographic variation in plumage, overall size

Arctic breeder, winters on coasts. Migrates coastally and inland; inland passage scarce over eastern Rockies and western Plains. Flocks often large, highly compact; wary, flush easily in tight formation. Both spring and fall flights average late.

Flocks noisy; harsh *chreee* call a common winter sound in coastal backwaters. Courtship song complex: alternating trills and short grating notes.

Curlew Sandpiper

Calidris ferruginea CODE 3

BREEDING FEMALE. *In breeding plumage, even dull female shows obvious rufous below; rufous on male is often much more extensive. Long, decurved bill is always prominent.* NEW JERSEY, MAY

NONBREEDING. Similar to Dunlin, but bill usually more decurved, pale eyebrow usually more prominent. Rump white (Dunlin has dark stripe down center). Legs longer than Dunlin.

ASIA, APRIL

JUVENILE. *Basic body plan as adult, but more scaly above and brighter overall than nonbreeding adult; prominent eyebrow and decurved bill always evident.*

EUROPE, SEPTEMBER

| L 8.5" | WS 18" | WT 2 oz | ♀ > ♂ |

▸ two adult molts per year; complex alternate strategy

▸ strong seasonal and age-related differences

▸ extent of rufous, especially on males, is variable

Siberian breeder; wanders widely to North America. Most records are from East Coast; has bred in Alaska. Favors open mudflats, sometimes with big groups of other shorebirds. Often wades into deeper water than other species in genus *Calidris*.

Call is a rolling *plirrup*, distinct from both Dunlin and Stilt Sandpiper. In large feeding assemblages, may give scratchier, harsher calls.

Stilt Sandpiper
Calidris himantopus   CODE 1

BREEDING ADULT.
Heavily barred chocolate and white below; some individuals are darker than shown here. Bright chestnut auriculars and at base of dark cap.
MANITOBA, JUNE

| L 8.5" | WS 18" | WT 2 oz | ♀ > ♂ |

▸ one adult molt per year; complex basic strategy

▸ strong age-related and seasonal differences

▸ extent of barring on breeding plumage variable

Arctic breeder; winters mainly south of our area. Migration largely east of Rockies: up the Plains in spring; more widely in fall, with small numbers to East Coast. Feeds like dowitchers: jabs rapidly into deep water. Flocks more dispersed than dowitchers.

Call a descending *tyurp*, with subdued quality. Pulsing song, given in flight, consists of squeaky notes interspersed with rising frog-like croaks.

NONBREEDING ADULT. *Like juvenile, but duskier and with plain upper-parts. In all plumages, foraging posture distinctive: with tail pointed upward, bird jabs rapidly into standing water.* TEXAS, MARCH

JUVENILE. *All Stilt Sandpipers are long-legged and slender, with a long, slender, slightly downcurved bill. Pale eyebrow prominent in all plumages. Juvenile is scaly above.*
NEW JERSEY, AUGUST

Buff-breasted Sandpiper
Tryngites subruficollis   CODE 1

| L 8.25" | WS 18" | WT 2.2 oz | ♂ > ♀ |

▸ one adult molt per year; complex basic strategy

▸ moderate age-related and weak seasonal differences

▸ individual variation in overall body size

Nests near arctic coasts, where males gather in communal courtship displays; winters on South American pampas. Migration confined to narrow belt through central Plains, especially spring; in fall, some drift to East Coast. A "grasspiper"; favors turf farms and prairie.

Call a short *crick*. Quiet, even on breeding grounds; simple song consists of dry *crick* notes. Spring migrants sometimes give partial courtship displays.

BREEDING ADULT.
Warm buff all over with blank stare. Underparts largely unmarked, upperparts stippled with black. Note structure: pot-bellied, short-billed, and relatively long-legged. ALASKA, JUNE

JUVENILE. *Scaly above; nonbreeding adult less so. Some nonbreeders are quite pale, as here; others brighter buff. Bare parts and blank stare as breeding adult.*
CALIFORNIA, SEPTEMBER

Short-billed Dowitcher
Limnodromus griseus CODE 1

ADULT BREEDING, WESTERN. *In breeding plumage, all Short-billed Dowitchers are variably orange-rufous. Breeding adults west of the Rockies are darker above, extensively barred below.* CALIFORNIA, MARCH

| L 11.5" | ws 19" | wt 4 oz | ♀ > ♂ |

- two adult molts per year; complex alternate strategy
- strong age-related and seasonal differences
- three subspecies show much plumage variation

Breeds across central Canada in three discrete clusters; favors dense muskeg with stunted spruces. Migrates overland to coastal wintering grounds; late spring migrant, early fall migrant. During winter, favors more-saline waters than Long-billed, but much overlap.

◄ Flight call a mellow *tu tu* or *tu tu tu*, recalling Lesser Yellowlegs; in flocks, becomes frenzied and can sound like Long-billed. Song loud and bubbly.

IN FLIGHT. *In all plumages, tail barred black and white with white barring as extensive as black barring; creates impression of white-tailed bird, especially at a distance.* TEXAS, SEPTEMBER

ADULT BREEDING, INTERIOR. *Extensive orange on underparts with reduced spotting and barring. Compared to Long-billed, Short-billed has a shorter bill that is not as perfectly straight.* MANITOBA, JUNE

ADULT BREEDING, EASTERN. *Least colorful subspecies of Short-billed. In all subspecies, relatively short bill can be obvious, as here, but there is some overlap with Long-billed.* NEW JERSEY, MAY

JUVENILE. *Similar to Long-billed, but scapulars and tertials are black with orange fringes and wavy orange internal patterns. Note that bill is not perfectly straight.* NEW YORK, SEPTEMBER

Long-billed Dowitcher
Limnodromus scolopaceus CODE 1

IN FLIGHT. *Dark bars on tail wider than white bars, the opposite of Short-billed. Flying away, tail appears dark and contrasts sharply with white wedge up back.* YUKON TERRITORY, MAY

BREEDING ADULT. *Extensively rufous below, extending to base of tail. Scapulars dark with broad white edges; overall effect is of high contrast above.* TEXAS, MAY

L 11"	WS 19"	WT 4 OZ	♀ > ♂

▸ two adult molts per year; complex alternate strategy

▸ strong seasonal and age-related differences

▸ bill length and extent of reddish coloration variable

Breeds on grassy tundra in Siberia and Alaska; arrives in lower 48 later than Short-billed Dowitcher. Favors fresh water in winter but overlaps with Short-billed, which prefers salt water. Much more likely inland in winter than Short-billed. Flocks large and compact.

Flight call a sharp *keek* or *k'keek*, distinct from Short-billed. In feeding flocks, however, both species give similar notes. Song rich and repetitious.

NONBREEDING ADULT. *Like Short-billed, but grayish wash on breast slightly more prominent, bill usually longer and straighter, "loral angle" from eye to bill usually shallower.* CALIFORNIA, DECEMBER

BREEDING ADULT. *If seen well, underwing pattern distinctive: axillaries are barred on Long-billed, but more diffusely marked and pale overall on Short-billed.* TEXAS, APRIL

JUVENILE. *Duller above than Short-billed; scapulars and tertials dark, lacking bright orange fringes and internal markings. Most Long-billed Dowitchers are noticeably long-billed.* CALIFORNIA, SEPTEMBER

Wilson's Snipe
Gallinago delicata CODE 1

IN FLIGHT. *From above, dark back shows prominent white stripes; relatively long tail is orange above. Flight erratic with wild twists and turns; long bill usually obvious.* MINNESOTA, MAY

| L 10" | ws 18" | wt 3.7 oz | ♀ > ♂ |

▸ two adult molts per year; complex alternate strategy

▸ weak differences between juvenile and adult

▸ overall warmth of ground color variable

Widely distributed; breeds in wet meadows. Medium-sized migratory flocks gather at edges of tidal creeks and prairie potholes. Inland, some winter well north, surviving wherever there are open creeks and warm seeps. Sits tight; difficult to spot until it flushes.

◀ Flushing birds give startling *skrayt* call. Nonvocal winnowing, produced by tail, is a spooky, far-carrying *ooh-hu-hu-hu-hu-hu-hu* given in flight, often at night.

ADULT. *Exceedingly long-billed. Juvenile similar, but median crown stripe slightly more prominent. Short-billed Dowitcher is less heavily barred below and shorter-tailed.* MONTANA, JUNE

ADULT. *Beak is surprisingly flexible; can be opened at tip while shut near base. Like other long-billed sandpipers, locates prey with sensory organs near tip of bill.* CONNECTICUT, MARCH

Common Snipe
Gallinago gallinago CODE 3

| L 10.5" | ws 18" | wt 3.7 oz | ♀ > ♂ |

▸ two adult molts per year; complex alternate strategy

▸ weak differences between juvenile and adult

▸ underwing pattern variable, subject to viewing angle

Eurasian counterpart of Wilson's Snipe; recently recognized as a separate species from Wilson's, in a return to older taxonomy. Habits and habitats much as Wilson's. Regular to western Aleutians, where breeding has occurred; rare elsewhere in Bering Sea region.

Songs and calls similar to Wilson's, but winnowing display lower-pitched, the result of Common having one fewer pair of tail feathers.

IN FLIGHT. *Like Wilson's, but subtle differences: pale underwing has two all-white bars; secondaries have white trailing edge; flanks less heavily barred. Juvenile slightly paler.* EUROPE, SEPTEMBER

ADULT. *Closely resembles Wilson's when standing. As with many other sandpipers, identification of Common and Wilson's Snipes is aided by studying in-flight field marks and listening to sounds.* EUROPE, JULY

American Woodcock
Scolopax minor CODE 1

ADULT. *Rotund and long-billed. Buff-orange below; splotchy black and gray above. In flight, wing uniformly gray on adult, subtly patterned gray and buff on juvenile.* CONNECTICUT, MARCH

IN FLIGHT. *Shows extensive buff below; wings plain gray above. Flushes from the edges of trails and clearings in forests or thickets, giving a short twittering in flight.* GEORGIA, FEBRUARY

ʟ 11"	ᴡs 18"	ᴡᴛ 7 oz	♀ > ♂

▸ one adult molt per year; complex basic strategy

▸ weak differences between juvenile and adult

▸ pale variants and red variants occasionally seen

Nests in early-successional habitats such as old fields and young forests. Winter habitat similar; birds at northern limit of winter range require warm seeps or bog edges. One of the earliest spring migrants; reaches latitude of Great Lakes by mid-February.

◀ Twilight courtship display is astonishing: male emits bizarre *eeent* calls from ground, then ascends to great heights, chirping wildly; then plummets to earth.

Wilson's Phalarope
Phalaropus tricolor CODE 1

BREEDING FEMALE. *Has black, white, and orange on head like other phalaropes, but prominent gray crown is unique. Breeding male is duller, often strikingly so.* MONTANA, JUNE

JUVENILE. *Pale and plain. Back is scalloped tan and gray; back of nonbreeding adult solidly pale gray. Long, straight, and exceedingly thin bill is distinctive in all plumages.* CALIFORNIA, AUGUST

IN FLIGHT. *Thin bill, small head, and overall pot-bellied shape create a distinctive impression. Rump white. Back and wings show little contrast, unlike other phalaropes.* CALIFORNIA, AUGUST

ʟ 9.25"	ᴡs 17"	ᴡᴛ 2 oz	♀ > ♂

▸ two adult molts per year; complex alternate strategy

▸ strong age-related, sex-related, and seasonal differences

▸ male alternate plumage more variable than female

Breeds inland in shallow wetlands at mid-latitudes; by early summer, disperses to hypersaline lakes in West to molt; then departs for wintering grounds in Bolivia and Argentina. The least pelagic phalarope, and more likely than other phalaropes to walk on mudflats.

Flight call a flat, low-pitched *frut* or *frup*. Intruders are chased from nest sites with a similar call, delivered loudly and incessantly.

Red-necked Phalarope

Phalaropus lobatus CODE 1

BREEDING FEMALE. *White chin set off by dark red neck and largely black face. Upperparts extensively gray; underparts mainly pale. Bill thin like Wilson's, but not quite as long.* ALASKA, JUNE

JUVENILE. *Black back has a few pale stripes. Prominent black patch behind eye is shared with Red, but lacking on Wilson's. Nonbreeding adult grayer with lower contrast overall.* CALIFORNIA, AUGUST

IN FLIGHT. *Compared to Wilson's, darker above and more contrasting. Similar to Red, but darker and more highly patterned in most plumages (juvenile here).* CALIFORNIA, AUGUST

L 7.75"	ws 15"	wt 1.5 oz	♀ > ♂

▸ two adult molts per year; complex alternate strategy

▸ strong age-related, sex-related, and seasonal differences

▸ extent and prominence of back streaking variable

Breeds on tundra, winters at sea. Like Wilson's, pauses at hypersaline lakes in West en route to wintering grounds. Feeding behavior of phalaropes, especially Red-necked and Red, is striking: foraging birds spin rapidly on water, creating deep vortex that pulls up food.

Call note, often given by birds in flight, a hard *pik*. Feeding flocks usually silent. Courtship song consists of alternating twittering and growling sounds.

Red Phalarope

Phalaropus fulicarius CODE 1

BREEDING FEMALE. *Completely brick-orange below; breeding male also extensively orange, but paler. Bill is conspicuously thick, opposite of the two other phalarope species.* ALASKA, JUNE

NONBREEDING ADULT. *Has black patch behind eye like Red-necked, but is paler overall: black in cap usually reduced and upperparts generally paler than Red-necked. Note thick bill.* CALIFORNIA, DECEMBER

IN FLIGHT. *Nonbreeding adult (the plumage most often seen at sea) is gray above; white wing stripe usually obvious. In a mixed-species flock, Red appears larger than Red-necked.* CALIFORNIA, DECEMBER

L 8.5"	ws 17"	wt 2 oz	♀ > ♂

▸ two adult molts per year; complex alternate strategy

▸ strong age-related, sex-related, and seasonal differences

▸ intensity and extent of yellow on bill variable

The most pelagic phalarope; unlike Wilson's and Red-necked, does not gather at inland lakes before migrating to sea. Inland sightings, often related to stormy weather, are uncommon in fall, rare in spring. On tundra breeding grounds, favors deeper water than Red-necked.

Flight call a sharp *pink* or *wink*, higher pitched and squeakier than Red-necked. On breeding grounds, combines loud screeching calls with softer warbling.

Ruff
Philomachus pugnax CODE 3

DISPLAYING MALE. *In full display, male puffs out feathers on head and neck ("ruffs"). Courtship plumes may be black, brown, white, red, or any combination thereof.* EUROPE, MAY

ADULT FEMALE. *Not nearly as spectacular as male. Grayish above; dusky or whitish below. Pot-bellied with slight bill and long, yellow legs.* ALASKA, MARCH

JUVENILE. *Distinctively buff with scaly upperparts, calling to mind Buff-breasted Sandpiper. All Ruffs are pot-bellied, thin-billed, and relatively long-legged.* EUROPE, SEPTEMBER

L 11"	WS 21"	WT 5 OZ	♂ > ♀

▸ three male molts per year; complex alternate strategy

▸ sexes differ greatly; much age-related, seasonal change

▸ males in breeding plumage exceedingly variable

Breeds in Eurasia, from northern Europe to eastern Siberia. A few reach North America every year, especially along coasts; has bred in Alaska. Famous in Old World for communal courtship displays in spring, sometimes performed in more subdued form in North America.

Rarely vocal; even in display, males are normally silent. Does not usually call in flight; feeding birds may give a quiet *gurg* or *gurrug*.

SHOREBIRD IDENTIFICATION

Shorebirds—especially sandpipers in the family Scolopacidae—are a well-known identification challenge. Beginners are often frustrated by the monotonous browns and grays of this family, and even expert birders are frequently thrown for a loop by familiar species. The core problem is that many sandpipers—even distantly related species—have fundamentally similar plumages. For example, the distantly related Greater Yellowlegs and Semipalmated Sandpiper are similarly streaked and spotted in gray and brown. But they differ greatly in bill and leg structure, in overall size and shape, in feeding behavior, in vocalizations, and in the timing and extent of their migrations. That basic lesson—namely, that non-plumage characters play an important role in identification—can be powerfully applied to closely related species, too.

Gulls, Terns, and Skimmers

FAMILY LARIDAE

50 species: 25 ABA CODE 1   9 CODE 2   7 CODE 3   4 CODE 4   5 CODE 5

Gulls, terns, and skimmers in the family Laridae present a well-known paradox: they are among the most familiar of birds, yet they are possibly the most difficult of birds to identify in North America. Anyone can recognize a "seagull," but even expert birders are often stymied by species-level identification of Herring Gulls and closely related species. Terns are nearly as familiar, but distinguishing among multiple species standing on a sandbar or foraging over the ocean can be exceedingly difficult. Identification of gulls and terns often requires paying careful attention to plumage—sometimes down to the detail of individual feathers. Also important in identification is an accurate assessment of body shape, overall size, and bill structure.

The taxonomy of this group is highly unstable: the terns were recently reorganized, the gulls are likely to be reorganized soon, and the status of the skimmers has changed over the years.

Terns and especially skimmers are inextricably linked to aquatic habitats, but gulls are more diverse in their habitat preferences. Western Gulls flourish at malls and fast food restaurants, Ring-billed Gulls flock to landfills, and Franklin's Gulls gobble up grasshoppers kicked up by combine harvesters. But other gull species are aquatic for most of their life cycles, and the mall-and-landfill gulls actually spend much of their time at lakes and reservoirs or out at sea.

Many species in the family Laridae are prone to vagrancy. Gulls are legendary in this regard, but terns, too, often wander far off course. Even the Black Skimmer—generally thought of as a highly specialized inhabitant of back channels in coastal salt marshes—routinely wanders inland, sometimes to locations more than a thousand miles from the coast. Wanderlust in the Laridae is so frequent that spatial and temporal patterns of vagrancy are well established; knowing when and where to expect a rarity is a large part of finding rare gulls and terns. Similarly, common species' daily and seasonal movements are regular and predictable, and it is beneficial to learn this aspect of the identification process.

Several species of gulls have adapted well to habitats modified by humans. These gulls are enjoying sustained population growth, coupled in some instances with rapid range expansion. Terns, however, being more strictly associated with marshland and coastal habitats, are at risk in many regions because of both human disturbance and habitat loss.

Ring-billed Gull
Larus delawarensis   CODE 1

IN FLIGHT. *Adult shows sharp contrast between black wing tips and gray of inner wing; in fresh plumage, black wing tip has one or two white spots.* COLORADO, NOVEMBER

BREEDING ADULT. *One of the smallest of the "large white-headed gulls." Fresh breeding adult shows yellow legs and yellow bill with black band. Mantle medium gray.* FLORIDA, FEBRUARY

L 17.5"   WS 48"   WT 1.1 lb

▸ two adult molts per year; complex alternate strategy

▸ strong age-related and moderate seasonal differences

▸ young birds variable, some resembling Mew Gull

Formerly persecuted, but has rebounded and is the most common gull in many regions, especially inland. A familiar sight around fast food restaurants and schoolyards. Nests colonially, usually at protected sites near open water. Range expanding and numbers increasing.

◀ Flight call a wailing *ay-eeee-ya.* Long call a steady series: *hyaw hyaw hyaw...,* with mocking quality. Also gives a deep, monotone *chochocho.*

JUVENILE. *Most are brown above, often "cleanly" patterned. Birds molting out of juvenile plumage (mainly in early fall) are highly variable.* COLORADO, JULY

IMMATURE, FIRST CYCLE. *Plumage "busier" than Mew Gull at same age; splotchy pattern of blacks, browns, and grays highly variable. Feet dull pink; bill usually pink with black tip.* NEW JERSEY, OCTOBER

IMMATURE, SECOND CYCLE. *Like adult, but wing tips all-black; bill less crisp, marked than breeding adul[t]. Gray mantle and yellow ey[e] of adult usually present by second cycle.* CALIFORNIA, OCTOBER

Mew Gull
Larus canus CODE 1

NONBREEDING ADULT. *Three-year gull. Smaller, slighter than Ring-billed. Dainty yellow bill usually unmarked on adult; second-cycle shows smudge or even "ring" like Ring-billed.* CALIFORNIA, JANUARY

IMMATURE, FIRST CYCLE. *Browner than Ring-billed Gull of same age, with lower contrast overall. Small bill and head give many individuals a "gentle" look. All ages are dark-eyed.* CALIFORNIA, JANUARY

IN FLIGHT. *Mantle darker gray than Ring-billed. Black wing tips have larger white spots than Ring-billed; contrast between mantle and trailing edge of wing fairly sharp.* BRITISH COLUMBIA, MARCH

JUVENILE. *Finely marked all over in gray and brown. Individuals still in extensive juvenile plumage are uncommon south of breeding grounds.* ALASKA, SEPTEMBER

16"	WS 43"	WT 15 oz	♂ > ♀

two adult molts per year; complex alternate strategy

strong age-related and moderate seasonal differences

three distinct subspecies in North America

American Mew Gull (viewed by some authorities as a species distinct from Old World populations) breeds in Alaska and western Canada. Winters mainly on Pacific coast; a few inland. Colony size usually small; some nest in trees. Flocks gather at ball fields, docks, and dumps.

CHICK. *Nestling plumage in most gull species is best thought of as a "work in progress"—the downy chick grows, becomes adult-size, and is capable of flight in a matter of weeks.* ALASKA, JUNE

Long call averages squeakier than Ring-billed, but much variation; note shrill yelping notes amid steadier cackling. Flight call a sharp, squeaky *keeeee-uh.*

California Gull

Larus californicus CODE 1

BREEDING ADULT. *Four-year gull. A medium-sized, fairly dark-mantled "large white-headed gull." Adult has yellow bill with variable black and red spots toward tip of lower mandible.* CALIFORNIA, JUNE

NONBREEDING ADULT. *Heavy smudging on head is typical. Long bill and long wings give California a lanky look. Seen closely, dark eye distinguishes it from adults of similar species.* CALIFORNIA, DECEMBER

L 21"	WS 54"	WT 1.3 lb	♂ > ♀

▸ two adult molts per year; simple alternate strategy

▸ strong age-related and moderate seasonal differences

▸ Great Basin breeders average darker-mantled than Northern

Breeds inland, often in large colonies, at large lakes in the western Plains and Intermountain West. Colonies are often within arid landscapes. Most move to West Coast in winter, but some remain inland. Range expanding eastward; inland wintering population increasing.

Long call variable, typically with a few grating notes. Flight call a somewhat modulated *klee*. Also gives a deep *chuchuchu* in flight.

IMMATURE, FIRST CYCLE. *Similar to Herring Gull of same age. Averages paler, especially neck and face; wings coarsely but variably patterned. Long bill pink at base, black at tip.* CALIFORNIA, OCTOBER

BLACK-TAILED GULL (Larus crassirostris). *Casual vagrant from Asia; increasingly sighted in recent years. Adult is fairly dark-backed; white tail has thick black band at tip.* ASIA, JULY

NONBREEDING ADULT. *In direct comparison and with good light, gull species can be separated by the darkness of their mantles. On lone individuals, however, assessment of darkness is problematic.* BRITISH COLUMBIA, FEBRUARY

Herring Gull
Larus argentatus   CODE 1

BREEDING ADULT. *Four-year gull. Mantle medium-gray; wing tips black with a few small white spots. Legs pinkish; bill yellowish with red spot toward tip of lower mandible.* NEW JERSEY, JUNE

| ∟ 25" | ws 58" | wt 2.5 lb | ♂ > ♀ |

- two adult molts per year; simple alternate strategy
- strong age-related and seasonal differences
- immense individual variation; species limits unclear

A notoriously variable gull; every individual in a flock may seem distinctive. In North America, breeds mainly in Canada and Alaska. Winters widely, with greatest concentrations near coasts; frequently seen out at sea. Breeding range expanding southward and inland.

Long call a fairly steady *hyoo hyoo hyoo hyoo*; notes richer and less strident than Ring-billed. Call notes average low and flat: *pooooo* and *pooo-ah*.

YELLOW-LEGGED GULL (Larus michahellis). *Casual stray to eastern seaboard. Distinguished from Herring by yellow legs and darker gray mantle, but always beware of "unusual" Herring Gulls.* EUROPE, MARCH

JUVENILE. *Resembles first-cycle, but darker and less coarsely patterned. Despite dramatic plumage differences, adult and older juvenile are basically alike in size and structure.* RHODE ISLAND, JULY

IMMATURE, FIRST CYCLE. *Coarsely patterned all over in chocolate, usually a little paler around face and throat. Tail and wing tips almost black. Bill usually all-dark; feet pink.* CALIFORNIA, DECEMBER

IMMATURE, SECOND CYCLE. *Paler than first cycle, with gray mantle feathers beginning to appear. Both first-cycle and second-cycle birds in flight show contrastingly pale inner primaries.* TEXAS, MARCH

IMMATURE, THIRD CYCLE. *By this stage, most have acquired the "fierce" yellow eyes and medium-gray mantle of adult. Most have black wing tips with little if any white spotting.* TEXAS, MARCH

Thayer's Gull
Larus thayeri CODE 2

BREEDING ADULT. *Four-year gull. Wing tips not as dark as Herring Gull, and legs usually brighter pink. "Gentle" expression created by rounded head, relatively slight bill, and dark eye.* NUNAVUT, JULY

JUVENILE. *Pale brown all over. At rest, primaries appear darker than body, but not as dark as Herring Gull. Variable; some are nearly as pale as Iceland Gull.* CALIFORNIA, FEBRUARY

IN FLIGHT. *Adult is readily identified in flight; compared to Herring, black in wing tips reduced and white extensive. Immature birds also show pale wing tips.* BRITISH COLUMBIA, FEBRUARY

L 23"	WS 55"	WT 2.2 lb	♂ > ♀

▸ two adult molts per year; simple alternate strategy

▸ strong age-related and moderate seasonal differences

▸ many individuals resemble Iceland and Herring Gulls

Breeds in high Arctic, west of Iceland Gull, except for zone of overlap on Baffin Island. Winters primarily on West Coast, but regular inland in small numbers. Some authorities question whether Thayer's and Iceland Gulls should be treated as separate species.

Vocalizations similar to Herring Gull. Long call perhaps more ringing, more strident than Herring. Call notes slightly purer, flatter, but overlap extensive.

Iceland Gull
Larus glaucoides CODE 2

NONBREEDING ADULT. *Wing tips white, with gray smudging, as here, in subspecies called "Kumlien's" Gull (L. g. kumlieni). Bare part colors like Herring, but build is daintier.* NEWFOUNDLAND, FEBRUARY

IN FLIGHT. *A few adults have pure white wing tips; these may be either strays from Europe, where Iceland Gull is paler, or Canadian breeders at the pale extreme.* MAINE, JANUARY

JUVENILE. *Dirty white all over. Separated from Glaucous Gull by smaller size, smaller head, smaller and darker bill, and relatively longer wings. Most are paler than Thayer's.* NEWFOUNDLAND, FEBRUARY

L 22"	WS 54"	WT 1.8 lb	♂ > ♀

▸ two adult molts per year; simple alternate strategy

▸ strong age-related and moderate seasonal differences

▸ extent and intensity of dark on wing tips variable

An arctic cliff-nester. In winter, disperses south along Atlantic coast and inland to Great Lakes; rare farther west. American Iceland Gull is geographically and morphologically intermediate between Thayer's and European Iceland Gulls.

Long call often with piercing quality: *eeerp eeerp eeerpeeya eeerpeeya.* All calls variable, but usually sound more shrill and intense than Herring.

Glaucous Gull

Larus hyperboreus CODE 1

ADULT. *Four-year gull. Larger and paler than Herring Gull. Pure white wing tips project slightly beyond tail at rest; bill distinctively large at all ages.* ALASKA, MARCH

JUVENILE. *Dusky white all over. Large bill is strongly two-toned with pink at base and black at tip.* ALASKA, MARCH

IN FLIGHT. *Most plumages strikingly white (worn second-cycle immature here). All appear broad-winged in flight with completely white wing tips.* ALASKA, JUNE

L 27"	ws 60"	wt 3 lb	♂ > ♀

▸ two adult molts per year; simple alternate strategy
▸ moderate age-related and seasonal differences
▸ hybridizes with Herring and Glaucous-winged Gulls

Nests widely along arctic shores, both singly and colonially. General-ist feeder, often predatory. South-ward movement in winter mainly coastal; but frequent inland, usually one or two birds in mixed-species flocks at reservoirs and along rivers.

Long call like other large gulls, but with subtle emphatic quality: *kaow kaow kaow kay-ah! kay-ah!* Call notes often two syllables: *heeyow* and *wuck-wuck.*

Glaucous-winged Gull

Larus glaucescens CODE 1

ADULT. *Four-year gull. Medium-gray mantle and wing tips are almost the same color, a distinctive pattern for gray-mantled gulls.* ALASKA, MARCH

HYBRID ADULT. *Hybrids with Western Gulls are frequent; called "Olympic Gulls" because many originate in Puget Sound area. Wing tips darker than pure Glaucous-winged.* CALIFORNIA, DECEMBER

HYBRID IMMATURE. *Pale gray-brown all over; primaries the same color as body. Hybrids with Glaucous Gull also occur. Browner overall than Glaucous, but note two-toned bill like Glaucous. Pure immature Glaucous-winged has more uniform bill.* CALIFORNIA, MARCH

L 26"	ws 58"	wt 2.2 lb	♂ > ♀

▸ two adult molts per year; simple alternate strategy
▸ strong age-related and moderate seasonal differences
▸ paler north than south; hybridizes with other species

Omnivorous and adaptable; popula-tion increasing. Typically an inshore species, seen a short distance inland at wharves and parking lots. Also along coast and at sea; sometimes strays far inland. Most individuals in Seattle, Washington, area are hybrids with Western.

Long call: *heeyoo heeyoo hoo hoo hoo hoo,* sounds mellow and sad. Call notes often with flat quality: *shrooya, hooo,* and *choo-choo-choo.*

Western Gull
Larus occidentalis   CODE 1

IMMATURE, FIRST CYCLE. *Wings, back, and tail dark; underparts variable. Large, dark bill has limited pale area at base.*
CALIFORNIA, JANUARY

L 25"	ws 58"	wt 2.2 lb	♂ > ♀

▸ two adult molts per year; simple alternate strategy

▸ strong age-related and weak seasonal differences

▸ southern birds are darker; hybrids frequent in north

BREEDING ADULT, "SOUTHERN." (L.o. wymani) *Four-year gull. The common dark-mantled gull of the West Coast, not quite as dark as Yellow-footed. Size and proportions similar to Herring.* CALIFORNIA, JANUARY

Often regarded as the southern counterpart of Glaucous-winged Gull, and hybridizes with it extensively in the Seattle, Washington, area; surprisingly, the two species are not closely related. More coastal than Glaucous-winged; less likely inland, more likely offshore.

IMMATURE, THIRD CYCLE. *Plumage similar to adult, but darker and duskier: tail has black streaks; contour feathers of body have variable smudging.*
CALIFORNIA, SEPTEMBER

Long call has choppy quality: *kee kee kee keeu keeu keeu...* Call notes varied; growling *kaow* a characteristic sound of West Coast beaches.

Yellow-footed Gull
Larus livens   CODE 2

L 27"	ws 60"	wt 2.8 lb

▸ two adult molts per year; simple alternate strategy

▸ strong age-related and weak seasonal differences

▸ legs of adult always yellow, but intensity variable

Breeds along Gulf of California during winter; a few hundred disperse north after nesting, reaching the Salton Sea in June; favors rock outcroppings along shoreline. Numbers dwindle in fall, but a few usually linger into winter. Numbers at Salton Sea increasing.

Calls lower-pitched, less variable than Western Gull (which is probably not a close relative of Yellow-footed). Listen for growling *cho* and *cha* elements.

BREEDING ADULT. *Three-year gull. Larger than Western. Yellow legs distinctive, but other dark-backed species can be aberrantly yellow-legged or appear yellow-legged in bright light.* CALIFORNIA, MAY

Slaty-backed Gull

Larus schistisagus CODE 3

NONBREEDING ADULT. *Four-year gull. Overall heft on par with Herring Gull. Mantle dark; legs bright pink. Most nonbreeding adults show a distinct, smudgy eye line unlike other dark-backed gulls.* ASIA, FEBRUARY

IMMATURE, FIRST CYCLE *Some are similar to Thayer's Gull, with pale dusky primaries and low-contrast, largely dark tail. Overall appearance is dark, splotchy, and often "messy."* BERING SEA, JUNE

L 25"   ws 58"   wt 3 lb   ♂ > ♀

▸ two adult molts per year; simple alternate strategy
▸ strong age-related and moderate seasonal differences
▸ adult mantle darkness and leg color variable

East Asian species that regularly reaches the Bering Sea region of Alaska and has bred there. Vagrant records in lower 48 have increased in recent years, doubtless because of the growing popularity of gull-watching, but possibly also reflecting a real population change.

Elements of long call average lower-pitched and more growling than most other large gulls; listen for baritone *kyaw* and *kyawp* notes.

Lesser Black-backed Gull

Larus fuscus CODE 3

NONBREEDING ADULT. *Four-year gull. Smaller than Great Black-backed; long-winged, small-headed. Legs and feet usually yellow. Mantle variable, but not so black as Great Black-backed.* FLORIDA, FEBRUARY

IMMATURE, FIRST CYCLE. *Small head and long wings impart a sleek look. Fairly pale head and underparts contrast with extensive dark on wings and tail. Bill completely black.* FLORIDA, FEBRUARY

L 21"   ws 54"   wt 1.8 lb   ♂ > ♀

▸ two adult molts per year; simple alternate strategy
▸ strong age-related and moderate seasonal differences
▸ actual and apparent darkness of mantle variable

A Eurasian breeder that winters in increasing numbers in North America. Present center of abundance is mid-Atlantic coast, but large flocks occur south to Florida and well inland in Pennsylvania; regular as far west as the eastern foothills of the Rockies.

Vocalizations average deeper and harsher than Herring Gull. Frequently heard calls are *skree-ya* and *skray-yuh*. In flight, a throaty *uh-chow-chow-chow.*

Great Black-backed Gull
Larus marinus CODE 1

SUBADULT, THIRD CYCLE. *Four-year gull. Most are huge and huge-billed. Adult has yellow bill with red spot; this individual has third-cycle bill characters but is otherwise adult-like.* FLORIDA, MARCH

IMMATURE, FIRST CYCLE. *Among "large white-headed gulls," one of the most distinctive in its first cycle: head nearly white; upperparts extensively checkered dark and white.*NEW JERSEY, SEPTEMBER

| L 30" | WS 65" | WT 3.6 lb | ♂ > ♀ |

▸ two adult molts per year; simple alternate strategy

▸ strong age-related and weak seasonal differences

▸ infrequently hybridizes with Herring Gull

Breeds mainly along Northeast coast and along Great Lakes. Sometimes seen far out at sea, but also goes far inland; strays are regularly noted west to the eastern foothills of the Rockies. Range expanding southward and inland; displacing Herring Gull in some places.

Calls deep and growling: *hooah* and *hoom*, descending in pitch and trailing off. Individual elements of long call have resonant quality: *hyew-ah* and *hyewa*.

KELP GULL (*L. dominicanus*). *Extraordinary stray from South America; has bred in Louisiana. Greenish legs help to separate it from other black-backed gulls.* MARYLAND, FEBRUARY

Heermann's Gull
Larus heermanni CODE 1

BREEDING ADULT. *Four-year gull. Combination of white head and dark body is unique among North American gulls. Bill blood-red; on most individuals, wings are uniformly dark.* CALIFORNIA, FEBRUARY

VARIANT ADULT. *A few adult Heermann's Gulls show a white wing patch, prominent in flight. This bird is in nonbreeding plumage, shown by its dusky gray head.* CALIFORNIA, SEPTEMBER

IMMATURE, FIRST CYCLE. *Chocolate-brown all over. Averages more smoothly and uniformly dark than any other first-cycle gull. Slender bill is extensively pale at base.* CALIFORNIA, FEBRUARY

| L 19" | WS 51" | WT 1.1 lb | ♂ > ♀ |

▸ two adult molts per year; simple alternate strategy

▸ strong age-related and moderate seasonal differences

▸ a few flash extensive white on upperwings

Breeds off western Mexico; migrates north into U.S. waters after breeding. Occurs along coast or well offshore. Seen at wharves and marinas; at sea, chases Brown Pelicans, stealing their food. Numbers increasing, and vagrants more frequently seen inland.

Long call is low and flat: *ah-wawawawa*. When chasing pelicans or other Heermann's Gulls, gives short, scratchy notes: *itch* and *ipipip*.

Laughing Gull

Larus atricilla CODE 1

BREEDING ADULT. *Three-year gull. One of the largest "small hooded gulls." Wings are dark gray with black tips; the tips lack white spotting. Bill entirely red in breeding plumage.* TEXAS, APRIL

NONBREEDING ADULT. *Head mainly white, with limited dusky smudging behind eye; eye arcs not as prominent as on Franklin's. Wings gray with mainly black tips; bill mainly black.* FLORIDA, FEBRUARY.

IMMATURE, FIRST CYCLE. *Duskier and "dirtier" overall than Franklin's of same age. Extensive gray-brown smudge across breast; ill-defined gray smudge on face and nape.* NEW JERSEY, SEPTEMBER

L 16.5"	WS 40"	WT 11 OZ	♂ > ♀

▸ two adult molts per year; complex alternate strategy

▸ strong age-related and seasonal differences

▸ bill color and extent of hood variable on winter adults

Inhabits beaches and salt marshes along the Atlantic and Gulf coasts; frequent vagrant far inland. Withdraws in winter from Northeast; regular in summer at Salton Sea. Common sight around cities and towns, but colonies susceptible to disturbance. Populations increasing.

Highly vocal; one of the telltale sounds of summer along the East Coast. Long call a rich, musical, honest-to-goodness laugh: *hahahahah haaa haaa haaa.*

Franklin's Gull

Larus pipixcan CODE 1

BREEDING ADULT. *Three-year gull. Black wing tips separated from gray inner wing by extensive white; compare Laughing. Eye arcs of Franklin's more prominent than Laughing.* TEXAS, APRIL

NONBREEDING ADULT. *Black on head reduced from breeding plumage, but not so much as Laughing. Rear of hood and white nape sharply delineated. Eye arcs remain prominent.* PERU, NOVEMBER

IMMATURE, FIRST CYCLE. *Compared to Laughing, black "half-hood" well-defined and eye arcs prominent. Underparts mainly white (Laughing extensively dusky).* CALIFORNIA, DECEMBER

L 14.5"	WS 36"	WT 10 OZ	♂ > ♀

▸ two adult molts per year; complex alternate strategy

▸ strong age-related and seasonal variation

▸ extent of rosy bloom on breeding adult variable

Breeds mainly on the prairies of south-central Canada; winters coastally and offshore near the equator. Requires extensive prairie marshes for breeding; colonies often large, but local populations vary from year to year. Large flocks gather on migration.

Noisy at colonies and sometimes in migratory flocks; otherwise fairly quiet. Long call *kah keeah! keeah! kah...* Call notes flat: *chull* and *uh-chupachup.*

Black-headed Gull

Larus ridibundus CODE 3

BREEDING ADULT. *Two-year gull. Larger than Bonaparte's; hood is dark brown, not black. Underwing dark with contrasting white outer primaries; similar pattern above, but lower contrast.* ASIA, JUNE

NONBREEDING ADULT. Similar to Bonaparte's, but bill usually retains substantial reddish. Wing pattern, the same as on breeding adult, is best distinction from Bonaparte's. CALIFORNIA, DECEMBER

IMMATURE, FIRST CYCLE. Upperwing pattern more diffuse than Bonaparte's; more dark overall. Immature has extensive red in bill; Bonaparte's bill is black. NEWFOUNDLAND, FEBRUARY

| L 16" | WS 40" | WT 9 OZ | ♂ > ♀ |

▸ two adult molts per year; complex alternate strategy
▸ strong age-related and seasonal differences
▸ bill length, body size, and bare part colors variable

Old World counterpart of Bonaparte's Gull; often joins Bonaparte's flocks in North America. Has recently colonized from Europe, where it is abundant. Fairly common in Newfoundland, especially in winter; a few breed. Uncommon but regular elsewhere in Northeast.

Calls average more "gull-like" than Bonaparte's; higher-pitched and more squealing. Most notes are monosyllabic but drawn-out: *keeer* and *kyeee.*

Bonaparte's Gull

Larus philadelphia CODE 1

BREEDING ADULT. *Two-year gull. In all plumages, note petite body structure. Black bill is thin and slight; feet usually red; mantle slightly darker than Black-headed Gull.* CALIFORNIA, MARCH

NONBREEDING ADULT. Black hood of breeding plumage replaced by smudge behind eye. From below, wing lacks extensive black of Black-headed Gull; above, note white wedge toward tip. COLORADO, NOVEMBER

IMMATURE, FIRST CYCLE. Upperwing mainly gray with black borders; underwing pale with dark trailing edge. Like Black-headed, has white tail with black terminal band. COLORADO, NOVEMBER

| L 13.5" | WS 33" | WT 7 OZ | ♂ > ♀ |

▸ two adult molts per year; complex alternate strategy
▸ strong age-related and seasonal differences
▸ some adults show a pale pink blush on breast

Nests in trees in Canada and Alaska, usually in forested lands near standing water. Winters coastally and on Great Lakes; widespread migrant inland, typically at large lakes and rivers. Feeds on small fish; not usually attracted to parking lots and garbage dumps.

Most calls are harsh and grating, lacking the squealing, anguished qualities of other gulls. Migrants usually quiet; sometimes give a tern-like *zheeeeerp.*

Little Gull

Larus minutus CODE 3

BREEDING ADULT. *Two-year gull. Tiny; noticeably smaller than Bonaparte's Gull. Legs red; bill black, thin, and slight. Wing tips entirely pale, unlike other "small hooded gulls."* EUROPE, JULY

NONBREEDING ADULT. *Black hood of breeding plumage reduced to black cap and smudge behind eye. Wings unique for a gull: short and rounded; black below with broad white trailing edge.* EUROPE, JANUARY

IMMATURE, FIRST CYCLE. *Black primaries and secondary coverts create M-pattern, similar to kittiwake. Short wings and buoyant manner of flight distinctive.* EUROPE, JANUARY

L 11"	WS 24"	WT 4.2 oz

▸ two adult molts per year; complex alternate strategy

▸ strong age-related and seasonal differences

▸ some adults show variable pink bloom on breast

Mainly Old World species; discovered breeding in North America in 1962. Regular but uncommon in winter in mid-Atlantic coastal region. Numbers possibly increasing here, but still low. Away from marshland breeding grounds, often associates with Bonaparte's Gull.

Typical call notes are short and clipped: *ko* and *kay*. Some are two syllables, with accent on second syllable: *k'ko* and *koko*.

Sabine's Gull

Xema sabini CODE 1

BREEDING ADULT. *Two-year gull. Small black bill has bright yellow tip; dark gray hood separated from white neck by black collar. Sabine's tends to ride high on the water.* ALASKA, JUNE

NONBREEDING ADULT. *Gray hood replaced by smudgy "half hood," suggesting Franklin's. Wing pattern always diagnostic: black-white-gray on adult, black-white-brown on juvenile.* NEW JERSEY, SEPTEMBER

JUVENILE. *Subadults inland in fall are usually still in juvenile plumage: extensively brown with finely scaled mantle. If identification is in doubt, wait for the bird to fly.* COLORADO, SEPTEMBER

L 13.5"	WS 33"	WT 6 oz	♂ > ♀

▸ two adult molts per year; complex alternate strategy

▸ strong age-related and seasonal differences

▸ body size and overall darkness are variable

High-arctic breeder; winters offshore in the tropics. In migration, common off West Coast, uncommon off East; regular inland, especially in West, on fall migration. Feeding behavior suggests phalarope: sits on water and plucks food from surface; also feeds on mudflats.

Calls have growling, tern-like quality; migrants sometimes give *zheeerp* and *wurf* notes; also excited *ch'ch'chick*. Long call: *zip zip zheerp zheerp...*

Black-legged Kittiwake

Rissa tridactyla   CODE 1

BREEDING ADULTS. *Three-year gulls. Plain overall: gray wings have black tips; yellow bill lacks markings; feet are black, unlike most other gulls in breeding plumage.* ALASKA, MAY

L 17"	WS 36"	WT 14 oz	♂ > ♀

▸ two adult molts per year; simple alternate strategy

▸ strong age-related and moderate seasonal differences

▸ individual variation in body size and bill length

NONBREEDING ADULT. *Acquires dark smudge behind eye in nonbreeding plumage; yellow bill remains unmarked. Black primary tips set off sharply against pale gray primaries.* CALIFORNIA, NOVEMBER

Nests on narrow ledges on steep sea cliffs; winters widely at sea, mainly in waters of North Temperate Zone. Avoids garbage dumps and parking lots, but is frequently attracted to ships. Numbers have increased recently, but essential prey are threatened by overfishing.

JUVENILE. *Boldly marked black and white overall. Upperwing has black bars forming prominent M-shape across mantle. Smudgy black band extends across base of nape. Bill black.*
CALIFORNIA, NOVEMBER

"Kittiwake" call variable, but typically with petulant quality: *daway!* or *iddy-ay*, often repeated. Other notes short, stuttering. Noisy at colonies.

Red-legged Kittiwake

Rissa brevirostris   CODE 2

L 15"	WS 33"	WT 13 oz	♂ > ♀

▸ two adult molts per year; simple alternate strategy

▸ moderate age-related and seasonal differences

▸ bare part colors of first-year birds variable

IMMATURE. *Two-year gull. Adults and older immatures have red legs and stubby yellow bill. Head white in breeding plumage; in nonbreeding plumage, vertical smudge forms behind eye.* ALASKA, JULY

The less common, less well-known counterpart of Black-legged Kittiwake. Breeds on steep sea cliffs on islands in Bering Sea, sometimes in colonies with Black-legged. Winter range poorly known; apparently mainly at sea in North Pacific. Populations possibly declining.

BREEDING ADULT. *Mantle darker than Black-legged Kittiwake. Secondaries and inner primaries of immature broadly tipped in white, creating a pattern like Sabine's Gull.*
ALASKA, JUNE

Vocal at colonies; generally silent at sea. Calls average higher-pitched than Black-legged. "Kittiwake" call a squeaky *chee-weekah* or *dee!-aha*.

Ross's Gull
Rhodostethia rosea CODE 3

BREEDING ADULT. *Two-year gull. Famous for its thin black collar and rosy wash below, but distinctive in other regards: tail wedge-shaped; wings long and white; bill black and tiny.* ASIA, JULY

| 13.5" | ws 33" | wt 6 oz |

» two adult molts per year; complex alternate strategy

» strong age-related and moderate seasonal differences

» extent and intensity of rosy bloom on breast variable

More common in North America than previously thought: found nesting at Churchill, Manitoba, in late 1970s; migrates off northern Alaska in the thousands; apparently winters in Bering Sea near limit of pack ice. Like Sabine's Gull, plucks food from surface of water.

Calls varied, often fairly musical. Around breeding grounds gives soft chattering and higher-pitched yelps; in winter may give short, rounded *kyew* notes.

NONBREEDING ADULT. *Black collar and rosy wash reduced, but nonbreeding adults retain distinctive characters: diminutive bill; long white wings; and overall paleness.* EUROPE, FEBRUARY

IMMATURE, FIRST CYCLE. *Black on wings contrasts sharply with overall whiteness; in flight, note M-pattern on upperwing and black tip on white tail.* JAPAN, DECEMBER

Ivory Gull
Pagophila eburnea CODE 3

| 17" | ws 37" | wt 1.4 lb | ♂ > ♀ |

» one adult molt per year; simple basic strategy

» strong differences between adult and first cycle

» juvenile plumage variable; adult bill color variable

Even among species of the high Arctic, Ivory Gull is exceptional. Breeds in desolate habitats of northern Canada; sometimes seen near North Pole. Most winter in Arctic Ocean; casual south of Maritime Canada. Uncommon even on breeding grounds; imperiled by global warming.

Calls varied, but strays in winter usually silent. A tern-like, descending *kee-ar* is characteristic. Other calls are more squealing and gull-like.

ADULT. *Two-year gull. Plumage completely white all year long. Eyes and feet black; bill yellow.* NORTHWEST TERRITORIES, JULY

JUVENILE. *White wings and tail variably flecked with black; face is smudgy black. Looks compact at rest, but long-winged and graceful in flight.* NOVA SCOTIA, JANUARY

Forster's Tern

Sterna forsteri CODE 1

BREEDING ADULT. *Pale gray wing tips fall short of tail tip. Feet orange. Bill extensively orange; mainly black in nonbreeding plumage. Bill is relatively thick for a Sterna tern.* TEXAS, APRIL

JUVENILE. *Brownish highlights change to gray and white as bird ages. Black eye patch distinctive; nonbreeding adult and older immature show same mark. Tail much shorter than adult.* NEW JERSEY, JULY

IN FLIGHT. *Primaries nearly all-white in breeding adult; outer tail feathers long. Nonbreeding adult and juvenile have less white in primaries, but primaries are not as dark as Common.* TEXAS, APRIL

L 13"   WS 31"   WT 6 OZ

▸ two adult molts per year; complex alternate strategy

▸ moderate age-related and seasonal differences

▸ darkness of plumage and overall body size variable

Nests in diverse freshwater or saltwater marshlands; patchy breeding range extends from coast to coast, north into prairie provinces and south to Gulf coast. Winters mainly in southern coastal plains. Conspicuous; plunge-dives for fish in large lakes and bays.

◀ Most calls harsh, for example loud, descending *zheeeeeeer*, given in flight. Harsh *cheer cheer cheer...* and stuttering *ch'ch'ch'* in feeding flocks.

Common Tern

Sterna hirundo CODE 1

BREEDING ADULT. *Pale, but darker overall than similar Forster's. Tail falls short of wing tips on bird at rest. Mainly red bill has thinner base than Forster's.* MINNESOTA, JUNE

IMMATURE. *All nonbreeding plumages show distinctive "carpal bar." Black on head is reduced compared to breeding plumage, but never forms the discrete eye patch of Forster's.* NEW JERSEY, SEPTEMBER

IN FLIGHT. *Adult in flight is conspicuously long-winged. Primaries have extensive black, especially toward tips. Most breeding adults show faint gray wash on underparts.* MAINE, JUNE

L 12"   WS 30"   WT 4.2 OZ

▸ two adult molts per year; complex alternate strategy

▸ moderate age-related and seasonal differences

▸ darker Old World subspecies a regular stray to Alaska

Breeds along East Coast and in much of southern Canada; has declined as a breeder in Great Lakes region. Noisy and highly aggressive at colonies. Widespread migrant inland, although uncommon throughout much of Intermountain West; migratory flocks sometimes large.

Calls variable; many with angry quality. Downward *keeeee-arr* is characteristic, sometimes shortened to *kya*. Also *ch'* and *graa* notes, often repeated.

Arctic Tern
Sterna paradisaea CODE 1

BREEDING ADULT. *Similar to Common Tern; unidentified birds are jokingly called "Commic" Terns. Tail longer than Common; extends past wing tips. Bill and feet blood-red.* ALASKA, JUNE

JUVENILE. *"Carpal bar" of nonbreeding plumages reduced or absent; prominent on Common. In all plumages, note peculiarly short legs and thin, short bill.* ALASKA, AUGUST

IN FLIGHT. *Wings are pale gray above with little patterning, and white below with a thin line of black along the tips of the primaries, visible at distance.* MANITOBA, JUNE

12"	ws 31"	wt 4 oz

- two adult molts per year; complex alternate strategy
- moderate age-related and seasonal differences
- extent and intensity of gray on breast variable

Fabled long-distance migrant; breeds as far north as northern Greenland, winters off Antarctic pack ice. Opportunistic; recently bred in Montana. Migration chiefly pelagic, but small numbers are annual inland, especially in fall. At sea occurs singly or in small flocks.

Vocal at colonies; attacks intruders with *kakaka* and *kshhh-ee-ar* calls; all year, gives downward *jeeee-ar*, higher than corresponding call of Common.

Roseate Tern
Sterna dougallii CODE 2

BREEDING ADULT. *A pale Sterna tern. Tail very long; usually projects well beyond wing tips. Thin bill mostly black; feet reddish orange.* NEW JERSEY, JUNE

JUVENILE. *Upperparts gray with black scalloping. All nonbreeding plumages have extensive black on head; "carpal bar" usually absent.* NEW YORK, AUGUST

IN FLIGHT. *Adult similar to Forster's: extensively white above; completely white tail is long and streamer-like. Many breeding adults show faint rosy bloom on underparts.* NEW YORK, JUNE

L 12.5"	ws 29"	wt 4 oz

- two adult molts per year; complex alternate strategy
- moderate age-related and seasonal differences
- rosy bloom on adult underparts is variable

Uncommon and patchily distributed over extensive, mainly tropical range; in our area, breeds from Long Island to Nova Scotia and in Florida Keys. Small numbers nest in mixed-species colonies; hybridizes with Common Tern. North American populations listed as threatened.

Typical call dry, distinctive: *k'chick* or *cheek.* In alarm gives downward *kyee*, not as throaty as Common or Arctic. Aerial courtship flights are impressive.

Gull-billed Tern

Gelochelidon nilotica CODE 1

L 14"   WS 34"   WT 6 OZ

▸ two adult molts per year; complex alternate strategy

▸ moderate age-related and seasonal differences

▸ California birds average larger than eastern birds

Globally widespread but patchily distributed. In our area, breeds on Gulf coast, on Atlantic coast north to mid-Atlantic states, and in southern California. Diet varied; fish and other animal matter. Does not plunge-dive. North American populations declining.

Typical call a two-syllable *kee-wick*, accented on second syllable; often extended to *k'diddy-wick* ("katydid"). Small flocks in salt marshes become frenzied.

NONBREEDING ADULT. *All plumages have thick black bill, more gull-like in shape than tern-like. Legs long, also contributing to a somewhat gull-like structure overall.* TEXAS, SEPTEMBER

BREEDING ADULT. *Long wings arc far back in flight. Note black cap on this breeding adult; nonbreeding adult and juvenile show smudgy black eye patch. Gull-like bill often visible from afar.* TEXAS, MAY

Sandwich Tern

Thalasseus sandvicensis CODE 1

BREEDING ADULT. *The smallest of the "crested terns," in some ways suggesting a Sterna tern. Breeding adult has long black bill with yellow tip; rear of crown shaggy.* TEXAS, MAY

L 15"   WS 34"   WT 7 OZ   ♂ > ♀

▸ two adult molts per year; complex alternate strategy

▸ moderate seasonal and age-related differences

▸ prominence of yellow at tip of bill variable

Nests in dense mixed-species colonies along Gulf and southern Atlantic coasts; populations north of Florida are migratory. Nests on beaches; even when foraging, does not go farther inland than inlets and outflows. Range expanding north.

Vocalizations varied. Most are short, with initial *k–* sounds: *kee!* and *k'k'k'k'* and *kr'r'r'r'* and *k'd'd'd'd*. High-pitched *keekeekee* given by female.

NONBREEDING ADULT. *All adults are pale gray above. Yellow-tipped bill retained in nonbreeding plumage, but black in cap is reduced. Immature similar to nonbreeding adult, but duskier.* FLORIDA, MARCH

ADULT IN FLIGHT. *Note narrow wings, very pale gray, with thin black edgings limited to the outermost primaries, both above and below.* TEXAS, MAY

Elegant Tern
Thalasseus elegans CODE 1

BREEDING ADULT. Bill distinctive: long, orange, and decurved. Crest averages shaggier on Elegant than other "crested terns." CALIFORNIA, JUNE

NONBREEDING ADULT. Black cap reduced. Adult has decurved orange bill all year; yellowish bill of subadult is straighter and similar to variant Sandwich Tern. CALIFORNIA, OCTOBER

IN FLIGHT. Wings are long and pointed; mainly white above. Many adults show pale peach or rosy bloom on underparts. Distinctive bill structure visible from afar. CALIFORNIA, MAY

| 17" | ws 34" | wt 9 oz | ♂ > ♀ |

▸ two adult molts per year; complex alternate strategy
▸ moderate age-related and seasonal differences
▸ pink flush across belly of adult variable

Nests among Heermann's Gulls and Caspian Terns in southern California and especially northwestern Mexico. After breeding, many move north along coast, regularly to San Francisco, rarely to British Columbia. Nesting success and extent of dispersal are related to El Niño.

Breeding flocks noisy; listen for descending *weer-ah* or *d'weer-ah*. Also gives *weer weer weer...* in frenzied series, and softer, stuttering *k'k'k'* notes.

Royal Tern
Thalasseus maximus CODE 1

BREEDING ADULT. A large tern with a substantial orange or red-orange bill; juvenile has yellow bill. In breeding plumage, crown is entirely black, often with wispy black plumes at nape. TEXAS, MAY

NONBREEDING ADULT. Black on crown significantly reduced; cap and forehead mainly white. Primaries of nonbreeding adult and subadult are extensively dark. NEW JERSEY, OCTOBER

ADULT IN FLIGHT. All plumages have a dark area on outer primaries- least extensive in breeding adult, but still prominent. TEXAS, MAY

| L 20" | ws 41" | wt 1 lb | ♂ > ♀ |

▸ two adult molts per year; complex alternate strategy
▸ moderate age-related and seasonal differences
▸ juvenile plumage and bare parts highly variable

Breeds on beaches in dense colonies. Forages over salt water, venturing farther inshore than Sandwich and Elegant Terns; occasional well inland after storms. Populations have expanded north along Atlantic coast, but numbers have apparently declined in California.

Most notes deep and growling: *runk* and *ronk* and descending *gr'r'r'aww*. Gives squeaky *cheek* at colonies; female gives high-pitched *yecyeeyee*.

Caspian Tern
Hydroprogne caspia CODE 1

ADULT IN FLIGHT. *Wingbeats deep and rowing; flight less buoyant than smaller terns. Wing tips dark, especially on underwing. Wing tips of juvenile are darker than adult.* CALIFORNIA, JULY

L 21"	ws 50"	wt 1.5 lb	♂ > ♀

▸ two adult molts per year; complex alternate strategy

▸ moderate age-related and seasonal differences

▸ bill color and extent of black on adult crown variable

Breeds widely but patchily in diverse habitats: salt marshes, barrier islands, and inland rivers and lakes; readily accepts restored wetlands and sites created by humans. Most withdraw to coasts in winter. Populations generally increasing.

NONBREEDING ADULT. *The largest tern, with an oversized, blood-red bill. Nonbreeding adult has smudgy black cap flecked with white; breeding adult has completely black cap.* CALIFORNIA, NOVEMBER

Adult vocalizations deep and hoarse: *ruk* and *rawk* and *ra-a-a-a-awk.* On migration, juveniles call to adults with squealing *eeee-ah* notes.

Least Tern
Sternula antillarum CODE 1

IMMATURE. *Small size usually diagnostic. Bill and feet black, with yellow highlights as bird ages. Often shows a carpal bar like Common Tern.* FLORIDA, JULY

L 9"	ws 20"	wt 1.5 oz

▸ two adult molts per year; complex alternate strategy

▸ moderate age-related and seasonal differences

▸ juvenile leg color and plumage variable

Breeds on coastal beaches and along major inland river systems. Departs by late summer for winter range in Central and South America. Nests placed in open habitats are susceptible to flooding, erosion, and predation; development and disturbance are additional threats.

BREEDING ADULT. *Our smallest tern. Bill and legs yellow. Black cap, nape and eye line contrast with white forehead.* CALIFORNIA, AUGUST

IN FLIGHT. *Wingbeats are flicking and choppy; foraging birds employ erratic twists and turns. From above, black wedge on outer wing contrasts with gray on rest of wing.* FLORIDA, JULY

Noisy at colonies and in feeding assemblages. Most notes sharp and clipped: *ki-dick* and *yip* and *yip-yip.* In attack, gives abrupt *jeerp.*

Black Tern
Chlidonias niger CODE 1

BREEDING ADULT. *Head, breast, and belly completely black; upperparts dark gray. A small tern, not much larger than Least.* NORTH DAKOTA, JUNE

NONBREEDING ADULT. *Plumage becomes more "tern-like": white head with irregular black cap or hood; upperparts paler.* CALIFORNIA, SEPTEMBER

L 9.75"   WS 24"   WT 2.2 OZ

▸ two adult molts per year; complex alternate strategy

▸ strong seasonal and moderate age-related variation

▸ transitional birds in late summer highly variable

Nests in loose colonies in freshwater marshes. Catches insects on the wing, sometimes joining flocks of swallows; also eats fish. Numbers fluctuate in response to local habitat conditions. Winters to our south; migrates at sea, singly or in small flocks, in late summer.

Most calls short and sharp: *kip* and *kick* and *kip-a-kip.* When attacking intruders, gives harsher *kek* or *keeer* notes, often repeated.

IN FLIGHT. *In all plumages underwings are uniformly gray. Flight erratic, with abrupt midair maneuvers reminiscent of Common Nighthawk.* NORTH CAROLINA, JUNE

JUVENILE. *Similar to nonbreeding adult, but averages browner. When bird is standing, note wavy brown-and-black on back.* NEW JERSEY, AUGUST

WHITE-WINGED TERN (C. leucopterus). *Eurasian species; casual visitor to North America. Similar to Black Tern, but note pale upperwings and contrasting black-and-gray underwings.* ASIA, MAY

Aleutian Tern
Onychoprion aleuticus CODE 2

L 12"   WS 29"   WT 4 OZ

- two adult molts per year; complex alternate strategy
- strong age-related and moderate seasonal differences
- amount of rusty in juvenile plumage variable

Widespread around coastal Alaska in summer; departs by August to begin trans-Pacific migration to wintering grounds in equatorial Pacific. In North America, often breeds in association with Arctic Terns. Numbers at colonies fluctuate from year to year.

Most calls soft and twittering, not nearly as harsh and throaty as Arctic Tern. Vocalizations more resemble small shorebirds: *preet* and *kiv* and *kiv-uk.*

BREEDING ADULT. *Traditionally thought of as similar to Arctic, but now known to be more closely related to Sooty and Bridled. Body gray overall; black cap broken by white forehead.* ASIA, MAY

IN FLIGHT. *Adult, as here, has pale underwing with thin dark trailing edge on secondaries; gray on underparts contrasts with lighter cheek. Juvenile scaly above and rusty overall.* ALASKA, JUNE

Sooty Tern
Onychoprion fuscatus CODE 2

JUVENILE. *Dirty dark brown all over with coarse white spotting on back and wings; spotting on wings most heavily concentrated on coverts.* TEXAS, MARCH

L 16"   WS 32"   WT 6 OZ

- two adult molts per year; simple alternate strategy
- strong age-related and weak seasonal differences
- adult underparts vary from pale gray to pure white

Worldwide in tropical waters; small numbers breed off Gulf coast. Wanders widely at sea in eastern North America; frequent inland after hurricanes. A master aerialist; aloft continually. Rarely lands on water; sometimes rests on debris. Catches flying fish in midair.

Vocal, especially around colonies. All day and all night gives scratchy, somewhat harsh *wick-a-wick* ("wide-awake") call, frequently doubled or trebled.

ADULT. *Nearly jet black above; darker than Bridled Tern. White patch on forehead extends back only to eye; compare longer, narrower patch of Bridled.* FLORIDA, MAY

IN FLIGHT. *From below, pale wing coverts contrast with dark primaries and secondaries; primaries and secondaries of Bridled not as dark.* FLORIDA, MAY

Bridled Tern
Onychoprion anaethetus CODE 2

JUVENILE. *Not nearly as dark as juvenile Sooty. Back and wings have wavy gray and brown marks. Adult face pattern (long, narrow extension of white forehead) shown by many juveniles.* FLORIDA, AUGUST

ADULTS. *Similar to Sooty Tern, but slightly paler above; upperparts often show a brownish tinge rather than black. White of forehead extends past the eye; white on Sooty ends at eye.* FLORIDA, MAY

L 15" WS 30" WT 3.5 OZ

▸ two adult molts per year; simple alternate strategy

▸ moderate age-related and weak seasonal differences

▸ extent and intensity of gray on underparts variable

A marine species, fairly common off Southeast coast. Dives for small fish, unlike Sooty Tern. Joins mixed-species assemblages; sometimes steals food from other species. Frequently rests on flotsam; Sooty seldom does.

At sea, sometimes gives yelping calls, *yep* or *yep yep*. Around colonies, gives a hollow rattle *k'r'r'r'r*, suggesting *Melanerpes* woodpecker.

IN FLIGHT. *Not quite as dark above as Sooty Tern; amount of white on tail more extensive than Sooty. From below, flight feathers pale gray; dark gray on Sooty.* NORTH CAROLINA, AUGUST

Brown Noddy
Anous stolidus CODE 2

L 15.5" WS 32" WT 7 OZ

▸ one adult molt per year; complex basic strategy

▸ moderate differences between juvenile and adult

▸ overall darkness variable, related to feather wear

Tropical species; breeding range extends to Dry Tortugas. Marine; disperses north casually to North Carolina. Feeding methods diverse: snags flying fish in midair, plucks fish from surface, and catches fish just below surface. Rests on flotsam or on water.

ADULT. *Dark chocolate brown all over, except on head. Pale gray cap blends into darker gray-brown of nape. Pale cap greatly reduced or lacking on subadult.* FLORIDA, APRIL

Fairly quiet away from colonies, but may give growling *ur-r-r-r-r*. In elaborate pair bonding, performs ritualize "nodding" display, chatters harshly.

Black Noddy

Anous minutus CODE 3

L 13.5"   WS 30"   WT 4 OZ

▸ one adult molt per year; complex basic strategy

▸ moderate differences between juvenile and adult

▸ overall darkness variable; foot color variable

In North American waters, the rare counterpart of Brown Noddy. Occurs in spring and summer; sometimes absent. Has been seen at or near Brown Noddy colonies on the Dry Tortugas; has attempted to breed. Feeds mainly by plucking small fish from the water's surface.

Vocalizations higher-pitched, not as gruff and growling as Brown Noddy: *trrrrrip* and *grrrrrip*. Also gives short chattering and croaking sounds.

IMMATURE. Entirely black except for white cap, which is set off sharply from black on rest of head. Adult plumage more like Brown Noddy, white cap blending with black nape. MIDWAY, MARCH

Black Skimmer

Rynchops niger CODE 1

JUVENILE. Mottled black and tan above. Distinctive bill structure acquired while juvenile plumage is held; younger birds do not yet have uneven mandibles. NEW JERSEY, AUGUST

BREEDING ADULT. Huge bill unique: lower mandible much longer than upper mandible. Plumage black above, white below. In nonbreeding plumage, cap and back are separated by white nape. TEXAS, MAY

L 18"   WS 44"   WT 11 OZ   ♂ > ♀

▸ two adult molts per year; simple alternate strategy

▸ strong age-related and moderate seasonal differences

▸ body size and bill length vary greatly

Feeding behavior extraordinary: open-mouthed, skims water with lower mandible; when tactile sensors detect a fish, bill snaps shut instantly. In perfect formation, small flocks forage in this manner. Roosts and nests in sociable colonies. Often feeds at dusk or night.

Generally vocal. Most common call is a ringing *ark!* or *arp!*— given singly or in series. Calls carry far; tone more pure than most tern vocalizations.

IN FLIGHT. Many other species engage in very brief bouts of surface-skimming while in flight, but only the Black Skimmer does so habitually, efficiently, and effortlessly. NEW JERSEY, JULY

Skuas & Jaegers

FAMILY STERCORARIIDAE

5 species: 3 ABA CODE 1   1 CODE 2   1 CODE 3

Long thought to be closely related to the gulls, the skuas and jaegers are now believed to be more closely allied with the puffins and murres in the family Alcidae. Well-known for their pugilistic behavior at sea, the skuas and jaegers are also bullies on their high-latitude breeding grounds. They are major predators of lemmings, seabird chicks, and other easy pickings of the Arctic and Antarctic regions. All three species of jaegers are regular in small numbers inland in North America, especially in the fall. The skuas, however, are almost never observed except offshore.

POLYMORPHISM

Identification of jaegers and skuas is famously difficult. The problem is that different species can look so similar, whereas individuals within the same species can look strikingly different from one another! What is the source of the confusion? It is not plumage differences between male and female or among different subspecies. Within a species, male and female jaegers and skuas are nearly identical; moreover, all of these species are "monotypic"—that is, they do not contain different subspecies. However, they do contain significant differences called color morphs.

It is natural to think of color morphs as discrete categories, and this approach is often useful in the field. For example, a first order of business when confronted with an adult Parasitic Jaeger is to assign it to one of three categories: dark-morph, intermediate-morph, or light-morph. Similarly, the dark and light morphs of Snow Goose are readily recognized in the field (the dark morph called Blue and the light morph confusingly called Snow). In truth, however, distinctions among color morphs are often fuzzy; a Parasitic Jaeger may fall in between the dark and intermediate morphs, or a Snow Goose may show traits of both the dark and the light morphs.

The condition of having two or more color morphs is known as "polymorphism," and it is genetically determined. The exact same gene is responsible for darkness in such disparate species as Parasitic Jaeger and Snow Goose. Polymorphism may have a geographic basis; for example, dark-morph Parasitic Jaegers are relatively more frequent on the Aleutian Islands than in north-central Canada. However, polymorphism may cut across geographically defined subspecies, as in the case of the Snow Goose: Both the "Greater" and the "Lesser" subspecies have dark and light morphs.

South Polar Skua
Stercorarius maccormicki CODE 2

INTERMEDIATE-MORPH ADULT. *Wings mainly dark below, with striking white flash at base of primaries. Intermediate and light morphs have paler head and underparts contrasting with wings. Stocky and robust, with broad wings and relatively short tail.* NORTH CAROLINA, MAY

DARK-MORPH ADULT. *Wing flash of skuas more prominent than on jaegers. Dark-morph South Polar often shows golden wash across nape. Subadult plumage averages more uniform.* CALIFORNIA, OCTOBER

L 21"	ws 52"	wt 2.5 lb	♀ > ♂

▸ two adult molts per year; simple alternate strategy

▸ weak differences between adult and subadult

▸ polymorphic; dark and light extremes are distinct

Breeds in Antarctic region; found along both of our coasts during warm months. Usually seen only out at sea; inland record from North Dakota is astonishing. Status in our area complicated by hybridization with Brown Skua; taxonomic questions are unresolved.

Largely silent in North American waters. During interactions in large feeding flocks, may utter soft, drawn-out grunts, but only rarely.

Great Skua
Stercorarius skua CODE 3

ADULT. *Very similar to South Polar; some cannot be identified to species. Great Skua tends to be more warmly colored, with more streaking and splotching underneath.* EUROPE, JUNE

ADULT. *Wing flash of both skua species is prominent both above and below. Note relatively warm ground color of underparts with dull splotching.* EUROPE, JULY

L 23"	ws 55"	wt 3.2 lb	♀ > ♂

▸ two adult molts per year; simple alternate strategy

▸ weak differences between adult and subadult

▸ color of adults shows extensive variation

Old World breeder in North Atlantic region; winters widely in small numbers off East Coast. Remains at least until May, when overlap with South Polar Skua is possible. Presumably wanders in the summer months to poorly studied waters off eastern Canada.

Like South Polar Skua, not likely to be heard in North American waters; may give soft, nasal *chench* when interacting in feeding flocks.

Pomarine Jaeger
Stercorarius pomarinus CODE 1

LIGHT-MORPH NONBREEDING ADULT. *Identification of subadult and nonbreeding adult is difficult. Important marks include bulky structure, thick bill, and relatively blunt-tipped central tail feathers.* HAWAII, MARCH

LIGHT-MORPH BREEDING ADULT. *Note hefty build and thick bill. Broad breast band often well-defined. Spoon-shaped central tail streamers are diagnostic if present and seen well.* CALIFORNIA, MAY

DARK-MORPH BREEDING ADULT. *Mainly dusky brown below; uniformity of coloration disrupted by usually prominent white flash at base of outer primaries.* NORTH CAROLINA, MAY

IMMATURE. *Most immatures are dark overall, lacking either warm tones or substantial pale areas. Bulky build, thick bill, and blunt-tipped tail are always useful marks.* CALIFORNIA, SEPTEMBER

L 18.5"	ws 52"	wt 1.5 lb	♀ > ♂

- ▸ two adult molts per year; simple alternate strategy
- ▸ strong age-related and seasonal differences
- ▸ polymorphic; light morph is most frequent

Nests on tundra; migrates mainly offshore to pelagic wintering grounds. Regular in small numbers inland in fall, generally later than Long-tailed and Parasitic Jaegers. Summer diet consists almost entirely of lemmings; at sea, eats squid and fish stolen from other birds; also consumes small sea birds and carrion.

Long call, heard on breeding grounds, repeated *wup* or *wuk* notes, changing speed midway through series. At sea, occasional *yip* and *wheep* and *gitchoo* calls.

LIGHT-MORPH BREEDING ADULT. *Bill large and thick, with prominent hook and relatively sharp gonydeal angle. Bill distinctively bicolored: pale at base, dark at tip.* NORTH CAROLINA, MAY

Parasitic Jaeger
Stercorarius parasiticus CODE 1

LIGHT-MORPH BREEDING ADULT. *Pointy central tail streamers, if present, an excellent mark. Cap paler and less extensive than Pomarine. Except for flash mark, upperwing uniform.* CALIFORNIA, SEPTEMBER

DARK-MORPH ADULT. *Even when streamers are absent, tail of adult is distinctively pointed. Note uniform upperwing; compare Long-tailed. Pomarine has heftier build.* NEW JERSEY, MAY

HEAD DETAIL. *With all jaegers, if close observation is possible, exact bill structure is important to note. Gonydeal angle is near the tip of Parasitic's long, slender bill.* ALASKA, JUNE

L 16.5"	ws 46"	wt 1 lb	♀ > ♂

▸ two adult molts per year; simple alternate strategy

▸ strong age-related and moderate seasonal differences

▸ polymorphic on a complex geographic basis

The most "falcon-like" of jaegers; seen in twisting flight swooping after seabirds, forcing them to disgorge food. On breeding grounds, robs eggs and hunts birds and mammals. Migrants are often seen from shore; regular in fall on Great Lakes, scarce elsewhere inland.

Long call a series of *hya* notes; slurred, higher-pitched than Pomarine Jaeger. In attack, gives sharp *wek* notes; infrequent high-pitched calls at sea.

MOLTING IMMATURE. *The broad-based wings and the triangular-shaped head, which seems notably small for the bird's size, can help to separate Parasitic from other jaegers in flight.* AUSTRALIA, NOVEMBER

LIGHT-MORPH BREEDING ADULT. *All jaegers flash white in the primary shafts, Long-tailed with just two or three, Pomarine with up to six, Parasitic (here) intermediate.* ALASKA, JUNE

Long-tailed Jaeger
Stercorarius longicaudus CODE 1

NONBREEDING ADULT. *Pale smoky gray below. Wing pattern usually diagnostic: above, gray coverts paler than dark flight feathers; below, uniformly dark with little if any wing flash.* CALIFORNIA, AUGUST

BREEDING ADULT. *Tail streamers very long, but often broken or missing. Black cap less extensive but more sharply delineated than other jaegers. More buoyant in flight than other jaegers.* ALASKA, JUNE

LIGHT-MORPH JUVENILE. *Most are conspicuously barred below; barring most prominent across vent and on underwing coverts. All morphs have cold or neutral ground color.* CALIFORNIA, AUGUST

HEAD DETAIL. *Bill slight and short. Upper mandible straight at base, curving downward toward tip; lower mandible also straight, showing slight gonydeal angle.* ALASKA, JUNE

15"	WS 43"	WT 11 OZ	♀ > ♂

· two adult molts per year; simple alternate strategy

· strong age-related and moderate seasonal differences

· juveniles are polymorphic, but adults are not

Breeds far north, favoring dry tundra and barrens. Winters farther south than other jaegers. Migrants at sea usually well offshore. Small numbers are seen inland in late summer; inland movement is earlier than other jaegers. Does not usually harass other birds.

LIGHT-MORPH BREEDING ADULT. *Smallest and slightest of the jaegers. Feeds on small mammals (lemmings and voles) during breeding season on tundra.* ALASKA, JUNE

Long call a series of downward notes, *kreer* and *krech*. In agitation, gives high, clipped *cheep!* Usually silent at sea.

Auks, Murres, and Puffins

FAMILY ALCIDAE

22 species: 11 CODE 1   8 CODE 2   1 CODE 3   1 CODE 4   1 CODE 6

In many respects the Northern Hemisphere counterparts of penguins, the alcids—as they are often called—are gregarious, mainly fish-eating, largely black-and-white seabirds. The relationship of the alcids to other taxa is unclear; their closest kin within the large order Charadriiformes appear to be the behaviorally and morphologically different jaegers and skuas. Within the family Alcidae, the most useful taxonomic boundaries for birders in the field are at the genus level; species within a genus are often confusingly similar, whereas species in different genera are rarely confused in the field.

All but two of the world's alcids have occurred in North America, and the family reaches its greatest diversity in the northwestern Pacific Ocean region, especially Alaska. Most alcid species breed in dense colonies on remote sea cliffs or offshore islands, and they feed on the open ocean. Alcids are rarely observed far inland. Two murrelets—the Ancient and, surprisingly, the Long-billed (which breeds only in Asia)—are regular inland in very small numbers in the late fall. All other species are best thought of as casual or accidental inland away from the breeding colonies. Most alcids undergo regular seasonal migrations, and several species disperse hundreds or even thousands of miles from the breeding colonies to winter feeding grounds far beyond the continental shelf.

What they lack in colorful plumage, alcids more than make up for in their outlandishly proportioned and colorful bills—think of the differences in bill structure among Razorbill, Black Guillemot, Dovekie, Rhinoceros Auklet, and Horned Puffin. Alcid bills play a role in courtship, but they are also related to feeding strategies. Most species are fish-eaters, but the two smallest species (the distantly related Dovekie and Least Auklet) strain plankton with their small bills.

Because of their colonial lifestyle, alcids both enjoy safety in numbers and are vulnerable to catastrophic events at the breeding colonies. In the past, hunting took a terrible toll on many alcid species; for example, the Great Auk, North America's only post-Columbian flightless bird, was exterminated by hunters. More recently, introduced predators—especially Arctic foxes on the Aleutian Islands—have been disastrous to colonies of alcids and other seabirds. Additional threats to alcids include oil spills and other forms of ocean pollution, collapse of regional fish populations caused by overharvesting, and possibly climate change.

Common Murre

Uria aalge CODE 1

BREEDING ADULT, BRIDLED MORPH. *Adult bill longer, straighter, and thinner than Thick-billed. Most individuals lack this bird's precisely delineated "bridle."* NEWFOUNDLAND, JULY

NONBREEDING ADULT. *Face acquires extensive white; thin black line extends back from eye. Immature similar to nonbreeding adult, but adult bill structure often not fully developed.* CALIFORNIA, OCTOBER

17.5"	ws 26"	wt 2.2 lb

- two adult molts per year; complex alternate strategy
- strong seasonal and weak age-related differences
- bridled morph in East; variable dark morph in West

One of the most numerous marine birds in the Northern Hemisphere, but populations are threatened by ocean pollution and climate change. Nests on sea cliffs; expert diver. Off West Coast, often seen along shore; off East Coast, sightings usually farther out at sea.

Colonies noisy. Amid general bustle, listen for loud growls, rattles, and moans. Juvenile at sea gives loud squeal: *rurrr urrr...* and *rip* and *r'r'r'r'rip.*

Thick-billed Murre

Uria lomvia CODE 1

BREEDING ADULT. *Thick-based bill has downward arch and horizontal white stripe. White of breast pinched off in sharp point on neck; border between breast and neck more rounded on Common.* ALASKA, JUNE

NONBREEDING ADULT. *Face whiter than breeding adult; lacks white behind eye and black line on cheek of Common Murre. Subadult similar, but bill often not fully developed.* CALIFORNIA, NOVEMBER

L 18"	ws 28"	wt 2.1 lb

- two adult molts per year; complex alternate strategy
- moderate seasonal and weak age-related differences
- Pacific birds larger, longer-billed, longer-winged

Nests on steep sea cliffs; generally more northerly than Common Murre, but considerable overlap. Foraging and social habits much as Common. Several colonies destroyed in 1900s were never reestablished. Formerly, but no longer, a regular transient on eastern Great Lakes.

Calls at colonies are guttural, faint or loud: deep *woo* and groaning *arrrr.* Gives series of muffled *woof* notes followed by rising *whoaaaaa!*

Razorbill
Alca torda CODE 1

BREEDING ADULT.
Bill laterally
compressed and
oversized; all black
except for thin,
precise, vertical,
white stripe toward
tip. Head black
except for thin white
line between eye
and bill. MAINE, JUNE

SUBADULT. Bill large, but not as distinctive as adult. All plumages are
black above and white below; subadult and nonbreeding adult have
dusky white smudge behind eye. NEW JERSEY, JANUARY

L 17"   WS 26"   WT 1.6 lb

▸ two adult molts per year; complex alternate strategy

▸ moderate age-related and seasonal differences

▸ bill structure and overall body size are variable

Breeds mainly in Iceland, Greenland; small, scattered colonies in
North America, south to Maine.
Winters at sea south to mid-
Atlantic; small numbers sometimes
at inlets and jetties. Propelled by
wings, dives after schools of fish;
flight surprisingly strong and agile.

At colonies, gives rising or falling low-pitched growls of
variable duration; for example *owwww-aww*. Juveniles at sea
sometimes whistle at adult male companion.

Dovekie
Alle alle CODE 2

BREEDING ADULT.
Black hood sharply
set off from pale
underparts.
Diminutive overall:
small-bodied and
stub-billed. In flight
appears pointier-
winged than most
other alcids.
NORTHWEST TERRITORIES,
JULY

NONBREEDING ADULT. Throat and neck mainly white; whitish behind
eye; immature similar but more smudgy. All Dovekies at rest show a
variable white wedge rising from breast. MASSACHUSETTS, MARCH

L 8.25"   WS 15"   WT 6 oz

▸ two adult molts per year; complex alternate strategy

▸ strong seasonal and moderate age-related differences

▸ extent of white variable, especially on juveniles

Most abundant North Atlantic alcid;
winters in huge flocks well offshore.
Regular south to mid-Atlantic, but
largest concentrations off eastern
Canada. Occasionally seen inland,
sometimes in large numbers, after
early-winter storms. Heavily hunted
by native peoples.

Vocal at remote breeding colonies; gives thin, strident *rrri-
chichichi...* and short, high-pitched *dree* and *pre* notes. Usually
silent at sea.

Black Guillemot
Cepphus grylle CODE 1

BREEDING ADULT.
Huge white oval on wing contrasts with otherwise jet black plumage. Feet bright red. EUROPE, JULY

NONBREEDING ADULT.
Body extensively white, usually with overall "dirty" aspect; subadult similar, but darker overall. Flight feathers of wing remain black in nonbreeding plumage.
NEW JERSEY, DECEMBER

IN FLIGHT. *White wing coverts, below and especially above, stand out on flying birds, even across great distances.*
EUROPE, AUGUST

L 13"   WS 21"   WT 15 oz

▸ two adult molts per year; complex alternate strategy
▸ strong seasonal and moderate age-related differences
▸ body size and extent of white vary geographically

Among East Coast alcids, the most likely to be seen from land; perches on rocks and cliffs, flies across inlets. In winter does not wander far south or out to sea, in contrast to other alcids. Although a common sight from New England north, world population is low.

Common call a pure-tone, high-pitched, slightly descending *peeeeeee*; a characteristic sound of the rocky shoreline of the Northeast.

Pigeon Guillemot
Cepphus columba CODE 1

BREEDING ADULTS. *Similar to Black Guillemot, but white oval patch on wing broken by black wedge.* CALIFORNIA, AUGUST

IMMATURE.
Body and neck predominantly white and pale dusky. Head paler and white on wing more extensive than juvenile.
CALIFORNIA, DECEMBER

JUVENILE. *Dusky gray-brown above in both guillemot species; oval on wing of adult lacking or indistinct on juvenile. Pigeon averages more extensively brown above than Black.*
ALASKA, FEBRUARY

L 13.5"   WS 23"   WT 1 lb

▸ two adult molts per year; complex alternate strategy
▸ strong seasonal and moderate age-related differences
▸ from south to north: larger, longer-winged, and whiter

West Coast counterpart of Black Guillemot. Nests in small colonies or as isolated pairs. Easily seen from coastal headlands and harbors. Feeds in shallow water inshore, diving straight to the bottom. Like Black Guillemot, frequently sighted but total numbers are low.

Calls similar to Black: *pweeeee* and *pwee-t't't't't't*; also shorter chatters. Vocal; often heard in sustained flight or when coming in for a landing.

Atlantic Puffin
Fratercula arctica CODE 1

BREEDING ADULT. Bill bizarre: laterally compressed, highly colorful, and exceptionally blunt. Pale face surrounded by black cap, nape, and throat. MAINE, JULY

NONBREEDING. Face grayer than breeding adult; bill smaller and duller. Plumage of nonbreeding adult similar; bill intermediate between subadult and breeding adult. VIRGINIA, FEBRUARY

IN FLIGHT. Wholly black underwing contrasts with white underparts. Other mid-sized East Coast alcids have pale gray or whitish underwings. NEWFOUNDLAND, AUGUST

L 12.5"   ws 21"   wт 13 oz

▸ two adult molts per year; simple alternate strategy

▸ strong seasonal and moderate age-related differences

▸ gradual increase in body size from south to north

Formerly hunted to near-extinction, but now an icon of the Northeast coast; a major tourism draw, benefiting from intensive reintroduction efforts. Except around colonies, sightings from land are rare. Flight is high and direct.

At colonies gives deep, rich, low moaning, suggesting a distant motor or cow: *urrrrrrr* or rising-then-falling *urrrrr-urrrp*. Generally silent at sea.

Horned Puffin
Fratercula corniculata CODE 1

BREEDING ADULT. Similar to Atlantic Puffin, but no range overlap. Bill even more outlandishly outsized than Atlantic, with substantial dull yellow at base; Atlantic mainly dark at base. ALASKA, JULY

NONBREEDING ADULT. A less gaudy version of the breeding adult: bill reduced in color and size; face dusky gray. Bill of subadult is smaller and less colorful. CALIFORNIA, MAY

IN FLIGHT. Underparts are white below in all plumages, a key point of distinction on distant flying puffins; Tufted Puffin underparts completely black. ALASKA, JULY

L 15"   ws 23"   wт 1.4 lb

▸ two adult molts per year; simple alternate strategy

▸ moderate seasonal and age-related differences

▸ body size averages greater from south to north

Western counterpart of Atlantic Puffin. Breeds colonially on northern sea cliffs, mainly Alaska; aggressive in colonies and at sea. After breeding, moves well out to sea; southern limit of nonbreeders' regular occurrence is poorly known. Numbers apparently declining.

Vocalizations similar to Atlantic Puffin; characteristic call a deep rumbling growl. Also gives a variety of higher, mainly short calls

Tufted Puffin
Fratercula cirrhata CODE 1

BREEDING ADULTS. *Wispy golden tufts protrude the length of the neck. Face unmarked white; face much duskier in nonbreeding plumage. Bill substantially orange throughout year.* ALASKA, JULY

SUBADULT. *Sooty gray all over, including underparts. Bill smaller than breeding adult, but impressively large; shows substantial orange or orange-yellow.* WASHINGTON, APRIL

IN FLIGHT. *All plumages very dark in flight; other puffins show extensive white below. Facial features usually prominent from afar.* ALASKA, JULY

15" ws 22" wt 1.7 lb

▸ two adult molts per year; simple alternate strategy
▸ strong seasonal and moderate age-related differences
▸ adult wing length and body mass variable

Breeds on steep, rocky sea cliffs; in vicinity of nest sites, can be glimpsed from shore. Otherwise, sightings are generally out at sea; after breeding, most head beyond shallow shelf waters for deep ocean. Some colonies have been wiped out by introduced predators.

Noisy in colonies and during interactions at sea. Characteristic call a deep, wavering growl: *purr-r-r-r-r-r.* Gives shorter, higher notes, too.

Rhinoceros Auklet
Cerorhinca monocerata CODE 1

BREEDING ADULT. *At a distance, looks dark and featureless. Close up, note elaborate breeding paraphernalia: two white plumes on face; pale plate protruding from upper mandible.* BRITISH COLUMBIA, JULY

JUVENILE. *A little paler below than above. Bill is stouter than most mid-latitude West Coast seabirds, but thinner than puffins. Nonbreeding adult similar, but bill thicker.* BRITISH COLUMBIA, AUGUST

IN FLIGHT. *Pale belly contrasts with dark underwing and breast. Appears robust—about the size of a puffin—in flight.* CALIFORNIA, JANUARY

L 15" ws 22" wt 1 lb

▸ two adult molts per year; simple alternate strategy
▸ moderate age-related and weak seasonal differences
▸ size of horn on adult bill is variable

A puffin-like seabird. Favors warmer waters and is more likely to be seen from shore in inlets and large harbors, or just offshore. Tends nest mainly at night. New colonies have recently been established in Oregon and California.

Vocal at colonies, less so at sea. Gives various groans and grunts, often in crescendo series: *woe woe whoa whoa!* Also gives sharper *weep* and *whip* notes.

Cassin's Auklet
Ptychoramphus aleuticus   CODE 1

L 9"	WS 15"	WT 6 OZ

- two adult molts per year; complex alternate strategy
- weak age-related and seasonal differences
- body mass and length increase from south to north

Nests widely on remote offshore islands; numbers have dropped where predators have been introduced. Tends to nest at night. Forms large flocks well offshore, but usually singly or in small flocks closer to shore. Migratory in northern portion of range, sedentary south.

Colonies noisy; birds at sea generally quiet. Many calls have distinct k– sounds: gasping *koooch koooch...* and hoarse *koook hoook...*

ADULT. *Dark and dumpy. Short white eyebrow exhibited by all plumages; slightly more prominent on breeders. Adults have pale eye; subadult eye usually dark.* CALIFORNIA, JUNE

IN FLIGHT. *Dark ashy gray all over, a little paler below. Mainly dark underwing flashes a paler gray patch.* CALIFORNIA, AUGUST

Ancient Murrelet
Synthliboramphus antiquus   CODE 2

L 10"	WS 17"	WT 7 OZ

- two adult molts per year; simple alternate strategy
- moderate seasonal and weak age-related differences
- body size averages larger in north

Young reared at sea; wanders widely after breeding, mainly in shallow waters just offshore. Rarely seen beyond continental shelf; often occurs at inlets and outflows, regularly in Puget Sound. A few are detected far inland every year, usually in late fall.

"Song" more elaborate than most alcids: begins with shorebird-like elements, *brrrrree-deet*, followed by twittering *bree* notes.

BREEDING ADULT. *Largest and most distinctive murrelet: head black with wispy white plumes (whence "ancient"); bill yellow and stubby; back gray; white underparts extend onto face.* CALIFORNIA, JANUARY

NONBREEDING. *In all plumages, contrast between gray back and black head usually obvious. White head plumes reduced or absent on subadult and nonbreeding adult.* CALIFORNIA, FEBRUARY

Xantus's Murrelet
Synthliboramphus hypoleucus   CODE 2

L 9.75"	WS 15"	WT 6 OZ

- two adult molts per year; simple alternate strategy
- weak differences between fresh and worn adults
- northern and southern races possibly distinct species

Nests on arid islands in southern waters; disperses north, well offshore, regularly to northern California, rarely to southern Canada. Occurs singly or in small numbers. Has been severely affected by introduced predators, but often responds well to predator removals.

All notes harsh, a ringing *cheep* and a short *chip*; calls often run together. Noisy at colonies; parent-offspring duos vocal at sea.

ADULT, NORTHERN. *Less black on face than on Craveri's: white extends to base of bill; white also extends a short distance in front of eye. Bill a little shorter than Craveri's. Southern population (not shown) has more white around eye; bill longer than northern.* CALIFORNIA, AUGUST

Craveri's Murrelet
Synthliboramphus craveri CODE 3

L 9.5"   WS 15"   WT 6 OZ

▸ two adult molts per year; simple alternate strategy

▸ weak differences between fresh and worn adults

▸ underwing variable, sometimes with white blaze

Southern counterpart of Xantus's Murrelet. Occurs regularly only off Baja California; nonbreeders range into California waters, rarely farther north. Northward dispersal depends on water temperature, influenced by El Niño; very scarce or absent in some years.

Vocalizations similar to Xantus's, but average thinner, shriller; in agitation, *chip* and *chee* notes run together in insect-like trill.

ADULT. *Dark on face is more extensive than Xantus's: black hood drops to just below eye level and extends to chin under base of bill. Bill longer and thinner than Xantus's.* CALIFORNIA, OCTOBER

IN FLIGHT. *Underwings dark; both subspecies of Xantus's have white underwings. Base of black hood curls out onto breast, forming a partial collar.*
CALIFORNIA, OCTOBER

Marbled Murrelet
Brachyramphus marmoratus CODE 1

MOLTING ADULT. *Fine wavy brown ("marbled") all over. Throat a little paler than face, but contrast low. Bill short; typically pointed upward on swimming bird.* BRITISH COLUMBIA, SEPTEMBER

NONBREEDING ADULT. *Dark above, white below. White on neck bulges back on nape; dark sides of swimming bird marked with white horizontal bar.* ALASKA, FEBRUARY

L 9.5"   WS 17"   WT 8 OZ

▸ two adult molts per year; complex alternate strategy

▸ strong seasonal and moderate age-related differences

▸ white on face variable in nonbreeding plumage

Northern birds nest on the ground; south of Prince William Sound, Alaska, nests high in conifers in old-growth forests. Difficult to glimpse around wooded breeding grounds, but fairly conspicuous otherwise: lone individuals frequent large bays, inlets, and breakers.

En route to and from arboreal nests, gives a ringing, pure-tone *keeeeer*, often repeated; once learned, a characteristic call of dense coastal forests.

IN FLIGHT. *Underwing smoky gray in all plumages. In nonbreeding plumage, as here, wedge of black extending from neck onto breast often prominent in flight.*
ALASKA, OCTOBER.

LONG-BILLED MURRELET (B. perdix). *Casual Asian visitor to North America, including sites well inland, especially in late fall. Nape extensively black in nonbreeding plumage.*
CALIFORNIA, AUGUST

Kittlitz's Murrelet
Brachyramphus brevirostris CODE 2

NONBREEDING ADULT. *Bill very small. Whiter overall than Marbled; black on head limited to crown. Subadult similar but duskier.*
BRITISH COLUMBIA, FEBRUARY

IN FLIGHT. *Underwing dark in all plumages. Breeding adult, shown here, paler below than Marbled. Belly and vent white; white outer tail feathers diagnostic, but difficult to see.* ALASKA, JUNE

L 9.5"	WS 17"	WT 8 OZ

- two adult molts per year; complex alternate strategy
- strong seasonal and moderate age-related differences
- geographic variation poorly known; possibly extensive

A little-studied seabird; potentially imperiled by climate change, pollution, and fisheries. Nests on ground at barren sites such as hilltops and glacier edges, up to 50 miles from sea. Breeders forage in glacial outflows; apparently moves out to open ocean in winter.

Apparently lacks long, wailing calls of Marbled Murrelet; most frequent call a short groan, *urrrn* or *urr-un*, fading at end. Also a nasal *ack*.

Parakeet Auklet
Aethia psittacula CODE 2

BREEDING ADULT. *Largest Aethia auklet. Bill orange, bulbous, and upturned; long white plume extends back from eye. Bill duller in nonbreeding plumage; black on subadult.* ALASKA, JULY

IN FLIGHT. *In breeding plumage, as here, resembles Rhinoceros Auklet; white belly, but otherwise dark. In nonbreeding plumage, white extends from belly all the way to throat.* ALASKA, AUGUST

L 10"	WS 11"	WT 11 OZ

- two adult molts per year; simple alternate strategy
- moderate seasonal and age-related differences
- bill color variable; facial plume sometimes incomplete

Breeds on rocky cliff faces like other northern alcids, but tends not to be gregarious, especially at sea, where usually seen singly. Extent of postbreeding dispersal little known; possibly well south far out at sea. Introduced predators have wiped out some colonies.

Noisy at nest and during interactions at sea; gives grating *chu* and *chee* notes, often building into an ascending trill: *chu chu chee chee ch'ch'ch'...*

Crested Auklet

Aethia cristatella CODE 2

L 10.5" ws 17" wt 10 oz

- two adult molts per year; simple alternate strategy
- moderate age-related and seasonal differences
- overall darkness of ground color variable

Gregarious: nests in dense colonies on steep sea cliffs; forages and migrates in huge flocks. Flight strong and straight. Details of seasonal movements poorly known; after breeding, most go to deep water at sea. Some colonies have been hit hard by introduced foxes.

Exudes a highly aromatic "perfume," readily detected by humans. Vocalizations diverse: goose-like honking, shorebird-like chattering, and dog-like yapping.

BREEDING ADULT. *Dark and chubby. Has upturned orange bill and long white plume trailing back from eye; also has long black plumes protruding from forehead.* ALASKA, JUNE

IMMATURE. *Compared to breeding adult, bill less colorful and facial and forehead plumes reduced.* ALASKA, JUNE

Whiskered Auklet

Aethia pygmaea CODE 2

L 7.75" ws 14" wt 4.2 oz

- one adult molt per year; complex basic strategy
- moderate age-related and seasonal differences
- body size increases from east to west in Aleutians

A scarce, little-known alcid of the Aleutian chain. Visits remote breeding colonies at night. Gathers in large assemblies offshore, but not far out to sea; sometimes singly or in small flocks close to shore. Introduced rats and foxes have caused local population losses.

Vocal around nest: gives a sad *ink* like a toy trumpet; also *buh-dink* or *biddy-wink* like a bicycle horn.

BREEDING ADULT. *Facial plumages more elaborate than other* Aethia *auklets. Small and dark; not much larger than Least Auklet. Immature similar to Crested, but smaller and paler.* PACIFIC OCEAN, JUNE

Least Auklet

Aethia pusilla CODE 2

L 6.25" ws 12" wt 3 oz

- two adult molts per year; simple alternate strategy
- strong seasonal and moderate age-related differences
- polymorphic; darkness of adult plumage variable

Although not closely related to Dovekie, Least Auklet is in many ways its Bering Sea counterpart: an abundant, small-bodied, voracious plankton feeder. Huge near-shore flights and at-sea foraging flocks often encountered. Nests on rocky cliffs and headlands.

Colonies or flocks at sea are audible at great distances. Calls varied, but most are harsh, nasal, and clipped: *cheer* and *churr* and *churn*.

BREEDING ADULT. *Small overall; bill stubby and reddish. All plumages have white throat. Immature and nonbreeding adult extensively white below; breeding adult variably dark below.* ALASKA, JULY

Pigeons & Doves

ORDER COLUMBIFORMES

The world's more than 300 extant species in the order Columbiformes are lumped together in a single family: Columbidae. Birds in this family are found worldwide, with the greatest diversity in Southeast Asia and the Australasian region. The Columbiformes are one of several orders of "near-passerines" that apparently are not closely related to the passerines; their placement just before the huge order Passeriformes is basically a matter of convenience. The terms "pigeon" and "dove" do not correspond to precise taxonomic groupings in our area; instead, these names are employed loosely to denote overall body size: Birds called pigeons tend to be larger than those called doves. On this note, the well-known Rock Pigeon was until recently called the Rock Dove; its name was changed partly because it is one of our larger columbids.

Pigeons and doves are upland species, typically found feeding on the ground or in trees. Singles or small flocks often gather in clearings or under bird feeders, where they look for fallen seed. In other settings, small flocks forage in the midstory or canopy for fruits and nuts. Pigeons and doves tend to feed slowly and methodically, and it is easy to overlook quietly foraging ground-doves just underfoot or even a flock of big Band-tailed Pigeons in a nearby tree.

Most North American pigeons and doves are either nonmigratory or so-called "partial migrants" that withdraw from only the northern portions of their range. However, several species are prone to vagrancy; the White-winged Dove, Common Ground-Dove, and Band-tailed Pigeon are recorded annually far out of range. Many species are strong fliers, with a distinctive "flattop" flight profile: the crown, back, and tail all form a straight line.

Population status of the pigeons and doves is a mixed bag. Some species—both native and exotic—are increasing in number and expanding their ranges. These include the Mourning and White-winged Doves, Inca and Ruddy Ground-Doves, and especially the Eurasian Collared-Dove. Others are in decline, such as the Band-tailed Pigeon, Common Ground-Dove, and Spotted Dove. The extinct Passenger Pigeon was, as far as we know, the most abundant bird species in North America. And the Dodo of the Indian Ocean island of Mauritius, a very large member of the order Columbiformes, is perhaps the most famous extinct species. The diverse fortunes of the pigeons and doves are, in general, closely related to human activities—among them changing land-use patterns, domestication and propagation, and deliberate eradication.

Rock Pigeon
Columba livia CODE 1

VARIANTS. *Dark, pied, brown ("red"), and other variants are common; note that wild-type adults are in the minority in this photo.* CALIFORNIA, OCTOBER

L. 12.5" WS 28" WT 9 oz

- one adult molt per year; complex basic strategy
- weak age-related and seasonal differences
- artificially selected color morphs vary greatly

The familiar city pigeon; occurs in the most heavily degraded and decrepit urban habitats imaginable, but can also be found nesting on remote sea cliffs and inland rimrocks. Powerful; capable of fast, sustained flight. Seen singly or in flocks of up to several hundred.

Cooing is low-pitched and syncopated, often building to a brief climax: *b'b'buh-buh-buh-roo! roo!* Effect is ventriloqual; cooing can carry far.

WILD-TYPE ADULT. *Plump and bullnecked. Neck iridescent green and purple, puffed out in male display. At rest, pale wings usually show two black bars.* CALIFORNIA, JANUARY

IN FLIGHT. *Wings pointed; Rock Pigeon in flight can be confused with falcons. Often soars briefly. Below, pale wings contrast with dark body; above, note white rump.*

BRITISH COLUMBIA, FEBRUARY

White-crowned Pigeon
Patagioenas leucocephala CODE 1

L. 13.5" WS 24" WT 10 oz

- one adult molt per year; complex basic strategy
- strong age-related and weak sex-related differences
- feeding birds often darkened or splotched by fruit

A fruit-eater; occurrence and movements in Florida and elsewhere closely tied to availability of food resources. Wary around people, but also attracted to lush plantings in residential and light commercial districts. Numbers have rebounded from earlier losses.

Cooing typically has three elements: *ooo! uh-ooh!*—usually repeated. Has whistled, breathy quality; higher-pitched than Rock Pigeon.

ADULT. *Extensive white on crown contrasts with otherwise dark blue-black plumage. Juvenile all dark; differs from variant dark Rock Pigeon by shorter wings, longer tail, and red bill.* FLORIDA, MAY

Band-tailed Pigeon

Patagioenas fasciata CODE 2

L 14.5"	ws 26"	wt 13 oz	♂ > ♀

▸ one adult molt per year; complex basic strategy

▸ strong age-related and weak sex-related differences

▸ interior birds paler and smaller than West Coast

Despite physical resemblance to Rock Pigeon, is associated mainly with remote mountain forests. Generalist in diet; finds food in natural settings, but also visits feeders. Does not usually venture far from wooded landscapes, even on migration. Populations in decline.

Cooing averages less complex than Rock Pigeon: rising *oooh* or *oo-ooh*, usually one syllable or slurred two syllables. Some calls closely match Rock Pigeon.

ADULT. *White hind-collar stands out on dark head. Iridescence on nape below hind-collar prominent in courtship display; muted otherwise. Bill yellow; eye dark.* CALIFORNIA, MARCH

JUVENILE. *Lacks the white hind-collar of adult. Upper parts have fine scaling, diminished or absent on adult.*
CALIFORNIA, JULY

IN FLIGHT. *In all plumages, outer half of tail paler (whence "band-tailed") than inner half.*
BRITISH COLUMBIA, SEPTEMBER

Red-billed Pigeon

Patagioenas flavirostris CODE 3

ADULT. *Extensive wine-colored hues, if seen well, are distinctive. Eyes and feet red; red bill has yellow tip. Juvenile is dull purple-gray all over.*
COSTA RICA, MARCH

L 14.5"	ws 24"	wt 11 oz

▸ one adult molt per year; complex basic strategy

▸ moderate age-related and weak sex-related differences

▸ extent of wine-colored gloss on head and breast variable

Fruit-eating tropical species that barely reaches our area in south Texas; extent of withdrawal in winter not perfectly known. Spends much of its time quietly feeding in dense treetops in thick woodlands; often difficult to spot. Numbers have declined with habitat loss.

Vocal throughout much of year, but tends to stay well-hidden while singing; cooing whistled, wavering, and breathy: *ooo-wuck-ah woooooo*, repeated.

Spotted Dove
Streptopelia chinensis CODE 1

L 12"   WS 21"   WT 6 OZ
- one adult molt per year; complex basic strategy
- moderate differences between juvenile and adult
- sometimes stained, especially below, from eating fruits

Introduced to Los Angeles from Eurasia around 1915; rapidly became established, spread into Central Valley and along coast. Numbers have declined recently. Prefers non-natural habitats such as tree-lined residential streets and agricultural windbreaks.

Vocal throughout much of year. Characteristic cooing a repeated *wunka-woo!* or *awunka!-woo*; has harsh and vibrant quality, audible over city noise.

ADULT. *Black nape has coarse white spotting. Juvenile, with plain nape, differs from superficially similar Mourning Dove by dark, rounded tail with white corners, prominent in flight.* HAWAII, OCTOBER

Eurasian Collared-Dove
Streptopelia decaocto CODE 1

L 13"   WS 22"   WT 7 OZ
- one adult molt per year; complex basic strategy
- weak differences between juvenile and adult
- hybridizes with exotic African Collared-Dove

Native to Eurasia; released in Bahamas in 1970s. Subsequent explosion across North America has been astonishingly rapid and successful. Invading populations favor small towns, suburbs, and ranches. Influence, if any, on native dove populations remains to be determined.

Cooing syncopated, usually consisting of three syllables; *hoo-hoo-hoo!* or *hoo-hoo!-ah*, repeated. Sings frequently, but unobtrusively.

ADULT. *Gray-tan, with black hind-collar. Larger and paler than Mourning Dove; in flight, tail is squared off, with broad white tips. Juvenile similar, but slightly scalier above.* COLORADO, MARCH

AFRICAN COLLARED-DOVE (S. roseogrisea). *Old World species; widely encountered in North America, but populations never become established. Smaller and paler than Eurasian Collared-Dove.* AFRICA, JANUARY

JUVENILES. *Pale tail tip contrasts with dusky undertail coverts; African Collared-Dove has white undertail coverts.* EUROPE, JULY

White-winged Dove
Zenaida asiatica CODE 1

ADULT. *Plain gray-tan like Mourning Dove, but larger; squared-off tail has broad white tip. White wing patch appears as thin white bar on bird at rest. Juvenile slightly scalier above.* ARIZONA, JULY

IN FLIGHT. *Below, black tail shows a broad band of blunt white tail tips. Above, white crescents on wing are striking, and contrast sharply with black flight feathers.* ARIZONA, AUGUST

L 11.5"	WS 19"	WT 5 OZ

▸ one adult molt per year; complex basic strategy

▸ weak differences between adult and juvenile

▸ paler and grayer in east; smaller in Texas

A flourishing species: population increasing; range expanding north in interior, also spreading along Gulf coast. Habitat change and warmer winters seem to be beneficial. Native to woodlands, but readily adapts to residential and commercial districts with planted trees.

Syncopated cooing, *ooh-coo!-ahoo!* ("Who cooks for you?"), is a characteristic sound of summer in towns and cities in the Desert Southwest.

White-tipped Dove
Leptotila verreauxi CODE 2

ADULT. *Plain overall. At rest, darker brown wings contrast with gray-brown body. Iridescence on neck can be prominent in courtship; muted otherwise. Juvenile slightly plainer.* TEXAS, NOVEMBER

ADULT. *In flight, appears plainer than either White-winged or Mourning Dove. Lacks bold white wing crescents of White-winged; tail has less white than either species.* TEXAS, MAY

L 11.5"	WS 19"	WT 5 OZ

▸ one adult molt per year; complex basic strategy

▸ weak differences between juvenile and adult

▸ color of bare parts, intensity of iridescence variable

Common in the tropics, barely reaches south Texas. Feeds quietly on the ground, where it eats fruits and seeds. Occurs in wooded habitats, especially near water. Prefers to hide in the shadows, along edges of trails and clearings. Wary, but can be closely approached.

Cooing is a mournful, drawn-out *ooh-wooooooo* or *ooh-wahooooo*, as if exhaled. Soft, but carries far; imparts a "tropical" feel to south Texas woodlands.

Mourning Dove
Zenaida macroura CODE 1

| L 12" | WS 18" | WT 4.2 oz | ♂ > ♀ |

▸ one adult molt per year; complex basic strategy
▸ moderate differences between juvenile and adult
▸ eastern birds larger and darker than western birds

Common to abundant throughout much of its extensive range; one of our most familiar birds. Adaptable; flourishes in large cities and in fragmented and degraded habitats, as well as in remote desert and ranch country. Absent only from extensively forested landscapes.

◀ Lilting *ooo-woo!-woo-woo* cooing is one of the first sounds of spring in many places; sounds soft, but carries far. Wing whistle is choppy but musical.

ADULT. *Gray-buff with long, pointed tail. Dark eye stands out on plain face. Wings have variable black spotting. Iridescent pink on neck prominent in display, muted otherwise.*
CALIFORNIA, NOVEMBER

ADULT IN FLIGHT. *White-tipped tail feathers prominent in short flights; in sustained flight, tail long and tapered.*
ARIZONA, AUGUST

JUVENILE. *Small and scaly; similar to ground-doves.* ARIZONA, SEPTEMBER

Ruddy Ground-Dove
Columbina talpacoti CODE 3

| L 6.75" | WS 11" | WT 1.4 oz |

▸ one adult molt per year; complex basic strategy
▸ moderate differences between male and female
▸ western birds duller than vagrants to south Texas

Habits and habitats generally as Common Ground-Dove, but current fortunes of the two species are reversed: Common generally declining, Ruddy increasing and pushing northward. Most U.S. records are of small numbers in fall; small colonies sometimes persist all year.

Cooing is a rapidly repeated *ga-yoop* or *oo-ooo-oop*, swelling midway through and then trailing off; lower-pitched, less obtrusive, and faster than Common Ground-Dove.

ADULT MALE. *Similar to Common Ground-Dove, but longer-tailed and a little larger; has dark bill (Common has red bill) and lacks scaling on breast and head. Female less reddish.* CALIFORNIA, OCTOBER

Inca Dove
Columbina inca CODE 1

ADULT. *Long-tailed like Mourning Dove, but noticeably smaller. Body has coarse scales all over, a little more prominent on male than female; a little less prominent on juvenile.* TEXAS, MAY

ADULT. *Bright rusty is conspicuous on spread wing. White border on long tail is usually concealed except in flight.* ARIZONA, APRIL

L 8.25"	WS 11"	WT 1.6 OZ

- one adult molt per year; complex basic strategy
- weak age-related and sex-related differences
- overall darkness of ground color variable

Expanding northward in mainly arid climates. Generalist in its habitats and food preferences, but with a preference for human companionship; favors towns, suburbs, and cities. Inconspicuous: hides in trees, walks slowly across lawns and rooftops.

Two-syllable cooing, *oooh-woop* ("no hope" or "whirlpool"), is repeated endlessly; in aggressive situations, sometimes expanded to three or four syllables.

Common Ground-Dove
Columbina passerina CODE 1

ADULT MALE. *Smaller than even the diminutive Inca Dove. Tail short and dark. Adult male scaly; more finely scaled than Inca Dove. Gulf coast birds more reddish than this individual.* ARIZONA, FEBRUARY

ADULT FEMALE. *Compared to male, less reddish and with more obscure scaling. Both sexes have dark spots on wings like Mourning Dove.* CALIFORNIA, MARCH

ADULT. *In flight or with wings outstretched, all plumages show extensive red on primaries and primary coverts.* TEXAS, MAY

L 6.5"	WS 10.5"	WT 1 OZ

- one adult molt per year; complex basic strategy
- moderate differences between male and female
- eastern birds are brighter and more reddish

Ranges across much of the southern tier of states. Population status tied to human activities: declining in many regions because of habitat loss, increasing in some areas with favorable agricultural practices. Nonmigratory, but vagrants widely noted north.

Cooing is a sharply rising *oor-ooo!* or *cuh-woo!*—repeated endlessly. Surprisingly loud for so tiny a bird; suggests a Sora, but lower-pitched and not as clear and sweet.

Parrots & Parakeets
ORDER PSITTACIFORMES

---

7 species: **4** ABA CODE 2   **1** CODE 3   **2** CODE 6

With well over 300 species worldwide, the parrots, parakeets, and allies are, on the whole, remarkably uniform. Most are extensively green, and nearly all have short, sharply decurved bills. The feet of all species in this order are zygodactyl, meaning they have two toes in front and two behind. With few exceptions, these foliage-colored, powerful-billed, agile-toed birds are at home in the treetops, where small flocks devour fruits and nuts. Parrots and parakeets are very popular as cage birds, with the result that captivity and escape into the wild are two major determinants of their population status worldwide.

DOCUMENTING EXOTIC BIRDS

The parrots and parakeets of North America in the early 21st century are completely different from the psittacid fauna of the 19th century. More than a century ago, Carolina Parakeets roamed in orchards and farmsteads in the Southeast, and Thick-billed Parrots irrupted into the high-elevation pinewoods of southeastern Arizona. Those species are gone from our area (the Carolina Parakeet extinct, and the declining Thick-billed Parrot restricted to Mexico), but many other species are now seen in the wild in North America.

Several species of psittacids appear to be well established here. Visitors to southern Florida encounter multiple species in the wild, and it is increasingly difficult to avoid seeing and hearing parrots in the Los Angeles area. Other places—Phoenix, Brownsville, even New York and Chicago—have parrots, parakeets, and their kin. Psittacids are easily recognized, and their screeching flight calls inevitably attract notice. Although they are wary, psittacids favor human companionship and often build nests in gardens, yards, and parks.

Yet the basic details of psittacid occurrence in North America are woefully undocumented. In regions where many species are found, the overall makeup of the psittacid community is poorly understood, and the population status is completely unknown in many cases. Birders have a splendid opportunity to fill in the gaps in our knowledge. On the one hand, there are questions with practical consequences. Understanding the status of the Red-crowned Parrot in Los Angeles would be of considerable benefit to managing native populations in Mexico, where the species is endangered. Monitoring the spread of Monk Parakeets could point toward correct decisions on whether to control populations of this potential pest. On the other hand, there are fascinating questions of basic science: Why are some species, such as the Black-hooded Parakeet apparently on the increase, whereas others, such as the Budgerigar, are on the decline? Questions like these are increasingly relevant to conservation biologists.

Monk Parakeet
Myiopsitta monachus   CODE 2

PEACH-FACED LOVEBIRD (Agapornis roseicollis). *Native to Africa; established in the Phoenix area. Small and stubby. Face washed reddish; tail sky blue.* AFRICA, JUNE

ADULTS. *Green above with blue in wings and tail. Belly yellowish; breast extensively gray with fine dark barring. Forehead pale gray on adult; darker gray on juvenile.* FLORIDA, MARCH

L 11.5"	WS 19"	WT 3.5 OZ

- ▸ one adult molt per year; simple basic strategy
- ▸ weak differences between adult and juvenile
- ▸ bluish avicultural variants are sometimes noted

South American native; established in North America by late 1960s. Other parrots nest in cavities, but Monk Parakeet builds huge stick nests on light fixtures and utility poles. Widely scattered in urban settings, especially parks; Chicago colony is legendary.

Chatters harshly like other parrots; most notes shrill and clipped. One note distinctive: a sharply rising *skyeet!*—calling to mind Western Scrub-Jay.

BLACK-HOODED PARAKEET (Nandayus nenday). *Native to central South America; well established in Florida. Extensive black on head contrasts with otherwise mainly green plumage.* FLORIDA, FEBRUARY

ROSE-RINGED PARAKEET (Psittacula krameri). *Native to Asia; breeding populations in Florida and California. Pale green with very long tail. Thin dark collar prominent; bill bright red.* FLORIDA, MARCH

Red-crowned Parrot

Amazona viridigenalis CODE 2

| L 12" | WS 25" | WT 11 OZ |

▸ one adult molt per year; simple basic strategy

▸ moderate differences between adult and juvenile

▸ sometimes hybridizes with Lilac-crowned Parrot

Native to a restricted region of northeastern Mexico, where it is endangered. Largest established population in North America is centered in Los Angeles metro area, where it is gregarious. Smaller populations elsewhere; birds in southern Texas may be wild vagrants.

Large mixed-species flocks perform amazingly loud screeching ensembles at dawn; Red-crowned leads the pack with its piercing *skee!up ratch-ratch-ratch.*

THICK-BILLED PARROT (Rhyncopsitta pachyrhyncha). *Breeds in Mexican pine forests; formerly strayed to southeastern Arizona. Forehead and eyebrow deep red; thick bill is black.* ARIZONA, FEBRUARY

ADULT. *Red on crown extensive in front of eye, diminished behind eye; juvenile lacks red behind eye. Red secondaries prominent in flight; usually reduced to a small patch on bird at rest.* TEXAS, MAY

LILAC-CROWNED PARROT. (Amazona finschi) *Western Mexican counterpart of Red-crowned; both species occur and interbreed in the Los Angeles area. Crown and nape pale purple-blue.* CALIFORNIA, FEBRUARY

Green Parakeet

Aratinga holochlora CODE 2

| L 13" | WS 21" | WT 8 OZ |

▸ one adult molt per year; simple basic strategy

▸ weak differences between juvenile and adult

▸ extent of red feathering on breast and chin variable

Native to Mexico. Currently established in southern Texas, where status is unclear; many birds are probably of introduced stock, but some may be wild strays from Mexico. Population increasing in Texas; small introduced population in Florida also apparently increasing.

Gives a variety of chattering and screeching notes: *eek* and *cheek*; rather tern-like. Feeds silently for long periods, then highly vocal in flight.

ADULT. *A truly green parakeet; unlike most other parrots, lacks splashes of red and other colors on wings or head. Adult may have a few red flecks on underparts; juvenile all green.* TEXAS, NOVEMBER

White-winged Parakeet
Brotogeris versicolurus   CODE 2

ADULT. *At rest, yellow secondary coverts are more prominent than white secondaries; in flight, white secondaries are conspicuous. Green of juvenile a little darker than adult.* FLORIDA, JANUARY

YELLOW-CHEVRONED PARAKEET (B. chiriri). *Like White-winged, has yellow secondary coverts, visible at rest. Secondaries are dark (white on White-winged).* FLORIDA, APRIL

L 8.75"   WS 15"   WT 2.1 OZ

▸ one adult molt per year; simple basic strategy

▸ weak differences between juvenile and adult

▸ may hybridize with Yellow-chevroned Parakeet

Native to South America, where enormous flocks roost in Amazonian clearings. Established in Florida, but population declining there; also present in California. Status uncertain, because species was merged until recently with Yellow-chevroned as "Canary-winged Parakeet."

Flight call is a clear, somewhat musical *cheep* or *cleep*. Flocks in chorus whip themselves into a frenzy of high-pitched chatter.

In their native ranges in the vast Amazon basin of South America, White-winged and Yellow-chevroned Parakeets have limited contact. In Florida, in contrast, introduced populations of the two species co-occur, and hybridization has been reported. An analogous situation occurs in California, where Red-crowned and Lilac-crowned Parrots interbreed, even though native populations in Mexico do not overlap. This "secondary contact" between pairs of closely related species provides valuable insight for scientists studying speciation.

Budgerigar
Melopsittacus undulatus   CODE 3

L 7"   WS 12"   WT 1 OZ

▸ one adult molt per year; simple basic strategy

▸ moderate differences between juvenile and adult

▸ blue, yellow, green, white, and mixed-color variants

Abundant in native Australia. Established and formerly numerous in Florida, but numbers have declined by more than 99%; competition with House Sparrows thought to be responsible. Sociable; visits feeders, readily uses bird boxes. Popular cage bird; escapes can be seen anywhere.

Along with rough chattering, characteristic of parrots, gives pleasing liquid notes: *gleer-ip* and *cheer-up*. Vocal when perched, unlike many other parrots.

WILD-TYPE ADULT. *Notably small; about the same mass as House Sparrow. Wild type is green below, finely barred yellow-and-black above. Juvenile duller overall; has barred forehead.* FLORIDA, MARCH

Cuckoos & Allies
ORDER CUCULIFORMES

This medium-sized order consists of about 150 species worldwide and is represented in North America by a diverse array of three cuckoos in the New World genus *Coccyzus*, two cuckoos in the Old World genus *Cuculus*, the beloved Greater Roadrunner, and two strange birds called anis. The order Cuculiformes is part of a higher taxonomic assortment of "near-passerines" with unclear relationships to each other and probably only distant relationships to the passerines (perching birds). Fittingly, the Hoatzin of South America—perhaps the strangest bird in the world—has been placed by different authorities within, adjacent to, or not even close to, the order Cuculiformes.

Despite the diversity of North America's handful of species, several general patterns can be noted. All cuculids are medium to large birds, slender in build, and long-tailed. They have fairly simple but often diagnostic vocalizations. With a good look, identification is often straightforward; in many instances, however, these birds stick to dense vegetation. Behavior and geography can be secondarily important in identification. At the same time, vagrancy has been recorded in most species.

Despite their generally unremarkable colors and unspectacular patterns, our seven regularly occurring species exhibit a remarkable array of behaviors and life history characteristics, ranging from highly solitary to surprisingly social, from highly migratory to virtually sedentary. American cuckoos and allies are mainly insect-eaters, with the Greater Roadrunner doubling as a vertebrate-eater and anis being part-time vegetarians. Their North American ranges run the gamut from widespread to peripheral.

Population status and conservation issues are predictably diverse. Numbers of American cuckoos (*Coccyzus*) are intrinsically variable, rising and falling with natural fluctuations in their food supply; at the same time, all three species may be undergoing long-term population declines related to habitat loss. One of the anis (Smooth-billed) is in sharp decline in our area, but the Greater Roadrunner is expanding its range. The range limits of the roadrunner are determined partly by thermal conditions, and the slow warming of the continent may be benefiting the species. Numbers of Old World cuckoos are in general decline, apparently because of population losses in species whose nests they parasitize. Our Yellow-billed and Black-billed Cuckoos only rarely lay eggs in other birds' nests.

Yellow-billed Cuckoo
Coccyzus americanus CODE 1

ADULT. *Wings and tail long. From below, tail black with large white spots; undertail pattern duller on juvenile. Large white spots on tail feathers visible even from behind.* TEXAS, APRIL

JUVENILE. *All adults and many juveniles have largely yellow bills, but some juveniles have all-dark bills; separated from Black-billed by more white on tail, throat, and breast.*
CALIFORNIA, SEPTEMBER

L 12"   WS 18"   WT 2.3 OZ

▸ one adult molt per year; complex basic strategy
▸ moderate differences between juvenile and adult
▸ western birds slightly larger and grayer above

Favors dense vegetation, where it forages slowly. Western subspecies restricted to high-quality riparian habitat. Late spring migrant (especially western subspecies) and early fall migrant. Numbers vary in response to caterpillar outbreaks.

◂ Song a clucking series that slows toward end: *ka ka ka kaw kaw kow kow kowl kowl... kowlp... kowlp.* Flight call, sometimes given by nocturnal migrants, *k'k'k'kulp,* descends in pitch.

Black-billed Cuckoo
Coccyzus erythropthalmus CODE 1

ADULT. *Distinguished from Yellow-billed by black (not yellow) bill and red (not yellow) orbital ring. Tail long like Yellow-billed, but mainly gray beneath, with small white tips.* TEXAS, JULY

JUVENILE. *Lacks red orbital ring; tail mainly gray below. Wings usually have some rufous. Breast and throat dusky; juvenile Yellow-billed is white on breast and throat.* NEW JERSEY, SEPTEMBER

L 12"   WS 17.5"   WT 2 OZ

▸ one adult molt per year; complex basic strategy
▸ moderate differences between juvenile and adult
▸ extent of rufous in wing of juvenile variable

In many respects, the northern counterpart of Yellow-billed. Generally seems less common than Yellow-billed, owing perhaps to its even more reclusive behavior. Sticks to treetops in disturbed woods, favoring forest edges, clearings, and streamsides.

Song a series of soft-sounding but far-carrying *cucu* or *cucucu* notes, evenly spaced. Flight call, sometimes given by nocturnal migrants, a rising *k'k'k'kick...*

Mangrove Cuckoo
Coccyzus minor　CODE 2

12"　ws 17"　wt 2.3 oz

» one adult molt per year; complex basic strategy
» moderate differences between juvenile and adult
» vagrants to western and northern Gulf coast darker

Like other *Coccyzus* cuckoos, skulks in arboreal vegetation. Occurs in namesake mangrove groves, but also in other vegetation. Movements of Florida birds not well understood; vagrants to Texas and northern Gulf coast wander from Mexico, not Florida.

Song a slow croaking series of *caw* notes, speeding up toward end. Call note a low-pitched *hvib*, easily overlooked.

ADULT. *Brown above, buff below. Undertail has more black than Yellow-billed; juvenile tail pattern muted. Thick bill is black above, yellow below. Gray crown contrasts with dark mask.* FLORIDA, APRIL

Common Cuckoo
Cuculus canorus　CODE 3

RED-MORPH FEMALE. *Only females have red morph; pattern of barring on underparts same as gray-morph female. Barred plumage, pointed wings, and long tail create falcon-like flight profile.* EUROPE, JUNE

GRAY-MORPH ADULT MALE. *All males and some females are blue-gray all over; darker above than below. Belly pale with fine dark barring; barring extends to breast on female.* EUROPE, APRIL

13"　ws 22–26"　wt 4–5 oz

» one adult molt per year; complex basic strategy
» strong differences between male and red-morph female
» females dimorphic (red or gray); males are not

Eurasian species that visits western Alaska in small numbers in late spring and early summer. In Old World, a solitary skulker in the treetops. Migrants to Alaska often stick to dense low vegetation if available.

The famous cuckoo-clock song of this bird, heard mainly on nesting grounds, is only rarely given by nonbreeding migrants to North America.

ORIENTAL CUCKOO. *(C. optatus) Asian species; casual to Bering Sea, mainly in fall. Similar to Common Cuckoo, but averages brighter buff below and darker gray above. Gray-morph adult, here; only female has red morph.* ASIA, MARCH

Smooth-billed Ani
Crotophaga ani CODE 3

ADULT. Bill lacks grooves and averages larger than Groove-billed. Body larger but tail proportionately shorter than Groove-billed; grackles have different bill shape and flight style. BELIZE, DECEMBER

JUVENILE. Both anis have reduced bills as juveniles, lacking grooves and protrusions from upper mandible; most juveniles not identifiable to species. BELIZE, DECEMBER

L 14.5"   WS 18.5"   WT 3.7 OZ   ♂ > ♀

▸ one adult molt per year; complex basic strategy

▸ moderate differences between adult and juvenile

▸ size of upper mandible of adult variable

Found in a variety of brushy or wooded habitats, often at trashy sites. Individuals or small, disorganized flocks clamber within or at edges of vegetation, sometimes venturing into the open. In recent years, numbers have dropped sharply.

Vocalizations variable; generally tinnier than Groove-billed. Commonly heard *tea-lick* call rises toward end. Some Great-tailed Grackle vocalizations similar.

Groove-billed Ani
Crotophaga sulcirostris CODE 2

ADULT. Usually separable from Smooth-billed by range. Smooth-billed averages larger-billed; Groove-billed has diagnostic but faint grooves. Bill of juvenile smaller; lacks grooves. TEXAS, JUNE

ADULT. Both anis convey an unkempt look: long tail is often dipped or flipped about slowly; wings often drooped or held away from body. TEXAS, DECEMBER

L 13.5"   WS 17"   WT 3 OZ   ♂ > ♀

▸ one adult molt per year; complex basic strategy

▸ moderate differences between adult and juvenile

▸ bill shape and depth of grooves in bill variable

Both anis differ from other North American cuckoos in their relatively gregarious behavior; less furtive than cuckoos. Communal mating system (females tend each other's eggs) is notable; away from nest sites, anis forage in loose aggregations.

Squeaky *tee-hoe* or *tijo* given singly or with other sharp, whistled notes such as *tweeek tijo*. Some vocalizations of Great-tailed Grackle similar.

Greater Roadrunner
Geococcyx californianus CODE 1

ADULT. *Unmistakable. Large overall, with long tail, sturdy legs, long neck, and shaggy crest, Streaked brown and white; "racing stripe" behind eye often prominent.* NEW MEXICO, AUGUST

JUVENILE. *Compared to adult, glossy highlights diminished; ground color of adult warmer buff or cinnamon than juvenile. This bird is sun-basking.* NEW MEXICO, NOVEMBER

ADULT. *Usually shows glossy highlights, especially in tail and wing. Crest often flattened, as here.* ARIZONA, DECEMBER

23"   ws 22"   wt 13 oz

one adult molt per year; complex basic strategy

weak differences between adult and juvenile

red-white-and-blue "racing stripe" behind eye variable

"classic" view, seen running across esert wash or along street. In norning, sometimes observed un-basking; throughout the ay, may be seen energetically ursuing lizards and rodents. therwise wary; most often seen etreating to or hiding behind cover.

ournful song, heard in late winter and early spring, a series of ur to six descending *hooo* notes. Rapid clicking sound made y bill is heard infrequently year round.

FORAGING. *Swift-footed hunter, capturing rodents, lizards, and smaller snakes; often seen perching on low shrubs and trees to swoop and chase prey.* NEW MEXICO, MAY

Owls and Nightjars

ORDERS STRIGIFORMES AND CAPRIMULGIFORMES

31 species: 11 ABA CODE 1   13 CODE 2   2 CODE 3   1 CODE 4   4 CODE 5

In many people's minds, these orders are the two most obviously nocturnal avian groups in North America. Owls and nightjars are placed next to each other in most taxonomic arrangements, a reflection of their presumed evolutionary closeness. However, recent molecular evidence calls into question the evolutionary cohesiveness (monophyly) of the nightjars, and suggests that the swifts and hummingbirds (order Apodiformes) may be closely allied with the owls and nightjars. All North American nightjars fall within the family Caprimulgidae, whereas owls are split into two families: the Barn Owl in the family Tytonidae, and all the rest (called "typical owls") in the family Strigidae.

All North American owls and nightjars are at least partially nocturnal, but there is much variation in the daily activity patterns of different species. Burrowing and Snowy Owls are as active by day as they are by night; pygmy-owls hunt by day and court during twilight; Great Horned Owls court during twilight, but hunt mainly at night; and screech-owls and Flammulated Owls are primarily nocturnal. Nightjars, too, are diverse in their daily schedules. Chuck-will's-widows and Whip-poor-wills are active primarily at night, whereas nighthawks frequently make aerial excursions in broad daylight. Day or night, voice is an important component in identifying owls and nightjars.

All of these species are predatory, their prey ranging in size from moths to porcupines. The huge-mouthed nightjars catch aerial insects (especially moths) on the wing, and the Elf and Flammulated Owls are mainly insectivorous. Most of the other owls, however, depend importantly on mammals and birds as prey. In several instances, an owl species appears to be the nocturnal equivalent of a diurnal species—a phenomenon known as niche partitioning. For example, the Barred Owl may be the nocturnal equivalent of the Red-shouldered Hawk, and the Great Horned Owl is in many respects the nocturnal counterpart of the Red-tailed Hawk. Especially curious is the Flammulated Owl, which fills the nighttime niche of small, migratory, insectivorous passerines.

The population status of owls and nightjars is difficult to assess, but several species appear to be declining, for various reasons. For example, Spotted Owls are decreasing at least partly because of competition with invasive Barred Owls and loss of habitat; Burrowing Owls have disappeared from districts where prairie dogs have been eradicated; and Common Nighthawks appear to have suffered losses from nest predation by American Crows.

Barn Owl

Tyto alba CODE 1

ADULT MALE. *Eyes black; face white; facial disk well-defined. Extensive white below distinctive, but all owls can appear white in headlights.* ARIZONA, FEBRUARY

ADULT FEMALE. *More buff on breast and flanks than male, with heavier spotting. In flight, appears smaller overall but more bullnecked than Great Horned Owl.* NEVADA, MAY

| 16" | ws 42" | wt 1 lb | ♀ > ♂ |

▸ one adult molt per year; simple basic strategy
▸ moderate differences between male and female
▸ extent and intensity of buff on undersides variable

One of the most widely distributed birds in the world; uncommon in North American range. Has declined in and withdrawn from areas in which traditional agriculture has been phased out. Nests in silos, abandoned outbuildings, cliffs, caves, and bird boxes.

◀ Call striking: a ghoulish scream *kssshhHHHHHhhhht*, ending abruptly. Also a squeakier *ksheeeeert*. Some calls of Great Horned Owl similar.

Great Horned Owl

Bubo virginianus CODE 1

ADULT. *Large and powerful; talons impressive; barred below, mottled above. When alert, "horns" prominent and white on breast conspicuous. Some are darker or paler than shown here.* FLORIDA, FEBRUARY

FLEDGLING. *Pale and "fuzzy." Younger juveniles are noisy and usually seen in presence of adults. Immature birds after juvenile molt similar to adult; just a little "fuzzier."* CONNECTICUT, APRIL

| L 22" | ws 44" | wt 3 lb |

▸ one adult molt per year; complex basic strategy
▸ moderate differences between juvenile and adult
▸ extensive geographic variation in color and size

Widespread and successful; in many regions, the most familiar and conspicuous owl. Powerful predator; takes surprisingly large prey. Hunts mainly at night, but may be seen at any time of day. When roosting, often attracts the attention of jays, magpies, and other birds.

◀ Females and males duet, *hu huhuhuhu hoo hoooo! hoo*, female higher-pitched. Also gives a shrieking *schkeeer!*—harsher and shorter than Barn Owl.

Eastern Screech-Owl
Megascops asio   CODE 1

GRAY-MORPH ADULT. *Small, but appears long-winged in flight. "Ears" often hard to see. Differs from Western in having pale bill and lacking prominent vertical streaks on breast.* CONNECTICUT, APRIL

RED-MORPH ADULT. *Most frequent in eastern portion of range. Like gray morph, has pale bill and little or poorly defined breast streaking. "Ears" sometimes raised.* CONNECTICUT, MAY

L 8.5"	WS 20"	WT 6 OZ	♀ > ♂

▸ one adult molt per year; complex basic strategy

▸ moderate differences between juvenile and adult

▸ complex geographic variation in color

A cavity-nester that adapts well to human presence; can be surprisingly common around farms and even in city parks. Requires woodlands, but generally avoids deep forests; favors groves, stream edges, and shelterbelts. Strictly nocturnal, and thus often overlooked.

◀ Primary song is a descending, pure-tone whinny: *w'he!he'he'he'he'hu'hu'hu'*. Secondary song a long, slow trill all on one pitch *huhuhuhuhuhu*.

Western Screech-Owl
Megascops kennicottii   CODE 1

ADULT. *Differs from Eastern Screech-Owl in having dark bill and fairly prominent vertical breast streaks; size and proportions about the same as Eastern.* CALIFORNIA, APRIL

ADULT. *Most are cold gray, but some are more pale gray-brown, as here, or darker rusty-brown; does not have red morph as Eastern.* CALIFORNIA, MAY

L 8.5"	WS 20"	WT 5 OZ	♀ > ♂

▸ one adult molt per year; complex basic strategy

▸ moderate differences between juvenile and adult

▸ geographic variation in size and overall color

A nonmigratory, cavity-nesting owl of western woods; especially characteristic of riparian zones in arid country. Patchily distributed; common at many spots, but also surprisingly absent from numerous landscapes. Has expanded east into shelterbelts and small towns.

◀ Primary song a mellow series, *wu... wu... wu wu wuwuwuwuwu*, with little drop in pitch. Secondary song like Eastern, but with short pause toward beginning.

Whiskered Screech-Owl

Megascops trichopsis CODE 2

ADULT. *Slightly smaller than Western Screech-Owl, with oddly small feet. Streaked below like Western, but streaks not as prominent; bill dark like Western.* ARIZONA, MAY

| L 7.25" | WS 17.5" | WT 3.2 OZ | ♀ > ♂ |

▸ one adult molt per year; complex basic strategy

▸ moderate differences between juvenile and adult

▸ rufous and brown morphs occasionally reported

An owl of tropical mountains; can be fairly common within limited North American range, and is perhaps expanding here. Generally found at higher elevations than Western Screech-Owl. Aggressive and curious; often ventures into campgrounds and outskirts of towns.

Primary song like Western Screech-Owl, but tempo does not increase. Secondary song an irregular series, *wu hu... hu hu huhu hu... huhuhu...*, like Morse Code.

Flammulated Owl

Otus flammeolus CODE 2

ADULT. *Smaller than a screech-owl, with soulful black eyes; "ears" often inconspicuous. All have reddish (flammulated) highlights, some more so than shown here.* ARIZONA, MAY

| L 6.75" | WS 16" | WT 2 OZ |

▸ one adult molt per year; complex basic strategy

▸ moderate differences between adult and juvenile

▸ size and wing length increase from south to north

Ecologically speaking, the nocturnal equivalent of Neotropical migrant passerines: small, insectivorous, and highly migratory. In most of range, a bird of extensive ponderosa pine forests; in especially arid regions, like Great Basin, ranges into spruce-fir woods.

Song a series of well-spaced hoots: *hooop* or *h'hooop* or *h'hu'hooop*. Hoots are surprisingly low-pitched, resembling the sound of blowing across a bottle top.

Elf Owl
Micrathene whitneyi　CODE 2

ADULT. *Very small; lacks "ears." Pale gray with variable cinnamon highlights, especially on female. Most females more cinnamon than shown here, but sexes not separable in the field.* ARIZONA, APRIL

L 5.75"	WS 13"	WT 1.5 OZ

- one adult molt per year; complex basic strategy
- weak differences between male and female
- grayer brown, less cinnamon in Texas than Arizona

Best known as inhabitant of saguaro "forests" in Arizona, but also ranges into mixed riparian groves, woodlands, and even into developed areas, where it catches large arthropods attracted to lights. Populations unstable, possibly declining, around edges of range.

Primary call a short, falling series of high-pitched yapping notes: *chew chewp! chip chip chip.* Also gives single squeaky notes; for example, *weeecha!*

Northern Pygmy-Owl
Glaucidium gnoma　CODE 2

ADULT. *Rotund and very long-tailed; tail is dark with conspicuous white bars. Crown dark with fine white spots. Some are grayer, others redder, than shown here.* COLORADO, FEBRUARY

L 6.75"	WS 12"	WT 2.5 OZ	♀ > ♂

- one adult molt per year; complex basic strategy
- moderate differences between juvenile and adult
- polymorphic: grayish, reddish, and intermediate

A fearsome predator of western coniferous forests; some movement into lowland and broadleaf woods in winter. Most active at dawn and dusk, but frequently also by day; rarely detected at night. Fairly common, but often overlooked; sometimes mobbed by small passerines.

Gives widely spaced, pure-tone whistles, *whew* or *wheep.* Southeast Arizona birds give a two-note whistle, in rapid series. Calls of all populations variable.

Ferruginous Pygmy-Owl
Glaucidium brasilianum CODE 3

ADULT. *Rusty all over; where ranges overlap, Northern Pygmy-Owl usually grayer. Tail is dark with conspicuous russet bars. Dark crown has fine white streaks; Northern has white spots.* TEXAS, MAY

ADULT. *Even at a distance, pygmy-owls are recognizable by their long tails and small stature overall; both species perch in the open in broad daylight. Juvenile darker than adult.* TEXAS, MAY

| L 6.75" | WS 12" | WT 2.5 OZ | ♀ > ♂ |

▸ one adult molt per year; complex basic strategy
▸ moderate differences between juvenile and adult
▸ Arizona birds average paler and grayer than in Texas

Numerous in tropics; uncommon at northern fringes of range in Arizona and Texas. Habitat does not usually overlap with Northern Pygmy-Owl: uses mesquite groves in Texas, low riparian woods in Arizona. Arizona population declining because of wildfires and urban sprawl.

Song a series of *pik* notes; more strident, not as pure-tone, as Northern Pygmy-Owl. Repeated rapidly and at length, like Northern Saw-whet Owl.

Burrowing Owl
Athene cunicularia CODE 1

ADULT. *Long-legged with intense, staring, yellow eyes. Brown above, with coarse white spots; barred and spotted brown and white below.* FLORIDA, MARCH

JUVENILE. *Unmarked below; has long legs and piercing countenance of adult. Wing has broad buff patch, usually lacking on adult.* ARIZONA, JUNE

| L 9.5" | WS 21" | WT 5 OZ |

▸ one adult molt per year; complex basic strategy
▸ strong differences between juvenile and adult
▸ Florida birds average slightly darker than western

A highly diurnal owl of open country; nests in small colonies, often in association with prairie dogs. Favors native prairies, but also accepts airports and golf courses. Two discrete populations: expanding north in Florida, declining at periphery of range in West.

In many situations, gives yapping *schrick* and *chrick* calls, often in series. Male gives a melancholy crowing, *woo hoooo*, with quality of Gambel's Quail.

Short-eared Owl

Asio flammeus CODE 2

ADULT. *A medium-large owl. Males pale gray to white below with dark streaks. Females average warmer buff below, but variable.*
MASSACHUSETTS, JANUARY

ADULT. *"Ears" tiny; hard to see even on an alert bird. Like most owls, cryptic plumage allows it to roost unmolested by day.*
CONNECTICUT, FEBRUARY

IN FLIGHT. *Wings pale below; darker above with broad, light buff patch on primaries. Wing beats shallow and choppy; with brief glimpse, can suggest Northern Harrier.* MASSACHUSETTS, FEBRUARY

L 15"	WS 38"	WT 12 OZ	♀ > ♂

- ▸ one adult molt per year; complex basic strategy
- ▸ weak differences between male and female
- ▸ Caribbean vagrants to Florida darker and smaller

Inhabits wide-open country; often abroad by day, especially on overcast days or late in the afternoon. Slow, butterfly-like flapping is distinctive. Forms large winter roosts. Widespread, but declining in Northeast; success depends on availability of small mammal prey.

Generally quiet, but gives yelping notes throughout year: *rrap* and *crip crape!* and *crep creep!* In striking display flight, gives audible wing-clapping.

Long-eared Owl

Asio otus CODE 2

ADULT. *When roosting, assumes elongated posture; Great Horned can appear similar, but lacks black vertical bars above and below eye. Like Short-eared, female has more buff below.*
CALIFORNIA, JANUARY

IN FLIGHT. *Similar to Short-eared, but not as boldly marked; buff patch above diminished, wings not as pale below. Manner of flying jerky and abrupt, like Short-eared.*
CALIFORNIA, JANUARY

L 15"	WS 36"	WT 9 OZ	♀ > ♂

- ▸ one adult molt per year; complex basic strategy
- ▸ weak differences between male and female
- ▸ western birds average paler than eastern

Best known for roosting close to tree trunks in conifer groves; roosts are often communal and can be closely approached. Nests in various woodland settings; feeds in open country, hunts only at night. In Intermountain West, highly dependent upon riparian habitat.

Although silent at daytime roosts, can be highly vocal otherwise, especially at dusk. Gives various hooting and yapping calls, some quite complex.

Barred Owl

Strix varia CODE 1

ADULT. *A large, chunky owl. Upper breast densely barred; remainder of underparts loosely streaked. Eyes black; bill yellow.*
CONNECTICUT, MARCH

N FLIGHT. *Wings and tail heavily barred. Usually seen in short-distance flight; appears bulky, with broad wings and rounded tail, in flight.*
ONNECTICUT, JANUARY

| 21" | ws 42" | wт 1.6 lb | ♀ > ♂ |

▸ one adult molt per year; complex basic strategy
▸ moderate differences between juvenile and adult
▸ hybridizes with Spotted Owl where ranges overlap

ives in deep woods, especially arge tracts. Favors bottomlands: swamps and floodplains. Generalist predator; active into midmorning and again by late afternoon. Especially common in South; rapidly nvading far West, where it poses a hreat to Spotted Owl.

Primary hooting is a far-carrying baritone *hu-hoo-huhoo hu-hoo-hu-hooaw*, last note descending. Gives various other calls, many with wild, maniacal quality.

Spotted Owl

Strix occidentalis CODE 2

ADULT, "NORTHERN." (S. o. caurina) *Averages darker than "Mexican," an adaptation to the less arid landscapes it inhabits. Spotting is less coarse than "Mexican."* OREGON, AUGUST

IMMATURE, "MEXICAN." (S. o. lucida) *In all plumages, belly has coarse white spots; Barred has coarse brown streaks on pale belly. Tail tips of immature white, pointed; blunt and browner in adult.*
ARIZONA, FEBRUARY

| L 17.5" | ws 40" | wт 1.3 lb | ♀ > ♂ |

▸ one adult molt per year; complex basic strategy
▸ weak differences between juvenile and adult
▸ overall color and extent of spotting vary geographically

Perhaps the most politically controversial bird species in North America; main focus is on "Northern" Spotted Owl of the Pacific Northwest, but all subspecies are associated with old-growth forests, usually with conifers. Tame and curious; flushes only a short distance.

Typical hooting a four-part series, *hoo... huhoo! hoo*, with pauses after first and third notes. Various others calls average less maniacal than Barred Owl.

Northern Saw-whet Owl

Aegolius acadicus CODE 2

ADULT WITH PREY. *Small, short-tailed, and bullheaded. Pale underparts have coarse rusty streaks; dark upperparts have coarse white spots.*
MINNESOTA, JANUARY

JUVENILE. *Belly unpatterned warm buff; upperparts dark chocolate with reduced white spotting. Dark head has contrasting white patch between eyes.* ALBERTA, SEPTEMBER

L 8"	WS 17"	WT 3 OZ	♀ > ♂

▸ one adult molt per year; complex basic strategy

▸ strong differences between juvenile and adult

▸ birds off coast of British Columbia are darker

Fairly common but infrequently detected; strictly nocturnal on breeding grounds, reclusive on wintering grounds. Migratory, with major passages noted at specialized monitoring operations in Great Lakes region; extent and regularity of dispersal not fully known.

Primary song a monotonous series of short whistles, repeated two or three times per second; easily imitated. Raspy "saw-whet" song rarely heard.

Boreal Owl

Aegolius funereus CODE 2

ADULT. *Similar to Northern Saw-whet, but pale-billed, larger overall, and darker; color of most individuals colder than Northern Saw-whet, but some have substantial russet.*
MINNESOTA, FEBRUARY

JUVENILE. *Darker below than Northern Saw-whet; belly and lower breast gray-brown. White between eyes usually has lower contrast than Northern Saw-whet.* ALASKA, AUGUST

L 10"	WS 21"	WT 4.7 OZ	♀ > ♂

▸ one adult molt per year; complex basic strategy

▸ strong differences between juvenile and adult

▸ intensity of russet hue variable

Nests in tree cavities in spruce-fir forests, where it eats small mammals. Probably numerous, but infrequently seen; favors dense stands, often in remote regions. Mainly sedentary, but a few wander far from breeding grounds in winter; roosting birds tame, approachable.

Primary song a rapid, tremulous series, *wuhuhuhuhuhuhuhuhu,* rising throughout or rising then falling; winnowing of Wilson's Snipe can be mistaken for Boreal Owl.

Northern Hawk-Owl

Surnia ulula CODE 2

| L 16" | WS 28" | WT 11 OZ | ♀ > ♂ |

▸ one adult molt per year; complex basic strategy
▸ weak differences between juvenile and adult
▸ bare parts variable; influenced by diet and blood flow

In many respects, more *Accipiter*-like or falcon-like than owl-like. Often abroad by day; scans for prey from conspicuous perches. Takes large prey such as ptarmigan, hares. Fearless; easily approached. Reasons for periodic irruptions southward are not well understood.

Song a rapid trill, *pupupupupupupupupu...*, similar to, but longer and more even in pitch than Boreal Owl. Also gives a screeching *shhhrrreek!*—ending abruptly.

IN FLIGHT. *Heavily barred beneath. Long tail is conspicuous; wings are broad-based with pointed tips. Hawk-like overall, but typically bullnecked like an owl.* MINNESOTA, JANUARY

ADULT. *A medium-large owl; long-tailed and densely barred below. Pale gray face framed by dark vertical bars. Juvenile similar, but a little darker.* MINNESOTA, JANUARY

Great Gray Owl

Strix nebulosa CODE 2

| L 27" | WS 52" | WT 2.4 lb | ♀ > ♂ |

▸ one adult molt per year; complex basic strategy
▸ weak differences between juvenile and adult
▸ overall color of facial disk varies from brown to gray

A bird of the boreal zone; roosts and feeds at edges of bogs and clearings, and along logging roads. Active at any time of day and night. Slight but annual movement southward, with strong irruptions in some years; southern strays in winter are often starving.

Song a simple series of deep *hoooo* or *hooo-ooo* notes, widely spaced. Individual notes with slight wavering quality; song does not carry well.

ADULT. *Large, bullheaded, and long-tailed. Facial disk has faint concentric circles; base of facial disk has broken white crescent. Adult a little browner than juvenile.* MINNESOTA, JANUARY

Snowy Owl
Bubo scandiacus CODE 2

ADULT MALE. *Large and mainly white. All have at least some dark barring below and dark spots or splotches above, diminished on adult male.* NEW YORK, FEBRUARY

IMMATURE. *Intensity and extensiveness of dark markings correlated with age and sex; birds this dark are usually immature.* NEW YORK, JANUARY

L 23"	ws 52"	wт 4 lb	♀ > ♂

▸ one adult molt per year; complex basic strategy

▸ moderate age-related and sex-related differences

▸ extent of black barring and splotching variable

An owl of open country: breeds in the high-arctic, where it hunts lemmings; winters regularly in northern U.S. Extent of movement southward varies from year to year. Wintering birds are often seen by day, when they strike a lethargic pose on dunes and outbuildings.

Song, heard only on breeding grounds, a series of hoots, with quality of Great Horned Owl. Defends winter territories with shrill whistles and rough barking.

Common Nighthawk
Chordeiles minor CODE 1

ADULT. *At daytime roosts, all nighthawks assume horizontal position on branch or ground. White patches at bend of wing and in primaries often evident on resting birds.* TEXAS, MAY

IN FLIGHT. *All plumages have prominent white patch across primaries. Male (here) has white subterminal tail band, lacking in female; white throat patch reduced on female.* NEW JERSEY, JUNE

L 9.5"	ws 24"	wт 2.2 oz

▸ one adult molt per year; complex basic strategy

▸ moderate differences between male and female

▸ much geographic variation in overall color and paleness

Widespread but generally declining; American Crows and other nest predators may be at fault. Hunts from dusk to dawn, as well as by day during overcast periods. Especially conspicuous in cities and towns, where it nests on flat roofs and hunts in brightly lit districts.

◀ Display is one of the marvels of summer evenings: male gives piercing *beeent*, repeated; then plunges toward earth, breaking with booming, non-vocal *hooooom*.

Lesser Nighthawk

Chordeiles acutipennis CODE 1

AT REST. *Low-contrast mottled brown plumage, especially, as here, on female and juvenile. Compared to Common, short-winged and long-tailed; wings at rest extend only to tail tip.* TEXAS, MAY

IN FLIGHT. *Outer primary short, making wing appear rounded. White bar relatively close to wing tip. Male (here) has more prominent throat patch and subterminal tail band than female.*

CALIFORNIA, JUNE

9"   WS 22"   WT 2 OZ

- one adult molt per year; complex basic strategy
- moderate differences between female and male
- intensity of buff ground color somewhat variable

A desert nighthawk; characteristic of vast lowland creosote bush stands. Common Nighthawk is usually at higher elevations, but much overlap. Lesser flies low to ground, barely clearing tops of desert shrubbery. Wing beats shallower than Common; sometimes seen in flocks.

Sounds are utterly unlike Common Nighthawk. Two songs: long-duration, toad-like trill on one pitch; berserk laughter, *w'hee!haw'w'hay!weng!whoa.*

Antillean Nighthawk

Chordeiles gundlachii CODE 3

ADULT FEMALE. *Slightly smaller and shorter-winged than Common. At rest, pale gray tertials visible. Male has white throat patch and subterminal tail band, reduced or lacking on female.* FLORIDA, MAY

IN FLIGHT. *Like Common, flashes prominent white bars on underwing; position of subterminal tail band on male (as here) similar to Common. Best separated from Common by call.*

BAHAMAS, MAY

L 8.5"   WS 21"   WT 2 OZ

- one adult molt per year; complex basic strategy
- moderate differences between male and female
- extent of rufous in ground color variable

Formerly treated as subspecies of Common Nighthawk, but found to nest side-by-side without interbreeding. Molt timing differs, as may wintering ecology; more study needed. Otherwise, habits much as Common; hunts at dusk, on cloudy days, and after rainstorms.

Display call a scratchy, kaytdid-like *diddy d'zick!*—sometimes shortened to *d'zick!* Nonvocal booming sound given in dive is similar to Common but softer.

Whip-poor-will
Caprimulgus vociferus CODE 1

ADULT FEMALE, "EASTERN." (C. v. vociferus) *Mottled gray-and-russet; gray scapulars often stand out. Base of chin has thin buff crescent on female; crescent is thicker and brighter white on male.* NEW YORK, NOVEMBER

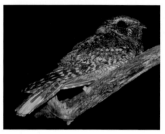

ADULT FEMALE, "ARIZONA." (C. v. arizonae) *Similar to "Eastern." Females of both populations have buff tail corners (white in males), prominent in flight; sometimes visible, as here, when roosting.* ARIZONA. MAY.

L 9.75"   WS 19"   WT 2 OZ

- ▸ one adult molt per year; complex basic strategy
- ▸ moderate differences between male and female
- ▸ "Eastern" and "Arizona" may refer to separate species

Well known if for no other reason than its evocative name; in many regions, has become scarce. Throughout range, favors dry woods: declining eastern population in deciduous or mixed woods, "Arizona" in pine-oak forests. Cryptic; roosts and nests on forest floor.

A well-known voice of the night. Eastern birds give honest-to-goodness *whip poo-oor will!*—clear and sweet. Song of "Arizona" burrier.

BUFF-COLLARED NIGHTJAR (C. ridgwayi). *Casual visitor to arid canyons of southeastern Arizona and southwestern New Mexico. Thin buff collar encircles neck.* ARIZONA, MAY

Chuck-will's-widow
Caprimulgus carolinensis CODE 1

ADULT. *Larger and usually brighter rufous than Whip-poor-will. Wings long, often extending to near tail tip on bird at rest.* FLORIDA, APRIL

ADULT. *Many are strongly rufous, but a sizable number are more grayish as here. Female has buff tail corners; male has white; difference usually discernible only in flight.* FLORIDA, APRIL

L 12"   WS 26"   WT 4.2 lb

- ▸ one adult molt per year; complex basic strategy
- ▸ moderate differences between male and female
- ▸ ground color varies from reddish to gray-brown

Nocturnal; flies low, hunting for flying insects. Also hunts on ground, plucking up large insects and even frogs. Occurs in various dry forest types: pine, live oak, and mixed deciduous. In regions of overlap with Whip-poor-will, tends to occur in more-open habitats.

Song a loud, pulsing *chuck! widow widow*, repeated. Also gives a sudden *quok!*—like Black-crowned Night-Heron—and a puffin-like *awrrrrr.*

Common Poorwill
Phalaenoptilus nuttallii CODE 1

ADULT. *Small-bodied, large-headed nightjar. At rest, gray overall with a bit of rufous in wings. Eye-shine is flaming orange; often detected in headlights.* CALIFORNIA, MAY

ADULT MALE. *In both sexes, wings flash extensive rufous in flight, or in aggressive display. Male has white tail corners (buff in female). All nightjars have huge mouths.* NEW MEXICO, MAY

L 7.75"   WS 17"   WT 2 oz

▸ one adult molt per year; complex basic strategy
▸ moderate differences between male and female
▸ darker in humid regions, paler in arid regions

Common in arid habitats throughout much of the West; occurs in low deserts and grasslands, but also well into mountains. Frequently seen on back roads around sundown. In response to environmental stress, such as a cold snap, goes into a depressed metabolic condition known as torpor.

Song a soft, mournful *poor will! up*, repeated. Second note higher and louder; third note inaudible except under still conditions or at close range.

Common Pauraque
Nyctidromus albicollis CODE 2

ADULT MALE. *A few birds in Texas may be as brightly colored as this one. Male has extensive white in outer tail feathers; female has limited white in tail.* MEXICO, OCTOBER

ADULT FEMALE. *All individuals are notably long-tailed. Brown eye patch contrasts with overall gray of rest of face. Most birds in Texas are fairly grayish overall.* TEXAS, NOVEMBER

L 11"   WS 24"   WT 2 oz

▸ one adult molt per year; complex basic strategy
▸ moderate differences between male and female
▸ hue of brownish gray ground color variable

Resident along wooded edges, especially mesquite groves, in south Texas. Sits on the ground, waits for flying insects, then flies up a short distance to catch prey. Especially active around dawn and dusk; at night, sometimes ventures into large open areas.

Full song consists of rapid, stuttering introductory notes followed by brief outburst: *p'p'p'p'p'wee!oh*. All but *p'wee!oh* is often omitted or inaudible.

Swifts and Hummingbirds

ORDER APODIFORMES

32 species: 11 ABA CODE 1   7 ABA CODE 2   2 ABA CODE 3   5 ABA CODE 4   7 ABA CODE 5

The swallow-like swifts and the unique hummingbirds seem at first glance to have little in common, but their close relationship is well established. Two structural similarities are key. First, both swifts and hummingbirds have diminutive feet; the name Apodiformes derives from Greek words meaning "without feet." Second, both families have long, arcing primaries that are well-suited to complex, midair maneuvers. With deficient feet but superior powers of flight, swifts and hummingbirds are adapted for feeding on the wing—although the two families have come up with two distinctly different methods of aerial foraging.

The relationship of the Apodiformes to other orders is unresolved. One possibility, suggested by molecular evidence, is a fairly close alliance with owls and nightjars.

The distribution of swifts and hummingbirds is strongly constrained by available habitat. Swifts require vertical surfaces for nesting, an extreme example being the Black Swift, which nests only behind "bridal veil" waterfalls with live moss. The Chimney Swift has been almost completely won over by chimneys and other artificial structures. Although hummingbirds do not face the same nest-placement constraints that swifts do, their distribution is importantly determined by availability of food resources. The recent proliferation of records of western hummingbirds in the Southeast is thought to reflect a

real winter range shift (as opposed to increased detection by hummingbird enthusiasts), enhanced by increasing numbers of sugar-water feeders in yards and gardens.

All of our hummingbirds and swifts are migratory. In some instances, annual movements are impressive, as with trans-Gulf migrants like the Chimney Swift and Ruby-throated Hummingbird. In other instances, annual movements are more restricted, as with several hummingbird species that withdraw in winter from the Desert Southwest to the mountains or west coast of Mexico. Hummingbirds and swifts are physiologically susceptible to daily and even hourly changes in thermal conditions and are adept at handling these changes. Several species are able to enter into a condition of reduced metabolic activity called "torpor."

Conservation and management of the Apodiformes is inextricably linked to availability of resources. Chimney Swifts have expanded west with the human population increase in the region, whereas the Vaux's Swift—a close relative that still relies heavily on natural cavities for nesting—has declined with the cutting of forests in the Pacific Northwest. The adaptable Anna's Hummingbird has expanded and is still expanding its range, feasting on exotic plantings wherever it goes. But the closely related Costa's Hummingbird may be declining due to loss of desert habitat that supports the flowering plants on which it depends.

Chimney Swift
Chaetura pelagica CODE 1

Vaux's Swift
Chaetura vauxi CODE 1

ADULT. *Flight style and wing shape similar to Chimney Swift; Vaux's is a smaller bird with a paler throat, but be aware of the effects of lighting.* CALIFORNIA, SEPTEMBER

ADULT. *In most of its range, the only swift; distinguished as such by its long, swept-back wings and rapid, twittering flight. Immature has slightly paler body than adult.* MARYLAND, JUNE

ADULT. *Rump the same color as rest of upperparts (rump lighter on Vaux's), but be aware of the effects of lighting.* TEXAS, MAY

ADULT. *Rump and neck slightly paler than wings, upperparts more uniform on Chimney. Juvenile has slightly paler primaries and secondaries—a further subtle contrast with Chimney.* CALIFORNIA, SEPTEMBER

L 5.25"   WS 14"   WT 0.8 oz (23 g)

▸ one adult molt per year; complex basic strategy

▸ weak differences between adult and immature

▸ slight individual variation in overall darkness

Seen almost exclusively in flight; occurs singly or in small flocks during the day, sometimes in huge swirling flocks above evening roosts. Nests in chimneys or other artificial vertical surfaces. Occurs in many habitats, but favors cities and towns.

◂ Call a simple *chip*, often run together in a musical twittering; notes flatter and less shrill than Vaux's. Courting pairs glide in unison, wing tip to wing tip.

L 4.75"   WS 12"   WT 0.6 oz (17 g)

▸ one adult molt per year; complex basic strategy

▸ weak differences between juvenile and adult

▸ contrast between rump and back variable

The wilder, western counterpart of Chimney Swift; some nest in chimneys, but many retain ancestral preference for hollow trees. Regional declines may be linked to cutting of old-growth forests. Seen in flight, usually high above coniferous or mixed forests.

Calls harsher, shriller, and higher-pitched than Chimney Swift: *chent* and *chewp*. With comparative experience, differences between the two species are obvious.

Black Swift
Cypseloides niger CODE 2

ADULT. *Large and dark; engages in dramatic, midair dives. Even at a distance, appears notably dark; tail often splayed out. Body feathers of juvenile flecked with white.* COLORADO, JUNE

AT REST. *Like all swifts, clings to vertical surfaces by bracing itself with tail. Swifts are rarely seen perched, and very few people have seen Black Swifts except in flight.* CALIFORNIA, AUGUST

L 7.25"   WS 18"   WT 1.6 OZ

▸ one adult molt per year; complex basic strategy

▸ moderate differences between juvenile and adult

▸ averages larger from south to north, but variable

Patchily distributed; absent from many landscapes. Breeds only amid mossy "bridal veil" waterfalls, where running water lathers vertical surfaces. Nest sites are ancestral, occupied every year. Forages high in the sky, often far from nests; rarely seen on migration.

Call a simple *chik*, usually repeated more slowly than Chimney and Vaux's Swifts. Sometimes these notes run together in a trill, like other swifts.

White-throated Swift
Aeronautes saxatalis CODE 1

ADULT. *Black breast with white center creates "vested" look. Throat and flanks white; tail long and forked. Can be confused with Violet-green Swallow. Juvenile slightly duller.* CALIFORNIA, JANUARY

ADULT. *Similarity to Violet-green Swallow (especially uncolored juvenile) is strongest from above. Wings of White-throated Swift swept back; tail tapered except when banking or soaring.* OREGON, AUGUST

L 6.5"   WS 15"   WT 1 OZ (28 g)

▸ one adult molt per year; complex basic strategy

▸ weak differences between juvenile and adult

▸ body size averages larger from south to north

Inhabits foothills and mountains, especially around cliffs and outcroppings; also nests and forages amid tall buildings in cities. Gregarious; occurs in small flocks most of the year, sometimes in large flocks during migration. Breeding and wintering ranges expanding.

Notes shrill and harsh, *jeer* and *jeet*, resembling Canyon Wren. Often given in series, becoming louder and then fading; effect sometimes ventriloqual.

Ruby-throated Hummingbird

Archilochus colubris CODE 1

ADULT MALE. *Green above; gray-green below, with little if any rufous. Forked tail is relatively long.* TEXAS, MAY

ADULT MALE. *As on most hummingbirds, color of the "gorget" (throat) of Ruby-throated depends highly on lighting. Gorget of Ruby-throated appears blazing red in good light; dark otherwise.* TEXAS, MAY

3.75"	WS 4.5"	WT 0.12 oz (3.8 g)	♀ > ♂

- one adult molt per year; complex basic strategy
- strong age-related and sex-related differences
- extent of green, especially on underparts, variable

The only hummingbird that breeds in eastern North America. Common in forest clearings, orchards, and gardens. Less common in forests, but does occur well into interior. Migrants conspicuous at flowers and feeders; breeders less conspicuous, often catch small insects.

Call a short, high *ch'*, given in direct flight or when perched. In aggressive encounters, a fast series: *ch... ch'ch'ch'ch'*. Humming of wings high-pitched.

IMMATURE MALE. *Throat often flecked with a few bright red feathers. Outer tail feathers of female and immature black with white tips; slightly less pointed than on Black-chinned.* TEXAS, SEPTEMBER

Until the late 20th century, the "conventional wisdom" was that all hummingbirds in eastern North America were Ruby-throated, unless proven otherwise. It is now known that any eastern hummingbird after mid-October is actually more likely to be a western species than it is a Ruby-throated. Fortunately, most late-season hummingbirds find their ways to feeders, where they can be observed at length, photographed, and sometimes banded.

FEMALE. *Green and gray; usually shows limited buff on underparts. Bill shorter, tail longer, and wings pointier than female Black-chinned, but field separation very difficult.* ONTARIO, JULY

Allen's Hummingbird

Selasphorus sasin   CODE 1

Rufous Hummingbird

Selasphorus rufus   CODE 1

ADULT MALE. *Extensively rufous; gorget broad. Typical males have much rufous on back and crown. Immature male has diminished red on throat, greener upperparts.* CALIFORNIA, MARCH

ADULT MALE. *Extensively rufous. All have green backs and green on crown; male Rufous usually orange above from crown to tail, but some can be extensively green like Allen's.* LOUISIANA, DECEMBER

JUVENILE MALE. *Very similar to Rufous and some cannot be identified. Tail, if seen well, is different: outer tail feathers more pointy and tapered than Rufous. Adult female similar.* CALIFORNIA, JULY

FEMALE. *Back green, but tail feathers have extensive orange at base; compare female Broad-tailed. Outer tail feathers broader and more rounded than Allen's.* CALIFORNIA, SEPTEMBER

| L 3.75" | WS 4.25" | WT 0.11 oz (3 g) | ♀ > ♂ |

▸ one adult molt per year; complex basic strategy

▸ strong age-related and sex-related differences

▸ extent of orange variable, especially on upperparts

Southern counterpart of Rufous. Movements fascinating: returns to coastal California by early January; begins "fall" migration to restricted Mexican wintering grounds in mid-May; a few stray to East Coast. Separate population on Channel Islands is apparently sedentary.

Call note a harsh, descending *tsick*, given singly or in rapid series. In chase, gives chattering series of *s'ticka* notes. Humming of wings metallic.

| L 3.75" | WS 4.5" | WT 0.1 oz (3.4 g) | ♀ > ♂ |

▸ one adult molt per year; complex basic strategy

▸ strong age-related and sex-related differences

▸ extent of rufous, especially on back, highly variable

Leaves breeding grounds in Northwest woods very early, with protracted "fall" migration well underway by early July; males migrate before females. Increasingly frequent in fall and winter in East Coast. Brief spring migration mainly through far West. Aggressive.

◀ Call a harsh *chep*. In chases, gives distinctive series: *zzee zickity zick*, all notes with buzzing quality; also gives single *zzzik* notes.

Black-chinned Hummingbird

Archilochus alexandri CODE 1

ADULT MALE. *Male gorget, if seen well, is distinctive: mainly black, but with broad purple base. Green above and gray-green below. Immature male has irregular purple flecks on throat.* CALIFORNIA, MAY

JUVENILE FEMALE. *Very similar to female Ruby-throated. With good view of perched bird in profile, wing tips appear broad (more pointed in Ruby-throated) and bill is longer.* COLORADO, JULY

L 3.75"	ws 4.75"	wt 0.12 oz (3.3 g)	♀ > ♂

▸ one adult molt per year; complex basic strategy
▸ strong age-related and sex-related differences
▸ overall body size smaller in Texas than elsewhere

Western counterpart of Ruby-throated Hummingbird. Habitat generalist, ranging from hot deserts to montane forests. Most numerous in southern portion of range, flourishing both in remote desert washes and in large cities. Increasingly sighted in winter along Gulf coast.

Call a flat *zip* or descending *zoop*, both lower-pitched than Ruby-throated. Other sounds, including humming of wings, average lower in pitch than Ruby-throated.

Broad-tailed Hummingbird

Selasphorus platycercus CODE 1

ADULT MALE. *Green above, green and gray below, with limited buff on flanks. Rose-colored gorget more extensive than Ruby-throated. Throat of immature male flecked with rose.* ARIZONA, AUGUST

ADULT FEMALE. *Tail is impressively large; female Rufous has more orange in tail feathers, female Calliope averages less and is distinctively shorter-tailed.* COLORADO, AUGUST

L 4"	ws 5.25"	wt 0.13 oz (3.6 g)	♀ > ♂

▸ one adult molt per year; complex basic strategy
▸ strong age-related and sex-related differences
▸ extent of russet on flanks and tail variable

The predominant breeding hummingbird of the southern Rockies; numerous in foothill canyons, mountain meadows, even at tree line. Local dispersal tied to retreat of snow pack, availability of blooming flowers. Copes with environmental stress by going into torpor.

Male in flight gives shrill trilling, created by wings; a common sound of summer in the Rockies. Gives *chew* and *chip* notes, not as modulated as Rufous.

Calliope Hummingbird

Stellula calliope   CODE 1

ADULT MALE. *Tiniest North American bird; short-tailed and relatively short-billed. Green above; green and white below with little buff. Gorget consists of long, coarse, magenta streaks.* MONTANA, JULY

ADULT FEMALE. *Dumpy overall; wing tips extend to tip of tail; bill short, straight, and thin. Female has dull buff wash on flanks, but little buff in tail; compare genus Selasphorus.* COLORADO, AUGUST

L 3.25"	WS 4.25"	WT 0.1 oz (2.7 g)	♀ > ♂

- ▸ one adult molt per year; complex basic strategy
- ▸ strong age-related and sex-related differences
- ▸ extent of buff on wings and flanks variable

Breeds in western mountains, generally to the north and west of Broad-tailed. Even more than Broad-tailed, subject to thermal stress; local distribution closely tied to microclimatic conditions. Increasingly reported in East in winter, possibly indicating a range shift.

Vocalizations average higher, thinner, and weaker than Broad-tailed and other hummingbirds. Call a thin *tsip*. Humming of wings very high-pitched, like a fly.

Anna's Hummingbird

Calypte anna   CODE 1

ADULT MALE. *Green above; green and gray below with little or no buff. Male has more red on head than any other North American hummingbird; red usually extensive on face and crown.* CALIFORNIA, APRIL

ADULT FEMALE. *Large hummingbird, relatively short-billed. Most females show limited red on gorget; immature males show more. Most have a prominent white patch behind or over the eye.* CALIFORNIA, MARCH

L 4"	WS 5.25"	WT 0.2 oz (4.3 g)

- ▸ one adult molt per year; complex basic strategy
- ▸ strong age-related and sex-related differences
- ▸ extent of red on throat variable in females, subadults

Aggressive and increasing. Range has extended north along Pacific coast, east into Desert Southwest. Prospers in urban settings with eucalyptus and other exotic plantings. Extent and nature of migration unclear and probably changing; tied to availability of resources.

◀ Song distinctive, *breezy breezy breezy...*, like tiny scissors; given by perched male. Male in display gives hard *snip!*—remarkably loud for so small a bird.

Costa's Hummingbird

Calypte costae CODE 1

ADULT MALE. *Like closely related Anna's, color on head extensive; intense purple on Costa's. Otherwise similar to Anna's: green above; green and gray below with little if any buff.* CALIFORNIA, MARCH

ADULT FEMALE. *Smaller overall and smaller-tailed than Anna's and Black-chinned; bill relatively longer and slightly decurved. Half of all females and most immature males show some purple on gorget.* CALIFORNIA, MARCH

3.5"	ws 4.75"	wt 0.11 oz (3.1 g)	♀ > ♂

- one adult molt per year; complex basic strategy
- strong age-related and sex-related differences
- extent of purple on female throat variable

A surprisingly little-known species. Breeds in Mojave and Sonoran Deserts. Movements poorly understood; probably related to availability of blooming plants. Occurs in washes, springs, and riparian areas; also visits feeders, typically at periphery of cities and towns.

Calls varied, generally unlike closely related Anna's Hummingbird. Song distinctive, but inaudible except at close range: a thin, high-pitched *suh-weeeeeee!*

Magnificent Hummingbird

Eugenes fulgens CODE 2

ADULT MALE. *Large and dark. Green above; green and black below. Gorget is green; cap is dark purple, appearing black except in good light.* ARIZONA, APRIL

ADULT FEMALE. *Large and green; bill impressively long. Rump mainly green (female Blue-throated has extensive copper on rump). Immature male similar, but more extensively green below.* ARIZONA, APRIL

L 5.25"	ws 7.5"	wt 0.25 oz (7 g)	♂ > ♀

- ▶ one adult molt per year; complex basic strategy
- ▶ strong age-related and sex-related differences
- ▶ extent of bronze and golden highlights variable

Tropical species that migrates north to Arizona; arrives by early April. Familiar visitor to feeders and lowland canyons, but also ranges well up into mountains; may range farther north in mountains than is currently known. Feeds in dry pine woods; nests near streams.

Call a hard *suh-tick* or *tsick*, fairly rich; recalls Dark-eyed Junco or Lincoln's Sparrow. In aggression, gives series of thinner, higher *kyoo* notes.

Blue-throated Hummingbird

Lampornis clemenciae   CODE 2

ADULT MALE. *Larger-bodied but shorter-billed than Magnificent; wingbeats slower than other hummingbirds. Dull green above, dull gray below. Male gorget dull blue, flecked with gray.* ARIZONA, APRIL

ADULT FEMALE. *Both sexes show dark ear-patch framed by white stripes. Large tail has white tail corners: prominent when tail is flared; visible even when tail is folded, as here.* ARIZONA, APRIL

L 5"	WS 8"	WT 0.27 oz (7.6 g)	♂ > ♀

- one adult molt per year; complex basic strategy
- moderate differences between male and female
- white in tail always present, but extent variable

Like Magnificent, a tropical species that migrates north to Southeast Arizona. More closely associated with broadleaf vegetation than Magnificent, favoring sycamores, oaks, and maples. Readily accepts manmade structures for nesting; increasingly reported during winter.

Vocalizations average higher than Magnificent. Call a rising *swee!* Song a whispered jumble of sweet, very high-pitched notes, like a tiny Brown Creeper.

Broad-billed Hummingbird

Cynanthus latirostris   CODE 2

ADULT MALE. *Dark overall. Gorget deep blue, often appearing black; blue limited in gorget of immature male. Long, decurved bill shows extensive red. Dark tail has deep notch.* CALIFORNIA, JUNE

FEMALE. *Fairly pale; dusky green above, mainly gray below. Dull individuals, as here, show limited red in bill and a faint white line behind eye; many females more prominently marked.* ARIZONA, JULY

L 4"	WS 5.75"	WT 0.1 oz (2.9 g)

- one adult molt per year; complex basic strategy
- strong age-related and sex-related differences
- brightness of bill and extent of auriculars variable

Tropical species that migrates north to Arizona and New Mexico, arriving mid-March. Mainly at low elevations: low canyons, adjacent washes, even in riparian habitats far from mountains. Generally associated with broadleaf vegetation: syca-more, cottonwood, and mesquite.

Call a husky *chick* or two-syllable *chivvik*, the latter sounding surprisingly like Ruby-crowned Kinglet; flatter than primary call of White-eared.

White-eared Hummingbird

Hylocharis leucotis CODE 3

ADULT MALE. *Thick white line behind eye more prominent than any Broad-billed. Gorget variable green and blue; often appears all-dark. Bill extensively red like Broad-billed.* ARIZONA, JULY

ADULT FEMALE. *Similar to female Broad-billed, but white line behind eye more prominent on White-eared; Side of breast has green spots; breast largely unmarked gray on Broad-billed.* COLORADO, JULY

| 3.75" | WS 5.75" | WT 0.12 oz (3.3 g) |

- one adult molt per year; complex basic strategy
- strong age-related and sex-related differences
- extent of green and white on underparts variable

Tropical species that reaches our southwestern mountains annually but in small numbers. Bulk of wanderers to North America may be in pine forests at fairly high elevations, where less likely to be detected than in popular low-elevation canyons. Possibly increasing here.

Call a rich *chink* or *tick*; differs from husky, often two-syllable note of Broad-billed. Identification by voice perhaps an underutilized method of detection.

Berylline Hummingbird

Amazilia beryllina CODE 3

ADULT. *Similar to Buff-bellied Hummingbird, but their North American ranges are widely separated. Back, head, and breast green; wings, rump, and tail rusty.* MEXICO, SEPTEMBER

IMMATURE. *Duller and grayer than adult, but rusty in wings and tail. Sexes are similar in genus* Amazilia, *unlike other North American hummingbird genera.* MEXICO, AUGUST

| L 4.25" | WS 5.75" | WT 0.16 oz (4.6 g) | ♂ > ♀ |

- one adult molt per year; complex basic strategy
- moderate differences between immature and adult
- extent of green and gray on underparts variable

Tropical species that barely reaches the mountains of the Desert Southwest; most records well into the summer. Presumably annual in small numbers; has bred here. Usually seen around feeders; probably spreads out into canyons and mid-elevation oak forests.

Song complex: *zrrrick zr'doo zr'doo...*, harsh and scratchy. Calls variable, mainly short, high-pitched, and modulated: *zzzrick* and *zzzree*.

Lucifer Hummingbird
Calothorax lucifer   CODE 2

ADULT MALE. *Green above, including crown. Gorget purplish like Costa's, but purple absent from crown and face. Forked tail appears tapered on perched bird. Bill strongly decurved.* ARIZONA, JULY

ADULT FEMALE. *In all plumages, bill long and decurved. A green-and-buff hummingbird: long line behind eye usually has buff tints; underparts and rump also show much buff.* ARIZONA, JULY

| L 3.5" | ws 4" | wt 0.11 oz (3.1 g) | ♀ > ♂ |

▸ one adult molt per year; complex basic strategy

▸ strong age-related and sex-related differences

▸ extent and intensity of bronzy tones variable

Among our "Mexican" humming-birds, distinctive for occupying open, arid habitats often far from woodlands. Life cycle of North American populations closely tied to flowering of agaves; ocotillo another good indicator. Not strongly inclined to visit feeders.

Call is a dry *chip*, sometimes run into a short series or slow trill. Song a wheezy series; reminiscent of Anna's Hummingbird, but shorter and softer.

Buff-bellied Hummingbird
Amazilia yucatanensis   CODE 2

ADULT. *Belly extensively buff; head, breast, and upperparts green; gorget has blue highlights. Similar Berylline Hummingbird has more rusty in wings. Immature duller.* TEXAS, NOVEMBER

GREEN VIOLET-EAR (Colibri thalassinus). *Casual stray—especially to Texas—from the tropics. Mainly green, with purple-blue highlights on face, breast, and tail.* COSTA RICA, JANUARY

| L 4.25" | ws 5.75" | wt 0.13 oz (3.8 g) | ♂ > ♀ |

▸ one adult molt per year; complex basic strategy

▸ weak differences between adult and immature

▸ brightness and extent of buff on belly variable

Tropical hummingbird that reaches Texas, where fairly common; favors brushy areas, parks, and residences. Present mainly March–August. In fall, some withdraw south to Mexico, but others disperse well north and east along Gulf coast; a few remain in south Texas.

Call a sharp *tek*. Territorial "song" a long series of *tek* notes. Gives a diversity of other vocalizations, mainly shrill and squeaky.

Violet-crowned Hummingbird
Amazilia violiceps   CODE 2

| L 4.5" | ws 6" | wt 0.19 oz (5.5 g) | ♂ > ♀ |

▸ one adult molt per year; complex basic strategy

▸ moderate differences between immature and adult

▸ extent of bluish highlights, mainly on crown, variable

Tropical species; barely reaches North America in Desert Southwest. Strongly associated with low-elevation broad-leaf groves, especially sycamores in canyons and washes. Patchy; rare in many areas, but locally common. Recent increase in winter records.

Call a hard *check*, given singly, in slow series, or in fast series. Song soft and inaudible except at close range: shorter, simpler than Anna's.

ADULT. *Mainly brown above, with greenish highlights. White below, including throat. Bill red and slightly decurved. Blue crown shown by both sexes, but reduced or lacking on immature.* MEXICO, JULY

Trogons and Kingfishers

ORDERS TROGONIFORMES AND CORACIIFORMES

---

5 species: 1 ABA CODE 1   3 CODE 2   1 CODE 4

These visually striking birds are best represented in Tropical America, Africa, Southeast Asia, and the Australasian region. Our North American species frequently employ a sit-and-wait foraging strategy, perching on a snag and watching for prey. North American kingfishers are strictly fish-eaters, with prey being caught in awkward-looking but efficient face-first plunge-dives. Trogons sally out to catch wasps, and other flying insects; less often they eat fruit. Although in separate orders, trogons and kingfishers are reckoned by many authorities as similar to each other and are often placed in the same "superorder."

BIOGEOGRAPHY

Even the most spectacular hotspots of avian diversity harbor only a small fraction of the world's more than 10,000 bird species. The countries with the richest bird faunas—for example, Colombia and Peru—harbor less than twenty percent of those species. At smaller scales (a preserve, a watershed, a mountainside), the total is considerably less; in all but the most extreme cases, such sites harbor far fewer than 500 species.

Most birds are surprisingly hardy, as proven by the thriving bird communities of deserts, oceans, and polar regions. A few species are found nearly worldwide, such as the Peregrine Falcon and Barn Swallow. But the overwhelming majority are restricted to relatively small geographic regions.

The study of spatial and temporal distribution of species and populations is called biogeography. Scientists studying biogeography are drawn to questions of process: What limits or restricts the geographic range of a particular species or population? At small spatial scales, these limiting factors are often biological. The distribution of Belted Kingfishers across the state of Oklahoma may be determined by the abundance of prey and availability of nest sites.

At larger spatial scales, range limits may have more to do with historical and geographical happenstance. It takes time for populations to become established, and large-scale distributional shifts may operate across greater time scales. Quite possibly, the mountains of southern Colorado could support Elegant Trogons. Quite possibly, the marshes of Oklahoma could support Ringed Kingfishers. Equally plausibly, currently established populations of trogons in Arizona or kingfishers in Texas could be extinguished by catastrophic natural or unnatural events. The lesson is that current range limits are set not only by biological conditions, but also by a dynamic mosaic of long-term regional and global changes involving glaciers, plate tectonics, oceans, and the atmosphere.

Elegant Trogon
Trogon elegans   CODE 2

ADULT MALE. *Large and long-tailed. Blue-green above, extensively red below. Head dark; eye ring red; bill bright yellow. Wing coverts gray; conspicuous on bird at rest.* ARIZONA, JUNE

ADULT FEMALE. *Less colorful than male. Red limited to undertail coverts; extends up to lower breast on male. Green of head and upperparts on male replaced by gray on female.* ARIZONA, JUNE

L 12.5"   WS 16"   WT 2.5 oz

▸ one adult molt per year; complex basic strategy

▸ strong age-related and sex-related differences

▸ intensity of greenish coloration depends on light

The only species in its order that regularly reaches North America, Elegant Trogon is perhaps the most famous Southwest specialty. Arizona population is migratory; occurs in lush lowland canyons, usually with sycamores. Locally threatened by intense eco-tourism.

Male song a series of harsh *ruff* notes. Each element descending in pitch and slightly slurred; entire series trails off. Other notes rough, unmusical.

JUVENILE. *Gray-brown overall with coarse spots on wings. All plumages show long, bronzy tail; all sit quietly in shade trees, body leaning forward and tail pointing down.* ARIZONA, AUGUST

EARED QUETZAL (Euptilotis neoxenus). *Casual stray from Sierra Madre to Southeast Arizona; has bred. Larger than Elegant Trogon; small-headed; gray-billed; broad tail white beneath.* ARIZONA, DECEMBER

Belted Kingfisher
Ceryle alcyon CODE 1

ADULT MALE. *Lacks rufous on flanks shown on female; note white collar, short tail, and heavy bill.* TEXAS, NOVEMBER

ADULT FEMALE. *All are large-headed with a shaggy crest. Female has blue breast band and rufous on flanks and belly.* FLORIDA, FEBRUARY

L 13"   WS 20"   WT 5 OZ

- one adult molt per year; complex basic strategy
- moderate differences between male and female
- western birds average larger and longer-billed

Widespread and adaptable, but needs clear water for diving; frequents both tiny ponds and huge bodies of water in freshwater or brackish habitats. Winter distribution limited by ice-free water, breeding distribution by availability of vertical earthen banks for nesting.

Common call, most often given in flight, a long, loud rattle: *ch!k'k'k'k'...*, exceedingly rough. When perched, gives a softer, shorter rattle: *ch'rrrrr...*

IN FLIGHT. *Broad blue wings flash extensive white at base of primaries. Even with crest flattened, as in flight, appears bullheaded.* TEXAS, NOVEMBER

HOVERING. *All individuals look for fish while hovering in midair, sometimes high above the water. Prey are captured at the water's surface in messy splash-dives.* CALIFORNIA, JANUARY

ADULT MALE. *Linings of underwing unmarked white. On most adult females, rufous on flanks and belly extends to axillaries.* BRITISH COLUMBIA, NOVEMBER

Ringed Kingfisher
Ceryle torquatus CODE 2

ADULT MALE. *Similar to female; huge, bicolored bill always prominent. Male largely rufous below; lacks thick blue breast band of female, but may show broken half-collar.* TEXAS, NOVEMBER

ADULT FEMALE. *Considerably larger than Belted Kingfisher; bill enormous, even for a kingfisher. Female belly and undertail coverts entirely rufous; compare Belted.* TEXAS, NOVEMBER

L 16"   WS 25"   WT 11 OZ

‣ one adult molt per year; complex basic strategy

‣ moderate differences between male and female

‣ extent and intensity of rufous on underparts variable

The conspicuous "King Kong Fisher" of southern Texas. Numbers apparently increasing; range expanding into Texas hill country. Habits similar to Belted, but tends to occur in especially open habitats, choosing high perches, and plunge-diving from more than 30 feet high.

Common call a harsh, unmusical rattle like Belted, but averages lower, longer, and much slower: *cha... cha... cha... cha... cha...*

Green Kingfisher
Chloroceryle americana CODE 2

ADULT MALE. *Like all kingfishers, appears large-headed with an oversized bill. Male is green above; white below with broad rusty breast band.* ARIZONA, NOVEMBER

ADULT FEMALE. *Lacks rufous breast band of male, but shows thin, broken bands of green across belly and upper breast. Crest usually less conspicuous than larger kingfishers.* TEXAS, DECEMBER

L 8.75"   WS 11"   WT 1.3 OZ

‣ one adult molt per year; complex basic strategy

‣ strong differences between male and female

‣ much individual variation in extent of green and white

Befitting its small size, hunts inconspicuously and flushes unobtrusively along wooded edges of quiet waterways. Catches small fish in shallow dives. Suffered population losses in 20th century; currently recovering, spreading into Edwards Plateau in Texas.

Common call is a weak, doubled *tick tick*, repeated slowly. Call sometimes shortened to a single *tick*, also repeated; ventriloqual, easy to overlook.

Woodpeckers
FAMILY PICIDAE

25 species: 17 ABA CODE 1   5 CODE 2   1 CODE 4   1 CODE 5   1 CODE 6

This familiar family demonstrates a powerful concept in evolutionary biology—that of a group of organisms whose common ancestor "figured out" long ago how to exploit an unoccupied ecological niche. Once the niche is "discovered," in an evolutionary sense, there follows impressive diversification by the descendants of the discoverers. Although woodpeckers are situated taxonomically before the passerines (perching birds), that placement is widely accepted as being provisional. Several authorities treat the woodpeckers as part of a lineage far removed from the passerines.

The well-named woodpeckers occur wherever there is wood to peck on—that is, in wooded districts nearly worldwide. In North America, virtually all tree communities host one or more woodpecker species. The "softwoods" (conifers) are attractive to many species, including some that are highly specialized—for example, White-headed Woodpeckers in ponderosa pine forests and Red-cockaded in mature pines afflicted with heartwood disease. The "hardwoods," especially aspen, provide food and nest sites to sapsuckers. Desert trees, including saguaros, host Gila Woodpeckers and Gilded Flickers. Diseased and dying trees are highly prized by Black-backed, American Three-toed, Hairy, and Lewis's Woodpeckers.

The simple vocalizations of woodpeckers are supplemented in many species by nonvocal drumming that serves the same function as song in most other kinds of birds. To the human ear, the general rule is that differences among species are hard to discern. But there are a few exceptions, such as the gradually slowing drums of sapsuckers and the powerful whacks of the Pileated Woodpecker.

North American woodpeckers, on the whole, are not impressively migratory. Most species are essentially sedentary, a reflection of their tight ecological and evolutionary associations with a particular tree species or tree community. The species that migrate are typically somewhat generalist—for example, the flycatching Red-headed Woodpecker and the anteating Northern Flicker. Sapsucker migrations are necessitated by the annual cycle of dormancy in their host trees.

Management and conservation issues for woodpeckers are closely linked with forest health. Most species require mixed-age stands with old, dying, or diseased trees that are not usually left after salvage logging. American forests have been so heavily transformed by forestry practices that simply "leaving them alone" will probably not be sufficient. Continued survival of the endangered Red-cockaded Woodpecker, for instance, will require active management of its pine forests.

Downy Woodpecker

Picoides pubescens   CODE 1

ADULT MALE, WESTERN. *All Downy Woodpeckers are pied above and white below; all have short bills compared to Hairy. Male has red on rear of crown; juvenile has red on forecrown.* CALIFORNIA, JANUARY

ADULT FEMALE, WESTERN. *Female lacks red cap. All plumages have white back; most have prominent black spots on the white outer tail feathers, lacking or reduced on Hairy.* CALIFORNIA, JANUARY

JUVENILE. *Both sexes have red on cap, sometimes extensively so; red on adult male is farther back on crown.* COLORADO, JULY

ADULT FEMALE, EASTERN. *White spots on wing coverts prominent; much reduced on Western. Some far-western birds, not shown, have sooty gray-white in place of white on underparts and head.* TEXAS, JUNE

L 6.75"   WS 12"   WT 1 oz (28 g)   ♂ > ♀

▸ one adult molt per year; complex basic strategy

▸ moderate age-related and sex-related differences

▸ extent of black and white on wings and tail variable

Widespread and familiar; inconspicuous but quietly industrious and approachable. Solitary except in late winter, when males ward off other males and pursue females. Feeds on smaller branches than other woodpeckers; fond of mullein stalks, where it probes for insects.

The geographic variation in the plumage of the Downy and Hairy Woodpeckers is similar across North America: in the East, both species show extensive white spotting on the wings and scapulars; in the Interior West, this spotting is reduced; and on the Pacific Slope, both species sometimes show extensive dusky below. This similar variation in plumage may indicate how regional environmental differences can create similar patterns in two different species.

Gives a soft *pik*, often while perched or foraging; short, descending whinny of *pik* notes, typically when flushing. Drum is soft, slow, short, and even.

Hairy Woodpecker

Picoides villosus CODE 1

ADULT MALE, WESTERN. *Plumage similar to Downy; geographic variation mirrors Downy, with western birds having reduced spotting on wings. Bill is longer than Downy.*
CALIFORNIA, JANUARY

ADULT FEMALE, EASTERN. *Plumage similar to Downy; note white spots on wing coverts on eastern birds. Rangewide, white outer tail feathers usually lack black spots shown by Downy.*
MINNESOTA, JANUARY

| 9.25" | ws 15" | wt 2.3 oz | ♂ > ♀ |

▸ one adult molt per year; complex basic strategy
▸ moderate age-related and sex-related differences
▸ ground color and overall patterning highly variable

Often regarded as the larger counterpart of the Downy Woodpecker, but apparently not its closest relative. Habitat broadly overlaps with Downy, but there are differences: in West, more closely associated with pinewoods and burns; in East, with larger forest tracts.

◀ Vocalizations similar to Downy, but sharper and harsher. Call note an explosive *peek!* Whinny a variable series of sharp notes, often irregular in delivery.

Red-cockaded Woodpecker

Picoides borealis CODE 2

ADULT FEMALE. *Wings and back striped black-and-white. Face mainly white; tiny red "cockade" of male near rear of crown hard to see; juvenile has larger red spot on forecrown.*
LOUISIANA, MAY

ADULT MALE. *The amount of red shown near the rear of the crown on this individual, although limited, is typical.*
ALABAMA, MAY

| ʟ 8.5" | ws 14" | wt 1.5 oz |

▸ one adult molt per year; complex basic strategy
▸ moderate differences between juvenile and adult
▸ northern and inland populations average larger

Endangered species; requires extensive, mature, open pine stands maintained by frequent summer fires started by lightning. Breeds cooperatively in living trees; nest sites indicated by gooey resin. Local populations sedentary; do not adapt well to logging.

Call note sharper than Downy, with burry, whistled quality: *fshwink.* Drum dull and flatulent. At nest or while foraging, small groups chatter softly.

Ladder-backed Woodpecker

Picoides scalaris CODE 1

ADULT MALE. *A little larger than Downy, with longer bill and more red on crown. Back and wings striped black-and-white; underparts dirty white with black spots.* TEXAS, APRIL

ADULT FEMALE. *Paler overall, more white on wings and back, and more extensive white on face than Nuttall's. Almost no range overlap between the two species.* ARIZONA, DECEMBER

L 7.25" WS 13" WT 1 oz

- ▸ one adult molt per year; complex basic strategy
- ▸ moderate differences between male and female
- ▸ individual variation in amount of barring and spotting

In most of North American range, the characteristic small woodpecker of lowland habitats. Habitat overlaps with Downy, but Ladder-backed generally in sparser, drier habitats: washes, Joshua tree woodlands, and cactus gardens. Inconspicuous, but visits towns and cities.

Vocalizations similar to Downy. Call a flat *pek*, slightly sharper than Downy. Whinny starts more ringing than Downy, but concludes with rougher notes.

Nuttall's Woodpecker

Picoides nuttallii CODE 1

ADULT MALE. *Red on crown does not extend as far forward as on Ladder-backed. White bars on back and white stripes on face narrower than on Ladder-backed.* CALIFORNIA, JANUARY

ADULT FEMALE. *More contrastingly black and white than Ladder-backed, which often has a buff or dirty white ground color on the underparts and face.* CALIFORNIA, MAY

L 7.5" WS 13" WT 1.3 oz

- ▸ one adult molt per year; complex basic strategy
- ▸ moderate differences between male and female
- ▸ hybrids with Ladder-backed in narrow zone of overlap

California counterpart of more widespread Ladder-backed Woodpecker. Closely tied to dry oak woodlands, especially in northern portion of range; farther south, more likely to occur in mixed riparian woodlands. Declining health of oak woods poses threat.

Vocalizations distinct from Ladder-backed and Downy. Call note a stuttering, strident *p'tick*, often repeated. Whinny an even series, usually not trailing off; suggests Belted Kingfisher.

Arizona Woodpecker
Picoides arizonae CODE 2

ADULT MALE. *On both sexes, back is brown; tail black; barred and spotted below. Face mainly white with a large brown spot. Male has red patch on head.* ARIZONA, NOVEMBER

ADULT FEMALE. *Both sexes are about midway in size between Downy and Hairy; bill structure about the same as Hairy. Female lacks red on head.* ARIZONA, MAY

L 7.5"	ws 14"	wt 1.6 oz	♂ > ♀

▸ one adult molt per year; complex basic strategy
▸ moderate differences between male and female
▸ extent of white and amount of brown spotting variable

Tropical species that ranges north into the sky islands of Desert Southwest. Inconspicuous and reclusive; can be hard to find. Widely distributed, although thinly so, in dry forests, especially oak and pine oak; some wander into sycamore-walnut stands in canyons.

Vocalizations similar to Hairy. Call note a high *peech*, averaging raspier than Hairy. Whinny similar to Hairy, but duller, less piercing; suggests Ladder-backed.

White-headed Woodpecker
Picoides albolarvatus CODE 1

ADULT MALE. *All plumages have largely white head; rest of body black, except for white flash in primaries. Male has limited red at rear of crown.* CALIFORNIA, JULY

ADULT FEMALE. *Like male, but red on crown of adult male replaced by black. Juvenile of both sexes has more red than adult male, forming cap that extends forward almost to eye.* OREGON, JULY

L 9.25"	ws 16"	wt 2 oz	♂ > ♀

▸ one adult molt per year; complex basic strategy
▸ moderate age-related and sex-related differences
▸ bill size and tail length increase slightly southward

All aspects of life history closely tied to pine forests, primarily ponderosa pine; depends heavily on pine seeds for food, especially fall–winter. Rarely wanders from pinewoods. Numbers have declined due to clear-cutting, snag removal, and fire suppression.

Call note distinctive: a sharp, descending *p'tenk* or *piddy-denk*; has strange mechanical quality, suggesting a small drill. Call soft, but carries well.

Red-headed Woodpecker
Melanerpes erythrocephalus   CODE 1

ADULT. *Head is completely red, unlike any other woodpecker in its range. Black-and-white contrast prominent from afar on both perched and flying birds. Sexes alike.* TEXAS, APRIL

JUVENILE. *Head is gray-brown, unlike adult. Plumage more smudgy overall, but large white wing patches prominent in flight and at rest.* OKLAHOMA, OCTOBER

L 9.25"   WS 17"   WT 2.5 oz

- ▸ one adult molt per year; complex basic strategy
- ▸ strong differences between adult and juvenile
- ▸ western birds larger, with tinge of red on belly

A feisty, flycatching, food-storing woodpecker. Distribution is complex: nearly absent from some landscapes, common in others; migratory in parts of range, sedentary elsewhere. Food availability affects annual movements.

◀ Call a rising *queerv*; shriller than Red-bellied, suggests Gray-cheeked Thrush flight call. In social situations, gives stuttering series of *uh-chuh!* notes.

Black-backed Woodpecker
Picoides arcticus   CODE 2

ADULT MALE. *All plumages are glossy black above; whitish below with variable black barring. Yellow cap of male can be hard to see; yellow replaced by black on female.* CALIFORNIA, JUNE

ADULT FEMALE. *Size and structure similar to Hairy, a species that frequently accompanies Black-backed in recent burns. Black-backed has three toes, but this feature is hard to see.* CALIFORNIA, JUNE

L 9.5"   WS 16"   WT 2.5 oz

- ▸ one adult molt per year; complex basic strategy
- ▸ moderate differences between adult male and female
- ▸ individual variation in black-and-white patterning

Best known as specialist on recent burn areas; also attracted to old, dying, and diseased stands. Occurs widely but thinly throughout montane coniferous forests. Mainly sedentary; some wander south in winter. Locally threatened by fire suppression and salvage logging.

Call a dry, blackbird-like *cleck*. Also gives a fast, high-pitched rattle, resembling Marsh Wren. Drum consists of a few slow taps, followed by faster tapping.

American Three-toed Woodpecker
Picoides dorsalis   CODE 2

ADULT MALE, NORTHWESTERN. *Back barred black-and-white; sides strongly barred. All males have yellow caps; cap streaked black-and-white in female.* BRITISH COLUMBIA, MAY

ADULT MALE, EASTERN. *Has less white on back than other populations; some resemble Black-backed. Head of Black-backed more solidly dark; back completely black.* MAINE, JULY

L 8.75"   WS 15"   WT 2.3 oz

» one adult molt per year; complex basic strategy
» moderate differences between adult male and female
» much geographic variation in amount of white

Like Black-backed, an uncommon inhabitant of coniferous forests. Prefers spruce, but scarce even there. Opportunistically invades extensive burn areas. Quiet; difficult to locate. Specializes on bark beetles; Black-backed on wood-boring beetles.

Call intermediate between Black-backed and Hairy, a raspy *chetch.* Rattle soft and squeaky. Drum distinctively changes tempo midway through.

All but two of our woodpeckers have four toes (two extending forward, and two extending backward in an arrangement called "zygodactyl"). The American Three-toed and Black-backed Woodpeckers have three toes, two facing forward and one facing backward. Despite the difference, these two species climb trees in a manner similar to other woodpeckers.

ADULT FEMALE, INTERMOUNTAIN WEST. *In southern Rockies, more resembles Hairy than Black-backed; females especially similar. Back nearly white, but sides barred black-and-white.* COLORADO, JULY

Red-bellied Woodpecker

Melanerpes carolinus　CODE 1

ADULT MALE. *All plumages barred black-and-white above; white or pale buff below, with or without red on belly and vent. Scarlet on male extends from bottom of nape all the way to bill.* TEXAS, APRIL

L 9.25"	WS 16"	WT 2.2 OZ	♂ > ♀

▸ one adult molt per year; complex basic strategy

▸ moderate age-related and sex-related differences

▸ individual variation in amount of red on underparts

Noisy, showy, and successful; expanding north and west. Recovery of northeastern forests and proliferation of bird feeders have probably encouraged range expansion. In much of core range, the most abundant woodpecker; in newly colonized regions, quickly becomes common.

◀ Call note a muffled and descending *cheerf*; given singly, a few times, or in muffled series. In spring and summer gives a rising, tremulous *queer-r-r.*

Woodpeckers in the genus Melanerpes *are generally more opportunistic feeders than other woodpeckers. Red-bellied Woodpeckers, for example, pluck fruit from trees, snag insects from foliage, and sally short distances for large flying insects. Enterprising individuals will even capture lizards and small birds.*

JUVENILE. *Head gray-tan with little if any red. All plumages show white on tail; Golden-fronted has all-black tail feathers.* TEXAS, APRIL

Gila Woodpecker
Melanerpes uropygialis CODE 1

ADULT MALE. *Back, wings, and tail barred black-and-white. Head and underparts gray-tan; adult male has small red cap.* ARIZONA, MAY

ADULT FEMALE. *Differs from adult male in lacking red cap. Juvenile resembles adult female.* ARIZONA, MAY

Golden-fronted Woodpecker
Melanerpes aurifrons CODE 1

ADULT MALE. *Head tricolored: red cap, orange nape, yellow in front of eye. "Zebra-backed" like Red-bellied, but tail is black; central tail feathers of Red-bellied mainly white.* TEXAS, NOVEMBER

ADULT FEMALE. *Head has yellow on nape and in front of eye; head of juvenile unmarked, like Red-bellied. In all plumages, all-black tail contrasts with white rump.* TEXAS, NOVEMBER

L 9.25"	WS 16"	WT 2.3 OZ	♂ > ♀

▸one adult molt per year; complex basic strategy
▸moderate differences between adult male and female
▸individual variation in tail pattern and crown color

Southwestern equivalent of Red-bellied and Golden-fronted. Noisy and conspicuous; characteristic of saguaro woodlands, but also common in urban and riparian environments. Clearing of riparian woodlands has caused declines in northern and western portions of range.

Throughout year, gives series of gull-like yaps. Also gives a rolling *queer-r-r*, slightly less musical than Red-bellied, slightly faster than Golden-fronted.

L 9.5"	WS 17"	WT 3 OZ	♂ > ♀

▸one adult molt per year; complex basic strategy
▸strong sex-related and moderate age-related differences
▸occasional hybrids with Red-bellied in zone of overlap

Replaces Red-bellied in dry brush-lands and semi-open woodlands of central Texas; especially character-istic of mesquite groves. Numbers apparently increasing and range expanding northward. Conspicuous; often seen feeding on cactus fruits and in pecan orchards.

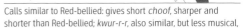

Calls similar to Red-bellied: gives short *choof*, sharper and shorter than Red-bellied; *kwur-r-r*, also similar, but less musical, less ringing.

Acorn Woodpecker
Melanerpes formicivorus CODE 1

ADULT MALE. *Extensive black in plumage contrasts with yellow-white on face and gray-white belly. Red on head of male extends from upper nape to above the eye.* CALIFORNIA, NOVEMBER

ADULT FEMALE. *Similar to male, but reduced red on head does not reach all the way to eye; female is shorter-billed than male.* ARIZONA, AUGUST

L 9"   WS 17.5"   WT 3 OZ

▸ one adult molt per year; complex basic strategy
▸ moderate differences between adult male and female
▸ Pacific-slope birds larger than Southwest and Texas

Occurs in noisy, conspicuous, highly social family groups. Defends stored food in communal caches called granaries; young raised cooperatively. Occurs primarily in oak woodlands; also pine-oak woods. Mainly resident, but individuals and local populations wander.

Gives diverse rattles, chatters, and yelps: some musical, quail-like; others harsh, parrot-like. Easily learned vocalization is rising series of *Jacob!* notes.

Lewis's Woodpecker
Melanerpes lewis CODE 1

ADULT. *In good light, exquisite: pink-bellied, red-faced, gray-collared, and green-backed. In poor light or at a distance, appears distinctly dark. Sexes alike. Flight direct with deep wingbeats.* CALIFORNIA, JULY

JUVENILE. *Lacks the pink and red highlights of adult. Underparts gray, extending onto sides of neck. Dark above; not as glossy as adult.* OREGON, JANUARY

L 10.75"   WS 21"   WT 4 OZ

▸ one adult molt per year; complex basic strategy
▸ strong differences between adult and juvenile
▸ individual variation in width of gray collar

Patchily distributed in broken woodlands; locally absent from many seemingly suitable swaths of habitat. Small groups gather in open ponderosa pine forests, riparian groves, and recent burns. Foraging method distinctive: in prolonged glides, sallies for flying insects.

Quiet compared to other members of genus. Calls varied, mainly soft: include shrill *jeeeer* like distant tern; chattering, squirrel-like series of *jee* notes.

Yellow-bellied Sapsucker

Sphyrapicus varius   CODE 1

ADULT FEMALE.
Throat usually white. Paler than other sapsuckers: red and black markings on head less prominent than on other sapsuckers; white barring on back and wings extensive.
CONNECTICUT, JUNE

ADULT MALE. *Has less red on head than other sapsuckers. Red on throat is broadly bordered in black. Back barred black-and-white; most have at least as much white on back as black.* CONNECTICUT, JUNE

L 8.5"   WS 16"   WT 2 OZ

• one adult molt per year; complex basic strategy
• strong age-related and moderate sex-related differences
• hybridizes with Red-breasted and Red-naped

Conspicuous in northern forests, especially early-successional hardwood stands; inconspicuous in winter, when coniferous groves are favored. Drills sap wells in horizontal rows; other species, especially Ruby-throated Hummingbird, are attracted to diggings.

Drum distinctive: starts out with three to five explosive elements, then trails off quickly. Call, heard year-round, a descending squeal, suggesting Red-shouldered Hawk.

The Yellow-bellied, Red-naped, and Red-breasted Sapsuckers are closely related. In certain respects, this complex has the characteristics of a single species with strong clinal variation—ranging from palest in color in the east (Yellow-bellied) to darkest in the west (Red-breasted).

JUVENILE. *Duller than female, and more diffusely patterned. All plumages show a long, broad stripe on the wing. Juvenile feathers retained longer into winter than Red-naped.* NEW JERSEY, OCTOBER

Red-naped Sapsucker
Sphyrapicus nuchalis   CODE 1

ADULT MALE. *Head has three discrete red patches: crown, nape, and throat. Red on throat more extensive than Yellow-bellied; bordered above by white, not by black.*
COLORADO, JULY

HYBRID. *All possible hybrid pairings involving the three sapsucker species are observed. Red-naped × Red-breasted, shown here, frequently encountered on migration in Great Basin.* CALIFORNIA, OCTOBER

ADULT FEMALE. *Less red on nape than male; red on throat reduced, too. Body has more black overall than Yellow-bellied. Juvenile similar to Yellow-bellied but darker.*
COLORADO, JULY

L 8.5"   WS 16"   WT 2 OZ

‣ one adult molt per year; complex basic strategy

‣ strong age-related and moderate sex-related differences

‣ individual variation in extent of red, mainly on nape

Rocky Mountain counterpart of Yellow-bellied Sapsucker. Common in summer in montane forests, especially aspen groves; widespread in winter in lowlands. Most behaviors similar to Yellow-bellied: drills sap wells that are visited by Rufous Hummingbird and other species.

Calls and drum essentially indistinguishable from Yellow-bellied; perhaps slightly quieter and more subdued overall than Yellow-bellied, but differences minor.

Red-breasted Sapsucker
Sphyrapicus ruber   CODE 1

ADULT MALE. *In northern half of range, head virtually all red, wings and back mainly black; farther south, head less red, upperparts more white. Intermediates, as here, are frequent.*
WASHINGTON, JUNE

ADULT FEMALE. *Even females of the paler southern subspecies, as here, show much red on breast; wings and back darker than other sapsuckers. Juvenile darker than Red-naped.* CALIFORNIA, JUNE

L 8.5"   WS 16"   WT 2 OZ

‣ one adult molt per year; complex basic strategy

‣ strong age-related and weak sex-related differences

‣ two subspecies; northern has more extensive red on head

Breeds from sea level to high mountains of Pacific slope in diverse forest types, but with preference for coniferous woods. Least migratory of the sapsuckers; uncommon east of the Cascades and Sierras. Declining in British Columbia, possibly increasing in California.

Sounds average slightly duller than Yellow-bellied and Red-naped. Call a little deeper, less squealing than those two species; drum not quite as emphatic.

Williamson's Sapsucker
Sphyrapicus thyroideus CODE 1

ADULT MALE. *Mainly black; has white "sapsucker patch" in wing, characteristic of genus. Head has two white stripes on face and red on chin; pale belly washed with yellow.* COLORADO, JULY

ADULT FEMALE. *So different from male that it was long thought to be a separate species. Head tan; body barred black-and-white; belly lemon yellow. Juvenile similar but browner overall.* CALIFORNIA, JULY

9"	ws 17"	wt 2 oz

▸ one adult molt per year; complex basic strategy

▸ strong differences between male and female

▸ bill longer and broader in western populations

Widespread and locally common in Intermountain West; compared to Red-naped and Red-breasted, breeding grounds average higher-elevation, with more of a spruce-fir component. Range spread west and south in 19th and 20th centuries, but now experiencing local declines.

Call note does not descend as sharply as other sapsuckers. Drum irregular like other sapsuckers, but does not slow down as fast.

Gilded Flicker
Colaptes chrysoides CODE 2

ADULT MALE. *Like "Red-shafted" subspecies of Northern Flicker, has red malar and lacks nuchal crescent. Breast patch larger than Northern; black bars on back less prominent than Northern.* CALIFORNIA, APRIL

ADULT FEMALE. *Like female "Red-shafted," lacks malar and nuchal markings. All Gilded Flickers have yellow in wings and tail, prominent in flight but often hard to see on bird at rest.* ARIZONA, APRIL

L 11"	ws 18"	wt 4 oz

▸ one adult molt per year; complex basic strategy

▸ moderate differences between adult male and female

▸ infrequent hybridizes with Northern in zone of overlap

Replaces Northern Flicker in saguaro and Joshua tree forests of Sonoran and Mojave Deserts. Also enters riparian habitats, where a few come into contact with Northern Flicker in higher-elevation pine and oak woods. Apparently nonmigratory; range limits not fully known.

Vocalizations much as Northern. Call a sweet *keeer*, subtly less disyllabic than Northern. Song variable; many variants indistinguishable from Northern.

Northern Flicker
Colaptes auratus CODE 1

ADULT MALE, "YELLOW-SHAFTED." (C. a. auratus) *Yellow in wing and tail prominent in flight, sometimes prominent on bird at rest. Male has black malar and red crescent (nuchal) on nape.* FLORIDA, MARCH

ADULT MALE, "RED-SHAFTED." (C. a. cafer) *Wing and tail salmon; distinct from "Yellow-shafted" but be aware effects of lighting. Male has red malar on largely gray face; lacks nuchal crescent.* UTAH, JULY

ADULT FEMALE, "YELLOW-SHAFTED." (C. a. auratus) *Lack black malar of male. All plumages have coarse black spots on pale undersides, black bars across tan upperparts.* FLORIDA, APRIL

ADULT FEMALE, "RED-SHAFTED." (C. a. cafer) *The dullest flicker; face gray, lacking both malar patch and nuchal crescent. Colorful wing and tail markings of all flickers are often obscure at rest.* UTAH, JULY

L 12.5"   ws 20"   wt 4.6 oz

▸ one adult molt per year; complex basic strategy
▸ moderate differences between adult male and female
▸ "Red-shafted" and "Yellow-shafted" groups distinctive

Common and conspicuous; often seen flying from ground to nearby tree. Terrestrial ants are important in diet, but also feeds on tree trunks. Adapts well to urban environment, but populations are in overall decline. "Yellow-shafted" is more migratory than "Red-shafted."

◀ Song a variable series of *wick, wake,* or *wake-up* notes. Call a down-slurred *kee-ur,* more distinctly two syllables in "Yellow-shafted." Drums on gutters and eaves.

INTERGRADE. *Many Red-shafted × Yellow-shafted combinations occur; this individual has the red malar of "Red-shafted" and a hint of the red nuchal crescent of "Yellow-shafted."* COLORADO, NOVEMBER

Pileated Woodpecker
Dryocopus pileatus CODE 1

ADULT FEMALE. *Crested and long-necked like male, but red on crown reduced; red malar of male replaced by black on female.* TEXAS, APRIL

ADULT MALE. *Substantially larger than any other North American woodpecker. At rest, body nearly all-black. Long neck and red crest distinctive. Male has more red on head than female.* TEXAS, APRIL

| ∟ 16.5" | ws 29" | wt 10 oz | ♂ > ♀ |

→ one adult molt per year; complex basic strategy
→ moderate differences between adult male and female
→ western birds average grayer, Florida birds darkest

Despite requirement of extensive tracts of old trees, a resourceful species that has adapted well to fragmented landscapes. Numbers stable or increasing throughout much of range; expanding into recovering forests in Northeast. Wary; can be difficult to glimpse.

IN FLIGHT. *Often flies higher and over greater distances than other woodpeckers; flight is more level, less undulating. Extensive white flashes at base of primaries.* FLORIDA, MARCH

◄ Frantic call a rapid series of *wuk!* notes; speeds up, then slows down. Also a slow series of well-spaced *wap* notes, flicker-like but slower. Drum powerful.

Tyrant-Flycatchers

FAMILY TYRANNIDAE

46 species: 24 ABA CODE 1   10 CODE 2   1 CODE 3   2 CODE 4   9 CODE 5

The huge order Passeriformes, comprising well over half the world's bird species, is divided into two suborders: the suboscines and oscines. This separation is based in part on anatomical differences that affect the way the birds vocalize. The suboscines are a mainly tropical American group represented in North America almost exclusively by the tyrant-flycatchers. The oscines include all the rest of the world's passerines—the species we commonly call "songbirds" or "perching birds." Of the 46 species of suboscines recorded in North America, all but two are in the huge family Tyrannidae, which numbers an amazing 400 species in the New World (more than are found in most entire orders). The two other suboscines recorded in our area are the Rose-throated Becard, rare in southeastern Arizona and southern Texas, and the Masked Tityra, accidental in Texas. These are probably close relatives of species in the subfamily Tyranninae, but their exact taxonomic placement has not been established.

The tyrant flycatchers are well represented in North America, with the greatest diversity occurring in the southern United States. Only half a dozen species regularly reach Alaska.

Most species require both a prominent perch from which to scan for prey and open sky in which to forage for aerial insects. Flycatching—sallying out from a perch and snagging a flying insect—is a foraging technique employed by many species, but other methods are used as well. Phoebes sometimes glean insects from the surface of ponds, *Empidonax* flycatchers occasionally glean arthropods from cobwebs, and kingbirds often pounce on grasshoppers in clearings.

Because most tyrant flycatchers are highly insectivorous, annual movements are strongly influenced by the availability of prey. Only the hardy phoebes remain at mid-latitudes in the winter, and even those are scarce in most regions with more than a few weeks of subfreezing temperatures in winter. Although most passerines migrate heavily by night, tyrant flycatchers have a strong diurnal component to their spring and fall passages. Nocturnal migration occurs, too, with most species moving silently—in contrast to most other nocturnal-migrant passerines.

Populations of tyrant flycatchers in North America are relatively secure. Several species (for example, the Willow and Yellow-bellied Flycatchers) face the perils associated with long-distance migration to the tropics, and others have declined locally (for example, the Buff-breasted and Vermilion Flycatchers). Others, though, have benefited from habitat clearing and other changes to the landscape, and their ranges appear to be expanding; these include the Black Phoebe, Great Kiskadee, and Tropical and Gray Kingbirds.

Northern Beardless-Tyrannulet

Camptostoma imberbe CODE 2

L 4.5"	WS 7"	WT 0.3 oz (9 g)

▸ two adult molts per year; complex alternate strategy
▸ weak differences between adult and juvenile
▸ Texas birds smaller, more greenish than Arizona birds

Unlike most flycatchers, picks insects from bark and foliage. Inhabits lowlands with dense vegetation, such as mesquite thickets and riparian scrub. Scarce overall in North America, but locally common. Winter status unclear; many remain, but are quiet and hard to find.

Song a series of clear whistles, third or fourth note often accented: *pee pee peee! pee pee...* Sharply descending *pee-uck* apparently is given only by female.

ADULT. *Like a tiny Ash-throated Flycatcher. Gray above, with puffy crest; crest reduced on juvenile. Bill orange below; wing bars buff; face shows faint eye crescents and eyebrow.* ARIZONA, JULY

Acadian Flycatcher

Empidonax virescens CODE 1

L 5.75"	WS 9"	WT 0.46 oz (13 g)

▸ two adult molts per year; complex alternate strategy
▸ weak differences between juvenile and fall adult
▸ extent of yellow in wing bars and on belly variable

Inhabits old-growth broadleaf woods, often choosing dark ravines and shady hillsides. Favors woodland streams throughout range; also uses bald cypress swamps in south, tamarack swamps in north. Has recently expanded range northward into recovering forest landscapes.

Song an explosive *pee!-chup* or *peech!-eeyup* ("pizza!"). Call a sharp *peezt!* or flatter *peek*. Birds on territory give a rapid series of soft *pip* notes.

ADULT. *Depending on light and feather wear, ground color varies from grayish to greenish. Wing bars prominent; eye ring thin. Throat gray; breast washed with olive; belly pale. Juvenile a little browner.* TEXAS, MAY

Yellow-bellied Flycatcher

Empidonax flaviventris CODE 1

ADULT. *Most adults are decidedly yellow below, more than any other Empidonax. Yellow extends to throat. Bill broad-based and orangish below. Juvenile not as yellow below as adult.* ONTARIO, MAY

L 5.5"	WS 8"	WT 0.4 oz (11 g)

▸ two adult molts per year; complex alternate strategy
▸ weak differences between juvenile and fall adult
▸ extent of yellow, especially in wing bars, variable

Breeds in northern forest landscapes. Avoids deep woods; most common in bogs, as well as around clearings and forest edges. On breeding grounds, forages on or near ground; migrants often forage high in trees. Declining in southern portion of breeding range.

Song a sneezy *killunk*, similar to Least, but not as abrupt or emphatic; given at more widely spaced intervals than Least. Call a rising *tuwee* or *tuawee*.

Alder Flycatcher
Empidonax alnorum   CODE 1

ADULT. *Formerly considered conspecific with Willow Flycatcher, and difficult to separate in the field except by voice. Eye ring diminished, but usually a little more prominent than on Willow.* MINNESOTA, JUNE

ADULT. *Most show at least a tinge of green on the back; throat grayish. Wings appear medium to long in all plumages; bill broad-based and orange below. Juvenile has brownish wing bars.* NEW YORK, JULY

L 5.75"	WS 8.5"	WT 0.5 oz (14 g)

- ▸ two adult molts per year; complex alternate strategy
- ▸ weak differences between juvenile and fall adult
- ▸ plumage contrast on underparts variable

Widespread and common across extensive breeding range, yet seemingly scarce on migration. Breeds in clearings and along forest edges. Spring migration late and brief; fall migration early and apparently brief. Completes prebasic (fall) molt on wintering grounds.

Song *rrraybeet* or *rrraybee-oh*, first note exceedingly harsh. Usually distinct from Willow, but intermediate songs sometimes noted. Call an emphatic *pip!*

Willow Flycatcher
Empidonax traillii   CODE 1

ADULT. *A large Empidonax, with a broad-based, orangish bill. All are brownish above; relatively short-winged and long-tailed. Eye ring is typically weak or absent.* MONTANA, JUNE

JUVENILE. *Color and darkness of upperparts variable, but all are mainly brownish. Juvenile has brownish wing bars; whitish on adult. All plumages show short wings, weak eye ring.* CALIFORNIA, AUGUST

L 5.75"	WS 8.5"	WT 0.5 oz (14 g)

- ▸ two adult molts per year; complex alternate strategy
- ▸ weak differences between juvenile and fall adult
- ▸ "Southwestern" subspecies (*E. t. extimus*) paler than others

The southern equivalent of Alder Flycatcher. Western birds breed in riparian thickets across wide altitudinal gradients; eastern birds breed in old fields with dense shrubbery, often on poorly drained soils. Like Alder, migrates late in spring and early in fall.

◀ Usual song a burry *fitz-bew*, unlike Alder. Less commonly sings a rough *rrrip*, resembling first note of Alder's song. Call note a liquid *wip*.

Least Flycatcher
Empidonax minimus　CODE 1

ADULT. *The smallest eastern* Empidonax; *usually appears short-winged and small-billed. Eye ring bold; wings dark with bold white wing bars and black-and-white tertials.*
MINNESOTA, JUNE

IMMATURE. *Most are grayish overall, but some have olive on back. Boldly marked wings, prominent eye ring, and small size overall are good marks. Juvenile has brownish wing bars.*
CALIFORNIA, DECEMBER

5.25"　　ws 7.75"　　wt 0.4 oz (11 g)

two adult molts per year; complex alternate strategy

weak differences between juvenile and fall adult

fall birds, especially juveniles, variably yellowish

Nests widely in northern woods with hardwood component. Migration mainly east of the Rockies. Earliest Empidonax in spring, returns early in fall although with some lingering fairly late. Unlike most other members of genus, frequently sings in migration, even in fall.

Song a snappy, agitated *ch'bik*, often given in rapid series; pace of delivery more frantic than Yellow-bellied. Call, a dry *wit*, given frequently.

Gray Flycatcher
Empidonax wrightii　CODE 1

ADULT. *Long overall and long-tailed; many are decidedly gray, with weak plumage contrast. Wings are short, reflecting the relatively short migration of this species.* CALIFORNIA, MAY

NONBREEDING. *Bill, if seen well, distinctive in all plumages; long and narrow, with lower mandible showing orange base and black tip. Long tail usually shows white outer edges.*
ARIZONA, MARCH

L 6"　　ws 8.75"　　wt 0.45 oz (13 g)

▸ two adult molts per year; complex alternate strategy

▸ weak differences between juvenile and fall adult

▸ weak yellowish suffusion on belly variable

Tail-dipping behavior, recalling Eastern Phoebe, distinctive. Found in dry shrublands and woodlands of the Intermountain West: old-growth sagebrush stands, pinyon-juniper woodlands, and lower-elevation ponderosa pine woodlands. Threatened by habitat conversion.

Song a rough *brrrt*, followed by a long pause and then a rolling *seeloo*; second note sometimes omitted. Call a dry *wit*, similar to Least and Dusky.

Dusky Flycatcher
Empidonax oberholseri   CODE 1

ADULT. *Intermediate in some respects between larger Gray and smaller Hammond's. Wings fairly short. Most are yellowish below, gray-olive above, and pale-throated.* CALIFORNIA, JULY

ADULT. *Most have medium-contrast eye ring and wing bars; compare Least. Medium-sized bill has orange base blending into darker tip. Juvenile has more yellow below and on wing bars.* ARIZONA, FEBRUARY

L 5.75"   WS 8.25"   WT 0.4 oz (11 g)

▸ two adult molts per year; complex alternate strategy
▸ weak differences between juvenile and fall adult
▸ ground color and plumage contrast variable

Breeding habitat intermediate between high-elevation Hammond's and lower-elevation Gray. Common in extensive pine forests, but also uses aspen groves and montane wetlands. May benefit from logging. Unlike Hammond's, completes prebasic (fall) molt on wintering grounds.

Song variable: usually three widely separated elements, the last note higher and clearer than the first two, *seelit...breek...seep!* Call a dry *wit*, distinct from Hammond's.

Hammond's Flycatcher
Empidonax hammondii   CODE 1

ADULT. *A small, often colorful* Empidonax *flycatcher. Spring adults variable because of irregular prealternate (spring) molt, but all are distinctively small-billed and long-winged.* CALIFORNIA, APRIL

IMMATURE. *Most appear large-headed; eye ring often pinched back in "teardrop." Juveniles average more yellow below. Long wings and small bill usually distinctive.* CALIFORNIA, SEPTEMBER

L 5.5"   WS 8.75"   WT 0.35 oz (10 g)

▸ two adult molts per year; complex alternate strategy
▸ weak differences between juvenile and fall adult
▸ color and contrast variable on spring migrants

Breeds at higher elevations and higher latitudes than Dusky, but extensive overlap. Nest placed high in conifers, especially firs. On breeding grounds, frequently forages well up in trees; migrants in diverse microhabitats. Completes prebasic molt before fall migration.

Song fundamentally similar to Dusky, but there are differences: Hammond's rougher overall, with each song element disyllabic. Call a sharp *peek*.

Pacific-slope Flycatcher

Empidonax difficilis CODE 1

ADULT. *Nearly identical to Cordilleran; in areas where both species occur on the breeding grounds, most cannot be identified in the field.*
ARIZONA, OCTOBER

JUVENILE. *Both Cordilleran and Pacific-slope are short-winged. Juveniles of both species have buff wing bars in fall; adults average paler.*
ARIZONA, OCTOBER

5.5"   ws 8"   wt 0.4 oz (11 g)

▸ two adult molts per year; complex alternate strategy

▸ weak differences between juvenile and fall adult

▸ Channel Islands population may be separate species

Replaces Cordilleran in humid coniferous forests of Pacific coast region. Eastern limit of breeding range poorly known. Migrants apparently disperse through lowlands, in contrast to Cordilleran. Migration, especially in fall, appears to have easterly component.

In much of range, "position note" a rising *s'weet*, distinct from Cordilleran; unidentifiable intermediate notes heard in Oregon and possibly elsewhere.

Cordilleran Flycatcher

Empidonax occidentalis CODE 1

ADULT. *Eye ring usually pinched back behind eye, forming a "teardrop"; other Empidonax flycatchers sometimes show this effect. Often prominently crested.*
ARIZONA, JUNE

ADULT. *All are yellowish, some conspicuously; bill is mainly or entirely orangish below. Juvenile similar, but wing bars brownish.*
COLORADO, JUNE

L 5.5"   ws 8"   wt 0.4 oz (11 g)

▸ two adult molts per year; complex alternate strategy

▸ weak differences between juvenile and fall adult

▸ shape and prominence of eye ring variable

With Pacific-slope Flycatcher formerly lumped as "Western Flycatcher." Cordilleran occurs in various woodland settings, usually with moist component, such as riparian groves and shady canyons. Migration appears to be mainly through mountains and foothills.

"Position note" clearly disyllabic: *peet-seet*, like Acadian; less emphatic. Song has three well-spaced elements, thin and squeaky: *tsee...seedick...zweek.*

Buff-breasted Flycatcher
Empidonax fulvifrons　CODE 2

ADULT. *A small* Empidonax. *All are buff-breasted, even when worn, as here; immature averages brighter buff. Eye ring and wing bar white; bill small with orange lower mandible.* ARIZONA, MAY

ADULT. *With experience, observers begin to notice subtle but consistent differences in "gestalt" (overall impression) among tricky flycatchers—for example, the short bill and narrow tail of Buff-breasted.* ARIZONA, APRIL

L 5"　　　WS 7.5"　　　WT 0.3 oz (9 g)

▸ one adult molt per year; complex basic strategy

▸ moderate age-related and seasonal differences

▸ color of underparts variable in all plumages

Uncommon visitor to pine-oak landscapes of Southeast Arizona; preferred microhabitat is mountain meadows and riparian areas. Migratory in North America; arrives late March and early April. Formerly more numerous and widespread, but grazing and fire suppression caused habitat loss.

Song a rolling, cheery *p'chick chew*, rapidly repeated. Call a hard *chet*, more like "Audubon's" Yellow-rumped Warbler than like other *Empidonax* flycatchers.

Olive-sided Flycatcher
Contopus cooperi　CODE 1

ADULT. *Larger than wood-pewees. Most show high-contrast "vest" created by dark sides of breast, belly and flanks contrasting with otherwise white underparts and throat.* MINNESOTA, JUNE

ADULT. *Bullheaded and often crested; large bill fairly dark below. Wings exceedingly long. White ovals on sides of rump distinctive if visible, but often concealed.* BRITISH COLUMBIA, MAY

L 7.5"　　　WS 13"　　　WT 1 oz (28 g)

▸ one adult molt per year; complex basic strategy

▸ weak differences between juvenile and fall adult

▸ prominence of white ovals on sides of rump variable

All *Contopus* flycatchers tend to perch on bare branches, but Olive-sided takes it to an extreme: often seen at tip of highest branch of tallest tree in sight. Breeds in coniferous forests from sea level to near tree line. Migrates late in spring, very early in fall.

Song in three parts: dry opening note, then two whistled notes, *wip heeeee reee* ("Quick! Three Beers"). Call a rapid series of three or four liquid *pip* notes.

Eastern Wood-Pewee

Contopus virens CODE 1

ADULT. *Larger and longer-winged than* Empidonax *flycatchers. Bill extensively orange below (Western Wood-Pewee darker); wing bars whitish on adult, buff on juvenile.* TEXAS, APRIL

ADULT. *Very similar to Western Wood-Pewee. Eastern slightly brighter and greener. Breast of Eastern often broken by a pale vertical stripe; Western tends to have uniformly gray breast.* TEXAS, APRIL

L 6.25"   WS 10"   WT 0.5 oz (14 g)

▸ one adult molt per year; complex basic strategy

▸ weak differences between juvenile and fall adult

▸ extent and prominence of wing bars variable

Declining but still common in eastern woods. Favors extensive broadleaf forests, but also found in conifer groves. Preferred microhabitat is clearings and edges. Returns late in spring. High in trees during breeding season, in lower vegetation on fall migration.

◀ Song, given throughout the day, consists of a sweet *peeee-a-weeeee*, followed by a long pause, then a down-slurred *peeeee-urr*. Opening *p–* sound is explosive.

Western Wood-Pewee

Contopus sordidulus CODE 1

ADULT. *Both wood-pewees are long-winged and habitually perch in the open (*Empidonax *flycatchers typically perch in vegetation). Western has dark lower mandible and weak wing bars.* CALIFORNIA, MAY

ADULT. *Both wood-pewees often appear crested. Western averages darker and grayer, usually with a solid gray wash across breast. Lower mandible dark (usually orangish on Eastern).* CALIFORNIA, MAY

L 6.25"   WS 10.5"   WT 0.45 oz (13 g)

▸ one adult molt per year; complex basic strategy

▸ weak differences between juvenile and fall adult

▸ minor geographic variation in size and ground color

Common but declining in western woods, from lowland riparian groves to montane pine forests. Uncommon in deep woods, but penetrates into forest interior at campgrounds and along logging roads. Fall migration early; most are gone from mid-latitudes by early September.

◀ Song generally distinct from Eastern: a down-slurred *jeeeeer*, much harsher than Eastern. At any time of year, can give sweet *peeya* notes, recalling Eastern.

Greater Pewee
Contopus pertinax CODE 2

ADULT. *A large, dark pewee. Bill usually bright orange below (dusky on Olive-sided Flycatcher). Crest often conspicuous; Olive-sided and wood-pewees can also appear crested.*
BELIZE, FEBRUARY

ADULT. *Breast grayish, lacking the full "vest" of Olive-sided. Worn adult, as here, quite grayish; fresh adult browner. Juvenile has buff on belly and wing bars.* ARIZONA, JUNE

L 8"　　　WS 13"　　　WT 1 oz (28 g)

▸ one adult molt per year; complex basic strategy

▸ weak age-related and seasonal differences

▸ bill color and prominence of crest variable

Breeds in pine forests, especially those with oak understory. Ranges farther north than most of the "Mexican" species that reach into the Southwest. Perches high in trees, scanning for insects from exposed perches. Some postbreeding wandering east and west of range.

Song has five syllables: *deelip deeree! ah* ("José María"), opening *deelip* repeated several times before *deeree! ah.* Gives *pip* calls like Olive-sided.

Eastern Phoebe
Sayornis phoebe CODE 1

ADULT. *Dark above and pale below; long-tailed. Head rounded and dark, with little if any eye ring; wings sooty gray with wing bars faint or lacking.* TEXAS, NOVEMBER

JUVENILE. *Proportions, patterns, and behavior as adult, but has extensive light yellowish wash below. Fresh adults in fall also yellowish below; whitish below in spring and summer.* NEW YORK, SEPTEMBER

L 7"　　　WS 10.5"　　　WT 0.7 oz (20 g)

▸ one adult molt per year; complex basic strategy

▸ moderate age-related and weak seasonal differences

▸ individual variation in extent of yellow wash on belly

Tail-dipping behavior distinctive. In East, earlier spring migrant than other flycatchers. Nests in clearings, often near water. Breeding range has expanded west and south, probably because of willingness to nest on human structures, such as bridges, barns, and porches.

Song consists of two widely spaced and burry "phoebe" elements: *free-beezzz* and *fee-br'r'ree.* Call a hard *chip*, strongly down-slurred; heard all year.

Say's Phoebe

Sayornis saya　CODE 1

ADULT. *Gray overall with orange on belly; tail black. In flight, fairly long tail and long, pale wings suggest female Mountain Bluebird more than other flycatchers.* ARIZONA, FEBRUARY

JUVENILE. *Orange on belly like adult, but also shows orange wing bars; prominent yellow gape disappears as bird progresses through juvenile plumage.* ARIZONA, APRIL

L 7.5"	ws 13"	wt 0.75 oz (21 g)

▸ one adult molt per year; complex basic strategy

▸ weak differences between juvenile and adult

▸ individual variation in bill size and structure

Like Eastern Phoebe, a hardy species that migrates early in spring and nests on artificial structures. Also dips tail like Eastern. Seems attracted to desolate, arid landscapes; nests in mine shafts and on outbuildings. Range apparently expanding northward.

Song melancholy: *puh-weeeeer*, sometimes lengthened to *piddy-weeeeer* or shortened to *pweeeeer*. Alternate song a scratchy, modulated *pjzzzzeer*.

Black Phoebe

Sayornis nigricans　CODE 1

ADULT. *Head, breast, and upperparts black; belly and undertail coverts white. Structure and behavior similar to Eastern Phoebe.* CALIFORNIA, OCTOBER

JUVENILE. *Usually shows rusty wing bars; back and rump often have rusty-flecked feathers. Black-and white plumage otherwise like adult.* CALIFORNIA, JULY

L 7"	ws 11"	wt 0.7 oz (20 g)

▸ one adult molt per year; complex basic strategy

▸ moderate differences between juvenile and fall adult

▸ extent of white on wings and belly variable

Tightly associated with water, from sea cliffs to desert oases. Breeds on human structures, such as pump houses, dam gates, and cattle troughs. Population fairly sedentary, but many individuals wander erratically after breeding. Range expanding northward and inland.

Vocal throughout year. Song a whistled *ch'weee* or *ch'd'weee*, the two components often alternated. Song elements average shorter, more piercing than Say's.

Vermilion Flycatcher
Pyrocephalus rubinus   CODE 1

ADULT MALE. *Flaming crimson below; dark brown above. Head mainly crimson; brown extends up nape and to eye.* CALIFORNIA, APRIL

ADULT FEMALE. *Gray-brown above. Dirty white below with fine streaks on breast; bright apricot wash across vent. Head fairly pale with broad dark mask through eye.* ARIZONA, MAYV

JUVENILE. *Scaly gray-brown above; off-white below with fine spots and streaks. In all plumages, tail is short and dark, with white outer edges.* ARIZONA, JULY

L 6"   WS 10"   WT 0.5 oz (14 g)

▸ one adult molt per year; complex basic strategy

▸ strong age-related and sex-related differences

▸ intensity of color on underparts varies in all plumages

Inhabits oases and riparian habitats within arid landscapes. Perches in the open, usually close to the ground, on irrigation structures and mesquites. Locally common, but absent from many areas; has withdrawn from portions of former California range.

As though its gaudy plumage were not impressive enough, Vermilion also has a spectacular courtship display; gives *jip jibberdy joo* in high skylarking flight.

Great Crested Flycatcher
Myiarchus crinitus   CODE 1

ADULT. *Head gray, belly yellow, wings and tail with rufous highlights. Larger, larger-billed, and darker than Ash-throated; smaller and smaller-billed than Brown-crested.* TEXAS, JUNE

ADULT. *In most of range, the only flycatcher in its genus; where ranges overlap, caution warranted. Innermost tertial shows prominent wide white edge, conspicuous in rear view. Juvenile darker, duller.* TEXAS, JUNE

L 8.75"   WS 13"   WT 1.2 oz

▸ two adult molts per year; complex alternate strategy

▸ weak differences between juvenile and fall adult

▸ southern birds have slightly larger bills

Occurs in forested landscapes, especially where fragmentation is moderate. Sometimes in deep woods, but also orchards, cemeteries, even residential areas with sufficient vegetation. Spends much time up in trees. Range apparently expanding north, west, and south.

◀ Gives piercing *wheeep!* or rapid series: *weep weep weep*. Also a burry *brrt brrt brrt...* Erratic: sometimes strangely silent at dawn, yet vocal at midday.

Brown-crested Flycatcher

Myiarchus tyrannulus   CODE 1

Ash-throated Flycatcher

Myiarchus cinerascens   CODE 1

ADULT. *Small, small-billed, and pale. Throat light gray; belly pale yellow. From below, rufous in tail framed by brown edges and tail tip; tail of juvenile more extensively rufous.* CALIFORNIA, APRIL

ADULT. *Color scheme the same as other flycatchers in the genus: head gray, belly yellow, wings and tail rufous. Bill is notably large; wings are relatively short.* CALIFORNIA, MAY

ADULT. *On flycatchers in this genus, the tail, seen from below, can be diagnostic. Brown-crested has rufous down center to tip, with brown along edges; tail of juvenile has more rufous.* ARIZONA, MAY

L 8.5"   WS 12"   WT 1 oz (28 g)

▸ two adult molts per year; complex alternate strategy

▸ weak differences between juvenile and fall adult

▸ individual variation in bill size and bill length

Common in arid landscapes, from Joshua tree woodlands to montane pine forests. Also in lush riparian groves, where it overlaps with Brown-crested. Fall migration early; rarely seen on breeding grounds past early September. Annual in small numbers to Atlantic coast.

◀ Song a series of *kabrick* or *kabreer* notes. Often clipped to *breer* or *brick*; sometimes extended to three or more syllables, approaching Brown-crested.

LA SAGRA'S FLYCATCHER (M. sagrae). *West Indies species; casual stray to Florida. Even paler and smaller than Ash-throated, but bill larger and longer.* CARIBBEAN, MARCH

L 8.75"   WS 13"   WT 1.5 oz

▸ two adult molts per year; complex alternate strategy

▸ weak differences between juvenile and fall adult

▸ birds in West larger and larger-billed than Texas birds

Habitat overlaps with Ash-throated, but Brown-crested less tolerant of sparse, arid expanses. Favors extensive riparian woods, but also extends into saguaro "forests"; Ash-throated occurs in such habitats, too. Population generally healthy; range expanding in places.

Song a rolling *wip w'deer w'deer* ("what will do"), generally distinct from Ash-throated, but songs of both species variable. Call a hollow *witch*.

Dusky-capped Flycatcher
Myiarchus tuberculifer CODE 2

ADULT. *Smaller than Ash-throated; brighter yellow below. Secondary coverts edged rufous (white in other flycatchers in its genus). Adult tail has little rufous, but juvenile tail more.* ARIZONA, MAY

ADULT. *Bill structure and tail color can be two key marks in the identification of Myiarchus flycatchers. Dusky-capped has a thin bill, and its tail shows very little rufous below.* ARIZONA, MAY

L 7.25"   WS 10"   WT 0.7 oz (20 g)

▸ two adult molts per year; complex alternate strategy

▸ weak differences between adult and juvenile

▸ extent of rufous in wings and tail variable

Tropical species; range extends into Southwest mountains. Favors riparian areas with broadleaf trees, especially sycamore; nests in tree cavities or bird boxes. Migratory; most depart by mid-August. Throughout Southwest, a few vagrants seen each fall, mainly immatures.

Call a mournful *wheeeer* ("Pierre"), sweet and descending; given all day long. Also a dry *hwit*. Song buzzy and frenzied, with "Pierre" elements interspersed.

Gray Kingbird
Tyrannus dominicensis CODE 2

ADULT. *Larger-billed and paler gray above than Eastern, with black eye-mask; lacks white tail tip of Eastern. Pale below with faint yellow wash; juvenile slightly more yellowish.* CARIBBEAN, NOVEMBER

ADULT. *In profile, bill appears substantially larger than Eastern Kingbird, but beware effects of viewing angle. Gray has shorter wings than Eastern, corresponding to shorter migration.* FLORIDA, APRIL

L 9"   WS 14"   WT 1.5 oz

▸ one adult molt per year; complex basic strategy

▸ weak differences between juvenile and fall adult

▸ like Eastern, has restricted, variable red crown patch

Habits and habitats much as Eastern. Usually in dry habitats near water: cultivated fields, residential developments, and coastal woodlands. Shoreline development may harm local breeders, but population healthy overall; has extended range west along northern Gulf coast.

Call distinct from Eastern: a simple trill, *k'rrrr*, or slightly more complex, *k'k'k'rrrrr*. Both versions descending; shrill overall but musical.

Eastern Kingbird
Tyrannus tyrannus CODE 1

ADULT. *Dark gray above; head nearly black. Underparts white. Black tail is broadly tipped in white, prominent from above or below.* TEXAS, APRIL

JUVENILE. *Paler above and scalier than adult, with "dirtier" underparts; white tail tip sometimes obscure. Some individuals superficially resemble smaller Eastern Phoebe.* NEW YORK, SEPTEMBER

L 8.5" WS 15" WT 1.4 OZ

▸ two adult molts per year; complex alternate strategy
▸ weak differences between juvenile and fall adult
▸ variable red crown patch is difficult to see

Famously aggressive inhabitant of farmlands and open country. During breeding season, mobs crows and hawks; noisily protests human intruders. Relatively subdued during conspicuous diurnal migration in early fall; sometimes gathers in small, well-behaved flocks.

◀ All vocalizations are sputtering and buzzy. Song an incoherent *g'g' giddy giddy gizz! giddy...*, like a live wire. Call a down-slurred *zeek*, with sharp buzz.

IN FLIGHT. *On quivering wings, sallies out to ward off intruders; black tail with broad white tip usually conspicuous in flight.*
BRITISH COLUMBIA, JULY

MOBBING

When a bird (or group of birds) flies directly at a larger bird for the purpose of driving it away, the behavior is referred to as "mobbing." Mobbing typically happens during the breeding season, and the purpose of the behavior is to deter would-be predators. Species that habitually engage in mobbing behavior include the Eastern Kingbird and Red-winged Blackbird. The American Crow is involved at both ends: The species often mobs Red-tailed Hawks, but it is also frequently mobbed by grackles and other mid-sized passerines.

Thick-billed Kingbird
Tyrannus crassirostris CODE 2

ADULT. *From almost any angle, bill appears oversized. Uniform gray-brown above. Most adults in spring and summer show limited yellow below.* ARIZONA, JULY

IMMATURE. *Juveniles and fresh adults in fall have extensive yellow below; juvenile has variable rufous feather edgings on wing.*
CALIFORNIA, NOVEMBER

L 9.5"   WS 16"   WT 2 OZ

‣ two adult molts per year; complex alternate strategy
‣ moderate age-related and seasonal differences
‣ extent of yellowish wash, especially on belly, variable

During brief breeding season in southwestern canyons, spends most of time high in cottonwoods and sycamores, where it is showy and conspicuous. Does not arrive until late May or early June. Status in flux; numbers apparently increasing, range possibly expanding.

Call buzzy, like other kingbirds; usually with three or four syllables, such as *k'd'zeeeee*. Also gives a strange metallic *sooweek?*—with clear rising inflection.

Cassin's Kingbird
Tyrannus vociferans CODE 1

ADULT. *Darker gray above than Western, but tail not as dark; overall contrast lower than Western. Tail usually shows low-contrast white tip. Juvenile similar, but has buff wing bars.* ARIZONA, JUNE

ADULT. *From below, tail may show thin white edges, like Western. Good point of separation from Western is darker head and breast, contrasting strongly with white chin and malar.*
ARIZONA, MAY

L 9"   WS 16"   WT 1.6 OZ

‣ two adult molts per year; complex alternate strategy
‣ weak differences between juvenile and fall adult
‣ contrast between white chin and gray breast variable

Determinants of occurrence complex; habitat separation from Western not fully understood. Where ranges overlap, Western tends toward arid, grassy landscapes, Cassin's toward savanna-like landscapes. Cassin's arrives earlier in spring, leaves later in fall than Western.

Song similar to Western, but most elements are gruffer and more growling; accent often on last note. Emphatic *chi-boo!* given separately or as part of song.

Western Kingbird
Tyrannus verticalis CODE 1

IMMATURE. *One of the most frequent western vagrants to the East (stray to New York shown here); recognized by yellow belly, gray-brown upperparts, and black tail with white edges.* NEW YORK, NOVEMBER

ADULT. *Head and breast pale gray, with little contrast; belly lemon-yellow. Bill relatively small. Back and wings olive-gray; tail nearly black, contrasting from behind or above with back.* TEXAS, APRIL

8.75"	WS 15.5"	WT 1.4 oz

- two adult molts per year; complex alternate strategy
- weak differences between juvenile and fall adult
- feather wear significantly affects appearance of tail

Common and conspicuous around ranches, roadsides, and fencerows. Aggressive and adaptable; common in towns and cities. Often in the same landscapes as Eastern, but Western tends to occur in drier habitats. Leaves early in fall; most are gone by beginning of September.

IN FLIGHT. *Like Eastern Kingbird, flies out on quivering wings to attack intruders. On fresh adult, black tail shows prominent white edges.* COLORADO, JULY

Call sputtering, but not as buzzy as Eastern: *er er urk! uree* ("Arickaree"), typically with antepenultimate note accented. Often sings at night.

JUVENILE. *Similar to adult, but paler and scalier. Dark (nearly black) tail has white edges; tail contrasts with lighter back.* CALIFORNIA, SEPTEMBER

IN FLIGHT. *Tail becomes heavily worn prior to prebasic (fall) molt. On some individuals, the white tail edges completely disappear.* COLORADO, JULY

Tropical Kingbird
Tyrannus melancholicus   CODE 2

Couch's Kingbird
Tyrannus couchii   CODE 2

ADULT. *Larger-billed than Western; most show a weak eye mask, often lacking in Western. A little grayer above and thinner-billed than similar Couch's. Juvenile slightly browner.* ARIZONA, JULY

ADULT. *Back sllightly greener than Tropical; bill has broader base. Differences are slight; voice is often essential for identification. Juvenile like adult, but has more brown on wings.* TEXAS, MAY

L 9.25"	WS 14.5"	WT 1.4 OZ

▸ two adult molts per year; complex alternate strategy

▸ weak differences between juvenile and fall adult

▸ red crown feathering variable, sometimes lacking

A truly characteristic species of the tropics. Has recently become established in Texas; numbers appear to be increasing in Arizona. Regular in fall along West Coast. Adaptable; often in cities and towns. Microhabitat separation from other kingbirds imperfectly known.

Song indeterminate; buzzy and sputtering notes, often in a prolonged series. At any time of day, gives a descending *tiddaree* or *tiddy-ree.*

L 9.25"	WS 15.5"	WT 1.5 OZ

▸ two adult molts per year; complex alternate strategy

▸ weak differences between juvenile and fall adult

▸ hybrids with Tropical, known from Mexico, possible here

Formerly the southern Texas counterpart of Tropical Kingbird, though Tropical has recently invaded lower Rio Grande valley. Couch's usually in denser and more extensive woodlands than either Tropical or Western, but much overlap. Vagrants seen throughout southern U.S.

Song variable and burry; not as high-pitched and sputtering as Tropical. Two calls: a distinct *bzweeeeer*, emphatic at beginning; *kip* much like Western Kingbird.

KINGBIRD EXPANSION

Both the Tropical and Couch's Kingbirds are expanding their ranges in North America, and both are increasingly detected as vagrants far out of range. In most of western North America, yellow-bellied kingbirds seen after late September are potentially exciting—that's because the widespread Western Kingbird is mainly gone from western North America by late September. In particular, Tropical Kingbird is the most expected kingbird on the West Coast during the autumn months. Later in the fall, Western Kingbirds are ironically more likely to be found, for instance, in New Jersey than in California.

Scissor-tailed Flycatcher
Tyrannus forficatus   CODE 1

10–15"	ws 15"	wt 1.5 oz

- two adult molts per year; complex alternate strategy
- moderate age-related and sex-related differences
- tail length highly variable in adults of both sexes

Habits and habitats kingbird-like: conspicuous in open areas with perches. Feeds like Western Kingbird: hunts insects both in aerial chases and in perch-to-ground captures. Common within fairly limited breeding range; seen as vagrant throughout much of North America.

Song resembles Western, but individual elements average less harp, more mellow; accent often on last note. Common call a harp *kip*, similar to Western.

ADULT. *Amazingly long tail usually obvious; tail of female shorter than male, but still impressively long. Salmon wash on belly intensifies to hot pink on underwing.* BELIZE, NOVEMBER

FORK-TAILED FLYCATCHER (T. savanna). *Migratory tropical species; annual to North America. Like an Eastern Kingbird with an exceedingly long tail.* SOUTH AMERICA

JUVENILE. *Tail shorter than adult, but longer than kingbirds; pale upperparts and dark tail recall Western Kingbird. Salmon and pink tones of adult replaced by dull yellow on juvenile.* TEXAS, JUNE

Sulphur-bellied Flycatcher
Myiodynastes luteiventris   CODE 2

L 8.5"	ws 14.5"	wt 1.6 oz

- two adult molts per year; complex alternate strategy
- weak differences between juvenile and adult
- individual variation in paleness and overall size

Catches food aerially, but also gleans insects from foliage. Returns late in spring to nest in sycamores and walnuts in mountain canyons; most depart by early September. Unlike many other migrants to Southwest "sky mountains," Sulphur-bellied is a long-distance migrant.

Call a squeaky series, with isolated notes extremely high-pitched, like a rubber duck: *pee pwee pzweeeee! pwee...* Also a lower, more rolling *pwee didderee.*

ADULT. *Yellowish underparts have coarse, dark streaks; head has bold dark-and-white stripes. Tail rufous in all plumages; wing of juvenile has rufous highlights, reduced in adult.* ARIZONA, JULY

Great Kiskadee

Pitangus sulphuratus CODE 2

ADULT. *Larger, more colorful, and more contrasting (especially on head) than kingbirds. Black crown of adult has yellow patch, usually concealed; patch absent on juvenile.* TEXAS, NOVEMBER

L 9.75"	WS 15"	WT 2 OZ

▸ one adult molt per year; complex basic strategy

▸ weak differences between juvenile and adult

▸ intensity of yellow on underparts variable

Aggressive and conspicuous along wooded waterways. Feeding methods diverse: hunts flying insects in aerial chases, skims water's surface for tadpoles and small fish, plucks fruit from trees, visits feeders. Range expanding in North America; a few wander after breeding.

Varied song consists of *rick! kiss!-kadee!* elements, individual syllables piercing. Sings all day long; song lengthy at dawn, shorter later.

Rose-throated Becard

Pachyramphus aglaiae CODE 3

ADULT MALE. *Plumage distinct from flycatchers, but structure and erect pose similar. Sooty gray all over; crown darker. Throat suffused with dull rose in Arizona, darker pink in Texas.* ARIZONA, JULY

ADULT FEMALE. *Pale buff below; dark buff above in Arizona, brighter buff in Texas. Female has dark crown like male; crown darker in Texas birds. Juvenile similar to adult female.* TEXAS, JUNE

L 7.25"	WS 12"	WT 1 oz (28 g)

▸ one adult molt per year; complex basic strategy

▸ strong age-related and sex-related differences

▸ larger and more strikingly marked in Texas than Arizona

Summer resident in small numbers along well-wooded, middle-elevation streams in Arizona, especially in dense sycamore-cottonwood groves. Irregular in Texas; has bred. Spends most of time in canopy, where it is inconspicuous; spectacular hanging nest a good clue to its presence.

Generally quiet. Song high-pitched and fast=paced: *twee yeeeee... twee yeeeee...*, squeaky and descending. On returning to nest, gives a high-pitched chatter.

Shrikes

FAMILY LANIIDAE

3 species: 2 ABA CODE 1   1 CODE 4

Best known for their predatory ways, the shrikes are a fairly small family concentrated primarily in the Old World. Although shrikes are songbirds, they look and act like raptors. They have hooked upper mandibles like raptors, and they wear the black masks of a bandit; juveniles are finely barred below, further conveying a raptor-like appearance. Shrikes capture live prey—large insects, small reptiles and amphibians, and even small birds—and then, in one of the most diagnostic avian behaviors, they impale their prey on barbed wire or the thorn of a locust or hawthorn.

POPULATION DYNAMICS

Bird populations are dynamic for a variety of reasons. There are the annual population changes associated with migration and breeding, but there are also longer-term trends. In many regions, populations of the Loggerhead Shrike are in a sustained decline. Meanwhile, populations of wintering Northern Shrikes vary substantially from year to year; in some winters, the species is common south of its breeding range, but in other years, it is hard to find even a single individual.

During the past hundred years, most bird species in North America have either increased significantly in number or decreased, have either expanded their range substantially or have withdrawn from previously occupied regions, or have done some combination of both. Among North American bird species, a truly stable population—with no discernible change in number or geographic range—is surprisingly rare.

Birds are sensitive to environmental change, and many of their behaviors are what biologists term "plastic." When conditions change, populations shift. Species such as the Double-crested Cormorant, Eurasian Collared-Dove, and Cave Swallow (among many others) are in the midst of remarkable range expansions that may be related to changing environmental conditions.

Birders have long been of central importance to the documentation of avian population change; efforts such as the Christmas Bird Count and the Breeding Bird Survey have depended almost entirely on the contributions of large numbers of amateurs. One new initiative with immense potential is eBird (www.ebird.org), a user-friendly, checklist-driven, customizable database that is dramatically improving our knowledge of the status and distribution of North American birds.

Loggerhead Shrike

Lanius ludovicianus   CODE 1

ADULT. *Broad black mask, black tail, and black wings contrast with gray upperparts. In flight, flashes limited white in tail and wing. Superficially resembles Northern Mockingbird.* WYOMING, JUNE

IMMATURE. *Similar to adult, but breast faintly barred; adult has plain gray-white breast. Bill bulbous and hook-tipped, unlike Northern Mockingbird.* CALIFORNIA, APRIL

L 9"	WS 12"	WT 1.7 OZ

- ▸ one adult molt per year; complex basic strategy
- ▸ weak differences between juvenile and adult
- ▸ gray on back variable; extent of white in tail variable

Impales insects, birds, mammals on thorns and barbed wire. Hunts in open country with scattered perches: old pastures, roadsides, and abandoned orchards. Populations declining in much of range; biocides and decline of traditional agriculture probably to blame.

Variable song consists of simple, repeated elements: for example, squeaky *k'weet!* or mellow *brrrt* like Baird's Sandpiper. Call a nasal, irritated *rinnn*, repeated.

Northern Shrike

Lanius excubitor   CODE 1

ADULT. *Larger, longer-billed, longer-tailed, and paler than Loggerhead. Black mask narrower, especially in front of eye. Adult faintly barred below; juvenile and immature more heavily barred.* NORTH DAKOTA, FEBRUARY

IN FLIGHT. *Wings black with white flash at base of primaries; tail black with small white corners. Flash patterns of wings and tail similar in Loggerhead.* EUROPE, JANUARY

L 10"	WS 14.5"	WT 2.3 OZ

- ▸ one adult molt per year; complex basic strategy
- ▸ moderate differences between adult and subadult
- ▸ western birds larger and paler, with more white in tail

Habits similar to Loggerhead; prey average larger, prey base tilted more toward vertebrates. Tame, easily approached; perches on fences and shrubs. Southward movements in winter irregular; nearly absent some years. Even in "invasion" years, not especially common.

Calls heard in winter: repeated *enk* like a miniature Peregrine, shrill *br'r'r'eet* like police whistle. Sometimes sings softly on sunny winter days.

Vireos

FAMILY VIREONIDAE

16 species: 10 ABA CODE 1; 3 CODE 2; 1 CODE 3; 2 CODE 5

Not long ago, the vireos appeared immediately before the wood-warblers in North American field guides. It made sense: The vireos are small, arboreal, migratory, insectivorous passerines that are just a little duller and slower than most wood-warblers. The vireos were a perfectly logical "warm-up" to the hyperactive and gaily colored wood-warblers. We have since learned that, taxonomically, the vireos do not belong anywhere near the wood-warblers; rather, they are part of a large Australasian lineage that is also represented in North America by the shrikes and corvids. The delineation of this group of 1,000+ species was one of the biggest shakeups ever in avian taxonomy, and it represents one of the crown jewels of the molecular approach to understanding evolutionary relationships among birds. The radiation of these Australasian passerines is a fascinating counterpart of the better known marsupials—a group of related mammals with diverse parallels elsewhere in the world. Analogously, the Australasian passerines consist of birds called "wrens," "robins," "warblers," "orioles," and "flycatchers"—but which are not closely allied with the Afro-Eurasian and New World versions of those taxa.

The word "vireo" is a neo-Latin word meaning "I am green," and several vireos from the tropics are called "greenlets." All of our species are small and mainly greenish or grayish. Except for Hutton's Vireo, North American vireos are migratory—many of them highly so. They are birds of complex plant communities. Most are forest inhabitants, and many spend the majority of their time well up in the canopy. A few are partial to thickets and edges, but even there they are inclined to hide within dense vegetation.

What vireos lack in visual splendor and conspicuousness, they more than make up for in vociferousness. One can easily hear a hundred Red-eyed Vireos in the course of a day's birding in the eastern deciduous forest, but it can be difficult to get a satisfactory look at even one of them. Even vireos of lower-stature vegetation can be maddeningly hard to see: Gray, Bell's, and White-eyed Vireos are adept at escaping visual detection, even though singing wildly all the while.

The conservation challenges presented by several vireo species are squarely related to habitat loss on the breeding grounds. For example, both the Black-capped Vireo and the "Least" subspecies of the Bell's Vireo are endangered, and both currently require aggressive habitat restoration for their survival. Interestingly, both species have been the beneficiaries of the relatively benign land-use practices associated with U.S. military installations.

Red-eyed Vireo
Vireo olivaceus CODE 1

L 6"	WS 10"	WT 0.6 oz (17 g)

▸ one adult molt per year; complex basic strategy

▸ weak differences between juvenile and adult

▸ much individual variation in size and shape

The most abundant long-distance migrant in North America. Spends most of time in canopy of broadleaf or mixed deciduous-coniferous forests. Greatest densities in interior of extensive forest tracts, but fragmented landscapes and mid-successional tracts also accepted.

◀ Incessant, monotonous song seems to answer itself: *chuwee? chuawee! ... cheerily? churro!* Call note a petulant, descending, nasal *chwaaaay.* Nocturnal migrants silent.

ADULT. *A large vireo; greenish above, pale below. Crown gray; broad white eyebrow has thin black borders. Superficially similar Tennessee Warbler has thinner bill, weaker face pattern.* TEXAS, APRIL

IMMATURE. *Similar to adult; eye is black or dark brown, but many adults can appear dark-eyed in poor light.* MINNESOTA, SEPTEMBER

ADULT. *Sings constantly; frequently cocks tail. Wings notably long. Red eye visible at close range; juvenile has dark brown eye.* CONNECTICUT, JUNE

Yellow-green Vireo
Vireo flavoviridis CODE 3

L 6.25"	WS 10"	WT 0.6 oz (17 g)

▸ one adult molt per year; complex basic strategy

▸ weak differences between adult and immature

▸ wanderers to California duller than Gulf coast birds

Uncommon across southern stretch of continent; rare breeder in south Texas. Presence here erratic: mainly in spring in Gulf coast region, annual in fall to coastal California. Split from Red-eyed Vireo in 1987; distributional status still being worked out.

Song a series of bright *cheep* notes; suggests House Sparrow. Reveals itself as close kin of Red-eyed Vireo with occasional *chiddy-lee* and *cheer-up* notes.

IMMATURE. *Like a large, low-contrast Red-eyed Vireo; wings slightly shorter, bill slightly larger. Underparts extensively yellow; eyebrow not as prominent as Red-eyed.* CALIFORNIA, SEPTEMBER

Black-whiskered Vireo

Vireo altiloquus CODE 2

L 6.25"   WS 10"   WT 0.6 oz (17 g)

▸ one adult molt per year; complex basic strategy
▸ weak differences between adult and juvenile
▸ black "whisker" (malar) can be faint or lacking

Caribbean species; replaces similar Red-eyed Vireo in southern Florida. Like Red-eyed, forages in treetops; coastal mangroves and hardwood forests favored. Spring migration early. Vagrants sometimes seen west along Gulf coast and north along Atlantic coast.

Song similar to Red-eyed, but individual notes less slurred, more two syllables: *sweet cheer... chip reeo...* Less mellow overall than Red-eyed.

ADULT. *Similar to Red-eyed, but duller and grayer. Adult has red eye; juvenile has dark brown eye.. Thin black "whisker" is a good point of distinction from Red-eyed.* FLORIDA, APRIL

Philadelphia Vireo

Vireo philadelphicus CODE 1

L 5.25"   WS 8"   WT 0.4 oz (12 g)

▸ one adult molt per year; complex basic strategy
▸ weak age-related and seasonal differences
▸ extent and intensity of yellow on underparts variable

A vireo of northern hardwood forests and parklands; especially characteristic of aspen woods. Breeding range contained within northern portion of Red-eyed Vireo's range; Philadelphia in younger woods, but much overlap. Logging and burning may benefit species.

ADULT. *A drab vireo, lacking wing bars. Thin dark line through eye is separated from gray crown by slightly arched pale eyebrow. Variable yellow below, strongest on center of breast.* TEXAS, APRIL

Song very similar to Red-eyed. Song of Philadelphia averages higher-pitched, more slowly delivered; much variation. Many cannot be separated by voice.

IN FLIGHT. *Although generally vocal, most vireos—including Philadelphia—do not have a "typical" flight call. Some give call notes in flight, however, and may even sing while flying.* LOUISIANA, APRIL

ADULT. *Some are very similar to Warbling; yellow beneath on Warbling stronger on flanks than on center of breast. In both species, juvenile and fresh fall adult more yellowish below. Dark lores of Philadelphia always a good mark.* TEXAS, APRIL

Gray Vireo
Vireo vicinior   CODE 2

L 5.5"   WS 8"   WT 0.5 oz (14 g)

- one adult molt per year; complex basic strategy
- weak differences between fresh and worn adults
- individual variation in size, shape, and ground color

Favors dry mid-elevation woods that birders rarely visit: neither in cool high-elevation forests nor lush lowland canyons; rather, in sparse juniper-pine woods, often in association with Black-chinned Sparrow. Rarely seen on migration. Habit of cocking tail distinctive.

Song suggests Plumbeous, but sweeter, simpler: *cheery... cheer-lee...* Lacks rich overtones and burry elements of Plumbeous. Call note simple and scratchy: *jig.*

ADULT. *Plain and gray; can be confused with more boldly marked Plumbeous Vireo, but habitats usually differ. Wing bars faint or lacking, depending on wear. Wings short; tail long.* CALIFORNIA, MAY

Warbling Vireo
Vireo gilvus   CODE 1

L 5.5"   WS 8.5"   WT 0.4 oz (11 g)

- one adult molt per year; complex basic strategy
- weak age-related and seasonal differences
- eastern and western birds may be separate species

Feeds in forest canopy; occurs in extensive forest tracts, but small fragments also accepted. Eastern birds more affiliated with riparian habitats than western, the reverse of the usual pattern in which western populations are more riparian-dependent.

◀ Song a rich warble, markedly different in eastern vs. western birds. Western gives a more herky-jerky song; eastern ends song with a strong up-slurred note.

Despite overall similarities in appearance, eastern and western populations of the Warbling Vireo exhibit consistent differences in vocalizations, habitat preferences, and molt strategies. The Warbling Vireo may consist of two "cryptic species"—species that differ more in behavior and ecology than in appearance.

ADULT, WESTERN. *Olive-brown above; pale gray below. Most are less yellow below than Philadelphia, but some juveniles and fall adults have much yellow on flanks and undertail coverts.* CALIFORNIA, MAY

ADULT, EASTERN. *A little larger than Western, but otherwise similar: both lack wing bars; both have gray cap, white eyebrow, and dusky gray stripe behind eye; both lack dark lores.* TEXAS, APRIL

Hutton's Vireo
Vireo huttoni   CODE 1

ADULT, PACIFIC-SLOPE. *Like Ruby-crowned Kinglet, small, stubby, and olivaceous, with white wing bars and broken eye ring. Hutton Vireo forages slowly; Ruby-crowned Kinglet frenetic.* CALIFORNIA, MARCH

ADULT, INTERIOR. *Both populations separated from Ruby-crowned Kinglet by bulbous bill with hooked upper mandible and absence of dark patch below lower wing bar. Juvenile more buff.* ARIZONA, APRIL

L 5"	ws 8"	wt 0.4 oz (11 g)

- one adult molt per year; complex basic strategy
- weak differences between juvenile and adult
- interior birds paler overall and longer-winged

Feeds in middle to upper reaches of canopy. Pacific-slope birds in oak woods south, spruce and Douglas-fir farther north; interior birds in oak or pine-oak woods. Pacific-slope birds sedentary; interior birds partially migratory. Nests early. Range has perhaps expanded.

Song colorless and persistent: insistent *chew* or *suawee* notes, best characterized by their endless delivery. Varied calls are dry and chattering.

Bell's Vireo
Vireo bellii   CODE 1

ADULT, MIDWEST. *Calls to mind a washed-out White-eyed Vireo: spectacles and wing bars reduced, but usually obvious; olive green above; pale below with extensive yellow on sides.* OHIO, MAY

ADULT, SOUTHWEST. *Western populations ("Arizona," V. b. arizonae, and "Least," V. b. pusillus) drab, with little yellow below and weak wing bars. "Least" nearly devoid of color. All Bell's Vireos are long-tailed and very small.* ARIZONA, APRIL

L 4.75"	ws 7"	wt 0.3 oz (9 g)

- one adult molt per year; complex basic strategy
- weak differences between adult and juvenile
- Midwest, Southwest, and California birds separable

Breeds in dense, shrubby vegetation: "Arizona" subspecies only in willow and tamarisk thickets in wet places; endangered "Least" Bell's Vireo in California along streams and in high-quality chaparral; Midwestern birds more broadly in brushy fields and forest edges.

◀ Heard more often than seen. Song is a scratchy *jaricka jaricky j'ree?* followed by a long pause and then *jaricka jaricky j'roo!*—as if answering itself.

White-eyed Vireo
Vireo griseus   CODE 1

ADULT. *Gray head marked with yellow spectacles and white chin. Eyes distinctively white (brown or gray in juvenile). Olive-green above with two bold wing bars. Flanks and undertail coverts yellowish.* TEXAS, APRIL

BILL DETAIL. *All vireos—including White-eyed—have a distinctive bill: fairly bulbous and broad-based with a downward hook at the tip of the upper mandible.* LOUISIANA, APRIL

L 5"   WS 7.5"   WT 0.4 oz (11 g)

▸ one adult molt per year; complex basic strategy

▸ moderate differences between juvenile and adult

▸ northern birds average larger and brighter

Southeastern counterpart of Bell's Vireo; skulks in dense, low vegetation. Favors overgrown pastures and streamside thickets; nests in multiflora rose and other shrubs. Unstable at northern edge of range; has expanded north in some regions but withdrawn from others.

Complex song variable, but usually with strong *chick!* notes at beginning and end, plus long, slurred note in middle, for example, *chick! adeeapuh weeeeeo chick!*

Yellow-throated Vireo
Vireo flavifrons   CODE 1

ADULT. *Head olive-yellow; throat bright yellow (duller in juvenile). Wing bars prominent; belly white. Distinct from other vireos, but colors and pattern suggest Pine Warbler.* TEXAS, MAY

ADULT. *Extension of primary feathers is longer than Pine Warbler's to assist in Yellow-throated's relatively greater distance of migration.* CONNECTICUT, JUNE

L 5.5"   WS 9.5"   WT 0.6 oz (17 g)

▸ two adult molts per year; complex alternate strategy

▸ weak differences between adult and juvenile

▸ individual variation in size and intensity of yellow

Breeds mainly in deciduous forests, especially bottomlands and swamps; highest densities around clearings and in forest fragments, but wooded landscape required. Numbers have increased and range expanded somewhat with reestablishment of eastern forests.

Song like Red-eyed, a series of slurred phrases. Distinguished by lazy, drawn-out quality: *eee-yaaaaa, zheer-aaay,* many elements with buzzy quality.

Plumbeous Vireo
Vireo plumbeus   CODE 1

L 5.75"   ws 10"   wt 0.6 oz (17 g)

▸ one adult molt per year; complex basic strategy
▸ weak differences between fresh and worn adults
▸ some are tinged yellow below, approaching Cassin's

Mainly in mountain pine forests, but some occur in broadleaf or mixed stands; favors warmer, drier forests than Cassin's and Blue-headed. Migrates later in fall than Cassin's; little temporal overlap. Population increasing; range expanding west and north.

◀ Song a series of lazy phrases: chuawee... chirrup... Burrier than Cassin's and Blue-headed, but any phrase can overlap completely with those two species.

ADULT. *Spectacles and wing bars prominent. Nearly devoid of color, but in fall can show yellow below and green in wings; compare Cassin's. White throat set off sharply from gray malar.* ARIZONA, MAY

Blue-headed Vireo
Vireo solitarius   CODE 1

L 5.5"   ws 9.5"   wt 0.6 oz (17 g)

▸ one adult molt per year; complex basic strategy
▸ weak differences between juvenile and adult
▸ darker-backed and larger in southern Appalachians

Increasing throughout range as forests regenerate, but generally scarce. Feeds in mid-story or canopy. Breeds in northern hardwoods community of maple, birch, and aspen; also in coniferous forests. Migrates earlier in spring, later in fall than other eastern vireos.

◀ Song richer and deeper than Red-eyed, with slower delivery: wheeya... clear-ah-lee... cheeer-yo... Some phrases grating and buzzy, for example, gzheer.

ADULT. *Sharper, more colorful than Cassin's Vireo. White throat set off sharply from blue-gray head. Yellow on flanks usually intense and extensive. Juvenile sliightly browner and drabber.* MAINE, JUNE

Black-capped Vireo
Vireo atricapilla   CODE 2

L 4.5"   ws 7"   wt 0.3 oz (9 g)

▸ two adult molts per year; complex alternate strategy
▸ moderate age-related and sex-related differences
▸ individual variation in intensity of black on head

Range-restricted habitat specialist; requires frequently burned, extensive brushlands. Especially favors shrub oaks. Has withdrawn from much of former range, but is locally common where habitat is protected and allowed to burn; cowbird control may help.

Song consists of buzzy introductory note followed by bubbly jumble: bzee... jiddy-juh-dee or bzee... juh-duh-duh jiddy.

ADULT MALE. *Small and secretive. Black head contrasts with broad white spectacles and white throat. Head grayish on female and juvenile; juvenile drabber and browner overall.* TEXAS, APRIL

Cassin's Vireo
Vireo cassinii   CODE 1

ADULT. *Some have relatively little color, more resembling Plumbeous than Blue-headed. Cassin's is slightly smaller; Plumbeous usually shows stronger contrast on the head.* CALIFORNIA, APRIL

ADULT. *Like a washed-out Blue-headed: colors a little drabber; white spectacles and chin do not contrast as strikingly with head. Slightly brighter and greener in fall.* CALIFORNIA, APRIL

L 5.5"	WS 9.5"	WT 0.6 oz (17 g)

▸ one adult molt per year; complex basic strategy

▸ weak differences between spring and fall adult

▸ variants resemble both Blue-headed and Plumbeous

Common, increasing in western forests; more inclined than Plumbeous Vireo to use broadleaf woods. Ranges down to sea level, but not especially high in mountains. Migrates early in fall. Strong easterly component to fall migration; regular to western Great Plains.

◀ Song sweeter than Plumbeous; similar to Blue-headed. Most phrases clear and sweet: *cheerah... clearly...* But some also buzzy: *zheerup... jurro...*

PATTERNS OF GEOGRAPHIC VARIATION

Many field guides and reference works split the North American avifauna into two major geographic sectors: East and West, with the dividing point roughly at the 100th meridian (just east of the Rocky Mountains). This division is useful, but can obscure the significant differences *within* the Western sector between the Intermountain West and the Pacific Slope. It can be more useful to consider North America as divided into three major regions of diversity: the East, the Intermountain West, and the Pacific Slope.

The "Solitary Vireo" complex is one of many examples of closely related land birds that separate into Eastern, Interior, and Pacific populations—the Blue-headed Vireo in the East, Plumbeous Vireo in the Intermountain West, and Cassin's Vireo on the Pacific Slope. The "Yellow-bellied Sapsucker" complex—consisting of the Yellow-bellied, Red-naped, and Red-breasted Sapsuckers—is another example. Other examples involve well-defined subspecies or subspecies groups within a complex currently recognized as a single species, among them the Downy Woodpecker, Hairy Woodpecker, Gray Jay, White-breasted Nuthatch, Hermit Thrush, Orange-crowned Warbler, and Pine Grosbeak.

A third category involves complexes that are currently classified as two species. Typically, it involves one geographically restricted species plus a more-widespread species with two well-defined subspecies. An example is the Virginia's/Nashville Warbler complex, with the Virginia's Warbler being restricted to arid woodlands in the Intermountain West and the Nashville Warbler consisting of well-defined eastern and far western populations.

Jays & Crows

FAMILY CORVIDAE

20 species: 12 ABA CODE 1   5 CODE 2   2 CODE 3   1 CODE 4

Wily and often conspicuous, the jays, crows, and their relatives are mainly birds of wooded landscapes. Relationships within the family Corvidae are not well established; the all-black crows and ravens (genus *Corvus*) seem to be a closely related taxonomic group, but the various species of jays, along with Clark's Nutcracker and our two species of magpies, are a taxonomic hodgepodge. The closest North American relatives of the family Corvidae are the shrikes and vireos, but the corvids in general are more closely related to the spectacular birds-of-paradise of New Guinea.

In terms of geographic distribution, the corvids are one of the most eclectic groups in North America. The Common Raven winters farther north than any other bird species, but several corvids—for example, the Brown Jay and Tamaulipas Crow—barely enter our area in south Texas. Several species are among the most range-limited in North America. The Yellow-billed Magpie is restricted essentially to California's Central Valley, and the Florida Scrub-Jay occurs only in central peninsular Florida. The Island Scrub-Jay takes the cake: It occurs nowhere else on the planet but Santa Cruz Island, one of the Channel Islands off southern California.

Many corvids are famously omnivorous. Western Scrub-Jays and Northwestern Crows are just as likely to gobble down an arthropod as they are a french fry, and Gray Jays are the well-known "camp robbers" that loiter about picnic areas and homesteads throughout their extensive range. Yet other species are closely tied to specialized food resources; Clark's Nutcrackers and especially Pinyon Jays undertake substantial movements in search of conifer seeds.

Sustained annual movements by corvids are relatively few. Many species are entirely sedentary, several are prone to periodic irruptions, and only a few (notably, the Blue Jay) have well-defined seasonal migrations. Corvids are very large as passerines go, and their flight style is distinct from that of most other passerines: On deep, slow wingbeats, they fly straight from point to point.

Populations of most species of corvids are healthy, despite some species' recent losses to West Nile virus. Several corvids are expanding their ranges, and a few are considered pests. American Crows, Black-billed Magpies, and others have been persecuted for hundreds of years—clearly with no ill effect on long-term population health. They and other corvids have long been acknowledged for their cleverness, and recent studies have revealed more intelligence than was formerly known. It is likely that most species will adapt well to environmental change in the 21st century and beyond.

Blue Jay
Cyanocitta cristata   CODE 1

ADULT. *Large, crested. Blue above; gray below with thin black necklace. Juvenile similar, but blue of adult replaced with blue-gray, and black replaced with black-brown.* CONNECTICUT, JANUARY

IN FLIGHT. *Tail blue with thin black bars; flashes white corners in flight. Primaries dark, but secondaries have white flashes. Migrants typically silent in flight.* NEW JERSEY, OCTOBER

L 11"   WS 16"   WT 3 OZ

▸ one adult molt per year; complex basic strategy

▸ weak differences between juvenile and adult

▸ rare hybrids with Steller's where ranges overlap

Showy and inquisitive. Occurs in diverse woodland settings, from deep woods to urban plantings; especially favors oaks. Present within range throughout year, but much annual movement; conspicuous diurnal migration can be strong. Range continuing to expand west.

◂ Calls include the following: harsh, raucous *jeer!* ("jay"); ringing *kway-kweedle*; repeated *clweer*; various soft chuckles; and imitations of other species.

Steller's Jay
Cyanocitta stelleri   CODE 1

ADULT, INTERIOR. *Head dark with prominent crest; underparts, wings, and tail blue. Populations in the Rockies have prominent white streaks on forehead and above eye.* COLORADO, FEBRUARY

ADULT, PACIFIC SLOPE. *Far-western populations slightly darker; facial stripes fewer, thinner, tending toward bluish. All juveniles have shorter crests, with diminished blue overall.*

CALIFORNIA, JULY

L 11.5"   WS 19"   WT 3.7 OZ

▸ one adult molt per year; complex basic strategy

▸ moderate differences between juvenile and adult

▸ extensive geographic variation in head plumage

Characteristic of western conifer forests; often the first species to greet a visitor to a campground or roadside pull-off. Common from sea level to tree line. A few wander into deserts and grasslands each fall; regular altitudinal movements difficult to discern.

Calls harsh and unmusical: a raspy *schweeeeek* or *schweeeeech*; also *shick shick shick shick*, rapidly delivered. Calls resemble Western Scrub-Jay, not Blue Jay.

Western Scrub-Jay

Aphelocoma californica   CODE 1

ADULT, "CALIFORNIA." (A. c. californica) *Strongly patterned in blue and white. White eyebrow prominent; white throat set off from whitish underparts by broken blue breast band.* CALIFORNIA, FEBRUARY

ADULT, "WOODHOUSE'S." (A. c. woodhouseii) *Subspecies of interior regions is drabber than "California"; eyebrow and breast band reduced. Juveniles of all populations grayish; all plumages are notably long-tailed.* COLORADO, MARCH

L 11.5"   WS 15.5"   WT 3 OZ

▸ one adult molt per year; complex basic strategy

▸ moderate differences between adult and juvenile

▸ "California" and "Woodhouse's" groups distinctive

"California" Scrub-Jay inhabits towns, farm districts, and oak woods in lowlands; "Woodhouse's" mainly in dry foothills with oak, juniper, and pinyon pine. "California" bold and aggressive; "Woodhouse's" relatively shy. "California" sedentary; "Woodhouse's" irruptive.

"California" gives screeching, shrieking *zhreeek!* "Woodhouse's" gives more subdued, more disyllabic *shreheck.* Both give *shruck shruck shruck* series.

Island Scrub-Jay

Aphelocoma insularis   CODE 2

ADULT. *Similar to "California" Scrub-Jay; slightly larger and bluer, with deeper blue on undertail coverts. Juvenile grayer and browner than adult.* CALIFORNIA, JULY

ADULT. *Isolated island populations are often larger than their mainland counterparts. Accordingly, Island Scrub-Jay is heavier, longer-billed, and larger overall than Western.* CALIFORNIA, SEPTEMBER

L 13"   WS 17"   WT 4 OZ

▸ one adult molt per year; complex basic strategy

▸ moderate differences between juvenile and adult

▸ prominence and extent of white on breast variable

Common on Santa Cruz Island, the only place where the species occurs; population less than 10,000, but stable. Occurs widely on Santa Cruz Island, but prefers oak and oak-chaparral habitats on east-facing slopes. Some are tame; hang around boat dock and greet tourists.

Calls similar to "California" Scrub-Jay. Species is completely isolated from mainland scrub-jays; thus, no selective pressure to evolve distinguishing calls.

Mexican Jay
Aphelocoma ultramarina   CODE 2

ADULT. Mainly blue above; back flecked with brown. Pale blue-gray below. Low-contrast overall; even dull "Woodhouse's" Scrub-Jay shows more contrast. ARIZONA, FEBRUARY

JUVENILE. Head browner and grayer than adult. Bill yellow; yellow in bill retained for about one year after bird has reached adult plumage.
ARIZONA, AUGUST

L 11.5"   WS 19.5"   WT 4.4 oz

- one adult molt per year; complex basic strategy
- strong differences between juvenile and adult
- Arizona birds more weakly patterned than Texas

Highly social, exceedingly sedentary; occurs in oak and pine-oak
woods, from foothills to middle
elevations. Permanent ancestral territories defended by noisy groups of
5–25 birds. Common within range,
but almost never irrupts into nearby
deserts and grasslands.

Call a rising *bzhink* or series of descending *bzheep* notes. Most
have electric-buzzer quality of "California" Scrub-Jay; Texas
birds more like "Woodhouse's."

Pinyon Jay
Gymnorhinus cyanocephalus   CODE 1

ADULT. Dull blue all over, intensifying on head; throat has pale streaks. Bill length variable; differences in bill morphology correspond to regional variation in pinecone morphology.
OREGON, AUGUST

JUVENILE. Duller than adult; blue limited except in wings and tail. In all plumages, notably shorter-tailed than Western Scrub-Jay
OREGON, AUGUST

L 10.5"   WS 19"   WT 3.5 oz

- one adult molt per year; complex basic strategy
- moderate differences between juvenile and adult
- geographic variation in bill size and shape

Highly social; found in arid woodlands of Intermountain West. Flocks
large; sometimes 500+ individuals.
Flocks in flight well dispersed;
sometimes 2 miles long. Nomadic;
wanders widely in search of cones
of pinyon and other pines. Usually
nests early in spring.

Musical call carries far: *raa-ah* and *raa-raa-raa* and *kya-raa*,
recalling Laughing Gull. Usually silent when feeding; flights in
flock sometimes noisy.

Florida Scrub-Jay
Aphelocoma coerulescens   CODE 2

L 11"	WS 13.5"	WT 3 oz

- one adult molt per year; complex basic strategy
- moderate differences between adult and juvenile
- some variation in size and paleness of plumage

Sedentary, habitat-restricted endemic to central Florida peninsula; requires frequently burned oak groves on well-drained, sandy soils. Breeding system cooperative; young adults forgo reproduction to help with rearing of younger siblings. Sometimes extremely tame.

Calls similar to closely related Western Scrub-Jay, but duller: *zhuh* or *zhu-u-uh*. Not likely to be confused with calls of sympatric Blue Jay.

ADULT. *The only scrub-jay in its range. Blue Jay is crested, shorter-tailed. Juvenile mainly brownish gray, with extensive blue only on wings and tail; juvenile Blue Jay much bluer.* FLORIDA, FEBRUARY

Green Jay
Cyanocorax yncas   CODE 2

L 10.5"	WS 13.5"	WT 3.5 oz

- one adult molt per year; complex basic strategy
- weak differences between juvenile and adult
- extent of blue in crown of juvenile variable

Despite outlandish garb, can be hard to see in shadows of dense woodland edges and old-growth mesquite stands. Reclusive during breeding season; small family groups forage stealthily in dense cover. More conspicuous in winter; visits feeders in towns and at campsites.

Calls consist of harsh notes, rapidly repeated: *jick jick jick* and *jaa jaa jaa* and *jail jail jail*. Also gives soft rattles and bell-like sounds.

ADULT. *Perched or feeding in vegetation, often appears dark. In flight, green-and-yellow tail and all-green back and wings are very bright. Juvenile has duller greens and yellows.* TEXAS, NOVEMBER

Brown Jay
Cyanocorax morio   CODE 3

L 16.5"	WS 26"	WT 7 oz

- one adult molt per year; complex basic strategy
- moderate differences between juvenile and adult
- bill color and extent of white on belly variable

Uncommon in south Texas woodlands; current concentration is well upstream of Rio Grande delta. Small flocks alternate periods of quiet industriousness with intervals of raucousness. Attracted to decrepit places.

Vocalizations frequent and insistent: a shrill, hawk-like *jay-ya*, given repeatedly by birds in chorus; also a shrill chatter, *kitchy kitchy kuh kitchy....*

ADULT. *Large and long-tailed like a magpie. Belly pale; rest of plumage dark brown. Bill of older adults black; bill of juvenile and younger adults yellow.* TEXAS, MAY

Gray Jay
Perisoreus canadensis   CODE 1

ADULT, EASTERN. *Gray above and below; head white with black crown and nape. All Gray Jays are fairly long-tailed and distinctively short-billed.* MAINE, JANUARY

ADULT, PACIFIC-SLOPE. *Head pattern about the same as birds farther east and north, but underparts brighter white.* OREGON, AUGUST

ADULT, INTERIOR. *Head paler than other populations; darkness on crown and nape reduced in both extent and intensity.* COLORADO, AUGUST

L 11.5"	WS 18"	WT 2.5 OZ

▸ one adult molt per year; complex basic strategy

▸ strong differences between juvenile and adult

▸ amount of black on head varies geographically

Widespread but generally inconspicuous in the boreal forest biome; especially partial to spruce-dominated landscapes. Frequent visitor to campsites and picnic areas; popularly known as "camp robber." Nonmigratory; wanders altitudinally but rarely out of habitat.

JUVENILE. *Dark sooty; bill is usually pale (darker on adult). Most have paler malar region.* OREGON, AUGUST

Uncharacteristically quiet for a corvid. Varied calls are soft and muted: mellow *kloop* and *kleep*; self-absorbed chatter, for example, *aaaa ratcha ratcha ratcha.*

Black-billed Magpie

Pica hudsonia CODE 1

ADULT. *Large and very long-tailed; boldly patterned in black and white. Wings and tail have variable blue and green glosses.* COLORADO, JUNE

JUVENILE. *Long tail feathers grow in gradually. Tail is broad and rounded at first, becoming longer and pointed as juvenile ages.* COLORADO, JULY

L 19"	WS 25"	WT 6 OZ

▸ one adult molt per year; complex basic strategy

▸ moderate differences between adult and juvenile

▸ tail length and shape variable in all plumages

A tame and inquisitive inhabitant of scrublands, forest edges, and riparian groves; frequents road kill and feedlots. Equally at home in remote wilderness and big western cities. Gathers in large flocks at dawn and dusk; otherwise, singly or in small flocks.

Call a short series, *mag mag mag* or *mick mick mick*, often given in chorus; like Steller's Jay but mellower. Also a soft *maa-aag?* and shrill *jay-yag?*

IN FLIGHT. *Flashes black and white in primaries and scapulars; visible from afar. Long tail also visible from afar. Wingbeats slow and powerful.* CALIFORNIA, OCTOBER

ADULT. *Both Black-billed and Yellow-billed Magpies frequently feed on carrion. They abandon it at the arrival of an intruder, and then quickly return.* COLORADO, JULY

ADULT. *With patient observation, corvid intelligence can be appreciated in the field. Black-billed Magpies engage in deception, food caching, and complex social interactions.* WYOMING, NOVEMBER

Yellow-billed Magpie
Pica nuttalli   CODE 2

L 16.5"	WS 24"	WT 5 oz

- one adult molt per year; complex basic strategy
- moderate differences between adult and juvenile
- tail length and shape variable in all plumages

Replaces Black-billed Magpie on Pacific slope of California; most numerous in the Central Valley, but also occurs in Sierra Nevada foothills and southern Coast Range. Locally common in oak savannas, pastures, and orchards, and along highway edges. Usually near water.

Repertoire of vocalizations essentially identical to Black-billed; short series of *mick* and *mag* may average slightly slower in delivery.

ADULT. *Bill yellow; bare skin below eye yellow. Juvenile less iridescent, more brownish. Black-billed in poor light or carrying food can appear yellow-billed.* CALIFORNIA, FEBRUARY

Fish Crow
Corvus ossifragus   CODE 1

L 15"	WS 36"	WT 10 oz

- one adult molt per year; complex basic strategy
- weak differences between adult and juvenile
- iridescent gloss variable, depending on light

Formerly associated with tidewater regions of Southeast. Still common coastally, but range has expanded far inland, mainly along major river systems, and is still expanding. Forms large flocks, often with American Crows. Adaptable and omnivorous; often in urban districts.

Call a nasal *cawr* or *car har* ("uh oh"); some notes of juvenile American Crow similar. Flocks and individuals are generally quieter than American.

ADULT. *Differences from American Crow are minor: bill and feet of Fish Crow smaller; wingbeats quicker; flight more erratic. Juvenile duller; eyes paler.* FLORIDA, MARCH

Tamaulipas Crow
Corvus imparatus   CODE 3

L 14.5"	WS 30"	WT 8 oz

- one adult molt per year; complex basic strategy
- weak differences between adult and juvenile
- extensive blue gloss variable, depending on light

Invaded Brownsville, Texas, area in late 1960s; hundreds were present by 1980s, followed by sharp decline in 1990s. Now rare and irregular in our area; fabled hangout at Brownsville dump is no longer reliable. Still shows up at other dumps and around outskirts of towns.

Croaking call low and flatulent: *frap* and *prut*. Sometimes given in steady succession at widely spaced intervals: *frap... frap... frap...*

ADULT. *Intermediate in length between American Crow and Great-tailed Grackle: smaller and lankier than American Crow, with extensive glossy blue in plumage. Juvenile duller than adult.* TEXAS, APRIL

Clark's Nutcracker

Nucifraga columbiana CODE 1

ADULT. *Large and sturdy-looking. Gray overall, with extensive black in wings and tail. Undertail coverts white on adult, gray-buff and often decidedly fluffy on juvenile.* CALIFORNIA, MARCH

IN FLIGHT. *Often seen swooping about rock outcroppings and alpine meadows. White in secondaries and outer tail feathers visible at great distances.* NEVADA, OCTOBER

L 12"	WS 24"	WT 4.6 oz

▸ one adult molt per year; complex basic strategy

▸ moderate differences between adult and juvenile

▸ extent and brightness of white on face variable

Inhabits rugged mountainous terrain, often up to and above tree line; during breeding season, rarely ventures below the ponderosa pine zone. Fall irruptions into desert and grassland are erratic: strong some years, nonexistent many others. Local populations unstable.

◀ Call a rising tremulous, far-carrying *ji-i-i-ick*. Also gives shorter *ick*, *juck*, and *ju-uck* calls, singly or in series; similar to Black-billed Magpie.

American Crow

Corvus brachyrhynchos CODE 1

ADULT. *Completely black; juvenile a little less glossy, more brownish. Stout-billed; at rest, wing tips are well short of tail.* CONNECTICUT, DECEMBER

IN FLIGHT. *Wings broad and rounded; rounded tail of even width. Flies with even, measured flaps. A large bird; more massive than the largest Great-tailed Grackle.* COLORADO, JANUARY

L 17.5"	WS 39"	WT 1 lb

▸ one adult molt per year; complex basic strategy

▸ weak differences between adult and juvenile

▸ eastern birds larger; south Florida birds larger-billed

Wily; widespread colonization of big cities in recent decades is a fascinating behavioral shift. Occurs singly, in small flocks, or in gigantic flocks; daily movements often extensive. Persecuted everywhere, but numbers increasing rapidly and range expanding quickly.

◀ Classic call a flat, familiar *caw* or *caw cawa*; variants approach the sounds of both Fish and Northwestern Crows. Many other calls, including a wooden rattle.

Northwestern Crow

Corvus caurinus   CODE 1

L 16"   WS 34"   WT 13 OZ

▸ one adult molt per year; complex basic strategy

▸ weak differences between adult and juvenile

▸ slight individual variation in overall body size

Replaces American Crow along Pacific coast from Puget Sound northward. Frequents beaches and intertidal pools; feeds on detritus and digs for clams. Also visits garbage cans and picnic areas. Some authorities believe that Northwestern is a subspecies of American.

Call harsh, grating: *graw* or vibrant *gra-a-aw*. Nasal like Fish Crow, but rougher. Some calls of Northwestern are similar to or indistinguishable from American.

ADULT. *Smaller than American, but difference slight; not safely separated from American under ordinary field conditions. Juvenile less black overall; has paler eyes.* ALASKA, MARCH

IN FLIGHT. *Both American and Fish Crows appear stocky and smoothly contoured in flight; wings broad and rounded; tail even-sided and rounded.* ALASKA, MARCH

Chihuahuan Raven

Corvus cryptoleucus   CODE 1

L 19.5"   WS 44"   WT 1.2 lb

▸ one adult molt per year; complex basic strategy

▸ weak differences between adult and juvenile

▸ white at base of neck feathers hard to see

Where range overlaps with Common Raven, habitats generally differ: Common in mountains, Chihuahuan in deserts and grasslands. However, both species wander and frequently visit the "wrong" habitat. More social than Common; forms large flocks after breeding.

Call like Common, but not as deep and growling: *cruck* and *cru-u-uck*. Some calls of Common identical. Often quiet, even when in large flocks.

ADULT. *Diagnostic white at base of neck feathers is rarely so visible as on this bird. Stout bill has nasal bristles extending at least halfway out along mandible. Juvenile duller.* NEW MEXICO, APRIL

Common Raven

Corvus corax CODE 1

ADULT. *World's largest passerine; long wings extend nearly to tip of tail. Nasal bristles extend only a short distance along upper mandible of long bill. Juvenile browner than adult.*
ARIZONA, DECEMBER

IN FLIGHT. *Soars competently for long periods of time; smaller crows soar briefly or not at all. Wings strikingly long and pointed; tail large and wedge-shaped.* CONNECTICUT, OCTOBER

L 24"	WS 53"	WT 2.6 lb

▸ one adult molt per year; complex basic strategy

▸ weak differences between adult and juvenile

▸ geographic variation in body size and proportions

Often seen in majestic soaring flight high above lonely rimrocks, but also seen rummaging about dumpsters. Mischievous: steals golf balls from driving ranges, peels labels off toxic waste drums, and removes rivets from aircraft. Range expanding south.

◀ Call deep and powerful, carries far: rolling *kr'r'r'uck* or emphatic *krock*. Also a melodious *whoop* or *whoop woop-woop*. Powerful wingbeats audible.

IN FLIGHT. *More than most other birds, engages in midair maneuvers that appear to be playful; much swooping, chasing, and choreographed soaring.* CONNECTICUT, OCTOBER

IN FLIGHT. *The immense size of the Common Raven (here mobbed by Red-winged Blackbirds) is usually obvious; the species is considerably larger than Red-tailed Hawk.* CALIFORNIA, APRIL

ADULT. *Common Ravens in the Mojave Desert, with relatively short bills and long nasal bristles, are more closely related to Chihuahuan Ravens than to other Common Raven populations.* CALIFORNIA, OCTOBER

Larks

FAMILY ALAUDIDAE

2 species: 1 ABA CODE 1   1 CODE 2

The ground-dwelling larks are a mainly Old World family represented in North America by the widespread Horned Lark and the range-restricted Sky Lark. Larks are especially characteristic of deserts and grasslands, and migratory flocks often assemble along ocean shores or in agricultural districts. Worldwide, larks are typically associated with desolate and rugged country, and so it is with our two species. The Horned Lark flourishes even at nuclear test sites in the arid deserts of the Great Basin, and Sky Larks occur in small numbers in remote, windswept, offshore islands in the northern Pacific region. Despite their superficial similarity to other grassland passerines (for example, pipits and longspurs), larks are considered to be in a distinct, well-defined, and taxonomically remote family within the large order Passeriformes.

SUBSPECIES

The subspecies concept is one of the most vexing problems for birders and professional biologists. Most authorities agree that a subspecies should be defined according to three basic criteria. First, a subspecies is a population that differs in some measurable way (size, color, behavior) from another subspecies. Second, the differences in these populations have a geographic basis; different subspecies may be geographically isolated from each other or their ranges may overlap only in a typically narrow "contact zone." Third, different subspecies are known or assumed to be genetically distinct; that is, their current evolutionary pathways are assumed to have some degree of independence from one another.

An immediate problem with the species concept is one of boundaries or limits. In many instances there is disagreement about whether well-defined populations correspond to separate subspecies or separate species. The seemingly endless dance of taxonomic "splitting" and "lumping" reflects this fundamental uncertainty. At the other end of the spectrum, there are many instances in which there is uncertainty about whether different populations of a species actually constitute separate subspecies, or whether they vary for other reasons. Other forms of variation below the species level include geographically widespread color morphs (polymorphism) and smooth geographic variation (called clinal variation).

Both North American larks are polytypic, meaning they have multiple subspecies. The Sky Larks introduced to Vancouver Island are of a European subspecies, whereas those that migrate naturally to and have bred on the Pribilof Islands are of a northeastern Siberian subspecies. In the case of the Horned Lark, more than 40 subspecies have been named worldwide, with close to 20 occurring in North America.

Horned Lark
Eremophila alpestris CODE 1

L 7.25"	WS 12"	WT 1 oz (28 g)

- one adult molt per year; complex basic strategy
- strong age-related and moderate sex-related differences
- complex, extensive geographic variation in plumage

Inhabits flat, barren expanses: from low deserts to alpine tundra, from arctic seacoasts to golf courses. Migratory tendencies vary among populations. Most form large flocks immediately after breeding. Common in most regions, but several populations in decline.

◀ Sweet, sputtering, accelerating song given day and night, in flight and perched: *cheer cheer cheeriddy-dee.* Call a downslurred *tseep* or *tsleep.* Flocks flush noisily.

ADULT MALE. *"Horns" distinctive, but often hard to see. All males have black breast band, mask, and "headband." Amount of white and yellow on face varies geographically.* COLORADO, FEBRUARY

ADULT FEMALE. *"Horns" absent or reduced. Facial pattern like male, but muted. In both sexes, color of upperparts varies geographically from reddish to brown to gray.* NEW HAMPSHIRE, OCTOBER

IN FLIGHT. *Pale overall and long-winged. Tail mainly black; longspurs, pipits, and other generally rarer winter birds have more white in tail.* COLORADO, DECEMBER

JUVENILE. *Variable; most are pale and washed out, with heavy scaling above. Some are surprisingly similar to Sprague's Pipit, a shorter-tailed and shorter-winged species.* ALBERTA, JULY

Sky Lark
Alauda arvensis CODE 2

L 7.25"	WS 13"	WT 1.4 oz

- one adult molt per year; complex basic strategy
- moderate age-related and seasonal differences
- Alaskan birds darker than Vancouver Island birds

Inhabits open short-grass habitats such as pastures and tundra. Two North American populations: on Vancouver Island, dwindling numbers of birds introduced from Britain; in Alaska, naturally occurring migrants from Asia. Has bred in small numbers on Pribilof Islands.

Famous aerial display a long series of sweet notes resembling Spotted Sandpiper interspersed with intricate, dry, warbled phrases. Flight call a low *clurrip.*

ADULT. *Intricately streaked above; breast buff, especially in fall, with fine dark streaking. Crest often raised. Juvenile scaly above; lacks crest.* ALASKA, JUNE

Swallows

FAMILY HIRUNDINIDAE

15 species: 8 ABA CODE 1   2 CODE 4   5 CODE 5

Familiar sights around meadows and marshes during the summer, swallows are best known for their habit of catching flying insects on the wing. Their foraging strategy calls to mind the swifts, but swallows and swifts are only distantly related. Swallows are relatively broad-winged, and they fly with languid wingbeats; swifts, in contrast, have strikingly narrow wings that arc far back, and they fly with unique twittering wingbeats.

Traditionally, the swallows have been treated as "primitive" and therefore placed toward the front of the passerine checklist. Molecular data, however, suggest that the swallows are embedded well within the bulk of "modern" passerine groups.

Swallows are restricted to habitats with abundant insect prey; they specialize on aerial insects such as mayflies, which, by and large, emerge earlier in the spring than caterpillars and other arboreal insects. As a result, swallows arrive on the breeding grounds about a month ahead of many forest insectivores, such as flycatchers and warblers. Swallows migrate chiefly in the daytime, and impressive flights may be observed in April and September. Fall flights of Tree Swallows, especially near the Atlantic coast, can be spectacular.

Most swallows are gregarious, especially during the breeding season, and many species have become accustomed to nesting on human structures. For example, Cliff Swallows place their densely packed adobe nests under highway culverts and on sides of buildings, and Bank Swallows often nest in the high walls of quarries or road cuts. Purple Martins have long been popular for dwelling in painstakingly crafted "apartments," and Tree Swallows returning from migration quickly commandeer bird boxes that are often intended for other species. Away from their nest sites, swallows frequently gather in mixed-species foraging flocks. Such assemblies are especially conspicuous during periods of stormy weather, when large flocks hover just above the surface of lakes and rivers, where the hungry birds search for insects.

Swallow populations are generally healthy, no doubt reflecting these species' willingness to coexist with humans. Besides profiting from additional nest sites provided by humans, swallows may be beneficiaries of global warming. Northern Rough-winged Swallows, in particular, have been returning earlier in the spring and wintering at higher latitudes in recent years, and several ornithologists have speculated that this behavioral shift may be a result of warmer winters. Major mortality events—usually related to inclement weather or to the perils of migration—are documented from time to time, but it is unclear whether such die-offs have long-term population consequences.

Barn Swallow
Hirundo rustica CODE 1

ADULT FEMALE. *Most females are shorter-tailed than males, with paler underparts, but tail length and plumage variable.* TEXAS, MAY

ADULT MALE. *Glossy blue above. Throat and forehead dull red; underparts orangish. Tail of male very long, appearing either deeply forked or tapered to a fine point.* TEXAS, APRIL

L 6.75"   WS 15"   WT 0.7 oz (20 g)

▸ one adult molt per year; complex basic strategy

▸ moderate age-related and sex-related differences

▸ vagrants to western Alaska have white underparts

A familiar summer sight: hawks insects all day long, perches on wires, and gathers mud from ground. Conspicuous nests are placed on human-made structures: under bridges, eaves, and porch roofs. Forages widely, but nest site often associated with water.

◀ Song a scratchy series with sweet slurred notes and a raspy trill: *wish-skwisha-wish... wheeer! schwisha skish bzhizzzzz swish...* Flight call: *see-yit.*

IN FLIGHT. *Tail can appear forked, as here, or tapered to a long point; note pale band across tail. Flight broad and graceful, with fewer twists and turns than other swallows.* ONTARIO, MAY

ADULT AND JUVENILES. *Although many passerines are secretive around their nests and young, swallows—including the Barn Swallow—are easily observed throughout the breeding cycle.* TEXAS, MAY

JUVENILE. *Compared to female, even shorter-tailed and paler below, with less-glossy upperparts; tail lengthens as bird progresses through juvenile plumage.* TEXAS, MAY

Cliff Swallow
Petrochelidon pyrrhonota CODE 1

ADULT. *Like Barn Swallow, glossy blue above with dull red on face. Tail much shorter than Barn Swallow; pale patch on forehead usually prominent.* UTAH, JULY

JUVENILE. *Duskier than adult, with more muted plumage pattern. Usually shows dull red on face, dark throat, and relatively pale nape.* UTAH, JULY

L 5.5"	ws 13.5"	wt 0.75 oz (21 g)

▸ one adult molt per year; complex basic strategy

▸ strong differences between adult and juvenile

▸ some birds in Southwest have chestnut foreheads

Highly social; breeds in tight colonies, ranging from huge aggregations under eaves of tall buildings to small clusters under culverts in remote deserts. Forages widely, especially near water, over lawns, and in other buggy habitats. Range has expanded far eastward.

◀ Twangy flight call begins sharply and trails off immediately, *chow!* or *chow!-wow*. Song a weak jumble with trilled *brrrrrt* notes interspersed throughout.

IN FLIGHT. *Tail squared off, unlike Barn Swallow. Forehead patch ("headlights") conspicuous in head-on flight.* WYOMING, JUNE

IN FLIGHT. *From above, rufous rump contrasts with steel-blue tail and back; Cave Swallow also shows this pattern.* UTAH, JUNE

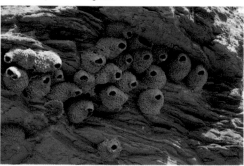

NEST COLONY. *Tightly packed nests are sometimes placed in natural settings, as here, but the species has been won over in most places by buildings, bridges, and other structures.* UTAH, SEPTEMBER

Cave Swallow
Petrochelidon fulva   CODE 1

L 5.5"   WS 13"   WT 0.5 oz (14 g)

▸ one adult molt per year; complex basic strategy
▸ strong differences between adult and juvenile
▸ birds in Florida deeper red than Southwest birds

Status rapidly changing: breeding population increasing and expanding in Southwest; late-fall wanderers to Northeast increasing dramatically in frequency and abundance. Expansion of breeding range is aided by this species' willingness to nest in bridges and culverts.

Flight call higher-pitched, more clipped than Cliff: *cha* or *chacha*. Song a sputtering series of short *ch'* and *cha* notes interspersed with weak trills.

Late-fall wanderers to the Northeast apparently come from the southwestern population of Cave Swallows, not from the closer Florida population. Any Petrochelidon swallow in the Northeast after mid-October is more likely to be Cave Swallow than Cliff Swallow.

ADULT. *Both Cliff and Cave have squared-off tail with orange rump. Most Caves have pale throat and dark forehead, the opposite of Cliff. Juvenile pale buff below and on throat.* TEXAS, MAY

IN FLIGHT. *Typical adult, as here, has throat paler than forehead; beware variant Cliffs with dark foreheads, as well as variant Caves with dark on throat.* TEXAS, JUNE

Northern Rough-winged Swallow
Stelgidopteryx serripennis   CODE 1

IN FLIGHT. *Wings fairly long but also broad-based; wingbeats relatively deep and slow, recalling Barn Swallow. Plumage low contrast overall.* NEW MEXICO, MARCH

ADULT. *Our dullest swallow; wholly gray-brown plumage above blends into diffuse wash across breast; otherwise pale gray below.* CALIFORNIA, JANUARY

L 5.5"   WS 14"   WT 0.6 oz (17 g)

▸ one adult molt per year; complex basic strategy
▸ moderate differences between adult and juvenile
▸ plumage less contrasting in north than south

Occurs widely in small numbers; forages over many habitats, especially ponds and rivers. Does not breed colonially. Nests in both natural and human-made cavities: crevices in walls and bridges, stream banks, and road cuts. Breeding and winter ranges expanding north.

JUVENILE. *Warmer than adult. Upperparts scalloped with buff; faint to prominent wing bars tinged with cinnamon. Throat also washed with cinnamon.* OREGON, JUNE

◀ Flight call a rough *fzzzap* or *fzzzup*. Song a slow series of notes similar to flight call. Quiet overall compared to other swallows; often forages silently.

Bank Swallow

Riparia riparia   CODE 1

L 5.25"	WS 13"	WT 0.5 oz (14 g)

▸ two adult molts per year; complex alternate strategy

▸ weak differences between adult and juvenile

▸ southern birds average paler, grayer, shorter-winged

Colonial; nests in earthen cavities in vertical surfaces near reservoirs, marshes, and other aquatic habitats. Uses both natural and artificial sites, such as river banks and quarries. Local abundance influenced both negatively and positively by human activities.

Guttural flight call, harsher than Cliff, often repeated quickly: *raa raa raa* and *razz razz razz*. Feeds noisily, unlike Northern Rough-winged.

ADULT. *Short-tailed; long-winged. Uniformly brown above. White below from chin to undertail coverts, except for broad brown breast band. Juvenile like adult, but has buff wing bars.* CALIFORNIA, JUNE

Tree Swallow

Tachycineta bicolor   CODE 1

JUVENILE. *Mainly brown above; pale below. Some show hint of breast band, as here, and can be confused with Bank Swallow; Tree has more extensive brown hood.* COLORADO, JUNE

ADULT MALE. *Upperparts blue or blue-green; in poor light appears dark above. Flight feathers of wing and tail dark brown. Underparts completely white.* CALIFORNIA, JULY

L 5.75"	WS 14.5"	WT 0.7 oz (20 g)

▸ one adult molt per year; complex basic strategy

▸ strong age-related, moderate sex-related differences

▸ sheen on upperparts variable, depending on light

Nests in cavities near water: uses abandoned woodpecker holes; readily accepts bird boxes. Forages over water. Arrives early in spring, departs late in fall; in cold weather, when insects are scarce, eats berries. Migrates by day; flocks in fall are sometimes enormous.

◀ Languid song consists of mellow warbles and lazy buzzes, for example, *drzzzt slweeweewa*. Call a liquid *tleewa*. Noisy around nest site; migratory flocks often quiet.

ADULT FEMALE. *Duller than male, but variable: some have little color above, as here; others, perhaps older females, are more similar to male.* CALIFORNIA, JULY

Violet-green Swallow
Tachycineta thalassina   CODE 1

ADULT MALE. *Colorful: head olive-green; back bright green; wings blue-green. White underparts extend onto much of face, as here, and well up onto sides of rump (prominent in flight).*
CALIFORNIA, JULY

ADULT FEMALE. *Duller than male; head smudged dusky olive. Juvenile even duskier, but usually shows some white on face and prominent white on sides of rump.*
CALIFORNIA, JULY

IN FLIGHT. *Above, white prominent on face and sides of rump. Overall pattern of dark and white similar to White-throated Swift, but flight style different.*
YUKON TERRITORY, MAY

5.25"	ws 13.5"	wt 0.5 oz (14 g)

- one adult molt per year; complex basic strategy
- strong age-related, moderate sex-related differences
- extent of white on flanks of juvenile variable

Breeding habitat generally distinct from Tree Swallow: Violet-green in montane forests, Tree around marshes and other waterways. Soars less than Tree. Nests solitarily or in small colonies. In bad weather, large flocks move to open water and gather just above surface.

Vocalizations drier than Tree Swallow. Call a simple *chivvit* or *heerit*. Song a series of *chiv* and *cheer* notes, often given at night in pinewoods.

Purple Martin
Progne subis   CODE 1

ADULT MALE. *Larger than other swallows. Dark blue all over; tail forked. Unlikely to be confused with other swallows, but flight profile of European Starling (also dark and glossy) similar.*
TEXAS, MAY

ADULT FEMALE. *Underparts gray with dusky spots and scaling; upperparts dark blue. Underparts of juvenile similar to adult female, but upperparts mainly gray.* TEXAS, MAY

L 8"	ws 18"	wt 2 oz

- ▸ one adult molt per year; complex basic strategy
- ▸ strong age-related and sex-related differences
- ▸ western females have whiter underparts and foreheads

Nest sites differ markedly between eastern and western populations: eastern birds nest only in human-made martin "apartments"; western birds nest in abandoned woodpecker holes or other natural cavities. Local populations crash in response to cold, wet weather.

Song alternates between rough and sweet notes, for example, *churf churf churf kleer kleer...* Vocalizations average richer, fuller, and deeper than other swallows.

Chickadees, Verdin, Long-tailed Tits, Nuthatches, and Creepers

FAMILIES PARIDAE, REMIZIDAE, AEGITHALIDAE, SITTIDAE, AND CERTHIIDAE

19 species: 15 ABA CODE 1   3 CODE 2   1 CODE 3

These are small, sociable, and industrious birds that probe for insects in woody vegetation. Relationships among the five families are unclear, but most taxonomists consider them to be fairly closely allied with one another. A complicating factor is their relationship to other groups that have traditionally been placed elsewhere in the avian family tree. For example, the Brown Creeper and the Verdin appear closely related to each other, according to one method of molecular analysis—but that same method also points to a close alliance between those two species and the wrens. Many taxonomists are hesitant to shuffle such well-established groups as the wrens, with the result that numerous small families in the middle of the passerine checklist are likely to be recognized for some time to come.

Most species in these five families habitually join mixed-species feeding flocks, and some of the most influential research on foraging behavior has been conducted on these birds. Chickadees often form the nucleus of feeding flocks, with nuthatches, creepers, and other species joining them on the periphery. One species that tends not to associate with chickadees—or any other species—is the Bushtit. However, it is hardly antisocial; tight-knit clans of Bushtits assemble soon after the breeding season and forage cooperatively throughout the winter.

Although primarily insectivorous, the members of this family do not withdraw to the tropics as do many other insectivorous North American passerines. Instead, they manage to exploit a wintertime arthropod fauna that is overlooked by many other species: eggs and egg-cases, galls and leaf-miners, and insects and spiders that spend the winter as adults in bark and leaf litter. Many of these birds supplement their diets with high-energy fruits and berries, and some—especially chickadees—flourish at bird feeding stations. The social tendencies of these birds are generally believed to be vital for communicating the whereabouts of critical food resources during the cold months.

The representatives of these five families enjoy generally healthy populations in North America, but there have been many regional population shifts in response to habitat alterations by humans. The nuthatches in particular are susceptible to forestry practices such as clear-cutting, stand thinning, and snag removal. Not surprisingly, given the thermal demands placed on small birds wintering at high latitudes, several species—among them Verdin and Carolina Chickadee—have been identified as apparent beneficiaries of climate warming.

Black-capped Chickadee
Poecile atricapillus CODE 1

L 5.25"   ws 8"   wt 0.4 oz (11 g)

▸ one adult molt per year; complex basic strategy

▸ weak differences between spring and fall adults

▸ hybridizes with Carolina Chickadee where ranges overlap

Familiar within range; industrious and tame, often visiting feeding stations. Occurs in woods of all sorts, from forest interior to small tracts in urban districts. Generally sedentary, but in some years irrupts south well into range of Carolina Chickadee.

ADULT. *Instantly recognized as a chickadee by its black cap, black bib, and white cheeks. Gray above; gray-and-buff below. Bill very small. Brighter overall in fall than spring.* MINNESOTA, JANUARY

◀ Song in most of range a simple, pure-tone, descending minor third: *wheee way-ay.* Call a husky, wheezy series: *fzicka bzee bzee bzee* ("chick-a-dee-dee-dee").

The range limits of the Black-capped and Carolina Chickadees are so closely matched that ecologists have hypothesized that the two species are in direct competition where their ranges meet. What may be at play here is a phenomenon known as "one-way competitive exclusion." Carolina Chickadees may exclude Black-capped Chickadees from the South, but Black-capped Chickadees flourish in the North, where Carolina Chickadees are at a disadvantage due to climate conditions.

ADULT. *Differs from Carolina Chickadee in a suite of minor characters: wing coverts and secondaries frosty; lower edge of bib sometimes ragged; cheek more extensively white.* MINNESOTA, JANUARY

Carolina Chickadee
Poecile carolinensis CODE 1

L 4.75"   ws 7.5"   wt 0.4 oz (11 g)

▸ one adult molt per year; complex basic strategy

▸ weak differences between spring and fall adults

▸ size and plumage vary weakly throughout range

Ecologically, the southeastern counterpart of Black-capped Chickadee; occurs in woodlands of all sorts. Narrow zone of contact between the two species drifting slowly but steadily northward. Wanderers almost never noted beyond well-defined northern limit of range.

ADULT. *Less white on wings and cheek than Black-capped; lower edge of bib smoother. After fall molt, can show white in wings like Black-capped. Grayer below in western portion of range.* TEXAS, APRIL

Song has four syllables: *whee way ee bay.* First and third syllables clipped in faster version: *w'way w'bay.* "Chickadee" call faster than Black-capped.

Mexican Chickadee

Poecile sclateri   CODE 2

ADULT. *General pattern like that of other chickadees. Black bib extensive; underparts mainly gray. Juvenile similar, but has slightly duller black cap.* ARIZONA, JULY

ADULT. *In fresh basic plumage (from fall into early winter), as here, all chickadees are crisply attired. As plumage wears (spring into summer), plumage contrast diminishes.* ARIZONA, DECEMBER

L 5"	WS 8.25"	WT 0.4 oz (11 g)

▸ one adult molt per year; complex basic strategy

▸ weak age-related and seasonal differences

▸ slight individual variation in plumage contrast

Limited in North America to Chiricahua Mountains of Arizona and to Animas and Peloncillo Mountains of New Mexico; the only chickadee in its range. Breeds in mountain forests, generally no lower than the ponderosa pine zone; a few descend to lower elevations in winter.

"Chickadee" call burry: *zh'ricka zhree zhree.* Also gives singsong *ch'rickadee? ch'rickadoo!* and descending, vireo-like calls: *p'titititi* and *zhraaaay.*

Mountain Chickadee

Poecile gambeli   CODE 1

FRESH ADULT, INTERMOUNTAIN WEST. *Black cap broken by broad white eyebrow. Birds in eastern portion of range have more buff below.* NEW MEXICO, DECEMBER

WORN ADULT, PACIFIC SLOPE. *Grayer below than Intermountain West birds; eyebrow not as prominent. In all populations, white eyebrow becomes muted in worn plumage, as here.* CALIFORNIA, JULY

L 5.25"	WS 8.5"	WT 0.37 oz (10.5 g)

▸ one adult molt per year; complex basic strategy

▸ moderate differences between spring and fall adults

▸ Pacific-slope birds grayer, especially on flanks

Mainly in dry mountain woodlands, from sparse juniper groves in foothills to spruce-fir stands near tree line. Where ranges overlap, Black-capped prefers riparian groves, Chestnut-backed prefers damp, shady forests. Minor winter dispersal to lower elevations.

Song varies geographically: steady *fee beee beee,* slurred *s'wee s'way,* and wavering *f'beee beee beee.* "Chickadee" call huskier than Black-capped.

Boreal Chickadee

Poecile hudsonica CODE 1

FRESH ADULT. *Cap brown; flanks bright rusty. Wings and tail solidly gray; Gray-headed usually has white in wings and tail. Cheeks grayish toward rear; Gray-headed has more white in cheeks.* MAINE, FEBRUARY

WORN ADULT. *As plumage wears, rusty in flanks diminishes and head markings become more muted, inviting confusion with rarer Gray-headed.* MAINE, JUNE

L 5.5" WS 8.25" WT 0.35 oz (10 g)

▸ one adult molt per year; complex basic strategy
▸ moderate differences between spring and fall adults
▸ geographic variation in extent of brown in plumage

Tame resident of boreal forests of all sorts: early and late successional stands, dense groves, and open parkland. Spruce and balsam fir especially favored. Annual elevational movements in mountainous portions of range; occasional irruptions well south of normal range.

Song unknown, but gives gargled series: *z'zwish ch'ch'ch'ch'ch'chee.* Calls harsh and raspy, like Mountain Chickadee, for example, *tsicka zzzeee zzzeee.*

Chestnut-backed Chickadee

Poecile rufescens CODE 1

ADULT, NORTHERN. *All have prominent reddish-brown on back and variable brown on cap and bib. In most of range, flanks warm chestnut. Overall plumage contrast lower on worn adults.* ALASKA, MARCH

ADULT, SOUTHERN. *From San Francisco southward, chestnut on underparts greatly reduced; most show just a little color on the vent and undertail coverts.* CALIFORNIA, DECEMBER

L 4.75" WS 7.5" WT 0.35 oz (10 g)

▸ one adult molt per year; complex basic strategy
▸ weak differences between spring and fall adults
▸ chestnut, especially on flanks, reduced southward

Inhabits wet, shady forests of the far West. Historically associated with humid conifer habitats, but has adapted to urban habitats and exotic hardwood plantings. Resident throughout range, but some drift upslope after breeding; winter wanderings related to snow cover.

Vocalizations diverse. Many are sweet and ringing, others buzzy: rapid *cheery cherry cheery*, emphatic *zzzeet*, and piercing *sweep* and *sweepit.*

Gray-headed Chickadee
Poecile cincta   CODE 3

L 5.5"	WS 8.5"	WT 0.44 oz (12.5 g)

- one adult molt per year; complex basic strategy
- moderate differences between spring and fall adults
- individual variation in facial patterning

Widely distributed and well studied in Eurasia, but scarce and little known in North America. Alaskan and Canadian population inhabits sparse, open woods; mainly spruce and willow groves along rivers and in foothills. Range limits and extent of vagrancy poorly known.

Like other "brown" chickadees (Chestnut-backed, Boreal), probably lacks a whistle-song. Call a twangy *dweer*, given singly or in steady series.

ADULT. *Compared to Boreal, has more white in cheeks, less rusty below, more white in wings, and longer, whiter tail. White in wing fades through spring and summer.* ALASKA, JUNE

Tufted Titmouse
Baeolophus bicolor   CODE 1

L 6.5"	WS 9.75"	WT 0.75 oz (21 g)

- one adult molt per year; complex basic strategy
- moderate differences between adult and juvenile
- hybrids with Black-crested Titmouse in zone of overlap

Energetic inhabitant of eastern forests; prefers deciduous stands. Versatile; flourishes in parks and backyards in towns and cities. Range has expanded north and is probably still expanding; climate change and proliferation of feeding stations likely involved.

◀Vocalizations highly varied. Song rich and repetitious: *peedo peedo peedo* and *peer peer peer*. "Chickadee" call a husky *see see djicka djicka.*

ADULT. *Larger than chickadees. Gray above; mainly white below. Crest usually prominent; black on forehead contrasts with rest of face. Flanks bright rusty.* KENTUCKY, FEBRUARY

ADULT. *Like all chickadees and titmice, Tufted Titmouse is acrobatic. It is often first detected by a sudden burst of motion.* NEW JERSEY, DECEMBER

JUVENILE. *Duller than adult; flanks usually unmarked and crest reduced. Forehead gray (forehead white on juvenile Black-crested).* OKLAHOMA, JULY

Bridled Titmouse
Baeolophus wollweberi   CODE 2

| L 5.25" | ws 8" | wt 0.37 oz (10.5 g) |

▸ one adult molt per year; complex basic strategy
▸ weak differences between juvenile and adult
▸ darker-backed in central than in southeastern Arizona

Inhabits oak and pine-oak woods, ranging from lowland riparian stands to mid-elevation forests; usually in denser, more luxuriant woodlands than Juniper Titmouse. After breeding, teams up with chickadees, nuthatches, and other small birds in rowdy feeding flocks.

Vocalizations varied. Most consist of rapidly repeated elements: dry *p'ch'ch'ch'ch* and liquid *dlee dlee dlee dlee.*

ADULT. *In size and plumage, intermediate between chickadees and titmice. Crested like other titmice, but black-and-white head pattern is more ornate. Juvenile slightly duller.* ARIZONA, FEBRUARY

Black-crested Titmouse
Baeolophus atricristatus   CODE 2

| L 6.5" | ws 9.75" | wt 0.75 oz (21 g) |

▸ one adult molt per year, complex basic strategy
▸ weak age-related and sex-related differences
▸ color of forehead ranges from white to cinnamon

Recently split from Tufted Titmouse, a return to older classification. Replaces Tufted in central Texas, where it is common in wooded habitats. Favors oak scrublands, but frequently wanders into riparian woods and residential districts. Habits much as Tufted.

Vocalizations recall Tufted. Song somewhat richer, often simpler: *reer reer reer* and *jare jare jare.* "Chickadee" call higher, more slurred than Tufted.

ADULT. *Face pattern the reverse of Tufted: crest black (gray in Tufted) and forehead pale (black in Tufted). Crest of female slightly more brownish, flanks of juvenile less rusty.* TEXAS, NOVEMBER

Oak Titmouse
Baeolophus inornatus   CODE 1

| L 5.75" | ws 9" | wt 0.6 oz (17 g) |

▸ one adult molt per year; complex basic strategy
▸ weak differences between spring and fall adults
▸ northern birds smaller, smaller-billed, and paler

With Juniper Titmouse, split from former "Plain" Titmouse. Common and conspicuous; inhabits warm, dry woodlands, especially oak and oak-pine. Maintains territories year-round; does not join winter foraging flocks. Sedentary; rarely noted out of range.

Songs varied, most with bright and cheery quality: *ridda ridda ridda* and *sleepa sleepa sleepa.* Calls also varied; for example, *r'zeea zeea zzzzz* and *cheeradoo!*

ADULT. *Plain gray-brown all over; Juniper Titmouse slightly paler and grayer. Both species are crested; Oak a little smaller. Worn birds, especially in summer, decidedly ratty.* CALIFORNIA, JANUARY

Juniper Titmouse
Baeolophus ridgwayi   CODE 1

| L 5.75" | WS 9" | WT 0.6 oz (17 g) |

▸ one adult molt per year; complex basic strategy

▸ weak differences between spring and fall adults

▸ northern birds average larger, grayer than southern

Common, but favors arid low-montane woodlands that birders rarely visit: juniper-sage transition zone, pure juniper stands, and pinyon-juniper forests. Some populations in western Great Basin, provisionally identified as Juniper Titmouse, inhabit dry oak forests.

Calls liquid and rolling like Oak, but not as cheery. Delivery faster: *clippidee clippidee clippidee* and *kl'dee kl'dee kl'dee* and *shick shick shick.*

ADULT. *Nearly identical to Oak Titmouse; slightly paler and grayer, and slightly larger. Like Oak, adults in worn plumage in summer are ratty-looking.* NEW MEXICO, DECEMBER

Verdin
Auriparus flaviceps   CODE 1

| L 4.5" | WS 6.5" | WT 0.25 oz (7 g) |

▸ one adult molt per year; complex basic strategy

▸ strong age-related, moderate sex-related differences

▸ amount and brightness of yellow on crown variable

Active, even at midday. Forages continuously, plucking arthropods from spiny desert shrubs; visits nectar feeders. Greatest densities in lush washes and riparian zones, but common even in monotonous creosote bush flats. Range expanding with desertification of Southwest.

◀ Vocalizations persistent and surprisingly loud. Call a bright, piercing, descending *tyeef;* song a short descending series, *tyee tyee tyee.*

ADULT MALE. *Very small; gray overall, with a sharply pointed bill. Most males have extensive yellow on head and blood red at bend of wing.* CALIFORNIA, APRIL

JUVENILE. *Almost completely gray. Distinguished from other small, gray, desert birds by pointy bill, active demeanor, and microhabitat.* CALIFORNIA, JULY

ADULT FEMALE. *Compared to male, most females have less yellow on head and diminished red in wings.* CALIFORNIA, NOVEMBER

Bushtit
Psaltriparus minimus CODE 1

MALE, INTERIOR. *Tiny; long-tailed and small-billed. All are basically gray. Interior birds have variable auriculars and plain gray crown. Adult male has dark eyes; yellow in female.* ARIZONA, FEBRUARY

FEMALE, INTERIOR. *Auriculars often indistinct, as here; many juveniles have extensive black in auriculars. A few adults, especially near Mexican border, have entirely black auriculars.* COLORADO, NOVEMBER

L 4.5"	WS 6"	WT 0.2 oz (6 g)

- one adult molt per year; complex basic strategy
- moderate sex-related and weak age-related differences
- much geographic and polymorphic variation

Almost always seen in roiling, energetic flocks of 10–25 birds, exceptionally 100+. Flies single file between shrubs. Inhabits washes, foothills, and scrublands; often in towns and cities, especially Pacific Slope population. Expanding north in eastern portion of range.

◀ Flocks chatter constantly; calls soft and high-pitched. Typical chorus consists of repeated *pee! p'pee!* elements interspersed with popping *zick* notes.

MALE, PACIFIC SLOPE. *Prominent cap is brownish and gray auriculars are indistinct, the opposite of Interior birds. Like Interior birds, male has dark eyes and female has yellow eyes.* CALIFORNIA, MARCH

ADULT AT NEST. *As if to make up for its paltry stature, the Bushtit builds an oversized nest— typically oversized and lined with live plant matter. Pacific-slope female shown here.* WASHINGTON, MAY

FEMALE, PACIFIC SLOPE. *Some are weakly patterned, but yellow eye of adult female always distinctive with close view. This individual is in molt, giving it a ratty appearance.* CALIFORNIA, SEPTEMBER

Red-breasted Nuthatch
Sitta canadensis   CODE 1

MALE. *Smaller than White-breasted, with slighter bill. Head has bold black-and-white stripes. Smooth blue-gray above; reddish below, usually extensively.* NEW MEXICO, DECEMBER

FEMALE. *Pattern similar to adult male, but colors duller: crown is blue-gray (black in male); underparts not as colorful as male. Immature slightly duller than adult.* CALIFORNIA, OCTOBER

IN FLIGHT. *All nuthatches are ovoid and pot-bellied, with a short rounded tail and rounded wings. All have white band on otherwise dark tail.* OREGON, APRIL

L 4.5"   WS 8.5"   WT 0.35 oz (10 g)

- one adult molt per year; complex basic strategy
- moderate differences between adult male and female
- reddish below variable, especially in adult male

Inhabits conifer forests during breeding season; uses other trees on migration and during winter, but still prefers conifers. Extent of fall dispersal varies greatly from year to year. Inquisitive and aggressive; responds readily to pishing and owl imitations.

All vocalizations nasal and petulant. Song a tinny *ehhnk... ehhnk... ehhnk*, with distinctive drawn-out quality. Call a whiny *ank*, usually in irregular series.

Pygmy Nuthatch
Sitta pygmaea   CODE 1

ADULT. *Small; blue-gray above and pale buff below. Dusky-brown cap and black eye stripe separates Pygmy from other nuthatches in its range.* CALIFORNIA, JANUARY

JUVENILE. *Plumage lower-contrast than adult: eye stripe is paler than adult; cap less richly colored. Worn adult in summer similar.* OREGON, AUGUST

L 4.25"   WS 7.75"   WT 0.4 oz (11 g)

- one adult molt per year; complex basic strategy
- moderate seasonal and weak age-related differences
- interior populations slightly larger and paler

Gregarious; usually in small flocks of about 5 individuals, but sometimes in flocks of 25+. Responds animatedly to pygmy-owl imitations. Found only in conifers, usually in mid-altitude yellow pine forests. Forages high in trees, often amid needles and smaller branches.

Call a piercing *peep!* or *pip!* Flocks especially noisy; often whip themselves up into frenzy of strident call notes. Some calls similar to Red Crossbill.

White-breasted Nuthatch

Sitta carolinensis   CODE 1

ADULT MALE, EASTERN. *All plumages have white face with black cap, blue-gray upperparts, and white underparts with rusty belly. Eastern birds relatively thick-billed and pale above.* CONNECTICUT, JUNE

ADULT FEMALE, EASTERN. *Females of all populations paler and duller than males; crown grayish (black in male). Immature male similar to adult female.* NEW HAMPSHIRE, NOVEMBER

ADULT MALE, INTERIOR WEST. *More similar to Pacific Slope bird than to Eastern; averages darker on the flanks than Pacific Slope.* ARIZONA, NOVEMBER

L 5.75"   WS 11"   WT 0.75 oz (21 g)

▸ one adult molt per year; complex basic strategy

▸ moderate differences between adult male and female

▸ geographic variation in calls, structure, and plumage

Occurs year-round in all forest types, with preference for older and more extensive groves. Little large-scale migratory movement, but local populations may disperse, especially in fall. Occurs singly or in small groups; active and industrious much of the time.

◀ Song of all populations a liquid series, *tu tu tu tu tu tu*. Call note varies regionally: flat *yank* (Eastern); stuttering *st't't't'* (Rockies); wavering *yiiirk* (far West).

Although widespread and successful, the White-breasted Nuthatch evidently prefers older, larger tracts of woodlands during the breeding season. The presence of breeding populations of this species may be a good indication of local forest health.

ADULT MALE, PACIFIC SLOPE. *Bill thinner than Eastern; black cap not as extensive; upperparts and flanks often darker.* CALIFORNIA, JULY

Brown-headed Nuthatch
Sitta pusilla   CODE 1

ADULT. *Differs from Pygmy in rufescent tones to brown head, diminished eye stripe, and less buff on underparts. Subadult and worn adult more dully marked.* LOUISIANA, APRIL

L 4.5"	ws 7.75"	wt 0.35 oz (10 g)

- one adult molt per year; complex basic strategy
- moderate seasonal and weak age-related differences
- smaller and paler on Florida peninsula than elsewhere

Southeastern counterpart of the Pygmy Nuthatch, but no range overlap. Like Pygmy, forages in noisy, small to medium-sized flocks that favor upper reaches of tall pines. Cooperative breeding system in which young adults forgo reproduction and help raise their siblings.

Song a nasal series with sharp opening note, followed by flatter notes in quick succession: *pwee! tu'tu'tu'tu.* Call note a single or doubled *pick,* given frequently.

Because of its heavy reliance on commercially valuable conifer forests, the Brown-headed Nuthatch is at risk in some locations. In particular, populations near the northern limits of the range are susceptible to decline or extirpation—and the prospects for local recovery are usually not good.

Brown Creeper
Certhia americana   CODE 1

ADULT. *Brown above; white below. Bill thin and decurved. Woodpecker-like tail pressed against tree trunk. Flight feathers of wing buff at base, forming broad chevron on bird at rest.* CALIFORNIA, MARCH

ADULT. *Eastern birds slightly paler and shorter-billed than western. Juveniles of all populations show slightly more buff overall; underparts weakly barred (unmarked on adult).* CONNECTICUT, APRIL

L 5.25"	ws 7.75"	wt 0.3 oz (9 g)

- one adult molt per year; complex basic strategy
- weak differences between adult and juvenile
- geographic variation in ground color and bill structure

Quiet but industrious; jerks its way up trees, but with smoother movements than woodpeckers. Often spirals upward around large trunk, then flies to base of nearby tree and repeats the process. Forages solitarily or in loose association with other species.

◀ All vocalizations high-pitched. Song is a wheezy jumble, sometimes fairly loud. Call note abrupt and buzzy: *dzwzzzz.* Flight call a short *szwit.*

Wrens and Dippers
FAMILIES TROGLODYTIDAE AND CINCLIDAE

10 species: 10 ABA CODE 1

The primarily terrestrial wrens are a distinctive group, and the highly aquatic American Dipper is perhaps the most distinctive bird species in North America. A close relationship between the families has been posited, but the taxonomic situation of both families is uncertain—especially that of the dippers. The wrens are a New World family, with greatest diversity in Mexico; only one species (the Winter Wren) reaches the Old World. The dippers are a small family (only five species) distributed widely in the Eastern and Western Hemispheres.

Wrens are portly, with long or short tails that are frequently cocked up or flipped about expressively. They occur in diverse habitats, but all species spend much time foraging in tight spaces—dense bramble patches, thick marsh vegetation, even amid canyon walls and rock outcroppings. When foraging, wrens often assume an odd, flattened profile. In overall structure, the American Dipper suggests an oversized wren: portly and bobtailed.

Many wrens observe surprisingly strict microhabitat boundaries. For example, Marsh and Sedge Wrens may occur in the same wetland complex, but the former prefers cattails in standing water, whereas the latter is more likely to be found in muddy meadows away from the water. Rock and Canyon Wrens often occupy the same canyon, but the former concentrates on dusty talus slopes, whereas the latter gravitates toward steep, often towering rock formations. The American Dipper is famously dependent upon rushing mountain streams; sightings even a few feet up on a bank or shore are extremely uncommon.

As if to make up for their drab appearance, wrens are highly prized for their vocal talents. The Winter Wren is rated by some as the world's greatest songster, and the Canyon Wren is one of the loveliest voices of the West. Male and female Carolina Wrens sometimes sing in a back-and-forth duet, and the outpourings of the Rock Wren call to mind a subdued Northern Mockingbird. The American Dipper, too, is a superb—and widely underappreciated—songster, frequently belting out a complex series of warbles and trills that may go on for several minutes.

Populations of North American wrens are mainly stable or increasing. Bewick's Wren is expanding within the Great Basin, and the Cactus Wren may be inching slowly northward with advancing desertification. In the East, Carolina Wren populations are pushing northward—set back periodically by strong cold spells. Local populations of Marsh and Sedge Wrens often decline or disappear due to habitat destruction, but both species are probably stable overall. The same can be said of the American Dipper—a species that has been proposed as a bioindicator of water quality throughout the West.

House Wren
Troglodytes aedon CODE 1

ADULT. *Small and stubby (Winter Wren even smaller and stubbier). Plain overall: brown above; paler gray-brown below. Eyebrow thin and pale.* OHIO, MAY

JUVENILE. *Warmer buff than adult (sometimes tending toward rufous); many show indistinct scalloping on throat and breast.* NEW JERSEY, JUNE

"BROWN-THROATED" WREN. (T. a. cahooni) *Ranges north from Mexico into mountains of Southeast Arizona. Warmer brown below than other North American House Wrens; eyebrow more distinct.* MEXICO, SEPTEMBER

L 4.75"   WS 6"   WT 0.4 oz (11 g)

▸ one adult molt per year; complex basic strategy
▸ weak differences between adult and juvenile
▸ "Brown-throated Wren" brighter and bolder

Flourishes in thickets and woodland edges; spends much time in dense shrubbery, but comes out for a view now and then. Common in towns and cities; readily accepts nest boxes. Also nests in outbuildings like barns and pump houses.

◀ Song is an outpouring of dry, bubbly notes; starts softly, then quickly erupts into rapid series. Call a harsh *chet*, like Common Yellowthroat, often in series.

Winter Wren
Troglodytes troglodytes CODE 1

WORN ADULT, EASTERN. *A tiny stub of a bird; short tail often cocked straight up. Compared to House Wren, darker overall, eyebrow usually more prominent, barring on belly stronger.* MAINE, JUNE

FRESH ADULT, WESTERN. *Averages darker than Eastern, often with reddish highlights. In all populations, juvenile has faint scalloping on breast and less-distinct barring on belly.* CALIFORNIA, OCTOBER

ADULT, BERING SEA. *Substantially larger than mainland populations; bill relatively longer. Plumage similar to Western, but usually without reddish highlights.* ALASKA, JUNE

L 4"   WS 5.5"   WT 0.35 oz (10 g)

▸ one adult molt per year; complex basic strategy
▸ weak differences between adult and juvenile
▸ extensive geographic variation in size, color, voice

Less confiding than House Wren. Even when just a few feet away, skillfully keeps out of sight. Usually stays close to ground, sometimes disappearing in large roots or clumps of moss. Western population possibly declining, Eastern apparently expanding.

Song an amazing series of complex, high-pitched trills; richest in eastern populations. Call a sharp *chimp*, usually doubled; drier in western populations.

Bewick's Wren
Thryomanes bewickii CODE 1

ADULT. *All plumages have long tail with white spots; tail constantly flipped sideways. Eyebrow always prominent. Juvenile a little paler overall than adult.* CALIFORNIA, APRIL

ADULT. *Ground color varies geographically; in eastern portion of range, some are reddish, suggesting Carolina Wren. Weak color polymorphism appears to exist within some populations.* TEXAS, APRIL

L 5.35"	WS 7"	WT 0.35 oz (10 g)

▸ one adult molt per year; complex basic strategy
▸ weak differences between adult and juvenile
▸ paler in Intermountain West than elsewhere

A wren of diverse scrubland habitats, from dry foothills in Intermountain West to lush urban parks in Pacific Northwest. Population expanding, especially northward, in Intermountain West; population has sharply declined and is still decreasing in much of East.

◀ Varied song often misidentified or simply unidentified. Most versions sound inhaled; for example, *hhheee hheer h'h'h'h'deee*. Call notes vary; most are raspy.

Carolina Wren
Thryothorus ludovicianus CODE 1

ADULT. *Reddish brown above; most are bright buff below. White eyebrow prominent. Warmer than even the brightest Bewick's; tail is slightly shorter and lacks white tips. Juvenile duller overall.* TEXAS, APRIL

ADULT. *Some are less colorful, especially in late spring and summer (before annual molt), but all show at least some buff below; Bewick's is always grayer below.* ONTARIO, MAY

L 5.5"	WS 7.5"	WT 0.75 oz (21 g)

▸ one adult molt per year; complex basic strategy
▸ weak differences between adult and juvenile
▸ individual plumage variation greater in southern birds

A characteristic voice of gardens and woodlands in the Southeast. Tolerant of humans, but wary; often hides in woodpiles and thickets. Range expanding steadily north-ward, with temporary withdrawals following harsh winters; westward expansion more erratic.

◀ Variable song is chanting and repetitious; for example, *chit mediator mediator mediator meet!* Varied calls include liquid *dihlip*, harsh rattle, and dry buzz.

Marsh Wren
Cistothorus palustris CODE 1

ADULT, WESTERN. *All populations brownish above and pale gray-brown below, with prominent eyebrow. Back is black with well-defined white streaks.* CALIFORNIA, APRIL

ADULT, EASTERN. *Similar to Western, but plumage shows higher contrast overall. Juveniles of all populations more smudgy than adult, with few if any white streaks on back.* NEW JERSEY, JULY

L 5"   WS 6"   WT 0.4 oz (11 g)

- ▸ two adult molts per year; complex alternate strategy
- ▸ moderate differences between adult and juvenile
- ▸ warmer and more boldly patterned in East

A voice in the cattails: sings at night, as well as through the hot midday hours; builds "dummy nests"—oval structures that are not actually used for egg-laying. Eastern birds usually nest in large marshes; Western birds accept golf course ponds and roadside sloughs.

◀ Song consists of several opening notes followed by harsh trill: *chik chick chu-u-u-u-u-u-r*. Song more liquid in West than East. Call note a soft *chuck*.

Sedge Wren
Cistothorus platensis CODE 1

ADULT. *Buff overall; spotting and streaking extensive, but plumage shows lower contrast than Marsh. Bill relatively small (species was formerly known as "Short-billed Marsh Wren").* NORTH DAKOTA, JUNE

ADULT. *Most adults are unmarked buff below, but some show weak barring on flanks. Juvenile has fainter streaking above and paler buff below than adult.* NORTH DAKOTA, JUNE

L 4.5"   WS 5.5"   WT 0.35 oz (10 g)

- ▸ two adult molts per year; complex alternate strategy
- ▸ weak differences between adult and juvenile
- ▸ individual variation in pattern on flanks

Secretive inhabitant of wet meadows. Local densities sometimes high, but species often absent from seemingly suitable habitat. Movements complex, poorly known; many populations apparently breed in north early, then breed again hundreds of miles south.

Song similar to Marsh, but sharper, more staccato, less liquid, and less musical: *chap chap chap ch'ch'ch'ch'ch'*. Call a sharp *chap*.

Rock Wren
Salpinctes obsoletus   CODE 1

ADULT. *Brown and buff above; gray below with buff on belly. Tail barred black-and-white below, with broad buff tip. Above, buff tip conspicuous in flight.* CALIFORNIA, APRIL

ADULT. *Some are mainly gray, with a little buff in tail. Plumage usually brightest in fresh basic plumage (early fall). Juvenile shows less plumage contrast overall.* CALIFORNIA, APRIL

| L 6" | WS 9" | WT 0.6 oz (17 g) |

▸ one adult molt per year; complex basic strategy

▸ moderate seasonal and weak age-related differences

▸ individual variation in darkness of upperparts

Breeds in rocky habitats from boulder-strewn tundra to arid foothills. Range overlaps broadly with Canyon, but Rock prefers different microhabitat: dusty, gravelly slopes, often with strong exposure to the elements. Entrance to nest paved with little trail of pebbles.

Song repetitive and jangling, suggesting a distant Northern Mockingbird: *chwee chwee chwee...chiss chiss chiss...jeery jeery jeery...* Call an explosive *ch'pwee!*

Canyon Wren
Catherpes mexicanus   CODE 1

ADULT. *White throat and breast contrast with otherwise largely rufous plumage; tail especially rufescent. Bill very long. Juvenile similar but has less spotting and barring.* CALIFORNIA, JULY

ADULT. *When foraging, often flattens itself out, enabling it to enter tight crevices and abutments.* ARIZONA, NOVEMBER

| L 5.75" | WS 7.5" | WT 0.4 oz (11 g) |

▸ one adult molt per year; complex basic strategy

▸ weak differences between adult and juvenile

▸ geographic variation in paleness of back and crown

Iconic species of rugged red-rock canyon country in the West; rarely seen away from steep rock faces. Scurries mouse-like up and down sheer surfaces, often detouring to explore a small cave or cranny; hard to catch a glimpse of, but clarion song is unmistakable.

Song a decelerating series of 10–15 clear whistles, each note descending in pitch: *dyeer! dyeer dyeer deer deer...* Call a shrill *beet*; song often ends with a few call notes.

Cactus Wren
Campylorhynchus brunneicapillus CODE 1

ADULT. *Our largest wren by far; size, plumage, and behavior suggest a small thrasher. All have black spots below, with a banded tail and prominent eyebrow.* TEXAS, APRIL

ADULT. *Most have buff on belly and dense spotting on breast; "San Diego" subspecies (C. b. sandiegensis) has less buff and more evenly distributed spots. Juvenile buff all over, with reduced spotting.* ARIZONA, APRIL

L 8.5"	WS 11"	WT 1.4 OZ

- ▸ one adult molt per year; complex basic strategy
- ▸ moderate differences between adult and juvenile
- ▸ "San Diego" has less buff below; even spotting on breast

An animated and sociable resident of hot, thorny deserts; ventures into towns and cities. Family groups forage for large arthropods in desert shrubs. Nest highly visible but amazingly difficult to access within thorny plants such as cholla and acacia.

Song, given right through the hottest parts of the day, is low and pulsing, like an old car trying to start: *chug chug chug chug...* Call note a growling *chut.*

American Dipper
Cinclus mexicanus CODE 1

ADULT. *Dark, rotund, and bobtailed. Bobs up and down in jerky movements on pale legs. Juvenile paler, especially on throat; wings and back scalloped with white; bill yellow.* UTAH, JULY

ADULT. *Easily observed in unique foraging behavior; this bird is beginning its dive into an icy stream.* COLORADO, JANUARY

L 7.5"	WS 11"	WT 2 OZ

- ▸ one adult molt per year; complex basic strategy
- ▸ moderate differences between adult and juvenile
- ▸ some subadults retain juvenile plumage into winter

Intimately connected with fast-flowing streams and rivers. Breeds from timberline down to sea level; withdraws in winter to limit of ice-free waters. Behaviors amazing: plops into frigid water, bobs on surface or submerges completely, and then repeats the process.

Song an endless series of halting warbles and trills; like a Winter Wren played slowly. Call a shrill *bzeet,* given in winding flight along streams and rivers.

Bulbuls
FAMILY PYCNONOTIDAE

1 species: 1 ABA CODE 2

The approximately 130 species of bulbuls are medium-sized and generally drab frugivores (fruit-eaters) native to Africa, India, and Southeast Asia. They are relatively uniform in appearance, and the Red-whiskered Bulbul is in a genus that comprises some 40 species. Several bulbul species have proved amenable to introduction, with the Red-whiskered having been established around Miami and Los Angeles, as well as in Australia and on certain oceanic islands. With a diet of berries and small fruits, the Red-whiskered Bulbul tends to forage slowly in dense arboreal vegetation—making it relatively easy to overlook. The overall population density is low in Florida and lower still in the Los Angeles area, adding further to the challenge of finding these birds.

Red-whiskered Bulbul
Pycnonotus jocosus CODE 2

ADULT. *Dark brown above; pale gray-brown below. Red "whisker" behind eye hard to see, but long crest and white cheek are prominent. Juvenile paler; lacks red on face.*
ASIA, JANUARY

L 7"   WS 11"   WT 1 oz (28 g)

- one adult molt per year; complex basic strategy
- moderate differences between adult and juvenile
- red on cheek and especially on crissum variable

Asian species that fares well in human-modified environments; has been widely introduced. Small population established in Miami area since early 1960s; a few in Los Angeles area since late 1960s. A fruit-eater; forages and roosts in well-vegetated gardens and parks.

Song consists of liquid *cheerlo* and *cheeralo* elements, recalling American Robin but slower and more repetitious. Call a descending *chinkalo*. All vocalizations rich and mellow.

ESTABLISHED EXOTICS

Proving that an exotic species is "established" in a particular region can be a challenge. The key is to demonstrate that a population is breeding in the wild and is self-sustaining. In the case of many exotic gamebirds and parrots, continued breeding in the wild may be the result of illegal releases or controlled introductions of captive stock. Once a population is "established," there is no guarantee that it will remain so forever. Populations of Red-whiskered Bulbuls, Spot-breasted Orioles, and other Florida exotics could be wiped out by changing landscaping practices or major hurricanes.

Kinglets and Old World Warblers & Gnatcatchers

FAMILIES REGULIDAE AND SYLVIIDAE

15 species: 4 ABA CODE 1   2 CODE 2   3 CODE 4   6 CODE 5

In the field guides of a generation ago, the kinglets and gnatcatchers were lumped together in the large family Sylviidae. On the surface, it was a reasonable alliance: Both the kinglets and the gnatcatchers are tiny, hyperactive, and insectivorous. But molecular studies of these two groups have raised major questions, and they are currently placed in separate families. The gnatcatchers remain in the family Sylviidae, a sprawling Old World assemblage of many look-alike species; the Arctic Warbler is a typical example. Meanwhile, the kinglets have been reassigned to their own family (Regulidae), which may have only distant affinities with the Sylviidae.

The kinglets, gnatcatchers, and Arctic Warbler have evolved a diverse suite of dispersal strategies related to their insectivorous lifestyles. The Arctic Warbler undertakes a long-distance migration from the insect-infested willow shrublands of western mainland Alaska to the hot jungles of Southeast Asia, and it is considered to be one of the longest-distance migratory insectivores wintering in the Old World. In contrast, large numbers of the diminutive Golden-crowned Kinglet winter in regions characterized by deep snow and long nights; one of the marvels of avian physiological ecology is that these thermodynamically challenged birds weighing a fifth of an ounce are able to find enough live insects to survive brutal winters in the Maine woods or in high-elevation forests in the Rockies. The Ruby-crowned Kinglet and Blue-gray Gnatcatcher are typical intermediate-distance migrants, withdrawing mainly to the southern United States and Mexico. And the California and Black-tailed Gnatcatchers are permanent residents in regions that have mild climates year-round.

Population health of North American regulids and sylviids is tied to human activities on the breeding grounds. For example, Golden-crowned Kinglets have expanded southward in recent years in the Northeast, a result of forest regeneration and proliferation of Christmas tree plantations. The California Gnatcatcher, in contrast, is listed as Endangered because of sharp population losses in the 20th century stemming from destruction of its required coastal sage-scrub habitat. Numbers of Black-tailed Gnatcatchers have declined locally with urbanization and other disturbances, but large-scale population changes have not been documented. Finally, it is worth remembering that some species occupy breeding ranges that have been largely unaffected, at least directly, by human activity; the breeding range of the Arctic Warbler is essentially uninhabited by humans, and the population is presumed to be stable overall.

Ruby-crowned Kinglet

Regulus calendula CODE 1

ADULT MALE. *Small and stubby, with distinctive olive-green cast overall. Red crown of male sometimes raised and prominent, but usually concealed.* CALIFORNIA, JANUARY

ADULT FEMALE. *Lacks red crown of male. Plumage similar to Hutton's Vireo, but black panel next to lower white wing bar is absent on Hutton's. Foraging actions different from Hutton's.* TEXAS, APRIL

L 4.25"   WS 7.5"   WT 0.25 oz (7 g)

▸ two adult molts per year; complex alternate strategy

▸ moderate differences between male and female

▸ male crown patch varies from prominent to invisible

Hyperactive; always fidgety. Flicks wings constantly. Forages at mid-level in trees, singly or in flocks with other species; does not usually flock with conspecifics, unlike Golden-crowned Kinglet. Breeds in spruce-fir forests; winters widely in lowlands.

◀ Call a scratchy *digit*, sometimes extends into long series. Song loud, long, and rich: *dear! dear! dear! diddy-dear diddy-dear ji-jiddy-jee ji-jiddy-jee.*

Golden-crowned Kinglet

Regulus satrapa CODE 1

ADULT MALE. *Even smaller than Ruby-crowned; more boldly patterned overall. Male sometimes raises bushy crest, exposing extensive orange.* MAINE, JUNE

ADULT FEMALE. *Head pattern distinct from Ruby-crowned: black eye line, white eyebrow; yellow crown (gray in juvenile) broadly outlined in black.* CALIFORNIA, NOVEMBER

L 4"   WS 7"   WT 0.2 oz (6 g)

▸ one adult molt per year; complex basic strategy

▸ moderate differences between male and female

▸ prominence of orange in male crown variable

Amazingly hardy; despite tiny size, winters at high latitudes. Surprisingly, eats only live insects in winter. Like Ruby-crowned, breeds in spruce-fir woods; expanding south into Christmas tree plantations in Northeast. Often in small flocks with conspecifics.

◀ High-pitched song consists of widely spaced opening notes, then a fast jumble: *see see see see s'd'gitchy-g'gitchy-g'see.* Call a fast series of 2–4 high *see* notes.

Blue-gray Gnatcatcher
Polioptila caerulea   CODE 1

BREEDING MALE. *Like a tiny Northern Mockingbird; long black tail edged in white. Soft blue-gray above; breeding male has black on forehead extending behind eye in thin line.* TEXAS, APRIL

ADULT FEMALE. *Lacks black on face of breeding male; blue-gray upperparts a little less intense. Nonbreeding male similar to adult female. All plumages have white eye ring.* FLORIDA, FEBRUARY

L 4.5"   WS 6"   WT 0.21 oz (6 g)

‣ two adult molts per year; complex alternate strategy

‣ moderate seasonal and sex-related differences

‣ in West, has less blue on back, more black on tail

Spritely inhabitant of eastern woodlands and western scrublands. Typically alone or only in loose association with conspecifics and mixed-species assemblages. Flips tail constantly; feeds in shrubs 5–10 feet above ground. Populations increasing; range expanding north.

◀ Calls frequently; a giddy, wheezy *bzweeeee* or *bzwaaaaay.* Song similar in timbre to call; a jumble of short notes, for example, *bzwee weea way'wah bees.*

Black-tailed Gnatcatcher
Polioptila melanura   CODE 1

ADULT MALE. *In most of range, the only gnatcatcher with extensive black on head. California darker overall; has more black in tail. Black-capped longer-billed with more white in tail.* ARIZONA, APRIL

ADULT FEMALE. *Similar to female Blue-gray: Black-tailed more uniformly brown above; tail has less white than Blue-gray. Nonbreeding male similar.* ARIZONA, FEBRUARY

L 4.5"   WS 5.5"   WT 0.2 oz (6 g)

‣ two adult molts per year; complex alternate strategy

‣ strong seasonal and sex-related differences

‣ birds in eastern portion of range average darker below

The desert gnatcatcher; especially fond of thorny plants. Greatest densities in washes, riparian thickets, and saguaro forests. During breeding season, Blue-gray Gnatcatcher at higher elevations; in winter and on migration, Blue-gray joins Black-tailed in the desert.

Calls varied. Many are scratchy, recalling Bewick's Wren; for example, *jit jit jit* and *jet jet jet.* Also gives a higher *jaaaaa,* resembling Blue-gray Gnatcatcher.

BLACK-CAPPED GNATCATCHER (*P. nigriceps*), *Adult Female. Casual visitor to Southeast Arizona; has bred. Tail white below; bill relatively long. Breeding male has black cap like Black-tailed.* ARIZONA, FEBRUARY

California Gnatcatcher

Polioptila californica   CODE 2

BREEDING MALE. *Darker overall than Black-tailed: ground color more brownish; black on head a little more extensive. Tail mainly black below; tail of Black-tailed more white below.* CALIFORNIA, MARCH

ADULT FEMALE. *Browner above than female Blue-gray; brown extends to lower belly and undertail coverts. Tail nearly all black below; eye ring thin or absent. Nonbreeding male similar.* CALIFORNIA, FEBRUARY

L 4.5"   WS 5.5"   WT 0.25 oz (7 g)

▸ two adult molts per year; complex alternate strategy

▸ strong seasonal and sex-related differences

▸ overall darkness variable, especially in females

Common the length of the Baja Peninsula; range reaches north into southern California, where restricted to waist-high coastal sage-scrub, a habitat that has been ravaged by development. Locally common but also absent from seemingly suitable swaths of habitat.

Most frequently heard call is distinct from Black-tailed and Blue-Gray Gnatcatchers: a long, drawn-out *eeeewwww*; exceedingly nasal.

Arctic Warbler

Phylloscopus borealis   CODE 2

ADULT. *Wings long; wing bars faint. Long, pale eyebrow prominent; bordered below by dark eye line, above by olive cap. Upperparts olive-brown in summer; olive-green in fresh fall plumage.* ALASKA, JUNE

ADULT. *Eyebrow often flares upward behind eye. Tail is broad, with squared-off or slightly notched tip.* ALASKA, JUNE

L 5"   WS 8"   WT 0.35 oz (10 g)

▸ one adult molt per year; complex basic strategy

▸ weak differences between spring and fall adults

▸ Aleutian vagrants yellower, larger-billed than breeders

The lone representative of the diverse Old World warbler family that regularly reaches North America. Widespread breeder well east into Alaska; winters in Southeast Asia. Breeds mainly in dwarf willow stands, sometimes in spruce woods. Active when feeding; flicks wings.

Song a loud, slow trill, suggesting Orange-crowned Warbler, but richer and lazier: *cheecheecheecheechee...* Call a loud, smacking *spitch*.

Thrushes

FAMILY TURDIDAE

29 species: 11 ABA CODE 1   3 CODE 2   4 CODE 3   6 CODE 4   5 CODE 5

These small to medium-sized birds are best known for their powers of song. Another unifying theme in the family Turdidae is bill structure: All of our species have "average" bills (fairly short, broad-based, and tapered), suitable for feeding on both insects and berries.

Most authorities consider the thrushes to be closely related to the Old World Flycatcher family Muscicapidae, seven species of which have reached Alaska as casual or accidental strays from Asia. At times, the family Turdidae has been subsumed within the family Muscicapidae, and further taxonomic instability may be expected in the years to come.

Thrushes are birds of terrestrial habitats. A number of species are associated with forested districts—for example, the Varied Thrush, Townsend's Solitaire, and the "spot-breasted" thrushes in the genus *Catharus*. Others are associated with open habitats: the Mountain Bluebird, Bluethroat, and especially Northern Wheatear. The American Robin—just about the most fabulously adaptable bird in North America—is nearly ubiquitous.

Among North American land birds, thrushes rank among the most competent of long-distance migrants. The Gray-cheeked Thrush winters in South America and migrates far north and west to taiga breeding habitats, with many individuals crossing the Bering Strait and penetrating well into Siberia! Two other remarkable migrants are the Bluethroat and Northern Wheatear, which migrate west across the Bering Strait in fall in exactly the opposite direction of eastbound Gray-cheeked Thrushes. Another highly migratory species is the American Robin, with huge coastal flights noted every fall and high-flying flocks heralding the passing of the seasons at inland locations continent-wide. Due to their generally strong powers of dispersal, the thrushes are prone to vagrancy; various western species wander all the way to the eastern seaboard every winter, and strays from outside the ABA Area routinely reach Alaska, maritime Canada, Texas, and Arizona.

With their widespread geographic distributions and generalist habits, populations of American thrushes remain fairly large. Several species are in decline, however. The *Catharus* thrushes are threatened by deforestation both on their boreal breeding grounds (where the paper industry poses a particular problem) and on their tropical wintering grounds (where clear-cutting for agriculture is generally unrestricted). In the West, the Western Bluebird and Varied Thrush may be adversely affected by modern forestry practices. On the flip side, the Eastern Bluebird is a well-known conservation success story; the species has benefited from a skillful public relations campaign, focusing in particular on the benefits of nest-box supplementation for locally threatened populations of this cheery emblem of America's farm country.

Northern Wheatear

Oenanthe oenanthe   CODE 2

BREEDING MALE, WESTERN. *A plump, short-tailed, long-legged ground dweller. Plumage shrike-like, but body structure very different. Western breeders have less buff on throat and breast.* ALASKA, JUNE

NONBREEDING MALE, EASTERN. *All individuals duller in nonbreeding plumage; all have black tails that flash white at base in flight. Eastern Canada population has more buff below.* EUROPE, SEPTEMBER

L 5.75"   WS 12"   WT 0.8 oz (23 g)

▸ one adult molt per year; complex basic strategy

▸ strong seasonal, sex-related, age-related differences

▸ eastern breeders larger and darker than western

Breeds in open, rocky, sparsely vegetated tundra, often with hilly aspect; North American and Eurasian breeders all winter in sub-Saharan Africa. Migration over Atlantic Ocean or Bering Sea, but vagrants noted throughout continent, mainly along Northeast coast in fall.

Flight call a soft, whistled *wheet.* Song a steady jumble of rich garbled notes and dry scratchy notes, interspersed with occasional twangy notes.

BREEDING FEMALE. *Black mask reduced. Nonbreeding female has more buff, with little black in wings and on face. Juvenile gray above with buff wings.* ALASKA, JUNE

Bluethroat

Luscinia svecica   CODE 2

L 5.75"   WS 9"   WT 0.7 oz (20 g)

▸ one adult molt per year; complex basic strategy

▸ strong seasonal and sex-related differences

▸ much individual variation in breast pattern

Skulks in vegetation on dwarf-shrub tundra; when alarmed, cocks tail like a wren. Forages actively: gleans insects from stunted birches and alders, forages in leaf litter, occasionally hawks after flying insects. Eastern limit of range, somewhere in Canada, unknown.

Song, given in flight or from perch, a series of *ch'pee!* notes, then dry bubble: *ch'pee! ch'pee! ch'pee! j'j'j'ch'ch'ch'j'j'jet.* Mimics other species.

ADULT MALE. *All plumages gray above, with extensive red in tail; all have boldly patterned head. Blue and red of male stunning; bright colors diminished on female, absent on immature.* ALASKA, JUNE

Siberian Rubythroat
Luscinia calliope CODE 3

L 5.5"   ws 10"   wt 0.7 oz (20 g)

▸ one adult molt per year; complex basic strategy

▸ strong age-related and sex-related differences

▸ some adult females have extensive pink on throat

Eurasian breeder; abundant east to Kamchatka Peninsula, but breeding range does not quite reach North America. Regular in spring in small numbers in western Aleutians; casual in fall throughout Bering Sea region. Habits similar to Bluethroat: furtive but active.

Call a harsh *rut* or short whistle. Like other species in genus *Luscinia*, an accomplished songster; wanderers to Alaska sometimes sing, but usually do not.

ADULT MALE. *Throat red; malar white. Throat of female and immature whitish or pinkish. All plumages lack bright rusty in tail of Bluethroat; all have relatively duller head markings.*
ASIA, JUNE

Eastern Bluebird
Sialia sialis CODE 1

ADULT MALE. *Blue above, with long wings. Orange below; orange creeps up onto throat and behind ears, unlike Western Bluebird. Southwestern subspecies paler.*
CONNECTICUT, JUNE

ADULT FEMALE. *Averages duller than male; throat usually white. Even the dullest of females have extensive blue in the flight feathers of the tail and wings.*
TEXAS, MAY

L 7"   ws 13"   wt 1.1 oz

▸ one adult molt per year; complex basic strategy

▸ strong age-related and sex-related differences

▸ extent and intensity of blue geographically variable

Fabled harbinger of spring in eastern North America. Nests early; often uses bird boxes. Occurs in open habitat with perches; forms loose flocks in winter. Feeding methods diverse: hawks for aerial insects, pounces on terrestrial arthropods, plucks berries from trees.

◀ Song consists of rich warbles preceded by a few scratchy notes: *p'ch'ch' churr churl churl*. Gives various scratchy calls and twangy flight call: *dwaya*.

JUVENILE. *Dark above and below with coarse white spots, often with a prominent eye ring. Blue in wings and tail variable, but always prominent.*
CONNECTICUT, JUNE

Western Bluebird
Sialia mexicana CODE 1

ADULT FEMALE. *Compared to male, indigo much reduced. Overall pattern the same as male: head uniformly dull blue-gray; back and scapulars have orangish wash.* CALIFORNIA, JUNE

ADULT MALE. *Indigo above; orange below. Orange does not creep up onto head as on Eastern, but instead wraps onto scapulars and back.* CALIFORNIA, MAY

L 7"	WS 13.5"	WT 1 oz (28 g)

▸ one adult molt per year; complex basic strategy
▸ strong sex-related and age-related differences
▸ extent of orange on back and scapulars variable

Breeds in forested landscapes, mainly in mountains: open pinewoods, burn areas, and camp-grounds. In winter, searches at lower elevations for juniper and mistletoe berries. Recent population losses due to snag removal and salvage logging in Northwest and elsewhere.

JUVENILE. *Like Eastern, dark above and below; coarse spotting becomes finer, as here, on older juveniles. Blue in wings and tail always prominent; white eye ring often conspicuous.* CALIFORNIA, AUGUST

Not as vocal as Eastern Bluebird. Vocalizations similar, but less ringing, less rich. Song disjointed and rambling: *p'choo... choof... churl... churl...*

Bluebirds and other blue-colored birds do not have blue pigments in their feathers. Blue coloration in birds is caused by tiny air cavities in the feather's structure. The size and location of the air cavities is amazingly precise, causing an interaction with light waves that produces blue colors.

ADULT MALE AND NESTLINGS. *All bluebirds are cavity nesters (natural or artificial), and all are easily observed at the nest site.* CALIFORNIA, JUNE

Mountain Bluebird
Sialia currucoides   CODE 1

ADULT MALE.
Glistening turquoise all over, especially above. Longer-winged and thinner-billed than other bluebirds.
COLORADO, APRIL

ADULT FEMALE.
Duller than male, but variably so. Some acquire reddish tones in fall, held through winter; differs from dull females of other species by bill and tail structure.
CALIFORNIA, JULY

JUVENILE. *Like other juvenile bluebirds, spotted below; spotting above reduced compared to other species. Relatively long wings and thin bill evident on older juveniles.*
OREGON, AUGUST

L 7.25"   WS 14"   WT 1 oz (28 g)

▸ one adult molt per year; complex basic strategy

▸ strong age-related and sex-related differences

▸ some females have ruddy breasts, especially in winter

Breeds in open country in cool landscapes, from sage-steppe to alpine tundra. Has benefited from diverse suite of human activities: forest clearing, grazing, and nest box supplementation. Diurnal migration in early spring, often through high elevations, is spectacular.

Diverse vocalizations similar to other bluebirds. Individual elements of rambling song average harsher than Eastern and Western: *chut* and *choot* and *cheerf.*

Townsend's Solitaire
Myadestes townsendi   CODE 1

ADULT. *Slim and long-tailed. At rest, shows white eye ring and small buff patch in wings. In flight, flashes white tail edges and extensive buff in wings.* NEW MEXICO, DECEMBER

JUVENILE. *Heavily spotted above and below with bright buff. Juveniles sometimes wander far from breeding grounds.* SOUTH DAKOTA, JUNE

L 8.5"   WS 14.5"   WT 1.2 oz

▸ one adult molt per year; complex basic strategy

▸ strong differences between adult and juvenile

▸ amount and intensity of buff in wing variable

Breeds mainly in montane conifer forests up to tree line. Often perches at tip of tree. Some winter just downslope from nest site; others disperse far eastward, at least a few years to east of the Mississippi. Fruiting junipers favored in winter.

◀ Song a leisurely, musical, remarkably long warble; some sing in winter. Call a rapid toot, similar to Northern Pygmy-Owl, but more metallic and with faster delivery.

Varied Thrush

Ixoreus naevius CODE 1

Rufous-backed Robin

Turdus rufopalliatus CODE 3

ADULT MALE. *Superficially similar to American Robin, with which it often associates in winter. Bluish above with bold orange eyebrow; orange below, with broad black breast band.* CALIFORNIA, NOVEMBER

ADULT. *Rufous below like American Robin, but rufous also extends to upperwing coverts and back. Streaks on throat more extensive than American. Female a little duller.* ARIZONA, NOVEMBER

L 9.25"	WS 16"	WT 2.7 OZ

▸ one adult molt per year; complex basic strategy

▸ weak differences between male and female

▸ individual variation in brightness, especially on back

Characteristic of Pacific slope of Mexico, where range apparently expanding. A few wander north each fall to North America; most records from Arizona, but ranging from California to Texas. Habits of strays to North America generally as American Robin in winter.

Call an abrupt *chutch*, suggesting Hermit Thrush. Song, rarely heard in North America, like American Robin, but less effusive: *early, urr-i-ly, url, urr-ily.*

ADULT FEMALE. *Dark bluish of adult male replaced with dull brown on female; juvenile duller, browner. In flight, all plumages flash long orange wing stripe above and below.* BRITISH COLUMBIA, JANUARY

L 9.5"	WS 16"	WT 2.7 OZ

▸ one adult molt per year; complex basic strategy

▸ moderate age-related and sex-related differences

▸ interior females average paler and grayer on back

Breeds in wet, shady, northwestern forests; apparently prefers old-growth stands. Shy and solitary in summer; occurs in low densities throughout much of breeding range. Migratory: most winter near Pacific coast; regular in winter far east of core winter range.

Song utterly bizarre: long, vibrant, metallic, breathy notes, spaced far apart: *zeeeeeeeng... zoiiiiiiing... zeeeeerng...* Call: *chook*, like Hermit Thrush.

AZTEC THRUSH (Ridgwayia pinicola). *Casual Mexican stray; most North American records from Southeast Arizona. Adult male boldly patterned in black-and-white; female, juvenile browner.* ARIZONA, JANUARY

American Robin
Turdus migratorius CODE 1

ADULT MALE. *Dark gray above, with nearly black head; bright rufous below. Prominence of white eye arcs and throat streaking variable.* OHIO, MARCH

IN FLIGHT. *Wings long; tail long and broad. Rufous underparts extend to underwing coverts. From above, assumes straight-line profile: wings, back, and head on same axis.*
BRITISH COLUMBIA, OCTOBER

FIELDFARE (T. pilaris). *Eurasian species; casual to Northeast in fall, to western Alaska in late spring. Upperparts gray, underparts intricately marked with dark chevrons.* EUROPE, JANUARY

ADULT FEMALE. *Averages paler than male; some are much paler, as here, but others are as dark as male. Both sexes generally paler westward, but variable.* CALIFORNIA, OCTOBER

L 10"   WS 17"   WT 2.7 OZ

▸ one adult molt per year; complex basic strategy

▸ strong age-related and moderate sex-related differences

▸ much individual and geographic variation in darkness

One of our most extraordinary species. Nests in downtown districts of huge cities, as well as in faraway wilderness. All aspects of behavior absorbing: birds at nest easily observed; winter flocks fascinating; diurnal and nocturnal migration may involve immense flocks.

◀ Song ebullient: *cheerily cheerio cheerup*, repeated endlessly in early morning. Calls include flat *tut* and sharp *cheep*. Flight call a piercing *zeeeert!*

JUVENILE. *Like most other thrushes, extensively spotted above and below. Spotting gradually diminishes as juvenile ages.* CALIFORNIA, JULY

Clay-colored Robin
Turdus grayi　CODE 3

L 9"　　　WS 15.5"　　　WT 2.6 OZ

▸ one adult molt per year; complex basic strategy
▸ moderate differences between juvenile and adult
▸ adult leg color variable: pink, green, brown-gray

Widespread in tropics, where it inhabits forests, edges, gardens, even large cities. Formerly thought to be rare winter stray to southern Texas; now known to be more common, with a few nesting pairs. Texas birds secretive; hide in tangles, sometimes wander onto lawns.

Song resembles American Robin but notes richer and deeper, highly slurred: *whirl-earl... leer-eerul... churl-earl...* Varied calls include distinct *zhigeero.*

ADULT. *Like a dull brown American Robin. Chocolate brown above; slightly paler below. Bill dull green; eye arcs of American Robin usually lacking. Juvenile spotted below.* TEXAS, NOVEMBER

Eyebrowed Thrush
Turdus obscurus　CODE 3

L 7.5"　　　WS 13"　　　WT 2.2 OZ

▸ one adult molt per year; complex basic strategy
▸ moderate differences between male and female
▸ amount of orange on breast and flanks variable

Breeds in dense spruce-fir and birch forests of Siberia; annual spring migrant through western Aleutians. Rare most years, but sometimes fairly common. Casual throughout Bering Sea region in fall. Strays to Alaska are active; feed on ground, flush frequently.

Calls include dull *chub;* flight call, a steady *zeee,* not descending like American Robin.

ADULT MALE. *Gray head has white eyebrow and broad white crescent below eye. Underparts orangish; upperparts brownish orange. Female paler below and less rufescent; throat whitish.* ASIA, FEBRUARY

Gray-cheeked Thrush
Catharus minimus　CODE 1

L 7.25"　　　WS 13"　　　WT 1.1 OZ

▸ one adult molt per year; complex basic strategy
▸ moderate differences between adult and juvenile
▸ markings on face variable but always faint

The only neotropical migrant passerine that regularly crosses the Bering Sea to breed in Siberia, but most breed in stunted boreal forests of North America. Migrates late in spring. Secretive on both spring and fall migration, but nocturnal migrants easily heard.

Song a series of nasal, buzzy, downslurred phrases: *bzee azeep azeee azooo.* Flight call, often heard at night, a buzzy *quee-yur.*

ADULT. *Cold gray-brown above; pale gray below, with heavy spotting on breast. Buff below limited; buff, if present, stronger on breast than throat. Long-winged. Juvenile spotted above.* TEXAS, APRIL

Hermit Thrush
Catharus guttatus   CODE 1

ADULT, EASTERN. *On all Hermit Thrushes, rufous tail contrasts with browner upperparts. Brown upperparts of Eastern have rufous highlights; flanks and breast have rufous wash.* NEW JERSEY, NOVEMBER

ADULT, INTERMOUNTAIN WEST. *The grayest and largest population, with relatively long wings. Gray-cheeked Thrush (rare migrant to southern Rockies) more similar to Hermit than Swainson's.* COLORADO, MAY

L 6.75"	WS 11.5"	WT 1.1 OZ

▸ two adult molts per year; complex alternate strategy

▸ moderate differences between adult and juvenile

▸ much geographic variation in size and redness

Stands quietly on ground, raises tail quickly, then slowly lowers it. Compared to other spot-breasted thrushes, migrates earlier in spring and later in fall, winters farther north; eats berries in winter. Much geographic variation in forest types preferred by breeders.

ADULT, PACIFIC SLOPE. *Brown upperparts have rufous highlights like Eastern, but flanks are grayish. Juvenile of all populations weakly spotted above.* CALIFORNIA, APRIL

Song like wind chimes: *erweeeeooo... eooweeaweeaweeaway.* Common call a hollow *chut.* Flight call, often heard on nocturnal migration, a pure *heee.*

Bicknell's Thrush
Catharus bicknelli   CODE 2

L 6.75"	WS 11.5"	WT 1.1 OZ

▸ one adult molt per year; complex basic strategy

▸ moderate differences between adult and juvenile

▸ color contrast between tail and body variable

Similar in most respects to Gray-cheeked; until recently, the two were treated as conspecific. Breeds in stunted fir forests in austere, wind-swept habitats; winters in Greater Antilles. Migratory path poorly known; apparently rare overland south of mid-Atlantic.

Song not as sharply descending as Gray-cheeked: *bzzaway, bzzawee, bzzawaaay.* Flight call a descending *peeeeer.*

ADULT. *Intermediate between Gray-cheeked and Hermit. More rufous above than Gray-cheeked, especially on tail and wings; wings shorter than Gray-cheeked. Juvenile spotted above.* NEW HAMPSHIRE, JUNE

Swainson's Thrush
Catharus ustulatus CODE 1

ADULT, EASTERN. *Thrushes in the genus* Catharus *are an underappreciated field identification problem; call notes, flight calls, and songs are of great help in the identification process.* MAINE, JUNE

ADULT, EASTERN. *The widespread population of Swainson's, sometimes called "Olive-backed"; (C. u. swainsonii) gray-brown above, with extensive buff on face and a bright buff eye ring. Juvenile spotted above.* MAINE, JUNE

L 7"   WS 12"   WT 1.1 OZ

▸ one adult molt per year; complex basic strategy
▸ moderate differences between adult and juvenile
▸ Pacific-slope birds more rufescent overall

Secretive but common. Widespread "Olive-backed" population breeds mainly in boreal conifer forests; "Russet-backed" population breeds down to sea level in dense broadleaf groves. Breeders more arboreal than other spot-breasted thrushes. Nocturnal migration spectacular.

◀ Song thin, sweet, upslurred: *urwee urwee urwee aweeaweeaway*. Call a liquid *pilp*. Flight call, often heard on nocturnal migration, a slightly rising *pweev*.

ADULT, PACIFIC SLOPE. *Sometimes called "Russet-backed" (C. u. ustulatus); mainly in lowland groves of far West. Brighter rufous above than "Olive-backed," with some rufous below; similar to Veery.* CALIFORNIA, APRIL

NOCTURNAL MIGRATION

Many passerines migrate at night, when they are invisible to birders. But that does not mean that nocturnal migrants cannot be identified, because a number of species give diagnostic flight calls as they fly overhead at night. The *Catharus* thrushes are an excellent starting point for those wishing to learn how to identify nocturnal migrants by voice; their low-frequency flight calls are loud and relatively long, with distinctive intonations and timbres. In many regions, the most frequently detected flight call is that of the Swainson's Thrush—a rich whistle that sounds like Spring Peepers (a common species of tree frog).

Veery
Catharus fuscescens   CODE 1

ADULT, EASTERN. *Warm reddish brown above; breast washed with warm buff. Spots on breast fainter than other* Catharus *thrushes. Darker western birds resemble western Swainson's Thrush.* CONNECTICUT, MAY

L 7"   WS 12"   WT 1.1 OZ

▸ one adult molt per year; complex basic strategy

▸ moderate differences between adult and juvenile

▸ western birds not as warmly colored as eastern

In most of range, breeds in damp deciduous woods; some Rocky Mountain populations in conifer bogs. Spends much time on ground, flipping over leaves for insects. Widely noted in woods on spring migration. Fall migration begins early; many are on the move by mid-August.

Distinctively downslurred song is nasal and buzzy: *d'veea veea veeur veeur veer*. Call a flat *vyurr*. Flight call rough, nasal, and downslurred: *veer-ick*.

The western populations of the Veery, Swainson's Thrush, and Hermit Thrush differ appreciably from their eastern counterparts. In the Pacific slope, for instance, Veeries can be confused with both Swainson's and Hermit Thrushes—a complication that does not usually present itself to birders in the East.

Wood Thrush
Hylocichla mustelina   CODE 1

ADULT. *Rufous-brown above; white below, with coarse white spots all over. Short-tailed, long-winged, sturdy-legged, and pot-bellied; suggests a large Ovenbird. Juvenile spotted above.* NEW JERSEY, JUNE

L 7.75"   WS 13"   WT 1.6 OZ

▸ one adult molt per year; complex basic strategy

▸ moderate differences between adult and juvenile

▸ southern birds slightly less rufous than northern

Common but declining in eastern deciduous forest, especially beech-maple; losses caused by forest fragmentation and cowbird parasitism. Habits much as *Catharus* thrushes: forages on forest floor; long-distance, trans-Gulf migrant; nocturnal flight call conspicuous.

◀ Song consists of flute-like phrase and tremulous buzz: *eeolay-eeeee* and *purpapeeta-peeeee*. Call explosive: *pit! pit! pit!* Flight call a buzzy *zeeerng*.

Research on the Wood Thrush in the late 20th century documented the distressing phenomenon of the population "sink." In small forest fragments in the Midwest, most Wood Thrush nests are lost to cowbird parasitism; these fragments act as "sinks" that drain regional populations of the species.

Babblers

FAMILY TIMALIIDAE

1 species: 1 ABA CODE 1

Restricted to dense brushland on the Pacific slope of North America, the Wrentit is hard to glimpse but impossible not to hear. Singles or pairs skulk in the undergrowth, coming into view to check out intruders and then quickly disappearing again. The taxonomic placement of the Wrentit has long been problematic, a fact that is reflected in its ambiguous name. It is fairly certain that the species is neither a wren nor a chickadee (chickadees are called "tits" in the Old World). Instead, it may be the lone representative in North America of the family Timaliidae (babblers), a large assemblage that reaches its greatest diversity in Southeast Asia. Babblers tend to be drab visually (although there are exceptions) but highly competent as vocalists; most are short-winged and non-migratory, with fairly weak powers of flight; and most are strong-legged species that live on or near the ground in densely vegetated habitats. All of the preceding applies well to the Wrentit, but the molecular evidence for an affinity with babblers is not airtight. In any event, the Wrentit is an evolutionarily distinctive member of the North American avifauna.

Wrentit

Chamaea fasciata   CODE 1

L 6.5"	WS 7"	WT 0.5 oz (14 g)

▸ one adult molt per year; complex basic strategy
▸ weak differences between adult and juvenile
▸ coastal and northern populations darker overall

Resident in chaparral along coast and in foothills; some get to 7,000+ feet on west slope of Sierra Nevada. Active but difficult to see; forages in dense shrubs, where it gleans arthropods from foliage. Sedentary; individuals wander little if at all from natal grounds.

ADULT. *Drab, long-tailed, and yellow-eyed. Juvenile a little scruffier. Compared to Bushtit, Wrentit is larger, with variable salmon wash below; Wrentit habitually cocks tail.* CALIFORNIA, OCTOBER

◀ Song as easy to hear as the songster is hard to see. Coastal chaparral country is defined by the Wrentit's "bouncing ball" song: *pip pip pip pipipipipipipi.*

Mimic-Thrushes and Starlings & Mynas

FAMILIES MIMIDAE AND STURNIDAE

---

13 species: 7 ABA CODE 1   4 CODE 2   1 CODE 4   1 CODE 5

The mimic-thrushes—consisting of the mockingbirds, catbirds, and thrashers—have traditionally been placed close to the true thrushes in the family Turdidae, in recognition of the similarities in plumage between the two groups. They are structurally distinctive, however, and certain lines of molecular evidence suggest that the mimic-thrushes are more closely allied with the starlings and mynas in the family Sturnidae.

The mimic-thrushes are a fairly small family (approximately 34 species) that occurs only in the New World, whereas the starlings and mynas are a medium-sized Old World family (approximately 115 species) with greatest representation in Africa and Southeast Asia. No doubt because of their talents as songsters, representatives of both families have long been popular with bird fanciers, and several species have become established in regions to which they are not native. The Northern Mockingbird has been established in Hawaii, and the European Starling is abundant in much of North America.

North American representatives of the families Mimidae and Sturnidae tend not to excel at migration. Five thrasher species that breed in the southern stretches of the continent are essentially non-migratory. The remaining species of mimids and sturnids are termed "partial migrants," meaning there is some overlap between their breeding and wintering grounds. The only species in these two families that routinely crosses the Gulf coast on migration is the Gray Catbird.

Several species of mimids and sturnids have succeeded wildly in North America. The European Starling is arguably the most successful bird species in North America, having flourished in virtually any habitat that provides cavities for nesting. The Common Myna, although not officially recognized on the American Birding Association Checklist, is well established in Florida and appears to be quickly increasing in number and expanding its range. Documentation of the rise of the Common Myna is of considerable significance to both ecologists and conservationists, but virtually nothing on the matter has been published to date; by simply recording and publishing their data on the species, birders in the field could make important contributions to our understanding of the natural history of the Common Myna. Population success stories in the families Mimidae and Sturnidae are not limited to exotic species. The native Northern Mockingbird has for many decades been expanding its range northward, in response to a combination of factors including local landscaping, regional habitat change, and perhaps climate change.

Northern Mockingbird

Mimus polyglottos CODE 1

ADULT. *Lanky and long-tailed. Medium gray above; pale gray below. Wings not much darker than back (compare shrikes), with conspicuous white "pocket handkerchief."* TEXAS, NOVEMBER

JUVENILE. *Eye dark (adult has yellow eye), bill yellowish, breast spotted. Spots on breast soon fade, and bill turns to black; dark eye retained well into winter.* TEXAS, APRIL

IN FLIGHT. *Flashes extensive white at base of primaries and in outer tail feathers, prominent from above and below.*

CALIFORNIA, NOVEMBER

ADULT. *Generalist diet includes the fruits of both native and exotic trees and shrubs—a trait that has "pre-adapted" the species for establishment beyond its original range.* TEXAS, NOVEMBER

L 10"   WS 14"   WT 1.7 OZ

▸ one adult molt per year; complex basic strategy
▸ moderate differences between adult and juvenile
▸ amount of white and black in wing variable

Highly animated. Prances about, flashing wings and tail; chases off perceived intruders of all sorts. Occurs singly or in tight-knit family groups in disturbed habitats: farm country, forest edges, cities. Also in deserts. Range expanding northward, especially in East.

◀ Prolific, high-fidelity mimic; single birds recognizably incorporate songs of 25+ species into seemingly endless song. Calls: harsh *chack* and grating hiss.

BAHAMA MOCKINGBIRD (M. gundlachii). *Casual Caribbean stray to Florida. Larger and browner than Northern Mockingbird, with black malar stripe. Adult, not shown, has streaking on belly.* FLORIDA, JANUARY

Sage Thrasher
Oreoscoptes montanus   CODE 1

L 8.5"	WS 12"	WT 1.5 OZ

- ‣ one adult molt per year; complex basic strategy
- ‣ moderate seasonal and age-related differences
- ‣ northern birds longer-tailed than southern

Breeding habitat largely restricted to sagebrush desert; some also nest in greasewood and bitterbrush. On migration and in winter, shrubs of any sort will do. Migrates early in spring, early in fall; fall migration has easterly aspect. Less furtive than other thrashers.

Song, often given at night, a long, disjointed series of sweet warbles and a few harsh notes: *urr whee chook weeur eer year chick urry whee-eer...*

FRESH ADULT. *Small and short-billed; in much of breeding range, the only thrasher. Gray-brown above; pale below. In fresh plumage, as here, has broad black spots below and buff flanks.* ARIZONA, FEBRUARY

JUVENILE. *Similar to adult, but upperparts more heavily streaked; wing bars usually more prominent.*
OREGON, AUGUST

WORN ADULT. *As plumage wears, spotting below becomes more diffuse, sometimes fading away almost completely. Can resemble Bendire's Thrasher, but is smaller and has paler plumage overall.*
CALIFORNIA, JUNE

Bendire's Thrasher
Toxostoma bendirei   CODE 2

L 9.75"	WS 13"	WT 2.2 OZ

- ‣ one adult molt per year; complex basic strategy
- ‣ weak differences between adult and juvenile
- ‣ slight individual variation in ground color

Uncommon. Inhabits deserts with scattered perches amid sparse grasses; Joshua tree forests and cholla groves are typical habitat. Feeds on ground like other thrashers, but also gleans vegetation for insects. Dispersal, especially following breeding, not well understood.

Song disjointed like Sage Thrasher, but faster and squeakier: *erweeka cheer cheek acheery-acheery jeerweeka.* Call a rough *chut.*

ADULT. *Similar to Curve-billed, but bill shorter and less curved; base of lower mandible pale. Juvenile slightly shorter-billed; can be confused with worn adult Sage Thrasher.*
ARIZONA, FEBRUARY

Curve-billed Thrasher

Toxostoma curvirostre CODE 1

ADULT, ARIZONA. *Dusky overall, with little spotting below; white in tail reduced. Division between eastern and western populations—perhaps separate species—is Southeast Arizona.* ARIZONA, MAY

ADULT, TEXAS. *Gray-brown above; pale below, with well-defined spots. Long tail has a bit of white in the corners. All adults have bills of medium length and curvature.* TEXAS, NOVEMBER

L 11"   ws 13.5"   wt 2.8 oz

▸ one adult molt per year; complex basic strategy

▸ moderate differences between adult and juvenile

▸ eastern birds have more distinct spotting below

Usually in dense vegetation, but sometimes runs through clearings or perches briefly on shrubbery. Avoids sparse desert; prefers yucca and mesquite stands, riparian tangles, plantings in cities. Eastern birds disperse somewhat after breeding, western birds tend not to.

Song burry and twangy, often with couplets: *chut-chut urpy-urpy chur cheer-cheer jurt.* Distinctive call a piercing *wit wheet!* or *wit wheet wheet!*

JUVENILE. *Bill shorter than adult; can be very similar to Bendire's, which is slightly warmer overall, with more extensive pale region at base of lower mandible.* ARIZONA, APRIL

California Thrasher

Toxostoma redivivum CODE 1

L 12"   ws 12.5"   wt 2.9 oz

▸ one adult molt per year; complex basic strategy

▸ weak seasonal and age-related differences

▸ darker and richer overall in northern parts of range

Favors coastal chaparral, where it hides in dense shrubbery and plucks grasshoppers and beetles from leaf litter; also in oak woodlands and in urban districts with sufficient plantings. Like other "curve-billed" thrashers, runs quickly through small clearings.

Disjointed song harsher, more growling than other thrashers: *uriwee-urkiweej... witchurra-witchurra... urr... purcha... urwee-urwee.* Call a rolling *churrrick.*

ADULT. *Large, dark, and long-billed; pale throat prominent. Crissal similar, but paler overall, and there is little range overlap; Worn adult ratty-looking; juvenile duller.* CALIFORNIA, JANUARY

Long-billed Thrasher

Toxostoma longirostre   CODE 2

L 11.5"	WS 12"	WT 2.5 OZ

▸ one adult molt per year; complex basic strategy

▸ weak differences between spring and fall adults

▸ tips of outer tail feathers sometimes buffy white

Habits much as Brown Thrasher, which it replaces as breeder in southern Texas. Greatest densities in brushlands of lower Rio Grande drainage, a habitat that has been largely destroyed; has expanded north in small numbers as mesquite and other shrubs invade northward.

Song is more raucous than Brown Thrasher: *chicker beat-dirty sweak urcheep...* Calls melodious *kleerk* or *kleerker* and dull *tsook*.

ADULT. *Darker and longer-billed than Brown; head of Long-billed distinctively gray. Underparts bold, with strong black streaks on all-white background; contrast eroded on worn adults.* TEXAS, NOVEMBER

Brown Thrasher

Toxostoma rufum   CODE 1

ADULT. *Large and long-tailed; rufous above with white wing bars; pale below with dark streaks. Long-billed Thrasher not as colorful. Juvenile has pale buff spots above, buff wing bars.* MICHIGAN, MAY

L 11.5"	WS 13"	WT 2.4 OZ

▸ one adult molt per year; complex basic strategy

▸ weak differences between adult and juvenile

▸ birds in western Great Plains larger and paler

Inhabits thickets and hedgerows east of Rockies. Eastern birds in forest clearings and edges, western birds in shelterbelts and wooded waterways. Migrates fairly early in spring, returns fairly late in fall. Range has expanded in all directions; still slowly expanding.

More than any other thrasher, song elements strongly doubled: *cheeper-cheeper... deerily-deerily... jeer-jeer... d'wee d'wee...* Call a harsh *stack*.

ADULT. *Large and long-tailed; rufous above with white wing bars; pale below with dark streaks. Long-billed Thrasher not as colorful. Juvenile has pale buff spots above, buff wing bars.* MICHIGAN, MAY

The Brown Thrasher is one of several dozen eastern land bird species that stray annually to western North America. A much smaller number of western species are annual vagrants to the East. In the East, the best season by far for western land bird vagrants is autumn; the same is true for eastern vagrants in the West, but the pattern is not as strong.

Crissal Thrasher

Toxostoma crissale   CODE 2

L 11.5"	WS 12.5"	WT 2.2 OZ

▸ one adult molt per year; complex basic strategy

▸ weak differences between adult and juvenile

▸ darker and grayer in eastern portion of range

Widely distributed in hot desert regions; greatest densities in washes and riparian thickets, but many occur well into desolate creosote bush stands. Some also range upslope into pinyon-juniper woodlands. Using feet, digs into and picks at soil for insects and spiders.

ADULT. *A long-billed desert thrasher. Most are paler than California; all are darker than Le Conte's. Undertail coverts (crissum) brick red. Juvenile has dull reddish cast overall.* ARIZONA, FEBRUARY

Song sweeter than California Thrasher, with many couplets: *durry-durry... ch'ree-ch'ree... breer... cheerip-cheerip.* Common call a distinctive *cheery churry.*

Le Conte's Thrasher

Toxostoma lecontei   CODE 2

L 11"	WS 12"	WT 2.2 OZ

▸ one adult molt per year; complex basic strategy

▸ weak differences between adult and juvenile

▸ birds in eastern portion of range average paler

Notable among even the desert thrashers for preferring the hottest, driest habitats; most occur in sere lowlands with scattered saltbush and shadscale. Uncommon, even in favored habitat. Terrestrial: digs pits in soil for arthropods; runs with tail held straight up.

ADULT. *Paler than other desert thrashers; sandy gray all over, with a pale apricot wash across undertail coverts. Juvenile has paler undertail coverts and slightly shorter bill.* CALIFORNIA, MARCH

Song alternating rich and thin, buzzy: *churwee url-churl... weet weet weet... urrrl... bzweet... churl-churl...* Call a rising *wheeerp!*

Gray Catbird

Dumetella carolinensis   CODE 1

L 8.5"	WS 11"	WT 1.3 OZ

▸ one adult molt per year; complex basic strategy

▸ weak differences between adult and juvenile

▸ western populations paler, especially on underparts

Skulks in shrubbery but less so than thrashers: often comes out for view; sometimes forages up in trees. In East, prefers forest edges and creeksides in farm country or suburbia; in West, associated with shelterbelts and riparian groves. Migrates silently at night.

ADULT. *Long-tailed and short-billed. Dark gray all over, with black cap and chestnut undertail coverts. Juvenile similar, but throat and sides have buff or rufous hints.* TEXAS, APRIL

◀ Song a complex series of short, thin, nasal notes: *aay-wa... chiddy... which... ur-beety...* Call—like a cat's meow—a nasal, querulous *nyewww?* or *nyaaay?*

European Starling
Sturnus vulgaris   CODE 1

NONBREEDING ADULT. *After fall molt, heavily spotted all over; spots wear off during winter, leaving glossy sheen of breeders. Bill usually dark in winter; yellow in summer.* COLORADO, SEPTEMBER

BREEDING MALE. *Both sexes black all over with intense iridescent highlights, especially greens and purples. Yellow bill has bluish base in male, pinkish base in female.* COLORADO, MARCH

JUVENILE. *Sooty brown all over. All plumages have distinctive shape: short-tailed and long-winged; portly overall and forward-leaning.* TEXAS, JUNE

IN FLIGHT. *Wings broad at base; wing tips pointy. Short-tailed. Flight profile unlike blackbirds; Purple Martin similar, but has languid, erratic flight typical of swallows.* PENNSYLVANIA, DECEMBER

L 8.5"   WS 16"   WT 2.9 OZ

▸ one adult molt per year; complex basic strategy

▸ strong seasonal and age-related differences

▸ adult bill color and juvenile plumage variable

Wildly successful introduced species. Characteristic of large cities and feedlots but occurs widely in natural habitats that supply cavities for nesting. Surprisingly wary. Forages on lawns and around dumpsters. Forms large, sociable roosts, especially in winter.

◀ Vocalizations varied; many are wheezy, others rattling. Calls include a hissing *sheeeer* and whistled *wheeeooo*. Imitates other species, especially Killdeer.

COMMON MYNA (Acridotheres tristis). *South Asian species; established and expanding in Florida. Plump; short-tailed. Black hood has variable yellow skin around eye; tail black-and-white.* HAWAII, MARCH

Wagtails & Pipits
FAMILY MOTACILLIDAE

---

10 species: 2 ABA CODE 1   2 CODE 2   2 CODE 3   2 CODE 4   2 CODE 5

These tail-pumping, ground-loving birds are at home in open, grassy habitats worldwide. Two genera occur in North America: the generally drab pipits in the genus *Anthus* and the brightly colored and boldly patterned wagtails in the genus *Motacilla*. Morphologically, vocally, and behaviorally, these two genera have much in common. Most species in both genera tend to be thin-billed and svelte overall, with long tails that are pumped constantly. Also, most of our motacillids engage in conspicuous diurnal migrations, giving loud and distinctive flight calls as they pass overhead. The evolutionary relationship of the wagtails and pipits to other families is not well established, with molecular and morphological evidence pointing to conflicting taxonomic affiliations.

For most North American birders, the widespread American Pipit is the flagship species of the family Motacillidae. Only one other species—the secretive Sprague's Pipit—breeds anywhere near the lower 48 states. Several species are casual or accidental, and one species surely ranks among the most bizarre of vagrants ever to have reached North America: Our single record of the Citrine Wagtail, a central Asian motacillid, comes not from Alaska, as one might have expected, but rather from central Mississippi.

Identification of wagtails and pipits— especially of birds out of range and away from breeding grounds—can be challenging, but behavioral and especially vocal cues can be of considerable use in the field. Outside the narrow time period during which it gives its mesmerizing courtship display, the Sprague's Pipit is nearly impossible to find if one does not recognize its flight call and microhabitat preference. Often, the first indication of a rare pipit—for example, a Red-throated or Olive-backed is its flight call.

North American populations of pipits and wagtails have been influenced by a diverse suite of environmental factors. The Sprague's Pipit appears to be in sharp decline, the result of degradation of the midgrass prairie habitat on which it specializes throughout its life cycle. Breeding habitat for the southernmost populations of American Pipit may be dwindling, too, with the steady shrinking in acreage of alpine tundra in the southern Rockies. But two species of wagtails may be recent beneficiaries of human modifications to the landscape. The elegant Eastern Yellow Wagtail flourishes in such inelegant landscapes as oilfields and roadsides, and recent studies have documented a range expansion in Alaska of the White Wagtail as humans expand their presence there.

White Wagtail

Motacilla alba   CODE 2

BREEDING MALE, "BLACK-BACKED." (M. a. lugens) *Back is all black; black on head and breast a little more extensive than "Gray-backed." In flight, wings mostly white (mostly dark on "Gray-backed").* ALASKA, JULY

BREEDING MALE, "GRAY-BACKED." (M. a. ocularis) *All adults have long, black tail with white edges. Breeding male has gray back; white face bordered by black crown, nape, chin, and throat.* ASIA, JULY

ADULT FEMALE. *Females of all populations have extensive black on breast; head paler and less contrasting than breeding male. Nonbreeding male similar.* ASIA, SEPTEMBER

L 7.25"   WS 10.5"   WT 0.65 oz (18 g)

▸ two adult molts per year; complex alternate strategy
▸ strong age-related and sex-related differences
▸ adult male "Black-backed" Wagtail distinctive

Habits similar to Eastern Yellow Wagtail, but more closely associated with sparsely vegetated seacoasts and river mouths. Population expanding as humans modify the Alaskan landscape. Most Alaskan breeders are gray-backed; "Black-backed" subspecies an uncommon breeder.

Flight call an abrupt *chivvip*. Call sharper: *tsweeweet*, suggesting Cordilleran Flycatcher. Song consists of two-noted phrases, repeated.

Eastern Yellow Wagtail

Motacilla tschutschensis   CODE 1

BREEDING MALE. *Bright yellow below; dark above. Black auriculars bordered above by white eyebrow, below by white chin. All plumages have long tail with white edges.* ALASKA, JUNE

ADULT FEMALE. *Paler below than breeding male; facial pattern similar but lower contrast. Nonbreeding male similar. Juvenile lacks yellow, but long tail with white edges distinctive.* ALASKA, JUNE

L 6.5"   WS 9.5"   WT 0.6 oz (17 g)

▸ two adult molts per year; complex alternate strategy
▸ moderate age-related and strong seasonal differences
▸ vagrants to western Aleutians larger, more yellow

Conspicuous on the tundra; catches insects animatedly, struts about, flushes frequently. Breeds in shrubby open areas, even those degraded by humans: roadsides, ditches, and drilling operations. Breeders arrive mainly in early June, depart by late August.

Flight call, given frequently, a ringing *psweep* or *pswee*, sometimes doubled to *tsee-veet*. Song vigorous: varied one-syllable notes repeated quickly.

Sprague's Pipit
Anthus spragueii   CODE 2

JUVENILE. *All plumages, especially scaly-backed juvenile, surprisingly similar to juvenile Horned Lark. Sprague's Pipit has pink legs, extensive white in tail, and plainer face.* CALIFORNIA, OCTOBER

L 6.5"   WS 10"   WT 0.9 oz (25 g)

▸ two adult molts per year; complex alternate strategy

▸ moderate age-related and weak seasonal differences

▸ extent and prominence of streaking and scaling variable

ADULT. *Sandy buff all over; brightest in fall. Broadly streaked above; below, just a few fine streaks. Black eye prominent on plain face. Tail has more white than American Pipit.* NORTH DAKOTA, JUNE

Uncommon and declining; furtive and cryptic. Courtship season brief. Spring migration in April, fall migration late September to October; migrants rarely detected. Easily overlooked in winter. Requires undisturbed midgrass prairie at all stages of life cycle.

Flight display astonishing for its duration and altitude; complex song sweet and twangy reminiscent of Veery. Flight call, *skweet*, often the best clue to the bird's presence.

ADULT. *Classic microhabitat, shown here, consists of tiny open patches amid much larger swaths of midgrass prairie.* ARKANSAS, FEBRUARY

Olive-backed Pipit
Anthus hodgsoni   CODE 3

L 6"   WS 10"   WT 0.7 oz (20 g)

▸ two adult molts per year; complex alternate strategy

▸ weak differences between spring and fall adults

▸ individual variation in extent of spotting

A tree-loving species that is rather out of its element in the treeless western Aleutians, where it is an annual migrant; casual elsewhere in Bering Sea region. Strays to Alaska tend to be secretive, sticking to dense cover. When out in the open, pumps tail vigorously.

Flight call a short, harsh *viss* or *veez*. Song, rarely heard in North America, consists of rapid trills alternating with soft, relaxed notes.

ADULT. *Olive-brown above; pale buff below with coarse spots. Brighter and bolder in fall. Eyebrow distinctive: white behind eye, yellow in front. Brown auriculars have white spot at rear.* ASIA, APRIL

American Pipit

Anthus rubescens   CODE 1

NONBREEDING ADULT. *Underparts more heavily streaked than in breeding plumage. All year, long dark tail has white edges, prominent in flight.* MINNESOTA, SEPTEMBER

BREEDING ADULT, SOUTHERN ROCKIES. *Long-tailed and long-winged, with sturdy black legs. Breeders are buff below; breeders in southern Rockies have reduced streaking below.* COLORADO, JUNE

L 6.5"	WS 10.5"	WT 0.75 oz (21 g)

▸ two adult molts per year; complex alternate strategy

▸ moderate differences between spring and fall adults

▸ geographic variation in color and pattern of underparts

Breeds on alpine and arctic tundra; winters in open areas with little vegetation, especially near water. Forms small to large flocks on migration and in winter. Diurnal migration conspicuous. At all times of year, struts around erratically, pumping tail as it goes.

◀ Flight call *tyip* or *tyippit*, sometimes extended to *pipipipit.* Repetitive song, given in towering display flight, a series of *cheer* and *cheedle* notes.

BREEDING ADULT, NORTHERN. *Widespread Northern population averages more heavily streaked below than in Southern Rockies.* MAINE, MAY

Red-throated Pipit

Anthus cervinus   CODE 3

L 6.25"	WS 10.5"	WT 0.75 oz (21 g)

▸ two adult molts per year; complex alternate strategy

▸ strong sex-related and moderate seasonal differences

▸ red on throat variable, especially on adult males

Erratic breeder in Bering Sea region; regular migrant, sometimes in medium-sized flocks there in early June and late August. Annual in fall in small numbers to California coast. Habits much as American Pipit: struts around on ground, perches on rocks, twitches tail.

ADULT MALE. *Face and breast extensively red. Most females not nearly as red; many males duller, too, especially in fall. All are dark above with pale streaks; belly pale with dark streaks.* ALASKA, JUNE

Flight call a high, thin, slightly descending *teeee* or *teeees.* Song like American Pipit, but elements more musical and delivered more rapidly.

Waxwings and Silky-Flycatchers

FAMILIES BOMBYCILLIDAE AND PTILOGONATIDAE

5 species: 3 ABA CODE 1   2 CODE 5

The North American representatives of these two families are sleek, crested, and erratic in their annual wanderings. The two families are closely related, and the silky-flycatchers are treated by some authorities as a subfamily within the Bombycillidae. Both families are small. The waxwings consist of only three species worldwide, and the silky-flycatchers of four. Their closest relatives appear to be two bird families that are even less diverse: the family Dulidae, which consists of a single species, the Palmchat, endemic to Hispaniola; and the family Hypocoliidae, also represented by a single species, the Hypocolius of the Middle East.

The three regularly occurring species in North America are the Bohemian and Cedar Waxwings in the family Bombycillidae and the Phainopepla in the family Ptilogonatidae. All three species engage in complex and poorly understood movements that are related to some combination of weather, food supply, and nest site availability. After a winter spent wandering widely in search of berries, most Cedar Waxwings migrate north in late May to breeding grounds in the northern U.S. and southern Canada; fall migration is erratic, with movements in some years as far south as southern Central America. Bohemian Waxwing dispersal following breeding is episodic, with large numbers reaching the southern Rockies in some years but few if any in most years; every year, at least a few flocks make it east all the way to maritime Canada. Oddest of all perhaps is the Phainopepla. Early in the spring, Phainopeplas defend breeding territories in low-desert mesquite groves infested with mistletoe; then they drift upslope in the early summer, undergo a personality change, and breed cooperatively in woodlands in the foothills.

Song in these three species is peculiar. In the waxwings, the call appears to double as the song, an arrangement that is unusual among passerines. Most of the time, Phainopeplas are heard to give only a simple call note, but the species also has a lengthy and complex song that sometimes incorporates high-fidelity imitations of other species.

Population status of the Phainopepla and the two waxwings is presumably related to food supply, and recent declines in the Phainopepla seem to have stemmed from destruction of berry-laden riparian groves in the Desert Southwest. Waxwings feasting on fermented berries sometimes become comically inebriated, but presumably with no long-term consequences for population health. Indeed, Cedar Waxwing populations appear to be on the increase, due in part to planting of exotic shrubs continent-wide and to reforestation of the Northeast.

Cedar Waxwing
Bombycilla cedrorum   CODE 1

ADULT. *Exquisite. Yellow tail band distinctive, even at a distance; red tips on secondaries prominent at close range. Brown-buff overall with yellow belly. Crested; black mask conspicuous.* TEXAS, MAY

JUVENILE. *Browner and duskier than adult; coarsely streaked below. Crest and eye mask often reduced compared to adult. Wings lack waxy red tips, but tail has prominent yellow tip.* WYOMING, SEPTEMBER

ADULT AND NESTLINGS. *Nesting is initiated later in Cedar Waxwings than most other species, probably reflecting the species' reliance on fruits that do not ripen until summer.* NEW YORK, JUNE

IMMATURE. *Body feathers as adult, but wings as juvenile: gray all over, lacking waxy red tips.*
NEW HAMPSHIRE, JANUARY

L 7.25"	WS 12"	WT 2 OZ

▸ one adult molt per year; complex basic strategy

▸ strong differences between adult and juvenile

▸ a few individuals have orange-tipped tails

Gregarious all year; occurs in small to medium-sized flocks in treetops. Flocks in flight bunch up tight. Gleans fruit from trees or flycatches. Regular, annual movements occur, but are hard to characterize; local populations wander in search of food at any time of year.

◀ High-pitched vocalization, heard year round, functions as both song and call: buzzy *zeeee* or slower, warbling *z'z'z'zee.* Similar to alarm calls of many other species.

FLOCK. *At any time of year, may be found in dense flocks, usually in treetops. Flocks are generally less noisy and active than flocks of Bohemian Waxwings.* NEW JERSEY, NOVEMBER

Bohemian Waxwing
Bombycilla garrulus CODE 1

ADULT. *Crested like Cedar, with prominent yellow tail tip. Larger and darker than Cedar; wings more intricately marked. Undertail coverts deep chestnut; white in Cedar.* MINNESOTA, JANUARY

IMMATURE. *Like adult, but lacks red on wings. Both sexes have black chin, more extensive in male. Juvenile, like juvenile Cedar, is streaked below.* NEW HAMPSHIRE, JANUARY

L 8.25" ws 14.5" wt 2 oz

▸ one adult molt per year; complex basic strategy

▸ strong age-related and weak sex-related differences

▸ number of red tips per wing on adults variable

Notorious for its wanderings. At least some movement eastward every fall; massive in some years, weak in others. Breeds in conifer forests; winters where there are berries. Eastern limits of breeding range and extent of regular easterly dispersal not well worked out.

Flocks amazingly chatty, with every member seeming to call at once. Compared to Cedar Waxwing, vocalization is more trilled, more musical, lower, and richer.

Phainopepla
Phainopepla nitens CODE 1

ADULT MALE. *Glossy black all over, with long, wispy crest. Eyes red. In flight, long tail especially prominent and primaries flash white.* CALIFORNIA, APRIL

FEMALE. *Sooty gray, but red-eyed and crested like male. Perched, shows limited white in wing; both sexes flash white in wing in flight, duller in female. Juvenile similar, but eyes browner.* CALIFORNIA, NOVEMBER

L 7.75" ws 11" wt 0.85 oz (24 g)

▸ one adult molt per year; complex basic strategy

▸ moderate age-related and strongsex-related differences

▸ slightly smaller in western portion of range

Alternates bouts of treetop foraging with twisting, erratic flight. Seasonal movements strange: nests early in spring in mistletoe-infested deserts, then nests in summer in low-montane woodlands. Desert nesters territorial; woodland nesters cooperative.

Common call a rising *wurt?* Song, given unpredictably and infrequently, a halting series of rich whistles and chatters; in distress, gives imitations of other species.

Olive Warbler and Wood-Warblers

FAMILIES PEUCEDRAMIDAE AND PARULIDAE

58 species: 41 ABA CODE 1   8 CODE 2   2 CODE 3   5 CODE 4   1 CODE 5   1 CODE 6

The wood-warblers are part of a large assemblage of mainly New World species called the nine-primaried oscines. Relationships among these species are unclear. The most important pattern seems to be that all of these groups (including the tanagers, sparrows, blackbirds, grosbeaks, and others) are closely related and have radiated quickly and recently. One species that seems to be distinct from the wood-warblers—and perhaps from all the nine-primaried oscines—is the Olive Warbler, currently recognized as the only member of the family Peucedramidae.

The wood-warblers are primarily insectivorous, a trait that is reflected in various aspects of their physiology and behavior: They are small and pointy-billed, they are active and largely arboreal, and they withdraw in winter from the North Temperate Zone. Foraging strategies differ greatly among species. Worm-eating Warblers poke into dead leaves for arthropods, Louisiana Waterthrushes glean from stream edges, Connecticut Warblers pluck food from the forest floor, and American Redstarts flash their wings and tail to startle prey. Even within the same tree, different foraging strategies are employed. In northern hardwood forests, for example, the Black-throated Green gleans from the upper surfaces of leaves, whereas the Chestnut-sided specializes on the undersides of leaves. And in boreal forests, the Cape May forages toward the tips of branches, whereas the Bay-breasted focuses its efforts closer to the center of the tree.

Migration is primarily nocturnal, with mixed-species flocks gathering by day to feed wherever there is leafy vegetation. More species are routinely encountered on migration in the East than in the West, but many eastern species regularly stray far west in both spring and fall. The majority of species sing during spring migration, especially as the season advances; in fall, most give only monosyllabic call notes that can actually be quite useful in the identification process.

Many wood-warbler species are of conservation concern. In particular, wood-warblers that breed in the eastern deciduous forest and boreal forest zones require large tracts of forest with few if any nest predators, Brown-headed Cowbirds (nest parasites), and White-tailed Deer (understory destroyers). Many species depend largely on caterpillars, and broad-spectrum insecticide treatment may severely depress wood-warbler populations at a regional scale. Meanwhile, deforestation in the tropics continues and poses a serious threat on the wintering grounds of most wood-warblers that breed in North America.

Olive Warbler

Peucedramus taeniatus CODE 2

ADULT MALE. *Size, structure, and plumage are warbler-like, but Olive Warbler is not a warbler. Head and breast orange; black mask prominent. Gray above with prominent white wing bars.* ARIZONA, JULY

ADULT FEMALE. *Pattern the same as male, but deep orange replaced by dull yellow; black mask reduced. Immature (especially female) duller still, with faint gray mask and buff-yellow.* ARIZONA, JUNE

5.25" ws 9.25" wt 0.4 oz (11 g)

- one adult molt per year; complex basic strategy
- strong age-related and sex-related differences
- intensity of color on head and breast variable

Inhabits pine forests; generally at high elevations, in ponderosa pine and Douglas fir zones. Feeding methods recall Pine Warbler: creeps and hops about. Mainly resident, but some annual movements by northern birds. Range appears to be expanding northwestward.

Song varied. Some versions ringing, like Tufted Titmouse: *Peer jeer jeer.* Others burrier and more rolling, like *Oporornis* warblers: *jurry jurry jurry.*

Tennessee Warbler

Vermivora peregrina CODE 1

ADULT MALE. *All are greenish above and pale below, with long wings and a short tail. Breeding male has much gray on head, with prominent white eyebrow; nonbreeding male less gray.* TEXAS, APRIL

ADULT FEMALE. *Drabber than male, especially first-fall female. Back greenish; underparts variable yellow; but undertail coverts bright white (undertail coverts yellow in Orange-crowned).* TEXAS, APRIL

L 4.75" ws 7.75" wt 0.35 oz (10 g)

- two adult molts per year; complex alternate strategy
- moderate seasonal, age-related, sex-related differences
- body size, especially bill length of male, variable

Classic "spruce budworm" warbler, susceptible to population swings associated with variation in food supply. Occurs mainly in early successional forests. Migrates fairly late in spring; migrates south in the midst of its fall molt, usual for eastern passerines.

Three-part song, chanted endlessly: *tecky tecky tecky tick tick tick tick tyew! tyew! tyew! tyew!* Call a sharp *tyick.* Flight call like Orange-crowned.

Orange-crowned Warbler

Vermivora celata　CODE 1

ADULT MALE. *All are small, short-winged, and pointy-billed. Orange crown, shown here, rarely visible in the field. Females and subadults average duller, sometimes substantially.* CALIFORNIA, MAY

IMMATURE, "EASTERN." (V. c. celata) *All plumages are short-winged and pointy-billed, with an obscure eyebrow and eye line. Eastern birds drab and gray; some have yellow limited to undertail coverts.* TEXAS, MARCH

ADULT, "PACIFIC SLOPE." (V. c. lutescens) *Spring males are yellow all over, brightest on undertail coverts and eyebrow. Female Yellow similar, but has longer wings and a blank facial expression.* CALIFORNIA, APRIL

L 5"	WS 7.25"	WT 0.35 oz (10 g)

- ▸ two adult molts per year; complex alternate strategy
- ▸ moderate age-related and weak sex-related differences
- ▸ geographic variation in yellow, especially on head

On its widespread breeding grounds, inhabits broadleaf woods, thickets, and conifers. Migrants occur in diverse habitats, for example, desert oases and open woodlands. More hardy in winter than most warblers; gleans berries and visits feeders.

ADULT, "INTERMOUNTAIN WEST." (V. c. orestera) *The most distinctively patterned Orange-crowned. Superficially resembles MacGillivray's, but structure and habits different. Gray hood with eye arcs.* CALIFORNIA, APRIL

FALL ADULT. *Even in fall, can show extensive yellow. Bright individuals in winter are sometimes mistaken for rarer species, such as Nashville and Yellow Warblers.* MEXICO, OCTOBER

◀ Variable song a loose trill, often with a pitch change midway through: *ch'ch'ch'ch'j'j'j'j'j'*. Call a short *tsiit*. Flight call slightly buzzy: *zzee.*

Nashville Warbler
Vermivora ruficapilla CODE 1

ADULT MALE, EASTERN. *Blue-gray head has prominent white eye ring, reduced in female and immature. Superficially similar to Connecticut, but size, structure, and behavior different.* NEW HAMPSHIRE, MAY

ADULT MALE, WESTERN. *All are yellow-green above and bright yellow below (including throat). Back of Western flecked with gray; tail-twitching of Western is conspicuous.* CALIFORNIA, APRIL

IMMATURE. *Dull, but adult pattern and colors evident: gray head with eye ring, greenish above, yellow below. Yellow throat washed with gray or white.* NEW JERSEY, SEPTEMBER

L 4.75"   WS 7.5"   WT 0.3 oz (9 g)
▸ two adult molts per year; complex alternate strategy
▸ weak seasonal, age-related, and sex-related differences
▸ western birds longer-tailed and brighter yellow on rump

Western birds nest on shrubby mountain slopes, eastern in second-growth woods. Both populations favor disturbed habitats such as recovering clear-cuts, not extensive old-growth. Western twitches tail, eastern does not. Fairly early circum-Gulf spring migrant in East.

Variable song two-parted and fast, not as piercing as Tennessee: *weeda weeda weeda weeda dididididi.* Call a nasal *pik.* Flight call a rising, buzzy *zwit.*

Virginia's Warbler
Vermivora virginiae CODE 1

ADULT MALE. *Like a Nashville with reduced color; yellow limited to undertail coverts and small patch on breast. Eye ring, tail-twitching, and small size good marks in all plumages.* CALIFORNIA, APRIL

IMMATURE. *Yellow undertail coverts contrast with otherwise gray plumage; adult female similar, but most have faint yellow on breast.* COLORADO, AUGUST

L 4.75"   WS 7.5"   WT 0.3 oz (9 g)
▸ two adult molts per year; complex alternate strategy
▸ moderate sex-related and weak age-related differences
▸ yellow on breast and red on head variable

Intermountain West counterpart of Nashville Warbler. Breeds in dry pine and oak forests on sunny slopes. Feeds by gleaning, mainly in low broadleaf vegetation; twitches tail like western Nashville. Fall migration early; some range east into western Plains.

Variable two-part song more languid than Nashville: *zwee zwee zwee zwee see see see.* Call a nasal *pik.* Flight call like Orange-crowned Warbler.

Colima Warbler
Vermivora crissalis CODE 2

ADULT. *Like an oversized and large-billed Virginia's. Plumage gray and brown, with conspicuous yellow undertail coverts. Gray head has white eye ring and reddish cap.* MEXICO, DECEMBER

ADULT. *All plumages can show greatly reduced brown, females and subadults more so on average than breeding males. Overall heft and large bill usually obvious.* TEXAS, MAY

L 5.75"	WS 7.75"	WT 0.35 oz (10 g)

- ▸ two adult molts per year; complex alternate strategy
- ▸ weak seasonal, age-related, sex-related differences
- ▸ individual variation in extent of brown in plumage

The bird of Big Bend National Park and environs; reaches North America only in the Chisos Mountains. Nests in oak and pine woods with grassy cover; most arrive in May, depart in September. Active; hops around while gleaning broadleaf vegetation for caterpillars.

Call a rapid trill, more like Orange-crowned than Nashville or Virginia's. Last note of trill distinct: *ch'ch'ch'ch'ch'ch'ch'chew!* Call an explosive *pisk!*

Lucy's Warbler
Vermivora luciae CODE 1

ADULT MALE. *Small and gray. Adult has red crown, reduced in female; both sexes have red rump. Fairly active; twitches tail frequently.* ARIZONA, APRIL

ADULT FEMALE. *Plain and drab; juvenile even duller. Immature female resembles dull Yellow Warbler, but Lucy's is shorter-winged and smaller; immature Verdin has shorter bill with broader base.* ARIZONA, APRIL

L 4.25"	WS 7"	WT 0.25 oz (7 g)

- ▸ one adult molt per year; complex basic strategy
- ▸ moderate age-related and sex-related differences
- ▸ extent of crown patch variable, especially on female

The desert warbler. Along with Yellow Warbler, Common Yellowthroat, and Yellow-breasted Chat, the only warbler that breeds in lowland deserts. Lucy's breeds in drier microhabitats than the others; mesquite and acacia bosques favored. Arrives early in spring, departs early in fall.

Rambling song descends, usually consists of two or three weakly differentiated segments: *chweea chweea cheea cheea cheeo.* Calls similar to Virginia's.

Blue-winged Warbler
Vermivora pinus CODE 1

ADULT MALE. *Face bright yellow with bold black eye line. Back and nape greenish; wings blue-gray with bold white wing bars. Underparts bright yellow.* TEXAS, APRIL

IMMATURE FEMALE. *Overall pattern and color the same as male and older female, but wing bars and black eye line reduced.* NEW JERSEY, OCTOBER

ADULT MALE HYBRID. *Many cannot be assigned to a tidy category. This individual has many characters of Golden-winged, but yellow wash on lower breast is more typical of Blue-winged.* CONNECTICUT, JUNE

"BREWSTER'S" WARBLER. *A frequently encountered hybrid; male blue-gray above and pale gray below, with yellow cap, black eye line, and yellow wing bars.* CONNECTICUT, JUNE

L 4.75"   WS 7.5"   WT 0.35 oz (10 g)

▸ one adult molt per year; complex basic strategy

▸ moderate sex-related and age-related differences

▸ hybridizes extensively with Golden-winged Warbler

Breeds at low to middle elevations in old fields, brushy marshes, and woodland edges. Probes curled leaves for insects. Range expanded northeast in 19th and 20th centuries, but forest recovery and urban sprawl may result in future population losses. Migrates fairly early in spring.

"LAWRENCE'S" WARBLER. *Rarely seen hybrid; male has striking black-and-yellow head, bright yellow underparts, and blue-gray wings with white wing bars.* CONNECTICUT, JUNE

Primary song a sighing *hhhhee bzzzzzz*, as if inhaled and then exhaled. Secondary song a long buzzy series. Call an abrupt *chik.* Flight call a rising *zeet.*

Golden-winged Warbler
Vermivora chrysoptera CODE 2

ADULT MALE. *Head boldly patterned in gray and black, topped off by lemon-yellow crown. Blue-gray wings have broad yellow panel. Plumage mainly gray otherwise.* CONNECTICUT, JUNE

ADULT FEMALE. *Like a washed-out version of male, but with overall pattern and color scheme evident; some immature females are even duller.* CONNECTICUT, MAY

L 4.75"   WS 7.5"   WT 0.35 oz (10 g)

▸ one adult molt per year; complex basic strategy
▸ moderate differences between male and female
▸ is being "genetically swamped" by Blue-winged

Habits much as Blue-winged: probes for insects in broadleaf vegetation in fragmented landscapes. Following population increases in 19th and 20th centuries, is now experiencing a worrisome decline. Losses are especially acute in southern and eastern portions of range.

Primary song a distinct *hree biss biss biss*, as if first note inhaled and subsequent notes exhaled. Secondary song a buzzy jumble. Calls like Blue-winged.

Chestnut-sided Warbler
Dendroica pensylvanica CODE 1

BREEDING MALE. *Cap lemon yellow; face boldly patterned black-and-white; flanks bright rusty; wing bars yellowish. Otherwise yellowish above and whitish below.* TEXAS, APRIL

FIRST SPRING FEMALE. *Some closely resemble adult male; others are much duller, as here. Even dull individuals show overall male pattern of yellow cap, rusty on flanks, and yellow wing bars.* CONNECTICUT, MAY

IMMATURE. *Yellow-green above; pale gray below. Face gray with white eye ring; cap lime green. Wing bars yellowish. Fall adults similar, but many retain traces of breeding plumage.* CALIFORNIA, OCTOBER

L 5"   WS 7.75"   WT 0.35 oz (10 g)

▸ two adult molts per year; complex alternate strategy
▸ strong seasonal, age-related, sex-related differences
▸ chestnut flank stripe longer on older birds

Formerly a rare inhabitant of burns and clearings. Has benefited greatly from the felling of eastern forests; flourishes in second-growth woods and edges of clear-cuts. Widely noted on migration. With tail cocked up, gleans for insects on undersides of leaves.

Primary song has emphatic ending: *dwee dwee dwee dwee dwee dweech!-ew*; secondary song more rambling. Call note a rich *chetch*. Flight call a buzzy *zzzhit*.

Northern Parula
Parula americana CODE 1

L 4.5" ws 7" wt 0.35 oz (10 g)

▸ two adult molts per year; complex alternate strategy
▸ strong sex-related and moderate age-related differences
▸ birds in western portion of range smaller and brighter

A treetop warbler; gleans insects from tips of twigs and foliage. Breeds in mature broadleaf or coniferous forests; highest densities along rivers and major streams. Where available, hanging *Usnea* and Spanish moss are used for nesting. Fairly early spring migrant.

Primary song rises, then ends abruptly: *bzzeeeeeeyip!* Secondary song halting: *b'zhwee wee weea weeeah*. Call a rich *jip*. Flight call a falling *zoo*.

BREEDING MALE. *Small and colorful. Mainly blue-gray upperparts with bold white wing bars and eye arcs. Chin yellow; breast yellow with bold black-and-orange breast band.* CONNECTICUT, JUNE

IMMATURE FEMALE. *Upperparts browner than adult; yellow below reduced. Bold wing bars and eye arcs, combined with olive-green back and yellow lower mandible, nearly always evident.* NEW YORK, SEPTEMBER

ADULT FEMALE. *Lacks breast band. In all plumages, two key marks are yellow lower mandible and olive-green back contrasting with blue-gray upperparts.* CONNECTICUT, MAY

Tropical Parula
Parula pitiayumi CODE 3

L 4.5" ws 6.25" wt 0.25 oz (7 g)

▸ two adult molts per year; complex alternate strategy
▸ moderate sex-related and weak age-related differences
▸ hybridizes with Northern Parula to unknown extent

Like Northern Parula, forages actively in forest canopy; more prone than Northern to glean from undersides of leaves. Occurs in oak or mixed riparian woodlands. Status in Texas unclear: extent of winter withdrawal not known; recent range shifts poorly understood.

Songs similar to western population of Northern Parula, but are more rambling; calls are also similar to Northern.

ADULT MALE. *Lacks eye arcs and dark breast band of Northern; yellow below more extensive. Female and immature duller, but always with yellow lower mandible and olive-green back.* TEXAS, APRIL

Yellow Warbler
Dendroica petechia   CODE 1

BREEDING MALE. *Most are completely yellow except for long red pinstripes on underparts; adult female lacks red pinstripes. Oddly blank look of face distinctive in all plumages.* TEXAS, APRIL

IMMATURE MALE. *Most are completely dull yellow, as here; a few are grayish yellow all over. Staring black eye prominent on plain yellow face. Longer-winged than Orange-crowned.* CALIFORNIA, SEPTEMBER

L 5"	WS 8"	WT 0.35 oz (10 g)

▸ two adult molts per year; complex alternate strategy

▸ strong age-related and moderate sex-related differences

▸ geographic variation in extent and pattern of rufous

Common and conspicuous in riparian and wetland woods; from tiny groves to extensive forests, from low desert oases to high mountain bogs. Nest placed low in broadleaf vegetation, but foraging often takes place at treetops. Nocturnal migration in August is impressive.

◀ Variable song consists of slurred opening notes and emphatic ending: *tswee tswee tswee tweesyweet!* Call a loud, bright *chip!* Flight call a scratchy *zziit.*

BREEDING MALE. *Most other* Dendroica *warblers have conspicuous white tail spots, but tail of Yellow lacks white.* CALIFORNIA, JUNE

"MANGROVE" WARBLER. (D. p. erithachorides) *Tropical subspecies, recently established in south Texas. Other subspecies in North America include "Golden" (D. p. gundlachi) of south Florida, with broader breast streaking.* TEXAS, MAY

IMMATURE FEMALE. *This individual is near the dull extreme for the species; as plain and dingy as Orange-crowned, but wings notably longer. Yellow and Orange-crowned differ in call note, face pattern, and bill structure.* CALIFORNIA, OCTOBER

Prairie Warbler
Dendroica discolor CODE 1

ADULT MALE. Dull yellow above; bright yellow below. Adult male has two thin black stripes on face, reddish streaking on back, and strong black streaks on sides. TEXAS, APRIL

IMMATURE FEMALE. Yellow below with white undertail coverts, like adult, but with only faint streaks on sides. Face pattern weak, but hint of adult markings usually evident. Tail-pumping always good to note. NEW JERSEY, SEPTEMBER

L 4.75"	WS 7"	WT 0.3 oz (9 g)

▸ two adult molts per year; complex alternate strategy

▸ strong sex-related and moderate age-related differences

▸ Florida population larger, with more white in tail

A tail-pumping inhabitant of old fields. Reaches greatest densities in sunny, shrubby oak and pine barrens. Nests near ground; forages in dense, low vegetation. Fairly early spring and fall migrant. Following earlier population increases, is now sharply on the decline.

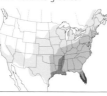

Song a harsh, rising series: *zzee zzee zzee zzee zzee....* Accelerates like Field Sparrow, but very different timbre. Call a bright *chip*. Flight call a buzzy *zeeu*.

Yellow-throated Warbler
Dendroica dominica CODE 1

ADULT MALE, INLAND. All have bright yellow on throat, breast; gray overall with bold black-and-white markings. Inland birds, sometimes called "Sycamore" Warbler, have all-white eyebrow. TEXAS, APRIL

ADULT MALE, COASTAL PLAIN. Eyebrow yellow in front of eye, white behind. All are long-billed, short-winged, especially in Atlantic coastal plain. Female and immature slightly duller. FLORIDA, NOVEMBER

L 5.5"	WS 8"	WT 0.35 oz (10 g)

▸ one adult molt per year; complex basic strategy

▸ moderate sex-related and weak age-related differences

▸ Midwest breeders have shorter bills and longer wings

Returns early in spring to nest in upland pine woods or riverside broadleaf forests; nests well up in canopy. Foraging method distinctive: working solitarily, creeps about trunks like Black-and-white Warbler and pokes long bill into crevices in bark and pine cones.

Song a clear, ringing series, often rising at end: *tswee tswee tswee tswee tswee tsweet!* Call a descending *chip*. Flight call also descending, a rich *tsoo*.

Palm Warbler
Dendroica palmarum CODE 1

BREEDING ADULT, "YELLOW." (D. p. hypochrysea) *Brown above; extensively yellow below with rufous streaks. Cap has extensive rufous; face yellow with rufous highlights.* MAINE, JUNE

NONBREEDING, "YELLOW." (D. p. hypochrysea) *Gray-brown above; extensive yellow below, brightest on undertail coverts. Broad yellow eyebrow prominent.* NEW YORK, OCTOBER

BREEDING ADULT, "WESTERN." (D. p. palmarum) *Duller than "Yellow." Paler below, with limited rufous; face has less rufous.* OHIO, MAY

NONBREEDING, "WESTERN." (D. p. palmarum) *Drab gray-brown, but always with bold yellow on undertail coverts and a broad white eyebrow; all Palm Warblers wag their tails constantly.* FLORIDA, FEBRUARY

L 5.5"	WS 8"	WT 0.35 oz (10 g)

- two adult molts per year; complex alternate strategy
- strong differences between spring and fall adults
- eastern Canada breeders considerably brighter

A ground-loving tail-wagger. Breeds in bogs ringed with conifers; winters, often well north, in brushy woods and coastal thickets; migrants found amid weedy fields and dunes. Spring migration early, especially in the case of "Yellow" Palm Warbler; fall migration late.

Song a loose, languid, buzzy trill: *zwees zwees zwees zwees...* Call note a hard *chick*, like Lincoln's Sparrow. Flight call a descending *zip*.

JUVENILE, "WESTERN." (D. p. palmarum) *Some are nearly devoid of color in first fall, but general facial pattern of breeding adult is present; note single rufous feather starting to grow in on crown.* ALBERTA, SEPTEMBER

Cape May Warbler
Dendroica tigrina CODE 1

BREEDING MALE. *Extensively yellow on head and underparts. Face has broad chestnut patch; underparts streaked black. White wing bars broad, sometimes coalescing in a single panel.* TEXAS, APRIL

IMMATURE MALE. *Drabber than breeding male, but with extensive yellow on face and below. All but the drabbest birds have yellow rump, prominent in flight or with wings drooped.* ILLINOIS, SEPTEMBER

BREEDING FEMALE. *Patterned as breeding male, but rusty on face reduced or absent; wing panel not as prominent.* OHIO, MAY

IMMATURE MALE. *Like most* Dendroica *warblers, Cape May flashes prominent white tail spots in flight. These spots are present in all plumages.* NEW JERSEY, OCTOBER

L 5"	WS 8.25"	WT 0.4 oz (11 g)

- two adult molts per year; complex alternate strategy
- strong sex-related, age-related, seasonal differences
- prominence of chestnut patch variable on breeding male

Nests and forages high in spruce, balsam fir, and other conifers. Densities variable; highest in areas infested with spruce budworm. Fall migration more easterly than spring. Forages toward outer branches, mainly by gleaning; also sallies out for flying insects.

All vocalizations high-pitched. Song, similar to Bay-breasted, a series of rising *swee* notes. Call a hard *dep*. Flight call a short, downward *zip*.

IMMATURE FEMALE. *Drab; closely resembles immature female Yellow-rumped. Cape May has green edges to flight feathers; bill thinner, longer, and more curved than Yellow-rumped.* ILLINOIS, SEPTEMBER

Grace's Warbler
Dendroica graciae CODE 1

ADULT MALE. *Similar to Yellow-throated, but smaller and not as boldly black and white. Eyebrow wider and shorter than Yellow-throated, with extensive yellow.* ARIZONA, JUNE

ADULT FEMALE. *Similar to male, but head has little black; immature female browner. Bright yellow throat and broad yellow eyebrow prominent in all plumages.* ARIZONA, JUNE

L 5"   WS 8"   WT 0.35 oz (10 g)

▸ one adult molt per year; complex basic strategy

▸ weak age-related and sex-related differences

▸ extent of yellow variable, especially on adult males

Inhabits sunny, open stands of mature pines in the mountains; characteristic of campgrounds and logging roads. Populations have declined in many regions because of logging and fire suppression, but breeding range is expanding north. Gleans insects from foliage.

Song is dry and colorless, a series of clipped notes, often speeding up toward end: *ch' ch' ch' ch'ch'ch'ch'.* Call a flat *chip.* Flight call a high *seet.*

Magnolia Warbler
Dendroica magnolia CODE 1

BREEDING MALE. *Bright yellow underparts with thick black streaks. Gray and black above; pale wing panel and broad black mask on face prominent. Tail from below broadly tipped in black.* MAINE, JUNE

IMMATURE. *Duller than breeding male; immature female especially drab. Most have eye ring and may suggest Nashville, but note streaking below and distinct black-tipped tail and wing bars.* NEW JERSEY, SEPTEMBER

L 5"   WS 7.5"   WT 0.3 oz (9 g)

▸ two adult molts per year; complex alternate strategy

▸ strong seasonal, age-related, sex-related differences

▸ brightness variable, especially in spring female

Fairly common on migration and breeding grounds; numbers may be increasing. Nests fairly close to the ground in boreal conifer forests, especially where there is dense second-growth vegetation. Forages alone, gleaning insects from broad-leaf and coniferous vegetation.

Song like a soft Hooded Warbler: *weeda weeda weedy!-o*; alternate song just a jumble of *weeda* notes. Call a distinct *clew-enk.* Flight call a buzzy *zeet.*

Yellow-rumped Warbler

Dendroica coronata CODE 1

ADULT MALE, "AUDUBON'S." (D. c. auduboni) *Yellow on throat does not wrap under auriculars, and is not prominent as on "Myrtle." "Audubon's" breeds in the Rockies, winters mainly in Southwest.* CALIFORNIA, APRIL

NONBREEDING, "AUDUBON'S." (D. c. auduboni) *Drab; all have yellow on rump, but some lack yellow on throat and sides. Face is blank compared to "Myrtle."* CALIFORNIA, MARCH

ADULT MALE, "MYRTLE." (D. c. coronata) *Gray and white, with flashes of yellow on crown, sides, and rump. White on throat wraps under black auriculars. "Myrtle" breeds north and east, winters widely.* MAINE, JUNE

NONBREEDING, "MYRTLE." (D. c. coronata) *Pale throat is white or tinged with yellow, wrapping under ear. Auriculars better defined than on "Audubon's."* NEW YORK, OCTOBER

L 5.5"   WS 9.25"   WT 0.45 oz (13 g)

▸ two adult molts per year; complex alternate strategy

▸ strong seasonal, sex-related, age-related differences

▸ "Audubon's" and "Myrtle" groups distinctive

Common throughout its extensive range. Breeds mainly in coniferous forests. Migrants widely noted, although not usually in dense forests. Both groups winter well north, "Myrtle" especially. Hunts insects both by foliage gleaning and flycatching; eats berries in winter.

◀ Song a loose, slow trill; changes pitch or loudness midway through. Call, often given in flight, a hollow *chep*, harder in "Myrtle." Flight call a rising *svit*.

JUVENILE. *Very dark and streaky; yellow absent except on rump. On breeding grounds, juvenile Yellow-rumped can outnumber any other species; rare away from breeding grounds.*

BRITISH COLUMBIA, JUNE

Kirtland's Warbler
Dendroica kirtlandii CODE 3

BREEDING MALE. *A large warbler. Blue-gray above; yellow below with black streaking on sides. Face shows prominent white eye arcs and black around base of bill.* MICHIGAN, MAY

ADULT FEMALE. *Duller than male. Separated from young Magnolia by larger size, larger bill, less black streaking below, low-contrast blue-gray plumage above, and constant tail-pumping.* MICHIGAN, JUNE

L 5.75"	ws 8.75"	wt 0.5 oz (14 g)

▸ two adult molts per year; complex alternate strategy

▸ moderate seasonal, age-related, sex-related differences

▸ plumage contrast on face variable on adult male

Tail-pumping inhabitant of extensive, early-successional jack pine forests in Michigan. Has always been rare; numbers dropped to perilous lows in mid-20th century, but have started to recover. Migrates late in spring, fairly early in fall; rarely seen on migration.

Song variable but usually distinctively rich, low, and liquid; for example, *ch' chip chip churry chip! chip!* Call a rich *chip.* Flight call a short *zzzt.*

Black-throated Green Warbler
Dendroica virens CODE 1

BREEDING MALE. *All are green-backed and yellow-faced, with black streaks below and yellow on vent. Breeding male has extensive black on throat and breast, reduced in other plumages.* CONNECTICUT, MAY

IMMATURE FEMALE. *All show at least a small amount of black on the breast; streaking below often reduced; but always a few streaks along flanks.* NEW JERSEY, OCTOBER

L 5"	ws 7.75"	wt 0.35 oz (10 g)

▸ two adult molts per year; complex alternate strategy

▸ strong age-related and sex-related differences

▸ breeders in southeast coastal plain smaller

Most nest in either coniferous or northern hardwoods forests; "Wayne's" Warbler subspecies (*D. v. waynei*) nests in coastal deciduous or cypress groves. Active: forages throughout day, gleaning insects mainly from upper surfaces of leaves; sings frequently on spring migration.

Buzzy song of 5 or 6 notes: lazy *zoo zeeee zoo zoo zee* or faster *zee zee zee zoo zee.* Call a hard, clipped *tik.* Flight call a short *tsee.*

Townsend's Warbler

Dendroica townsendi   CODE 1

ADULT MALE. *Boldly marked in black and yellow; note broad stripes on face, extensive black on throat, and yellow underparts with black streaking. Back greenish; wing bars prominent.* CALIFORNIA, APRIL

IMMATURE FEMALE. *Head pattern same as adult male, but yellow areas paler and black areas tinged with olive. Yellow reduced on underparts, with just a few streaks on flanks.* CALIFORNIA, AUGUST

L 5"	ws 8"	wt 0.35 oz (10 g)

▸ two adult molts per year; complex alternate strategy
▸ strong age-related and moderate sex-related differences
▸ birds that winter on Pacific coast are shorter-winged

In breeding season often in treetops in moist coniferous forests, especially those with Douglas-fir; also in drier forests. Insectivorous; gleans with equal proficiency from both broadleaf and coniferous vegetation. Fall migration more easterly than spring.

Variable song a fairly short series of wheezy notes: *zweea zweea zweea zwee zwee.* Call like Black-throated Green, but weaker; flight call similar.

Hermit Warbler

Dendroica occidentalis   CODE 1

ADULT MALE. *Face completely yellow with staring black eye. Black on head limited to small amounts on nape, throat, and rear of crown. Underparts whitish, with little if any streaking.* CALIFORNIA, APRIL

IMMATURE FEMALE. *Like breeding male, has much yellow on face. Crown and nape olive, not black; throat white or pale gray. In all plumages, whitish below with prominent wing bars.* CALIFORNIA, AUGUST

HYBRID. *A frequent hybrid combination has much yellow on the face (like Hermit), along with some yellow below (like Townsend's).* CALIFORNIA, AUGUST

L 5"	ws 8"	wt 0.35 oz (10 g)

▸ two adult molts per year; complex alternate strategy
▸ moderate age-related and sex-related differences
▸ frequently hybridizes with Townsend's Warbler

On breeding grounds, the southern counterpart of Townsend's Warbler. Nests in conifer forests, especially those with Douglas-fir; fall migration not as easterly as Townsend's. Preferred foraging zone averages lower in tree than Townsend's, but there is much overlap.

Variable song notoriously similar to Townsend's and Black-throated Gray; most dialects average faster and thinner. Call note and flight call like Townsend's.

Golden-cheeked Warbler

Dendroica chrysoparia CODE 2

BREEDING MALE. *Darker than Black-throated Green; eye line, crown, and back are black. Lacks greenish tones.* TEXAS, APRIL

ADULT FEMALE. *Female and immature are told from Black-throated Green by white vent and more yellow on face.* TEXAS, APRIL

L 5"   WS 7.75"   WT 0.35 oz (10 g)

▸ two adult molts per year; complex alternate strategy

▸ strong sex-related, moderate age-related differences

▸ amount of yellow variable, especially on breeding male

Rare overall and endangered, but locally common; has fared well on lands managed intensively for the species. Requires old-growth or mature second-growth oak-juniper woodlands in rugged terrain. Arrives on breeding grounds by mid-March; most are gone by late July.

Buzzy song usually of five or more elements, with flourish toward end: *bzz bzz bzz bzz bzza-wee! way.* Call a clipped *tik.* Flight call a short *tseet.*

Black-throated Gray Warbler

Dendroica nigrescens CODE 1

ADULT MALE. *Aptly named; gray above with a black throat. Differs from Black-and-white in lacking streaking above and in having only a few streaks below; foraging methods are different.* ARIZONA, APRIL

IMMATURE FEMALE. *Similar to adult male, but black areas of male replaced by gray; black on throat much reduced. All plumages have a yellow spot in front of eye.* CALIFORNIA, AUGUST

L 5"   WS 7.75"   WT 0.35 oz (10 g)

▸ two adult molts per year; complex alternate strategy

▸ moderate age-related and sex-related differences

▸ interior breeders paler, have more white on tail

Closely related to the four species in the "black-throated green" complex. Breeds in dry pine, pinyon-juniper, and pine-oak forests with brushy undergrowth. Feeds by gleaning, lower in trees than Townsend's and Hermit Warblers. Tame; some populations appear to be increasing.

Variable song buzzy, often intensifying toward end: *zwee zwee zwee zwee zweech! zweech!* Call note a dull *thik.* Flight call a short *see.*

Blackburnian Warbler
Dendroica fusca CODE 1

BREEDING MALE. *Blaze orange on face and throat. Black and white otherwise; prominent wing bars coalesce into an irregular white panel.* MAINE, MAY

IMMATURE. *Can be surprisingly similar to immature female Townsend's; Blackburnian always has a thin yellow or orange crown stripe and prominent pale bars on back.* VIRGINIA, JUNE

L 5" WS 8.5" WT 0.35 oz (10 g)

▸ two adult molts per year; complex alternate strategy
▸ moderate seasonal, age-related, sex-related differences
▸ sharpness of facial pattern variable on subadults

Breeds in diverse, usually mature, coniferous forest types: for example, *Usnea*-draped spruce woodlands, southern hemlock forests. Migrants use any woodlands. Forages in treetops, either by gleaning from leaves or by hovering at tip of vegetation and hawking for insects.

Song starts out high-pitched, becomes even higher-pitched: *tweewit tweewit tweewit tweeeee*. Call note a smacking *zick*. Flight call a short buzzy *zid*.

Cerulean Warbler
Dendroica cerulea CODE 1

BREEDING MALE. *Extensive pale blue above; white below with thin dark blue breast band. White wing bars prominent in all plumages.* CONNECTICUT, MAY

BREEDING FEMALE. *All plumages are long-winged and short-tailed. The dullest first-fall females are yellow-gray below with a prominent yellow-gray eyebrow; upperparts dull gray-green.* TEXAS, APRIL

L 4.75" WS 7.75" WT 0.35 oz (10 g)

▸ two adult molts per year; complex alternate strategy
▸ strong sex-related, moderate age-related differences
▸ some adult males are greenish-backed in fall

Forages high in canopy of mature broadleaf forests; methodically gleans insects from foliage. Nests in moist bottomland groves and along rivers. Migrates fairly early in spring, quite early in fall. Despite local range expansions, overall numbers in decline.

Rising, buzzy song suggests Northern Parula and Black-throated Blue Warbler: *zwee zwee zwee zweea zweea zweez*. Call a hard *chet*. Flight call a buzzy *zizt*.

Pine Warbler
Dendroica pinus   CODE 1

ADULT MALE. *Bright yellow throat; back and breast olive-yellow. Belly and undertail coverts white; white wing bars prominent. Superficially resembles Yellow-throated Vireo.* FLORIDA, FEBRUARY

IMMATURE FEMALE. *On very dull birds, as here, distinctive body structure is essential for identification: tail long, wings short, bill relatively thick.* NORTH CAROLINA, SEPTEMBER

L 5.5"   WS 8.75"   WT 0.4 oz (11 g)

- one adult molt per year; complex basic strategy
- moderate sex-related and age-related variation
- overall brightness variable in all plumages

Strongly affiliated with pines; even far-flung vagrants somehow manage to find a pine tree. A hardy species that winters well north, eating seeds and berries; consorts with flocks of Eastern Bluebirds and Chipping Sparrows. Migrates early in spring, late in fall.

◀ Song, heard year-round, a steady trill on one pitch; increases, then sometimes decreases in volume. Call a dull *yip*. Flight call a short, piercing *tseet*.

Bay-breasted Warbler
Dendroica castanea   CODE 1

BREEDING MALE. *Chestnut ("bay") on crown, throat, and flanks. Mainly black face bordered behind by large creamy patch.* TEXAS, APRIL

BREEDING FEMALE. *Toward end of first plumage cycle, begins to show pattern like male, but duller; older females in breeding plumage show much chestnut on sides; younger females, as here, do not.* TEXAS, MAY

L 5.5"   WS 9"   WT 0.45 oz (13 g)

- two adult molts per year; complex alternate strategy
- strong seasonal, age-related, sex-related differences
- extent of chestnut on adult male in fall variable

Breeds in spruce-fir forests, especially in districts infested with spruce budworm. Both migrants and breeders tend to forage in canopy, near center of tree in dense foliage. Migrates fairly late in spring; fall migration fairly early, generally preceding Blackpoll.

High-pitched song like Cape May Warbler, but notes slur downward: *s'weep s'weep s'weep s'weep*. Call note a rich *chip*; flight call a flat *seet*.

Blackpoll Warbler
Dendroica striata CODE 1

BREEDING MALE. *All plumages long-winged and streaky, with white wing bars; most have yellow-orange feet, white undertail coverts. Adult male has black crown, white cheek, and black malar.* TEXAS, APRIL

NONBREEDING. *Similar to Bay-breasted, but washed in pale greenish yellow, with faint streaking below. White wing bars prominent; undertail coverts usually white. Wings always long. Note yellowish feet.* NEW JERSEY, SEPTEMBER

BREEDING FEMALE. *Common but infrequently seen, as females tend to forage silently in dense treetop foliage; note streakiness, wing bars, yellow-orange feet, and white undertail coverts.* ILLINOIS, MAY

L 5.5"   WS 9"   WT 0.5 oz (14 g)

▸ two adult molts per year; complex alternate strategy
▸ strong seasonal and sex-related differences
▸ foot color variable; some have nearly black feet

Much of breeding range in spruce-dominated taiga, even taiga-tundra transition zone; at southern range limit (New England), breeds in stunted mountaintop forests. Migrates very late in spring, fairly late in fall. Substantial numbers migrate over Atlantic Ocean in fall.

High-pitched song a simple *si si si si see see si,* rising in loudness, then quickly trailing off. Call note like Bay-breasted; flight call also similar, but higher.

LONG-DISTANCE MIGRATION

The Blackpoll Warbler is one of the icons of long-distance migration. Weighing less than half an ounce, it routinely completes an immense biannual migration from its boreal forest breeding grounds to its South American wintering grounds and back again. The Blackpoll has been the subject of several classic investigations into the physiology of migration, including pioneering studies on celestial and magnetic navigation, and the metabolic requirements for long-distance flight.

Like so many long-distance migrants, Blackpoll Warblers are critically dependent upon high-quality stopover sites—especially groves and tangles that are not treated with insecticides. An acute crisis for long-distance migrants that travel by night involves the ever-proliferating array of wind turbines and cell phone towers; these obstacles to migration are largely invisible to nocturnal migrants, and huge numbers of warblers and other nocturnal migrants are killed by striking these tall structures.

Black-throated Blue Warbler

Dendroica caerulescens CODE 1

ADULT MALE. *Dark blue above, with bold white patch in wings. Face and flanks black; belly and most of breast white. Less seasonal variation than in other* Dendroica *warblers.* TEXAS, APRIL

FEMALE. *Olive-brown above; olive-yellow below. Dark face bordered by curved eyebrow and pale malar. Wing patch prominent on adult, reduced or absent on immature female (here).* NEW JERSEY, SEPTEMBER

L 5.25"   WS 7.75"   WT 0.4 oz (11 g)

▸ one adult molt per year; complex basic strategy

▸ strong differences between male and female

▸ southern males are darker above, especially on crown

Breeds in shady interior of maple/birch/beech forests, often with hemlock. Migrants use all woodland types; bulk of migration east of Ohio River valley. Forages actively, mainly in shrub and midstory layers, focusing efforts on undersides of broadleaf foliage.

Song decidedly raspy: *zeeer zeeer zeeer zeeeee*, last note higher and rising. Call note a smacking *tsick*, like Dark-eyed Junco; flight call a rising *sit*.

Black-and-white Warbler

Mniotilta varia CODE 1

ADULT MALE. *Coarsely streaked black-and-white all over; compare with adult male Blackpoll and Black-throated Gray. Long bill reflects habit of poking for insects in furrows of trunks.* TEXAS, MAY

IMMATURE FEMALE. *Head whiter than adult male: auriculars reduced and dusky (black in male); throat white or creamy (black in male). Adult female slightly darker overall.* NEW JERSEY, SEPTEMBER

L 5.25"   WS 8.25"   WT 0.4 oz (11 g)

▸ two adult molts per year; complex alternate strategy

▸ moderate sex-related and age-related differences

▸ extent of buff tones, especially on female, variable

Creeps jerkily, nuthatch-like, on trunks and large boughs; also gleans insects from leaves. Returns fairly early in spring; favors lush broadleaf woods, especially extensive tracts, for nesting. In winter, accepts all forest types, including mangroves and pinewoods.

High-pitched song a lazy series of two-syllable notes: *wee-see wee-see wee-see wee-see wee-see...* Call note a smacking *tseck*; flight call a thin, buzzy *zzzeee*.

American Redstart

Setophaga ruticilla   CODE 1

L 5.25"   WS 7.75"   WT 0.4 oz (11 g)

▸ two adult molts per year; complex alternate strategy

▸ strong age-related and sex-related differences

▸ extent of color variable, especially in subadults

Coquettishly spreads tail and droops wings; flashes colorful patches to flush prey. Foraging motions are animated, exaggerated: lunges after insects and somersaults through air. Breeds in various forest types; favors sunny edges and clearings. Common but declining.

◀ Last note of variable song often clipped, abrupt: *wees weesa weet!* or *seeta seeta seeta seet!* Call note a clear *chip*; flight call a sharp, clear *swee*.

ADULT MALE. *Mainly black and orange; belly white. Orange in wings and tail often displayed by perched birds.* TEXAS, MAY

IMMATURE MALE. *Plumage in spring and summer after year of hatching (first alternate plumage) is female-like, except for black splotching on face and usually orangish rather than yellowish patch on sides of breast.* TEXAS, MAY

ADULT FEMALE. *Orange of adult male is replaced by yellow. Head gray, upperparts mainly olive, underparts pale dusky. Immature female similar but duller; flashes yellow in wings and tail.* TEXAS, APRIL

Worm-eating Warbler

Helmitheros vermivorum   CODE 1

L 5.25"   WS 8.5"   WT 0.45 oz (13 g)

▸ one adult molt per year; complex basic strategy

▸ weak differences between spring and fall adults

▸ breast varies from rich cream to buff yellow

Feeding behavior distinctive; pokes into clusters of dead or dying leaves to extract spiders and insects. Breeds in extensive broadleaf forests, sometimes with extensive hemlock, often in hilly country. Uncommon; seen only in small numbers on migration.

Song a long, dry, steady trill; drier than Pine Warbler or even Chipping Sparrow. Call a clear, dry *chep*; flight call a short, level *szit*, often doubled.

ADULT. *Buff-olive all over, with warmer flush across breast. Head striped olive and black; bill long, spiky, and usually pale. Spring adult a little paler overall than fall.* TEXAS, APRIL

Prothonotary Warbler
Protonotaria citrea   CODE 1

ADULT MALE. *Head and breast entirely yellow; immature male duller. Black eye prominent on plain face. Wings and tail unmarked blue-gray. Tail distinctively short; bill distinctively large.* TEXAS, MAY

ADULT FEMALE. *Yellow areas duskier than on male. All plumages are large-billed and bobtailed, with blue-gray wings and tail. Bill conspicuously black in spring; paler in fall.* TEXAS, APRIL

L 5.5"   WS 8.75"   WT 0.6 oz (17 g)

▸ one adult molt per year; complex basic strategy

▸ moderate age-related and sex-related differences

▸ extent of yellow on belly and vent variable

Nests in tree cavities in swamps and other wet woodlands; prefers standing or slowly moving water. Migrants seek wooded waterways if present. Slowly gleans insects from foliage at shrub layer and midstory. Spring migration early; fall migration early.

Song a loose trill, like Swamp Sparrow: *sweet sweet sweet sweet...* Call note a soft *squip*; flight call rising and sweet like American Redstart, but more buzzy.

Swainson's Warbler
Limnothlypis swainsonii   CODE 2

ADULT. *Bill is pale, long, and spiky. Olive-brown all over, paler below than above. A very plain warbler, but long and broad eyebrow is prominent. Immature slightly more buff.* CARIBBEAN, MARCH

ADULT. *Some have a cold and neutral ground color, as here. All are short-tailed and pink-legged; eyebrow and long bill always conspicuous.* TEXAS, APRIL

L 5.5"   WS 9"   WT 0.7 oz (20 g)

▸ one adult molt per year; complex basic strategy

▸ weak differences between immature and adult

▸ individual and geographic variation in ground color

Secretive ground-dweller; forages in leaf litter and on fallen logs. Two breeding populations: one in dense swamps with extensive cane; the other in lush hemlock/rhododendron forests in southern Appalachians. Arrives fairly early in spring; fall migration protracted.

Striking song consists of rich, slurred notes: *ee dee yay ewee weet!* Call a rich *chap*, often given persistently. Flight call a ringing *teeeth*.

Northern Waterthrush

Seiurus noveboracensis   CODE 1

ADULT. *Most have extensive dull yellowish wash below; streaking below dense, usually extending onto throat. Long eyebrow usually narrows behind eye. Juvenile weakly spotted above.* ARIZONA, SEPTEMBER

ADULT. *Some are nearly as white below as Louisiana, but throat is spotted. Differences in song, call, and tail-pumping behavior are often important in identification.* NEW HAMPSHIRE, MAY

L 6"	ws 9.5"	wt 0.65 oz (18 g)

▸ one adult molt per year; complex basic strategy

▸ weak differences between immature and adult

▸ brightness and extent of yellow on underparts variable

Associated with calm water at all times of year; Louisiana Waterthrush prefers rushing streams. Pokes around in leaf litter or at water's edge. Pumps tail quickly; Louisiana does so more slowly. Breeds in bogs in northern and montane forests; widely noted on migration.

◀ Descending song usually with three distinct segments: *s'chee s'chee chew chew choop choop*. Call note a sharp *cheenk*; flight call a piercing, buzzy *zzzhip*.

Louisiana Waterthrush

Seiurus motacilla   CODE 1

ADULT. *Most are white below with diffuse brown streaks; throat usually devoid of streaks. Broad white eyebrow does not narrow appreciably behind eye. Legs often bright red.* CONNECTICUT, APRIL

ADULT. *Some are buff below; heaviest concentration on flanks; long eyebrow, extending without contraction behind eye, usually a good mark. Juvenile faintly spotted above, like Northern Waterthrush.* KENTUCKY, JUNE

L 6"	ws 10"	wt 0.7 oz (20 g)

▸ one adult molt per year; complex basic strategy

▸ weak differences between immature and adult

▸ some show streaking or spotting on throat

Closely tied to rushing streams; used as bioindicator of water quality. Even on migration, rarely found away from running water. Walks on ground, slowly pumping tail. Returns earlier in spring than any other neotropical migrant landbird. Fall migration very early.

Song ringing and jumbled, with downslurred elements: *s'wee s'wee s'deea weet weep*. Call note *check*, flatter than Northern; flight call *zweest*, less buzzy.

Common Yellowthroat
Geothlypis trichas   CODE 1

ADULT MALE. *Variable, but all adult males have extensive black mask, bordered below by bright yellow; upper border of mask variably whitish, yellowish, or bluish.* TEXAS, APRIL

ADULT FEMALE. *Yellow on throat ends abruptly at edge of face. All have long, sturdy legs; all tend to flip tail expressively, like a wren.* TEXAS, APRIL

L 5"   WS 6.75"   WT 0.35 oz (10 g)

▸ two adult molts per year; complex alternate strategy

▸ moderate age-related and strong sex-related differences

▸ much geographic variation in extent of yellow

Lurks in dense, brushy vegetation. Most western populations associated with water; eastern populations also in dry brushy fields and woodland edges. Flies in immediately to pishing. Migratory in most of North American range. Hardy; sometimes lingers into early winter.

◀ Song a series of scratchy, identical phrases: *witch-i-ty witch-i-ty witch-i-ty...* Call note a harsh *chack*; flight call a buzzy *zzzt*, like a spark.

IMMATURE MALE. *Yellow confined to smaller patch on throat than adult; black mask reduced in extent, mottled overall. In all plumages, note short brown wings and long, broad tail.* NEW JERSEY, SEPTEMBER

GRAY-CROWNED YELLOWTHROAT (*G. poliocephala*). *Casual stray from Mexico; some hybridize with Common. Larger and larger-billed than Common; all plumages have pale eye crescents.* TEXAS, MARCH

ADULT MALE. *Some are almost all-yellow below; variation in yellow related to diet, possibly to age, and especially to feather wear and geography.* NEW JERSEY, MAY

Ovenbird
Seiurus aurocapilla CODE 1

ADULT. *Portly and bobtailed. All are white below with prominent black streaks. Orange crown has black borders; most show white eye ring.* ILLINOIS, APRIL

IMMATURE. *Patterned like adult, but duller: upperparts lack greenish tones of adult; crown not as colorful.* VIRGINIA, SEPTEMBER

L 6" WS 9.5" WT 0.7 OZ (20 g)

▸ one adult molt per year; complex basic strategy
▸ weak differences between immature and adult
▸ back color ranges from greenish olive to nearly gray

Struts on leaf litter of mature broadleaf forests; sometimes walks along fallen logs or horizontal boughs. Numerous, but nest failure frequent in forest fragments and near edges. Feeds deliberately; turns over fallen leaves or plucks arthropods from soil surface.

Heard far more often than seen. Song a far-carrying *erteach erteach erteach...* or *teach teach teach...* Call note a dry *chick*; flight call a sharp *seet*.

MacGillivray's Warbler
Oporornis tolmiei CODE 1

ADULT MALE. *Colorful; greenish above and yellow below, with a blue-gray hood. White eye arcs prominent; black extensive in front of eye, usually reduced at base of hood.* CALIFORNIA, APRIL

IMMATURE FEMALE. *Hood dingy gray, including throat and breast; eye arcs prominent. Mourning shorter-tailed and longer-winged. Orange-crowned in Intermountain West can be similar.* CALIFORNIA, SEPTEMBER

L 5.25" WS 7.5" WT 0.4 OZ (11 g)

▸ two adult molts per year; complex alternate strategy
▸ strong sex-related and moderate age-related differences
▸ northern birds paler above and brighter below

Widespread and common in forested landscapes, especially in mountains. Flourishes in disturbed and early-successional habitats; may benefit from logging. A skulker, but readily responds to pishing. Commonly encountered on migration, unlike other *Oporornis* warblers.

Short, burry song dies off at end; for example, *burry burry burry bore*. Call note a dry *chat* like Common Yellowthroat; flight call a scratchy, rising *zzit*.

Mourning Warbler

Oporornis philadelphia CODE 1

ADULT MALE. *Similar to MacGillivray's, but lacks eye crescents; hood has more black at base, less black in front of eye. Female has paler hood, with no black at base or in front of eye.* TEXAS, MAY

ADULT FEMALE. *Plainer than male; hood does not contrast strongly with rest of body. Immature female duller, often with thin eye ring; hood indistinct, with yellow extending onto throat.* TEXAS, MAY

L 5.25"	WS 7.5"	WT 0.45 oz (13 g)

- two adult molts per year; complex alternate strategy
- strong sex-related and moderate age-related differences
- some show hints of white eye arcs and dark lores

Best known as a hard-to-find spring migrant through the East and Midwest; skulks in dense, low vegetation. Spring passage late. Breeds in disturbed areas: regenerating forests and edges of bogs or streams. Feeds on the ground or on low branches in undergrowth.

Two-part song drops at end: *churry churry choory choory* or *churr churr churr choory*. Call note a dry *chit*; flight call a rising *swit*.

Connecticut Warbler

Oporornis agilis CODE 2

ADULT MALE. *Large and long-legged. Blue-gray hood has prominent white eye ring; hood paler in female. In both sexes, hood paler than corresponding plumage of Mourning.* MINNESOTA, JUNE

IMMATURE FEMALE. *Hood brownish, poorly defined. Eye ring prominent; immature Mourning can show a thin white eye ring. Always appears large, long-legged, long-winged, short-tailed.* NEW JERSEY, OCTOBER

L 5.75"	WS 9"	WT 0.5 oz (14 g)

- two adult molts per year; complex alternate strategy
- moderate age-related and sex-related differences
- some males have paler hoods with olive flecking

A warbler much sought after; races north during spring migration in late May to nest in spruce-tamarack bogs and moist broadleaf woods. Migrants skulk in poorly drained weedy places. Spring migration up Ohio River drainage; fall migration more easterly.

Song rich and rollicking: *chipaweea chuppa cheechuppa*, with singsong quality. Call note a dull *fwit*; flight call buzzy like Yellow Warbler, but sharper.

Kentucky Warbler
Oporornis formosus CODE 1

ADULT MALE. Olive-green above; extensively yellow below. Face pattern vaguely suggests Common Yellowthroat, but Kentucky has reduced black mask and bold yellow spectacles. NEW JERSEY, SEPTEMBER

ADULT FEMALE. Face pattern similar to male, but black mask not as extensive. Immature female has limited black on face, but yellow spectacles prominent in all plumages. TEXAS, MAY

L 5.25"	WS 8.5"	WT 0.5 oz (14 g)

- two adult molts per year; complex alternate strategy
- moderate age-related and sex-related differences
- amount of black on head of adult female variable

Forages near ground in moist broadleaf forests; scratches with feet for arthropods and pokes bill into leaf litter. Favors forests with rich understory such as spicebush and sassafras. Absent from forests where understory is disturbed, for example by deer browsing.

Song a ringing, musical *chorry chorry chorry chorry chorry*; steady, does not trail off at end. Call note a wooden *thuck*; flight call a burry *zeek*.

Wilson's Warbler
Wilsonia pusilla CODE 1

ADULT MALE. *Dusky green above; bright yellow below, including face. Western populations average brighter, more orangish than Eastern. Black cap prominent on many males.* CALIFORNIA, APRIL

IMMATURE FEMALE. *Black cap of adult male replaced by dull olive; broad yellow eyebrow usually prominent. Wilson's is small and active; tail is relatively long, and often flipped about.* TEXAS, NOVEMBER

ADULT. *Many females show extensive black on cap, sometimes as much on this individual, which is probably not safely assignable to a sex.* CALIFORNIA, APRIL

L 4.75"	WS 7"	WT 0.25 oz (7 g)

- two adult molts per year; complex alternate strategy
- moderate age-related and sex-related differences
- extent of black cap on female highly variable

With tail cocked, forages in low vegetation. Abundant migrant through much of West, less so in East. Breeds in wet brushlands in mountains and along Pacific coast; favors beaver ponds, mountain meadows, and recovering clear-cuts. Migrants use riparian habitat.

Song a slow trill, often dropping at end: *chi chi chi chi chi chet chet*. Call a rising *wenk?* Flight call an abrupt, scratchy *chick*.

Hooded Warbler

Wilsonia citrina   CODE 1

L 5.25"   WS 7"   WT 0.4 oz (11 g)

▸ one adult molt per year; complex basic strategy

▸ strong sex-related, moderate age-related differences

▸ extent of cowl variable, especially on adult female

Feeds close to ground, employing diverse tactics: hover-gleaning, leaf-gleaning, and aerial hawking. Breeders require large forest tracts with mature trees: maple/beech/ oak forests in northern part of range, cypress/gum forests in south. Most populations declining.

Song clear and ringing, like an amplified Magnolia Warbler: *weechy weechy weechy-o!* Call note a rich and penetrating *chep*; flight call a short *chint*.

ADULT MALE. *Olive-green above; yellow below. Bright yellow face encircled in broad black cowl.* OHIO, APRIL

ADULT FEMALE. *Cowl never as extensive as adult male. Some females have splotchy black on throat and breast; others have little if any.* FLORIDA, APRIL

IMMATURE FEMALE. *All plumages have extensive white in long, broad tail; tail often spread, exposing white. Drabbest individuals, usually immature females, have cowl greatly reduced.* TEXAS, APRIL

Red-faced Warbler

Cardellina rubrifrons   CODE 2

ADULT MALE. *Medium gray above; pale gray below. Face of male is black and intense, brilliant red; female and immature average more red-orange, also of a very intense hue.* ARIZONA, MAY

L 5.5"   WS 8.5"   WT 0.35 oz (10 g)

▸ one adult molt per year; complex basic strategy

▸ moderate age-related and sex-related differences

▸ individual variation in red on face; not well studied

Habits and ecology suggestive of genus *Wilsonia*, to which Red-faced may be closely related. Feeds by flycatching and foliage-gleaning. Migrates to mountain conifer forests, where it nests in a small depression in the ground. Locally common, especially in New Mexico.

Song a bright, singsong series: *chippy weechy cheewa chee cheewa*; recalls Canada but sweeter, more languid. Call a harsh *chet*.

Canada Warbler
Wilsonia canadensis CODE 1

L 5.25"   WS 8"   WT 0.35 oz (10 g)

▸ two adult molts per year; complex alternate strategy
▸ moderate age-related and sex-related differences
▸ individual variation in extent of black on face, breast

Active and expressive: cocks tail and flicks wings as it hunts arthropods in understory and midstory; also forages on ground and by flycatching. Breeds in cool, moist forests with rhododendron and other shrubs; migrants pause in dense tangles of grape and honeysuckle.

Staccato song recalls Connecticut Warbler, but not as low and rich: *chip chippitty chip chippiwee.* Call note similar to Hooded; flight call a piercing *chit.*

BREEDING MALE. *Blue-gray above; bright yellow below. Breeding male has necklace of thick black streaks; thin white eye ring and yellow line in front of eye stand out on dark face.* CONNECTICUT, MAY

ADULT FEMALE. *Resembles male, but with less black on face and breast. Like other* Wilsonia *warblers, amount of black on adult female is variable.* FLORIDA, APRIL

IMMATURE FEMALE. *All individuals are long-tailed and solidly blue-gray above. The dullest show at least a trace of a necklace, and all have a prominent white eye ring.* NEW JERSEY, SEPTEMBER

Painted Redstart
Myioborus pictus CODE 2

L 5.75"   WS 8.75"   WT 0.35 oz (10 g)

▸ one adult molt per year; complex basic strategy
▸ strong differences between adult and juvenile
▸ individual variation in amount of white in wings, tail

Occurs near streams in pine-oak woods. Behavior recalls American Redstart, to which Painted is not closely related; both species flash wings and tail to startle prey. Feeding techniques varied: gleans foliage, hovers in front of foliage, even catches insects over water.

Song a disorganized, wiry jumble: *t'wee twee dweet twee dyee dyee d'weet.* Call note a rich, downslurred *syeer,* unlike any other North American wood-warbler.

ADULT. *Black above and red below, with extensive white in wings and tail. Juvenile lacks red below; underparts entirely black.* ARIZONA, MAY

Yellow-breasted Chat

Icteria virens CODE 1

ADULT MALE, EASTERN. *Large; the size of a* Catharus *thrush. Olive-brown above; bright yellow on throat and breast. All plumages long-tailed and large-billed, with white spectacles.* ONTARIO, MAY

ADULT FEMALE, WESTERN. *More orange-yellow below than Eastern, with slightly longer tail. Female has gray lower mandible; bill is all-black in male.* ARIZONA, SEPTEMBER

L 7.5"	WS 9.75"	WT 0.9 oz (25 g)

- one adult molt per year; complex basic strategy
- weak differences between adult male and female
- western birds longer-tailed, more orange-yellow below

Seems to delight in evading detection in dense tangles; forages on ground and in low vegetation. Eastern birds inhabit abandoned farmland, coastal thickets, and powerline rights-of-way. Western birds are generally associated with riparian and marshland habitats.

◀ Diverse vocal array unlike any other wood-warbler. Song endless and clownish: *gleerp chet chet peee! chegga... cheer cheer cheer... whoit! kook? gleerp...* Call note a growling *zheeemp.*

CASUAL STRAYS FROM MEXICO

These warbler species are currently classified by the American Birding Association as casual (Code 4)—not annual, but of somewhat regular occurrence in the ABA Area.

GOLDEN-CROWNED WARBLER (Basileuterus culicivorus). *At least 10 have wandered to south Texas. Olive-gray above; yellow below. Namesake golden crown broadly bordered in black.* CENTRAL AMERICA, FEBRUARY

RUFOUS-CAPPED WARBLER (Basileuterus rufifrons). *At least 25 have wandered to south Texas and Southeast Arizona. Head boldly patterned; tail long and flipped about, wren-like.* CENTRAL AMERICA, FEBRUARY

SLATE-THROATED REDSTART (Myioborus miniatus). *At least 10 have wandered to south Texas and Southeast Arizona. More extensively red below than Painted Redstart; lacks white wing panels.* MEXICO, DECEMBER

Tanagers
FAMILY THRAUPIDAE

6 species: 3 ABA CODE 1   1 CODE 2   2 CODE 4

The four tanager species that regularly reach North America are the northernmost representatives of a large tropical assemblage (more than 200 species) of wildly colorful forest birds. Our four species are closely related, and are placed in the same genus (*Piranga*). Their relationship to other tanagers, however, is not at all clear. Indeed, the tanager assemblage is likely "polyphyletic," meaning that it consists of groups that are more closely related to other families than they are to each other. Ongoing molecular studies indicate that some tanagers are probably best classified with cardinals and allies (Cardinalidae), that others belong with the sparrows (Emberizidae), and that still others are most closely related to the finches (Fringillidae). A major revision of the tanager family is foreseen by many avian taxonomists.

The four regularly occurring North American tanager species are migratory, arboreal, and largely clad in reds and yellows. All species exhibit a fair amount of variation in color intensity, the result of both parasites (for example, mites that degrade feather quality) and diet (red, orange, and yellow pigments originate in the birds' food). Tanagers rely heavily on insects during the nesting season, but all four species readily flock to fruiting trees if available. Their bills are stout and bulbous, recalling the bill morphology of various primarily tropical groups that specialize in fruit consumption.

Despite their glorious colors, tanagers can be frustratingly hard to see as they forage slowly in tall treetops. Their songs and call notes carry far, however, and are often the first clue to their presence. All four species frequently call in flight, too, and the flight call of the Scarlet Tanager is one of the characteristic sounds of nocturnal migration in the East, especially in early autumn. Tanagers seem especially susceptible to weather-related groundings during spring migration, with large numbers of starving individuals foraging on the ground during rainy or snowy weather.

The welfare of tanager populations in North America is tied to forest health. Declines in Scarlet Tanager populations are thought to be due in part to fragmentation of the eastern deciduous forest, and losses in Summer Tanager populations in the West are a result of destruction of the riparian habitats on which they depend. Numbers of the two tanager species that breed in western pine forests—Hepatic and Western—are relatively stable.

Scarlet Tanager
Piranga olivacea CODE 1

BREEDING MALE. *All individuals have pale bill with smoothly decurved upper mandible. Plumage of breeding male entirely bright red, except for black wings and tail.* TEXAS, APRIL

IMMATURE MALE, FIRST SPRING. *Flight feathers of wing have brown or olive tint (wings completely black on adult male). Note single green scapular retained from first-fall plumage.* TEXAS, MAY

IMMATURE MALE, FIRST FALL. *All Scarlet Tanagers in fall are greenish. Male has mainly black wings, tinged brown or olive in immature plumage; wings of female paler.* KENTUCKY, OCTOBER

ADULT FEMALE. *Same general pattern as adult male, but colors different: wings and tail brownish black; body feathers olive-green. Similar to Western, but back olive and wing bars absent.* TEXAS, APRIL

L 7"	WS 11.5"	WT 1 oz (28 g)

- ▸ two adult molts per year; complex alternate strategy
- ▸ strong age-related, sex-related, seasonal differences
- ▸ males vary in brightness, related to diet and parasites

Breeders favor extensive broadleaf tracts with continuous canopy. During nesting season, sticks to treetops, where it forages slowly or sits motionless. Migrants are more likely to come down into lower vegetation, such as fruiting shrubs.

◀ Song is loud and burry, consisting of rising and falling phrases. Call note is a far-carrying *chip-burrrr*; flight call is a soft, slurred *tuwee*.

ADULT FEMALE. *Drab female and immature tanagers can be confused with orioles, but tanagers are proportionately shorter-tailed, with shorter and more-bulbous bills.* TEXAS, MAY

Western Tanager
Piranga ludoviciana CODE 1

BREEDING MALE. *The only bird in North America with a cherry-red head and striking black-and-yellow body. Nonbreeding male and some breeding males have diminished red on head.* CALIFORNIA, MAY

ADULT FEMALE. *Extensive mustard-yellow on head and underparts. Good points of separation from Scarlet include gray back and white wing bars.* CALIFORNIA, MAY

IMMATURE MALE. *Shows variable red on head by first spring (first alternate plumage), but juvenile wing feathers retained.* CALIFORNIA, MAY

L 7.25"   WS 11.5"   WT 1 oz (28 g)

▸ two adult molts per year; complex alternate strategy

▸ strong age-related, sex-related, seasonal differences

▸ slightly larger in eastern portion of range

Western counterpart of Scarlet Tanager. On breeding grounds, favors pine forests in mountains, but some range into lowland broadleaf groves. Migrants widely noted, even in deserts and grasslands. Impressive groundings occur during cold, wet weather in spring migration.

◀ Song similar to Scarlet, but not as hoarse. Call distinct from Scarlet: a rolling, rising *piddywit*. Flight call more slurred, less disyllabic than Scarlet.

ADULT FEMALE. *Some females show limited yellow, and resemble drab female Bullock's Oriole; latter has pointy bill, shorter wings, and longer tail with extensive yellow.* CALIFORNIA, JUNE

FLAME-COLORED TANAGER (*P. bidentata*). *Casual to Arizona; many records involve hybrids with Western. Both sexes have dark auriculars and white wing bars; male flame-orange, female yellow.* ARIZONA, APRIL

Summer Tanager
Piranga rubra   CODE 1

ADULT MALE, EASTERN. *Uniform rosy-red all over, including wings and tail, with little plumage contrast. Some are paler and more yellowy than shown here, especially in Desert Southwest.* TEXAS, APRIL

ADULT FEMALE. *Uniformly olive-yellow. Distinguished from female Hepatic by pale bill (dark in Hepatic) and low-contrast plumage (Hepatic has prominent auriculars).* TEXAS, APRIL

L 7.75"	ws 12"	wt 1 oz (28 g)

▸ one adult molt per year; complex basic strategy

▸ strong age-related, sex-related, seasonal differences

▸ western birds larger-billed and paler than eastern

East of Texas, breeds in various forest types, usually with broadleaf component; sticks to forest canopy during nesting season. West of Texas, prefers riparian vegetation, especially where extensive. Eastern birds are prone to vagrancy; western birds less so.

Song similar to Scarlet, but longer and more musical. Distinctive call note is an explosive *picky tucky tuck*; flight call is a harsh *toovee*.

IMMATURE MALE, FIRST SPRING. *Coarsely patterned rose-and-yellowish plumage retained until early August; variable. Bill appears large in all plumages, especially in Desert Southwest.* TEXAS, APRIL

Hepatic Tanager
Piranga flava   CODE 2

L 8"	ws 12.5"	wt 1.3 oz

▸ one adult molt per year; complex basic strategy

▸ strong differences between adult male and female

▸ intensity and extent of red and yellow variable

In North America, closely tied to mountain pine forests in the Desert Southwest. May range into arid oak woods, but rarely into the low-elevation riparian forests favored by Summer Tanagers. Northern range limit poorly known; possibly fluctuates.

Song sweeter than other tanagers; suggests Black-headed Grosbeak. Call note a one-syllable *tuck*, unlike other tanagers. Flight call also a single syllable; harsher than Western.

ADULT MALE. *Brick red overall; color not as vibrant as Summer Tanager. Gray auriculars stand out; most have grayish hue to flanks, wings, and back.* ARIZONA, APRIL

ADULT FEMALE. *Brick-red of male replaced with yellow-orange on female, duller overall and less extensive. Grayish auriculars and large, dark bill evident on all individuals.* ARIZONA, MAY

Sparrows & Allies
FAMILY EMBERIZIDAE

60 species: 37 ABA CODE 1   10 CODE 2   3 CODE 3   3 CODE 4   7 CODE 5

These birds, sometimes referred to as "LBJS"—little brown jobs—are small, mainly brown, and often furtive. The emberizids are a primarily American group, with good representation from the arctic to the southern tip of South America. Most of the North American species are called sparrows, whereas the handful of species that make it to the Old World are called buntings. And to confuse matters further, North American buntings belong to the family Cardinalidae, whereas the Old World sparrows are assigned to the family Passeridae. These complexities notwithstanding, the bulk of the species in the family Emberizidae are closely related. Boundaries among several genera of New World sparrows are fuzzy, and there have been several instances of intergeneric hybrids in the family Emberizidae.

Sparrows have bills that are conical and fairly stout—an immediate tip-off that they are seed eaters. Not surprisingly, given their diet, many species occur in weedy fields and amid flowering forbs. Several are familiar visitors at feeding stations provisioned with millet and sorghum. Most are intermediate-distance migrants, and a few are essentially nonmigratory; since they do not have to forage for insects in the winter, there is relatively little pressure for them to migrate long distances in autumn to the insect-rich tropics.

As if to make up for their drabness, most sparrows produce remarkable vocalizations. The songs of many are simply beautiful: Cassin's, American Tree, Fox, Vesper, and White-throated are among the species that come to mind. Other species sing songs that are not usually characterized as beautiful, but that are nonetheless bewitching and evocative: The white-noise buzzing of Clay-colored, the odd hiccupping of Henslow's, and the hissing of Seaside are exemplary.

Geographic variation is extensive in the sparrow clan, and several currently recognized species may represent complexes of two or more species. Examples include the Sage, Seaside, Fox, Savannah, and White-crowned Sparrows. Paradoxically, the wildly variable Song Sparrow, with several dozen named subspecies, is not a candidate for a future split; the many populations freely interbreed with populations from adjacent regions, and thus do not behave as separate species.

Conservation priorities for the family Emberizidae focus heavily on those species that depend on grasslands and shrublands. Species in serious decline include Brewer's, Baird's, and Henslow's Sparrows, along with the Chestnut-collared and McCown's Longspurs. Also of concern are distinctive subspecies, among them Sage Sparrows in coastal California and Grasshopper and Seaside Sparrows in Florida.

White-collared Seedeater
Sporophila torqueola   CODE 3

ADULT MALE. *All are small and compact, with short wings and rounded tail; all have a thick, short, rounded bill. Male has black-and-white wings and variable black on face.* BELIZE, NOVEMBER

ADULT FEMALE. *Dull buff above, brighter buff below; wing bars not as prominent as male. Some individuals are quite drab, but are recognized by their body shape and small size overall.* MEXICO, JANUARY

L 4.5"   WS 6.25"   WT 0.35 oz (10 g)

▸ two adult molts per year; complex alternate strategy
▸ strong differences between adult male and female
▸ darkness of cap and paleness of rump variable on male

Small flocks pluck seeds from tall weeds and grasses in brushy fields, often near water. Formerly common year-round in Rio Grande valley; now occurs in just a few small flocks, mainly in summer. Habitat destruction and overuse of agrochemicals may have caused decline.

Song consists of a rich whistle or whistles followed by a trill that ends abruptly: *tweeee d'd'd'd'd'd'd'd' weech!you.* Call notes: descending *choo* and rising *chwit.*

Olive Sparrow
Arremonops rufivirgatus   CODE 2

ADULT. *Greenish above; gray below. Head gray with thin rusty stripes and thin white eye ring. Juvenile duller; olive-buff overall, with diffuse streaks above and below.* TEXAS, MAY

ADULT. *Structure, coloration, and behavior recall Green-tailed Towhee; adult Green-tailed distinctive, but dully marked immature resembles Olive Sparrow.* TEXAS, NOVEMBER

L 6.25"   WS 7.75"   WT 0.85 oz (24 g)

▸ one adult molt per year; complex basic strategy
▸ strong differences between juvenile and adult
▸ amount of plumage contrast on adult variable

Spends much time in dense cover, where it plucks food from open patches of bare ground and from low vegetation. Scratches with feet in leaf litter, like towhees. Range may be expanding north into grazing-induced chaparral that was formerly coastal prairie.

Heard much more often than seen. Song a distinctive "bouncing ball" series, slower and drier than Wrentit. Call note a sharp, dry *swit*, given frequently.

Eastern Towhee

Pipilo erythrophthalmus CODE 1

ADULT MALE. *A large, long-tailed, boldly patterned sparrow. Black above with white patch in wing; belly white; flanks and undertail coverts rufous. In most of range, adult has red eye.* OHIO, MARCH

ADULT FEMALE. *Browner above than male; variable, but even the dullest individuals have extensive rufous on flanks.* FLORIDA, FEBRUARY

IMMATURE. *Similar to adult female; back and breast often show faint, diffuse streaking.* FLORIDA, APRIL

ADULT MALE. *Most breeders in peninsular Florida are smaller than elsewhere, with white eyes. Eyes pinkish on coastal plain from northern Florida to Carolinas; red elsewhere.* FLORIDA, FEBRUARY

L 8.5" ws 10.5" wt 1.4 oz

▸ one adult molt per year; complex basic strategy

▸ moderate sex-related, strong age-related differences

▸ geographic variation in eye color and body size

Familiar inhabitant of edges and brushy fields; often visits feeders. Feeds on the ground, methodically working a small patch of leaf litter or fallen grain. In summer and winter, usually near or under dense cover; migrants more generalist. Some populations sedentary.

◀ Song consists of two short notes and trill: *peek year chee'e'e'e'e'e'* ("drink your tea"). Call notes a rising *schweeek* ("chewink") and buzzy *zweeeee*.

JUVENILE. *Dusky overall, with streaking below. Flight feathers of wings and tail adult-like: wings have white patch, and long tail has white corners, prominent in flight.* NEW YORK, SEPTEMBER

Spotted Towhee
Pipilo maculatus   CODE 1

ADULT MALE. *All have white in tail. Birds from arid Intermountain West, as here, usually show extensive white; other populations show reduced white.* ARIZONA, APRIL

ADULT MALE. *Like Eastern, black above and white below with extensive rufous on flanks. Differs from Eastern in having coarse white spots on back and wing coverts.* CALIFORNIA, OCTOBER

ADULT FEMALE. *Differences between the sexes not as pronounced as in Eastern; most females are slightly paler above than males.* CALIFORNIA, MARCH

IMMATURE. *Birds of the Pacific Northwest show fewer white spots on wings than other populations.* BRITISH COLUMBIA, NOVEMBER

L 8.5"   WS 10.5"   WT 1.4 OZ

▸ one adult molt per year; complex basic strategy

▸ strong age-related and weak sex-related differences

▸ extensive geographic variation in white spotting above

Habits generally as Eastern Towhee, with which it was formerly lumped as "Rufous-sided Towhee." Populations vary in subtle ecological details, but all are basically birds of shrubby habitats; dry foothills with dense broadleaf vegetation are especially favored.

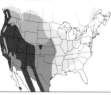

◀ Songs vary regionally: *whee whee ch'e'e'e'e'* in eastern Rockies; growling *churrrr* along Pacific coast; and other dialects. Call note a wavering *zhree-ee-eee.*

JUVENILE. *Dark overall, with streaking below. Wings usually show pale spots, like adult; tail long with white corners, also like adult.* CALIFORNIA, JULY

California Towhee

Pipilo crissalis   CODE 1

| L 9" | WS 11.5" | WT 1.5 OZ |

▸ one adult molt per year; complex basic strategy

▸ moderate differences between adult and juvenile

▸ darker in southern coastal California than elsewhere

A ground-loving inhabitant of brushy habitats of all sorts; active around hedges in urban settings, but also flourishes in coastal chaparral and oak woodlands with sufficient undergrowth. Wary but usually approachable; frequently wanders out to edges of lawns and paths.

ADULT. *Dusky gray-brown overall. Long-tailed with chestnut undertail coverts; muted chestnut highlights on face. Juvenile paler than adult, with even weaker facial pattern.*
CALIFORNIA, JANUARY

Song a sputtering series of sharp notes, erratically speeding up toward end: *peep peep peep-peep peep-peep-peep.* Call note a sharp *peep,* as in song.

Canyon Towhee

Pipilo fuscus   CODE 1

| L 9" | WS 11.5" | WT 1.9 OZ |

▸ one adult molt per year; complex basic strategy

▸ moderate differences between adult and juvenile

▸ averages larger and paler in north, darker in Texas

Occurs widely in arid landscapes. Especially characteristic of lowland washes with scrubby vegetation; also ranges upslope into pine-oak woodlands in rough, rocky terrain. Foraging techniques more diverse than other towhees; spends less time scratching in dirt for food.

Song a slow trill composed of *tyoo* notes; lower and more musical than songs of Abert's and California Towhees. Call note a liquid *tleeup.*

ADULT. *Similar to California; ranges do not overlap. Paler overall, with bolder markings on head and breast; many are peculiarly scruffy-looking. Juvenile browner, faintly streaked below.* TEXAS, APRIL

Abert's Towhee

Pipilo aberti   CODE 1

| L 9.5" | WS 11" | WT 1.6 OZ |

▸ one adult molt per year; complex basic strategy

▸ moderate differences between adult and juvenile

▸ western birds average paler and more cinnamon

Among towhees, occurs in the hottest and lowest deserts, but copes by restricting activities to high-quality riparian zones. Adaptable; as native cottonwood-willow groves have been lost, has shifted to such habitats as irrigation ditches lined with exotic salt cedar.

Song distinctly two-parted; slow at first, then an accelerating trill: *peep peep peep ch'ch'ch'ch'ch'churrr.* Call a sharp *peep,* resembling California.

ADULT. *Unlike Canyon, has extensive black around eye and base of bill. Habitats differ: Canyon usually in washes and foothills, not low desert. Juvenile warmer brown, streaked below.* ARIZONA, NOVEMBER

Green-tailed Towhee
Pipilo chlorurus   CODE 1

ADULT. *The smallest towhee. Greenish above; dull gray below. Head mainly gray with white on throat, on malar, and in front of eye. Crown rich reddish on male; slightly duller on female.* COLORADO, JULY

JUVENILE. *Often seen in the same habitat as Spotted; both are extensively streaked, but juvenile Green-tailed usually shows green in flight feathers of wings and tail.* CALIFORNIA, JULY

L 7.25"   ws 9.75"   wt 1 oz (28 g)

▸ two adult molts per year; complex alternate strategy

▸ strong age-related and weak sex-related differences

▸ brightness of crown variable on adult female

Breeds in cool habitats in Intermountain West. Common in high-quality shrub-steppe landscapes, especially sagebrush; also in disturbed areas of montane forest landscapes. More migratory than other towhees; in winter, occurs in desert washes and brushlands.

Song more complex and musical than Spotted; for example, *wh'seeee cheee ch'ch'ch'ch'ch'*. Song can resemble interior Fox Sparrow. Call note a mewing *nywaee*.

Cassin's Sparrow
Aimophila cassinii   CODE 1

ADULT. *Large; drab. Where range overlaps with Botteri's, note finely scalloped upperparts and bolder facial pattern of Cassin's. In flight, rounded tail flashes prominent pale corners.* ARIZONA, JULY

ADULT. *Sometimes confused with Brewer's, another drab scrubland sparrow with a thin white eye ring and thin black throat stripe. Brewer's much smaller, daintier, and smaller-billed.* ARIZONA, JULY

ADULT. *Even worn adults, as here, show weak spotting above and faint stripes on head. Juvenile is brighter buff overall with heavy streaking and strong facial pattern.* ARIZONA, JULY

L 6"   ws 7.75"   wt 0.7 oz (20 g)

▸ two adult molts per year; complex alternate strategy

▸ moderate seasonal and strong age-related differences

▸ uncommon rufous morph has richer upperparts

Inhabits arid shrub grasslands, where numbers fluctuate greatly from year to year, in part due to variation in rainfall. Most inhabit mixed-grass prairie with yuccas, mesquites, or chollas, but some populations flourish in shortgrass, others even in creosotebush desert.

Five-note song, given in shallow skylarking display, consists of two clear whistles, a descending trill, then two more whistles. Call note a high *sit*.

Botteri's Sparrow

Aimophila botterii CODE 2

ADULT. *Similar to Cassin's, but slight differences add up: Botteri's is longer-billed and plainer-faced; back has coarse streaks; wing at rest is fairly plain and gray.* ARIZONA, JULY

ADULT. *Tail averages darker than Cassin's; when flushing, shows less contrast in tail corners. Juvenile similar to juvenile Cassin's, but plainer-faced and longer-billed.* ARIZONA, JULY

L 6"   WS 7.75"   WT 0.7 oz (20 g)

▸ two adult molts per year; complex alternate strategy

▸ moderate seasonal and strong age-related differences

▸ Arizona birds redder and slightly darker than Texas

Habitat overlaps broadly with Cassin's Sparrow; prime habitat for Botteri's slightly more dense and lush. In Arizona, occurs in grasslands near woodlands; in Texas, occurs in coastal prairie with scattered shrubs. Feeds on ground, where it chases grasshoppers.

Song a complex series of strange, sputtering notes, often more than 5 seconds long, usually including an accelerating trill. Call note similar to Cassin's.

Rufous-crowned Sparrow

Aimophila ruficeps CODE 1

ADULT. *Grayish overall; all adults have black throat-stripe and rusty crown bordered below by thin white stripe. Juvenile has weaker facial pattern and is streaked below.* ARIZONA, JULY

ADULT. *Like other* Aimophila *sparrows, Rufous-crowned is large overall, large-billed, and plain-breasted. Sparrows with rusty crowns in other genera are smaller and less bulky.* CALIFORNIA, APRIL

L 6"   WS 7.75"   WT 0.65 oz (18 g)

▸ one adult molt per year; complex basic strategy

▸ moderate differences between adult and juvenile

▸ California birds smaller; upperparts less brownish

Throughout most of range, inhabits dry hillsides and canyons with scattered shrubs and grass; coastal California population has declined because of destruction of favored sage-scrub habitat. Range limits imperfectly known. Forages both on the ground and in low foliage.

Song dry and bubbly, like a House Wren, although not as spirited. Call note a nasal *dyew*, often given in rapid series and often incorporated into song.

Bachman's Sparrow
Aimophila aestivalis CODE 2

ADULT. *Cap rusty; eyebrow gray; back streaked; breast buff. Juvenile heavily streaked below, with buff wash across breast.* FLORIDA, FEBRUARY

ADULT. *Some Bachman's approach Swamp Sparrow in plumage, especially in Mississippi River drainage. Structure different: Bachman's is more bulky overall, larger-billed, longer-tailed.* FLORIDA, FEBRUARY

L 6"   ws 7.25"   wt 0.7 oz (20 g)

▸ two adult molts per year; complex alternate strategy

▸ strong differences between adult and juvenile

▸ western birds redder above, more buff underneath

A specialty of southeastern pine forests with a dense ground layer of grasses and forbs. Short-term and long-term movements mysterious: spring and fall migration poorly understood; apparently expanded northward in late 19th century, but has subsequently retreated.

Song consists of a sweet opening note, followed by one or two trills: *pweeeee ch'ch'ch'ch'ch' chip-chip-chip-chip.* Call note a high *chee.*

Rufous-winged Sparrow
Aimophila carpalis CODE 2

ADULT. *Similar to Rufous-crowned, but bill is shorter and mainly yellow. Most show two "whiskers" bordering malar region. Juvenile more buff, with streaking and spotting beneath.* ARIZONA, FEBRUARY

L 5.75"   ws 7.5"   wt 0.5 oz (15 g)

▸ two adult molts per year; complex alternate strategy

▸ strong differences between adult and juvenile

▸ rufous lesser wing coverts often concealed

Resident of Sonoran desert scrublands where thorn scrub is interspersed with bunchgrass; usually in flat or slightly hilly terrain. Onset of nesting is strongly tied to the start of summer monsoon rains. Long-term population status dynamic and unstable.

Most common song four-parted: two chips, an extremely high note, and a fast trill: *pip pip peeeee p'p'p'p'p'p'p'p'.* Also sings a simpler, accelerating trill.

All sparrows in the genus Aimophila *are prone to short-distance wandering. These movements may be pronounced in some years, but weak or absent in others. The causes of these wanderings are poorly understood, but may be related to the interplay between regional weather and food availability.*

Five-striped Sparrow

Aimophila quinquestriata CODE 3

ADULT. *Brown above; gray below. Head has bold black-and-white stripes. Juvenile duller, yellow below, with lower-contrast face pattern; some retain yellow through winter and spring.* ARIZONA, JULY

ADULT. *Plumage suggests Sage Sparrow, but Five-striped is larger overall, longer-tailed, and longer-billed than Sage.* ARIZONA, JULY

L 6"   WS 8"   WT 0.7 oz (20 g)

▸ two adult molts per year; complex alternate strategy

▸ strong differences between adult and juvenile

▸ amount of yellow on belly of subadult variable

Uncommon amid acacias and ocotillos of hot canyons and steep hillsides in south-central Arizona. Briefly conspicuous at beginning of breeding season, triggered by onset of monsoon rains; otherwise hard to find. Winter status unclear; at least a few remain.

Song consists of scratchy notes interspersed with musical trills: *ch' chweet-chweet-chweet chit chit chit chee-chee-chee ch'ch'*. Call note a low *churt*.

Chipping Sparrow

Spizella passerina CODE 1

BREEDING ADULT. *Head pattern distinctive: rufous crown, white eyebrow, black eye stripe. Adult gray-breasted all year; small body size and relatively long tail evident in all plumages.* MAINE, JUNE

NONBREEDING ADULT. *Black eye stripe, extending from bill to nape, dominates the head pattern; compare Brewer's and Clay-colored Sparrows.* CALIFORNIA, OCTOBER

JUVENILE. *Widely noted away from breeding grounds. Breast streaked, in contrast to subsequent plumages. Black eye stripe distinctive and petite build conspicuous, as in older birds.* OREGON, AUGUST

L 5.5"   WS 8.5"   WT 0.4 oz (11 g)

▸ two adult molts per year; complex alternate strategy

▸ strong seasonal and age-related differences

▸ western birds larger and grayer than eastern

Forages on ground, actively hopping and running. Western birds breed in dry, open woodlands; eastern birds are common breeders in towns and suburbs. Not especially hardy; rarely lingers north of core winter range. Outside breeding season, joins mixed-species flocks.

◀ Song a dry trill with uniform pitch and loudness. Call note a sharp *chip*; flight call, often heard during nocturnal migration in fall, a piercing, rising *seen?*

Clay-colored Sparrow
Spizella pallida   CODE 1

BREEDING ADULT. *Relatively high-contrast head markings: brown auriculars bordered above and below by black; eyebrow white; nape gray. Juvenile buff below with extensive fine streaks.*

MINNESOTA, JUNE

NONBREEDING. *Face less boldly marked than in breeding plumage, but underparts often richer buff. Brown rump (Chipping gray), prominent in flight, a good mark in all plumages.* NEW YORK, SEPTEMBER

L 5.5"   WS 7.5"   WT 0.4 oz (11 g)

- ▸ two adult molts per year; complex alternate strategy
- ▪ moderate seasonal and strong age-related differences
- ▪ extent of buff variable, especially in late summer

Usually in prairies mixed with low shrubs; may be the most numerous breeder in such habitat. Breeding range expanding east into regions with abandoned farms and reclaimed strip mines. Bulk of migration is through Great Plains, but regular vagrant to both coasts.

◀ Song a series of well-spaced buzzes that consist of white noise: *zzzz... zzzz... zzzz...* Call note like Chipping Sparrow; flight call a rising *sit.*

Brewer's Sparrow
Spizella breweri   CODE 1

ADULT. *The drabbest Spizella; like a washed-out Clay-colored. Dusky crown has faint streaks; auriculars lack prominent borders; malar stripe relatively well-defined.*

CALIFORNIA, JUNE

ADULT. *Like Clay-colored, eye line extends behind but not in front of eye (eye line of nonbreeding Chipping Sparrow extends all the way to bill). Many Brewer's show an eye ring.*

CALIFORNIA, OCTOBER

IMMATURE. *In all plumages, small and dainty overall like Chipping and Clay-colored. "Timberline" subspecies, not shown, is darker. Juvenile streaked below; face pattern like adult.*

CALIFORNIA, AUGUST

L 5.5"   WS 7.5"   WT 0.4 oz (11 g)

- ▸ two adult molts per year; complex alternate strategy
- ▪ strong age-related and weak seasonal differences
- ▪ "Timberline Sparrow" (*S. b. taverneri*) may be separate species

Most breed in sagebrush country; numbers in decline, but the species remains common in most areas. "Timberline Sparrow" breeds in stunted willows; southern limits of range, somewhere in Intermountain West, unknown. Winters in mixed-species flocks in desert shrublands.

Intricate song consists of multiple series of 4–10 notes each; sounds inhaled. Call note a sharp *chip*; flight call more level, more buzzy than Clay-colored.

Field Sparrow

Spizella pusilla CODE 1

ADULT. *Gray and rusty. Size and shape about the same as Chipping Sparrow, but Field has a more gentle countenance; note pink bill and white eye ring.* MINNESOTA, JUNE

ADULT. *Sometimes confused with American Tree: Field has lower-contrast markings overall; breast is unmarked (American Tree has central spot). Juvenile buff below with heavy streaking.* CONNECTICUT, MAY

L 5.75"	WS 8"	WT 0.45 oz (13 g)

▸ two adult molts per year; complex alternate strategy

▸ strong age-related and weak seasonal differences

▸ western birds grayer overall and longer-tailed

Inhabits overgrown pastures, woodland edges, and bramble patches. Some populations declining due to habitat conversion. Unlike Chipping Sparrow, does not tend to breed in congested and developed districts. Migrates fairly early in spring, returns fairly late in fall.

Song an accelerating series of sweet notes: *twee twee twee tee tee t't't't't'.* Call note a forceful *spit*; flight call a long, descending *seeewp.*

American Tree Sparrow

Spizella arborea CODE 1

ADULT. *Gray and rusty overall, with red crown and gray eyebrow. Plain breast has a central spot; juvenile, rarely seen south of arctic breeding grounds, is heavily streaked below.* NEW YORK, FEBRUARY

ADULT. *In winter, some resemble Field and Chipping: American Tree larger overall, with bicolored bill (dark above, yellow below); spot on breast of American Tree is usually diagnostic.* ILLINOIS, MARCH

L 6.25"	WS 9.5"	WT 0.7 oz (20 g)

▸ two adult molts per year; complex alternate strategy

▸ strong differences between adult and juvenile

▸ western breeders paler and slightly larger than eastern

Not a bird of trees, but rather of shrubby fields. Known to most birders as a winter visitor; small flocks feed on seed heads. Breeds north of tree line in arctic. One of the latest passerines to return in fall; departs early and unobtrusively from wintering grounds.

◀ Call note a tinkling *seerleerp*; flight call a short, ringing *tsweerp.* Song, sometimes given in late winter, a high, musical jumble, descending overall.

Black-chinned Sparrow

Spizella atrogularis CODE 1

ADULT MALE. *Washed heavily in dark gray, deepening to solid black on chin and around bill. Female and nonbreeding male show less contrast on head; both sexes have pink bill.* NEW MEXICO, APRIL

JUVENILE. *Gray overall like adult, but lacks black on face. All plumages are impressively long-tailed; all appear slight of build overall and small-billed.* CALIFORNIA, JULY

L 5.75" ws 7.75" wt 0.4 oz (11 g)

▸ two adult molts per year; complex alternate strategy

▸ moderate age-related, sex-related, seasonal differences

▸ northern birds average purer gray, with less brown

Breeds in arid scrublands of little-visited slopes below the ponderosa pine zone; shares habitat with Gray Vireo and Rufous-crowned Sparrow. Winters in desert scrublands and washes. Rarely detected on migration. Feeds on the ground and in low broadleaf vegetation.

Song an accelerating series similar to Field Sparrow, but drier; faster and more buzzy at end. Call note a weak *seet*; flight call a level *siss*.

Black-throated Sparrow

Amphispiza bilineata CODE 1

ADULT. *Uniform gray-brown above and gray below. Head gray with bold white streaks; throat jet black. Tail dark and somewhat rounded; flashes white corners in flight.* ARIZONA, FEBRUARY

JUVENILE. *Lacks black throat of adult; breast has dusky streaks. Similar to adult Sage, but Black-throated has prominent white eyebrow; Sage usually has central spot on plain breast.* ARIZONA, AUGUST

L 5.5" ws 7.75" wt 0.5 oz (14 g)

▸ one adult molt per year; complex basic strategy

▸ strong differences between adult and juvenile

▸ browner and smaller in west and north of range

One of the best-adapted desert bird species; specialized physiology enables it to go without water for many days. Often the dominant bird species in sun-baked creosote bush monocultures. Less common in colder, sagebrush-dominated deserts, where migratory.

Spiritless song consists of 1–3 notes followed by jangling trill: *kweet kweet ch'ch'ch'cheer*. Call note, given in long, halting series: a high *sip* or *see*.

Sage Sparrow
Amphispiza belli CODE 1

ADULT, INTERIOR. *Gray head has white eye ring. Upperparts brown; underparts pale with central breast spot and sparse streaks on flanks. Juvenile darker; has dense streaking below.* NEW MEXICO, NOVEMBER

ADULT, "BELL'S." (*A. b. belli*) *Colors and pattern similar to Interior, but head darker gray and back darker brown. "Bell's" has bold throat streak (indistinct on Interior birds).* CALIFORNIA, MARCH

L 6" WS 8.25" WT 0.6 oz (17 g)

▸ one adult molt per year; complex basic strategy

▸ strong differences between adult and juvenile

▸ "Bell's" Sparrow of California coast darker

Like a thrasher, runs from shrub to shrub across bare ground with tail cocked up. Widespread Interior populations breed mainly in sagebrush-dominated cold desert; "Bell's" subspecies is resident in coastal sage-scrub. In winter, gathers in large flocks in deserts.

Song low and halting, typically with abrupt break in middle: *chippaweezeea...chippawoozee-oo*. Tinkling call note, given frequently, sounds like Black-throated.

Lark Bunting
Calamospiza melanocorys CODE 1

BREEDING MALE. *Extensive white on secondary coverts contrasts with otherwise entirely black plumage. High-contrast plumage especially prominent in flight.* COLORADO, JUNE

NONBREEDING MALE. *Dark overall with extensive coarse streaking. Plumage mainly brown, but variable black is usually present on wings, face, throat, and underparts.* ARIZONA, DECEMBER

ADULT FEMALE. *Shows little black, but white wing patch is usually prominent. In all plumages, note large size overall, combined with short wings, short tail, and large bill.* COLORADO, JUNE

L 7" WS 10.5" WT 1.3 oz

▸ two adult molts per year; complex alternate strategy

▸ strong seasonal and sex-related differences

▸ fall molt prolonged; transitional birds variable

Numerous within fairly narrow range: territorial males prominent on shortgrass prairie; large flocks gather on migration and in winter. Tolerates grazing. Migrates mainly by day, often conspicuously; spring migration fairly late, fall migration quite early.

◀ Song, given in flight, a slow, pulsing series of low rich chirps and trills in series, like an exuberant Brewer's Sparrow. Flight call distinctive; a rising, musical *wurt?*

Lark Sparrow
Chondestes grammacus CODE 1

IMMATURE. *After molt out of juvenile plumage, immature acquires face pattern like adult, but not as colorful. In briefly held juvenile plumage has heavy streaking below.*
CALIFORNIA, AUGUST

ADULT. *One of our most distinctive sparrows: large overall, with a bold and colorful face pattern; plain breast has prominent central spot.*
CALIFORNIA, JUNE

L 6.5" ws 11" wt 1 oz (28 g)

▸two adult molts per year; complex alternate strategy

▸moderate differences between adult and immature

▸amount of streaking on back of adult variable

Inhabits small groves, open woodlands, and forest edges, often in areas with poor soil and overgrazing. Forages on the ground, preying on surprisingly large arthropods. Fall migration early and conspicuous; large flocks assemble by day. Eastern limits of range unstable.

Song, often given in conjunction with male strutting display, a variable jumble of trills that features abrupt changes in pitch and timbre. Flight call an abrupt *seek*.

ADULT. *All plumages have distinctive tail: long and rounded with extensive white. From below, tail appears mainly white; in flight, white forms bold corners around tail.* CALIFORNIA, OCTOBER

Vesper Sparrow
Pooecetes gramineus CODE 1

L 6.25" ws 10" wt 1 oz (28 g)

▸one adult molt per year; complex basic strategy

▸weak differences between spring and fall adult

▸prominence of rusty patch on wing coverts variable

Dwells on the ground in open, often dry, habitats: old fields, dirt roads, and shrub-steppe. Early spring migrant, fairly late fall migrant. Feeds actively, scratching vigorously with feet and pecking rapidly in dirt. Sharp declines in eastern portion of range.

ADULT. *Usually shows a thin white eye ring. Flanks tinged with buff; streaks on breast do not converge in central spot. White outer tail feathers, visible here, are prominent in flight.*
NEW JERSEY, OCTOBER

◀ Song is rich and low, often given in early evening: *deeer deer dzee d'd'd'd'd'*. Call note a low, hard *jip*; flight call a rising, buzzy *swit*

ADULT. *Larger and longer-tailed than Savannah Sparrow. Rufous at bend of wing more prominent on birds in worn plumage in summer (here) than in fresh plumage in fall.* MAINE, JUNE

Savannah Sparrow
Passerculus sandwichensis CODE 1

TYPICAL ADULT. *In much of range, breast finely streaked and belly white; front portion of eyebrow usually yellow. Juvenile duskier overall, with coarser streaking below.* QUÉBEC, JULY

TYPICAL ADULT. *Many are weakly patterned, showing diminished or obscure yellow in eyebrow. Short tail and slight bill usually a good point of separation from Song and Vesper Sparrows.* COLORADO, JUNE

ADULT, "LARGE-BILLED." (P. s. rostratus) *Breeds in northwestern Mexico; a few winter north to Salton Sea, and southern coastal California. Streaking diffuse; bill large. Many scientists suspect it is a distinct species.* CALIFORNIA, JANUARY

ADULT, "IPSWICH." (P. s. princeps) *Breeds only on Sable Island, Nova Scotia; winters on dunes along East Coast. Large overall and very pale; slight bill and short tail are typical of species.* NEW YORK, JANUARY

L 5.5"	WS 6.75"	WT 0.7 oz (20 g)

▸ two adult molts per year; complex alternate strategy

▸ weak seasonal and moderate age-related differences

▸ extensive geographic variation in size and plumage

Characteristic of open country in all but the extreme northern tip of the continent: found in hay meadows, marshes, and tundra. Approachable and conspicuous during breeding season: sings from fence posts and atop prominent shrubs. More reclusive in winter.

ADULT, "BELDING'S." (P. s. beldingi) *Another distinctive population; restricted to salt marshes in southern California; darker than most Savannahs. Relatively long bill adapted for foraging in mud.* CALIFORNIA, JANUARY

◀ Song a gasping *sip sip sip seeeee say* ("please, please, please, please! stay"). Call note a smacking *stick*; flight call a descending, slightly buzzy *tsip*.

Grasshopper Sparrow
Ammodramus savannarum   CODE 1

ADULT. *All plumages are flat-headed, large-billed, and short-tailed; body structure is stubby overall. Adult has unmarked plain breast; juvenile breast buff with fine dark streaks.*
MINNESOTA, JUNE

ADULT. *All adults have a white central crown stripe and white eye ring, prominent in good light on fresh birds; most have some buff on face, and yellow or yellow-orange in front of eye.* MONTANA, JUNE

ADULT. *Most adults are pale overall, as here, but Florida population is darker and southeastern Arizona population is redder.* MAINE, JUNE

L 5"	ws 7.75"	wt 0.6 oz (17 g)

▸ two adult molts per year; complex alternate strategy

▸ strong differences between adult and juvenile

▸ moderate geographic variation in brightness, contrast

Locally common breeder in fields with grasses of intermediate height; often co-occurs with the more generalist Savannah Sparrow. In prime habitat, breeds in semi-colonial clusters. Territorial males conspicuous; inconspicuous in fall and winter in weedy fields.

Two songs: a hiccup and a long buzz, *p'tup zeeeeeee*; a high, wiry jumble, *zh'weea-zeezy-swee-j'wee-sweejeeze*. Flight call a level, buzzy *zeezt*.

Henslow's Sparrow
Ammodramus henslowii   CODE 2

ADULT. *Like other* Ammodramus *sparrows, is large-billed, flat-headed, and short-tailed. Adult is buff below with extensive but irregular dark streaks; streaking reduced on juvenile.* OHIO, MAY

ADULT. *Color pattern separates Henslow's from other* Ammodramus *sparrows: head washed in greenish yellow; wings and back have extensive chestnut.* OKLAHOMA, JULY

L 5"	ws 6.5"	wt 0.5 oz (14 g)

▸ two adult molts per year; complex alternate strategy

▸ moderate differences between adult and juvenile

▸ rare eastern population darker, deeper chestnut

Breeds in tall grass; winters in fire-maintained savannas. Locally numerous breeder in high-quality grasslands, but absent from large swaths of degraded habitat. A declining species; poorly differentiated eastern population of Atlantic coast marshes may be extinct.

Song an odd, distinctive *fshlick*. Call note a sharp, smacking *fsick*, similar to song. Flight call a long, wavering, buzzy *seeeth*.

Baird's Sparrow

Ammodramus bairdii CODE 2

ADULT. *Flat-headed, large-billed, and short-tailed. Most adults have warm tan tones overall, with a few prominent streaks below. Juvenile less contrastingly marked; streaking reduced.* MONTANA, JUNE

ADULT. *Some have warm tones limited to face, as here; all adults have black spot behind eye, plus malar bordered in black.* NORTH DAKOTA, JUNE

L 5.5"   WS 8.75"   WT 0.6 oz (17 g)

▸ two adult molts per year; complex alternate strategy

▸ moderate differences between adult and juvenile

▸ streaking of breast on adult female variable

A much sought-after prairie specialty. Except when singing, hides in tall grass. Migrants rarely detected; even at core of winter range, hard to find. Flushes low, dives into deep grass, stays hidden. Populations naturally variable, but in overall decline.

Song consists of tinkling opening notes, then variable trill: *tslip tslip tslip tsaaaay*. Call note a downslurred *zleep*; flight call a short, high *si*.

Le Conte's Sparrow

Ammodramus leconteii CODE 1

ADULT. *All have basic* Ammodramus *body plan of large bill, flat head, and short tail. Adult shows orange-buff on face; juvenile diffuse yellow-buff all over and finely streaked below.* MINNESOTA, JUNE

ADULT. *Separated from similar species by thin white crown stripe bordered by black, diminished gray on nape, and tan-and-black streaks on back.* MINNESOTA, JUNE

L 5"   WS 6.5"   WT 0.45 oz (13 g)

▸ two adult molts per year; complex alternate strategy

▸ moderate differences between adult and juvenile

▸ intensity of yellow-buff on juvenile variable

Typical of genus *Ammodramus*, is hard to detect except when males are singing. Breeds in marshy meadows; winters in similar habitat, but also in drier weedy fields. Migrates mainly through eastern Plains; annual in smaller numbers well outside main migratory corridor.

Song like primary song of Grasshopper Sparrow, but long terminal buzz even thinner: *titi-zeeeeeee*. Call note a short *tsee*; flight call a long high *tzeee*.

Saltmarsh Sharp-tailed Sparrow

Ammodramus caudacutus CODE 1

L 5.25"	WS 7"	WT 0.7 oz (20 g)

- ▸ two adult molts per year; complex alternate strategy
- ▸ moderate differences between adult and juvenile
- ▸ southern birds darker above than northern birds

Nests in cordgrass and salt hay in Atlantic coast salt marshes north into New England, where population overlaps extensively with Nelson's Sharp-tailed. Numbers are locally unstable because of nest losses due to storms and high tides. Generally not as conspicuous as Seaside Sparrow.

Strange song unlike Nelson's; sputtering and whispered, a complex of thin trills, weak hiccups, and high buzzes. Flight call a high, piercing *seees*. Males rarely sing.

ADULT. *Orange or orange-yellow extensive on face, but reduced or absent below; heavy streaking on breast and flanks. Bill long. Juvenile buff overall, including underparts; heavily streaked below.* NEW JERSEY, JUNE

ADULT, HYBRID. *Where Nelson's overlaps with Saltmarsh in northern New England, hybrids occur. Color and pattern of head and breast intermediate, but variable, on hybrids.* MAINE, JULY

Nelson's Sharp-tailed Sparrow

Ammodramus nelsoni CODE 1

ADULT, "COASTAL." (A. n. subvirgatus) *Duller than Prairie population, with diffuse streaking below; bill longer than Prairie, an adaptation for probing in mud.* MAINE, JUNE

ADULT, "COASTAL." (A. n. subvirgatus) *All sharp-tailed sparrows (including Saltmarsh) have extensive gray on nape. Hudson Bay subspecies (A. n. alterus), not shown, is intermediate in brightness between Prairie and Coastal.* MAINE, JUNE

L 5"	WS 7"	WT 0.6 oz (17 g)

- ▸ two adult molts per year; complex alternate strategy
- ▸ moderate differences between adult and juvenile
- ▸ Atlantic coast population duller, less contrasting

Western population breeds in inland prairie marshes, central population in dense sedge bogs above high tide line of Hudson Bay, eastern population in salt marshes along Atlantic coast. All migrate stealthily to coastal marshes from mid-Atlantic southward.

Song a fading *k'zheeeee*; sounds like a hot poker being plunged into water. Call note a short, sharp *tik*; flight call like Le Conte's but shorter, sweeter.

ADULT, "PRAIRIE." (A. n. nelsoni) *More coarsely patterned than Le Conte's; burnt orange on face, breast, flanks, and undertail coverts. All juveniles rich buff-orange all over, with sparse streaking below.* TEXAS, MAY

Seaside Sparrow
Ammodramus maritimus CODE 1

ADULT, ATLANTIC COAST. *Averages duskier and less boldly marked than Gulf coast populations; even the duskiest birds show pale throat and a bit of yellow in front of eye.* NEW JERSEY, MAY

ADULT, "CAPE SABLE." (A. m. mirabilis) *Paler below and a little more olive above than other Seaside Sparrows. Restricted to small area of south Florida; breeds in freshwater marshes and flooded prairies* FLORIDA, MAY

ADULT, GULF COAST. *The largest, longest-billed* Ammodramus *sparrow. Underparts dark and dingy, with diffuse streaking; white throat and yellow in front of eye often prominent.* TEXAS, FEBRUARY

L 6"	WS 7.5"	WT 0.8 oz (23 g)

▸ two adult molts per year; complex alternate strategy

▸ moderate differences between adult and juvenile

▸ much geographic variation in pattern and darkness

A complex of well-differentiated, fairly isolated populations; most are restricted to medium-height cordgrass in salt marshes. Locally common; averages more confiding than typical of genus *Ammodramus*. Probes in mud with long bill; runs skillfully across mudflats.

Song like Nelson's Sharp-tailed but more complex: *pickaweea-zhheeeee*; only terminal *zhheeeee* audible at a distance. Flight call a long, piercing *seeeer*.

JUVENILE. *Paler than adult, especially below. Underparts dirty buff with fine streaking. Spiky tail, very long bill, and hulking demeanor distinctive in all plumages.* DELAWARE, JULY

Fox Sparrow
Passerella iliaca   CODE 1

ADULT, "RED." (P. i. iliaca) *The brightest Fox Sparrow; clad in "foxy" red and gray. All Fox Sparrows are large-bodied with at least some red in tail. All juveniles have underparts washed buff.* NEW JERSEY, DECEMBER

ADULT, "SLATE-COLORED." (P. i. schistacea) *Reddish coloration restricted mainly to tail and wings; head and back extensively gray. Underparts marked with irregular rows of discrete chevrons.* COLORADO, DECEMBER

ADULT, "THICK-BILLED." (P. i. megarhyncha) *Slightly less colorful than "Slate-colored," but otherwise similar in plumage; best distinctions are call note and substantially larger bill of "Thick-billed."* CALIFORNIA, JUNE

ADULT, "SOOTY." (P. i. unalaschcensis) *Dark and sooty overall, with only a hint of red in tail. Streaking below runs together in extensive, splotchy, dark wash across underparts.* OREGON, OCTOBER

L 7"	WS 10.5"	WT 1.1 oz

▸ two adult molts per year; complex alternate strategy

▸ weak differences between adult and juvenile

▸ much geographic variation in color and pattern

All populations breed in dense thickets: for example, "Red" in alders near spruce woods, "Thick-billed" in montane chaparral. All migrate early in spring, but timing of fall migration varies among groups. All feed by "double-scratch" method employed by towhees.

◀ Song variable; sweet in "Red" Fox Sparrow, drier in other groups. Call note of "Thick-billed" a sharp *deek*; other groups give a duller *tuck*.

ADULT, "RED." (P. i. zaboria) *Each of the four major groups consists of multiple subspecies. For example, western "Red" Fox Sparrows (here) are darker and duller than their counterparts in the East (top left).* CALIFORNIA, JANUARY

Song Sparrow
Melospiza melodia CODE 1

ADULT. *Most are fairly large, with a long, rounded tail; most have pale malar bordered below by thick black stripe. Streaking below coalesces into a central breast spot.* CALIFORNIA, JANUARY

ADULT. *Variation is clinal, such that genetically isolated populations are Infrequent. In any given area, individuals may vary in paleness and ground color.* COLORADO, DECEMBER

ADULT. *Several populations distinctive, such as large, dark, diffusely streaked birds from the Bering Sea. Also distinctive is a pale reddish subspecies of Desert Southwest.* ALASKA, MARCH

JUVENILE. *Buff overall. Finely streaked breast with buff ground color invites confusion with Lincoln's.* NEW HAMPSHIRE, JULY

L 6.25" WS 8.25" WT 0.7 oz (20 g)
- ▸ one adult molt per year; complex basic strategy
- ▸ moderate differences between adult and juvenile
- ▸ much geographic variation in size and color

Found in dense cover without continuous forest canopy. Many populations nest in gardens and parks near human habitation; others are specialized inhabitants of beach grass, salt marshes, or riparian areas. Migratory strategies variable. Some populations are declining.

◀ Song variable; most have clear opening whistles, followed by buzz or trill, then a few short notes. Call note a nasal *chemp*; flight call a high *seeet*.

ADULT. *The long rounded tail is jerked up and down in flight, a good mark on flushing birds.* ARIZONA, NOVEMBER

Swamp Sparrow
Melospiza georgiana CODE 1

BREEDING ADULT. *Variable gray and rufous overall; all have reddish in wings and tail, buff along flanks. Gray head has rusty cap, black line behind eye, and thin black malar.* NEW HAMPSHIRE, APRIL

NONBREEDING ADULT. *Colder and darker than breeding plumage; cap becomes brownish. Most have a distinct white throat, set off from gray wash across breast.* NEW HAMPSHIRE, OCTOBER

JUVENILE. *Finely streaked below. Similar to Lincoln's and Song, but wings and tail more reddish. After first fall molt, acquires immature plumage that resembles nonbreeding adult.* MANITOBA, AUGUST

BREEDING ADULT, MID-ATLANTIC COAST. *Darker and duller than other Swamp Sparrows. Also longer-billed, like coastal Ammodramus sparrows.* DELAWARE, MAY

L 5.75"   WS 7.25"   WT 0.6 oz (17 g)

▸ two adult molts per year; complex alternate strategy

▸ strong seasonal and moderate sex-related differences

▸ darker toward southern and eastern portions of range

Nests in diverse sorts of wetlands: cattail marshes, boreal bogs, and cedar swamps. Breeding populations locally abundant, but naturally unstable due to flooding. Winters mainly in wetlands, but migrants range into drier habitats where there is dense brush.

Song, a simple trill, is more leisurely and jangling than Chipping Sparrow: *chee chee chee chee...* Call note a hard *cheet;* flight call not as thin as Lincoln's.

Despite sounding simple to human ears, the song of the Swamp Sparrow varies considerably. Females are able to assess male dominance and "fitness" (a measure of genetic quality) by discerning the exact frequency and intonation of individual elements in the song. Differences in male song may also communicate information about local habitat quality.

Lincoln's Sparrow
Melospiza lincolnii   CODE 1

ADULT. *Plumage similar to Song, but markings crisper and finer. Washed buff-brown below with fine streaks; gray face has white eye ring. When agitated, raises crown in bushy crest.* CALIFORNIA, APRIL

ADULT. *Some closely resemble Song, but almost all show at least a thin eye ring and some buff across breast. Buff malar is bordered by black. Juvenile is more heavily streaked below.* TEXAS, APRIL

L 5.75"   WS 7.5"   WT 0.6 oz (17 g)

▸ two adult molts per year; complex alternate strategy

▸ moderate differences between adult and juvenile

▸ eastern breeders rustier-backed than elsewhere

Skulks in dense thickets, but easily lured into view with pishing. Breeds extensively throughout boreal forest zone; locally abundant amid low willows near bogs. Widespread on migration, especially in West; migrants occur anywhere there is sufficient cover.

◀ Bubbly song consists of multiple trills; common pattern is three trills, the middle one higher. Call note a junco-like *tsik*; flight call a fine, buzzy *zeee*.

Golden-crowned Sparrow
Zonotrichia atricapilla   CODE 1

BREEDING MALE. *Extensive black cap has variable lemon-yellow forecrown. Gray on face and chin extends smoothly to underparts. Bill yellowish, especially on lower mandible.* ALASKA, JUNE

NONBREEDING ADULT. *Most show dusky yellow forecrown bordered by dark brown. Some approach appearance of Pacific coast populations of White-crowned Sparrow.* OCTOBER, CALIFORNIA

IMMATURE. *Brownish overall; bill usually yellowish, especially lower mandible. Dusky brown cap has yellow or tan patch toward forecrown. Juvenile is streaked below.* CALIFORNIA, OCTOBER

L 7.25"   WS 9.5"   WT 1 oz (28 g)

▸ two adult molts per year; complex alternate strategy

▸ strong age-related and seasonal differences

▸ yellow on crown variable within any plumage

Breeds at forest-tundra transition. Common in winter along Pacific slope, east to foothills of eastern Cascades and Sierra; only a few penetrate farther east in Intermountain West. Closely related to White-crowned Sparrow.

Song consists of three pure whistles; first note typically highest, second note lowest, third note intermediate. Call note a flat *chup*; flight call like White-crowned.

White-throated Sparrow
Zonotrichia albicollis CODE 1

ADULT, WHITE-STRIPED MORPH. *Bright white throat contrasts with gray breast and cheeks. Broad white eyebrow bordered above and below by thin black stripes; central crown stripe white.* MAINE, JUNE

ADULT, TAN-STRIPED MORPH. *Duller overall than white-striped morph: throat, eyebrow, and crown stripe dusky; yellow in front of eye usually reduced.* MINNESOTA, SEPTEMBER

IMMATURE. *Widely encountered during winter months. Head pattern resembles tan-striped adult; dusky breast has variable streaking, sometimes coalescing in central breast spot.* CALIFORNIA, OCTOBER

JUVENILE. *Plumage briefly held; usually not seen away from breeding grounds. Heavily streaked below. Most show broad pale eyebrow and unmarked pale throat, suggesting adult.* BRITISH COLUMBIA, JULY

L 6.75"   WS 9"   WT 1 oz (28 g)

▸ two adult molts per year; complex alternate strategy
▸ strong differences between adult and immature
▸ tan-striped and white-striped morphs throughout range

One of the most abundant and characteristic species of the northern boreal zone; favors clearings, bogs, and beaver ponds. Winters in flocks in undergrowth in woodlots; frequently attends feeders.

White-striped morph White-throated Sparrows are dominant to the tan-striped morph in many social settings. The tan-striped morph would seem to be at a disadvantage, and might be expected to disappear as a consequence of being outcompeted by the white-striped morph. However, the tan-striped morph persists because of a "balanced polymorphism." The two morphs are maintained in nearly equal proportions because white-striped birds breed only with tan-striped birds and vice versa.

Two songs of sweet, wavering whistles: "Old Sam Peabody" and "Sweet Canada." Call note a sharp *chink*, harder than White-crowned; flight call a buzzy *zeezz*.

White-crowned Sparrow
Zonotrichia leucophrys CODE 1

BLACK-LORED ADULT. *Populations breeding in eastern Canada and Intermountain West have pink-red bill and black line (supraloral) extending from eye to forehead.* TEXAS, NOVEMBER

ADULT, "GAMBEL'S." (Z. l. gambelii) *Populations breeding in Alaska and surrounding regions, called "Gambel's," have pale supraloral area; bill not as reddish as bill of black-lored birds.* ARIZONA, FEBRUARY

ADULT, "NUTTALL'S." (Z. l. nuttalli) *Two Pacific coast races, including southern "Nuttall's," have pale supraloral, yellow bill, and underparts washed brownish (grayer in other populations).* CALIFORNIA, OCTOBER

IMMATURE. *Widely seen on migration and throughout the winter months. In all populations, black-and-white head stripes of adult are brown and gray in immature.* CALIFORNIA, OCTOBER

L 7"	WS 9.5"	WT 1 oz (28 g)

▸ two adult molts per year; complex alternate strategy

▸ strong differences between adult and immature

▸ geographic variation in bill color and face pattern

Breeds mainly in open habitats, from alpine meadows in southern Rockies to northern tundra to grasslands along Pacific coast. Common to abundant as migrant throughout much of West; less common in East. Often in large, loose flocks in winter and on migration.

◀ Song consists of one or more long sweet notes, followed by a short buzzy jumble. Compared to White-throated, call note is flatter, flight call less buzzy.

LEUCISTIC ADULT, "NUTTALL'S." (Z. l. nuttalli) *In most bird species, pure-white adults (albinos) are rare. Many leucistic (partially albino) individuals often survive to adulthood.* CALIFORNIA, NOVEMBER

Harris's Sparrow
Zonotrichia querula CODE 1

BREEDING ADULT. *Notably large; gray overall, pink-billed, and boldly patterned. Extensive black around base of bill extends up onto crown and down onto breast.* MANITOBA, MAY

NONBREEDING ADULT. *Black around bill is reduced; gray on face of breeding plumage becomes tan. Face pattern variable; some individuals are oddly reminiscent of House Sparrow.* NEW MEXICO, NOVEMBER

IMMATURE. *Shows traces of black on crown, around base of bill, and on breast; belly and lower breast mainly white. Large size and colorful bill always distinctive.* CALIFORNIA, OCTOBER

L 7.5"   WS 10.5"   WT 1.3 oz

▸ two adult molts per year; complex alternate strategy

▸ strong age-related and moderate seasonal differences

▸ amount of black on face of winter adult variable

One of only three bird species that breeds exclusively within Canada, but known to most birders as a winterer in shelterbelts and brush piles in the southern Plains; status at edge of winter range varies annually. Breeding and wintering ranges appear to be expanding.

Song simpler than other species in genus *Zonotrichia*; a few sweet whistles on same pitch. Call note a raspy *chinch*; flight call a steady, sibilant *teees*.

Dark-eyed Junco
Junco hyemalis CODE 1

ADULT MALE, "SLATE-COLORED." (*J. h. hyemalis*) *Dark gray (female gray-brown) with white belly. All Dark-eyed populations are small-bodied and pale-billed; flash white outer tail feathers in flight.* MINNESOTA, JANUARY

ADULT, "OREGON." (*J. h. oreganus*) *Variable, but all have black hood (gray in female) and pinkish back and flanks.* CALIFORNIA, JULY

L 6.25"   WS 9.25"   WT 0.7 oz (20 g)

▸ two adult molts per year; complex alternate strategy

▸ strong differences between adult and juvenile

▸ immense geographic variation in color and pattern

Widespread and common; feeds on ground in jerky motions. Breeds in northern and mountain forests. Winters along woodland edges or in thickets; frequently visits feeders. Where multiple populations co-occur in winter, they often sort out by subspecies; more study needed.

◀ Song varies locally and individually, but not among subspecies; all sing a loose trill, more musical than Chipping Sparrow. Call note a smacking *tick*.

ADULT, "WHITE-WINGED." (J. h. aikeni) *Larger than "Slate-colored," with white wing bars and more white in tail. Breeds in Black Hills; winters mainly in eastern Rockies.* COLORADO, MARCH

ADULT, "RED-BACKED." (J. h. dorsalis) *Bill bicolored (darker above, paler below). Throat pale gray; tertials and wing coverts tinged russet. Some give songs like Yellow-eyed Junco.* NEW MEXICO, JUNE

ADULT, "PINK-SIDED." (J. h. mearnsi) *Black in front of eye stands out on pale gray hood; pale russet wash on sides extensive, nearly meeting across lower breast. Breeds central Rockies.* COLORADO, DECEMBER

ADULT, "CASSIAR." (J. h. cismontanus) *Many hybrid pairings occur where ranges overlap; "Cassiar" Junco (breeds Canadian Rockies, winters widely) is intermediate between "Slate-colored" and "Oregon."* CALIFORNIA, NOVEMBER

ADULT, "GRAY-HEADED." (J. h. caniceps) *Mainly gray; mainly orange back forms high-contrast triangle on bird at rest. Black in front of eye contrasts with otherwise gray face. Breeds southern Rockies.* NEW MEXICO, DECEMBER

JUVENILE. *Abundant in late summer on breeding grounds, but rarely seen elsewhere. All are scruffy and streaky; all flash white outer tail feathers like adult.* CALIFORNIA, JULY

Yellow-eyed Junco
Junco phaeonotus　CODE 2

ADULT. Eye yellow; bill dark above, yellow below. Plumage similar to "Gray-headed" and especially "Red-backed" Dark-eyed Juncos, but wing coverts show extensive rusty. ARIZONA, DECEMBER

ADULT. All adults have staring yellow eyes; eye of juvenile starts dark, ordinarily becoming yellow before molt to first fall plumage.
ARIZONA, SEPTEMBER

L 6.25"	ws 10"	wt 0.75 oz (21 g)

- ▸ two adult molts per year; complex alternate strategy
- ▸ strong differences between adult and juvenile
- ▸ some resemble "Red-backed" race of Dark-eyed Junco

Replaces Dark-eyed Junco in the Sierra Madre, reaching North America in the sky islands of the Southwest; common around campsites and picnic areas. Species limits unclear: "Red-backed" Junco, currently placed within Dark-eyed, was formerly classified under Yellow-eyed.

Song generally distinct from Dark-eyed: consists of two or three discrete trills. Some "Red-backed" Juncos sing a similar song. Calls resemble Dark-eyed.

Smith's Longspur
Calcarius pictus　CODE 2

BREEDING MALE. Bright buff below extends to neck and nape; head and face boldly patterned in black and white. Bill thin. Outer tail flashes more white than Lapland.
MANITOBA, JULY

BREEDING FEMALE. Like a muted version of breeding male: buff below not as bright; face pattern not as bold. Nonbreeding male similar. All plumages fairly long-winged.
MANITOBA, JUNE

IMMATURE. Even the drabbest birds are warmer buff below than other longspurs; white wing bars often prominent. Thin bill, white tail edges, and fairly long wings always good marks.
CALIFORNIA, OCTOBER

L 6.25"	ws 11.25"	wt 1 oz (28 g)

- ▸ two adult molts per year; complex alternate strategy
- ▸ strong seasonal, age-related, sex-related differences
- ▸ individual variation in brightness of underparts

For most birders, a mid-continent specialty. Winters in cultivated fields, overgrazed pastures, and around airports, often in single-species flocks. Migrants are especially fond of stubble fields. On breeding grounds, occurs on flat terrain at forest-tundra transition.

Warbled song a mix of sweet and buzzy notes; short duration. Common flight call is a short rattle, higher-pitched than Lapland; also a sharp *pip*.

Lapland Longspur

Calcarius lapponicus CODE 1

BREEDING FEMALE. *Lacks black face and breast of male, but rusty nape, along with auriculars boldly framed in black, recalls facial pattern of male.* ALASKA, JUNE

BREEDING MALE. *Face and breast black; outlined by angled white-and-buff stripe. All plumages long-winged with stout bill; all flash white outer tail feathers in flight.* ALASKA, JUNE

L 6.25"	WS 11.5"	WT 1 oz (28 g)

▸ two adult molts per year; complex alternate strategy

▸ strong seasonal, age-related, sex-related differences

▸ in North America, eastern breeders paler than western

NONBREEDING MALE. *Retains a trace of black on breast; boldly framed auriculars and long wings always evident. Most individuals show rusty wing coverts, forming panel on bird at rest.* MINNESOTA, OCTOBER

Breeds on wet tundra, where it is typically the most abundant passerine; in some regions, it is the *only* breeding passerine. Forms huge flocks in core wintering range in Great Plains, often keeping company with Horned Larks; much scarcer in winter toward coasts.

◀ Song soft and ringing, lacking strong cadences. Flight calls, the most commonly heard vocalizations for most birders, include a dry rattle and harsh *pweer*.

Away from their breeding grounds, female and immature long-spurs can be difficult to identify. Wing length, bill structure, and tail pattern are important marks. Bear in mind that even the drabbest of longspurs typically shows a hint of the bold patterns of breeding males.

IMMATURE MALE. *Diffusely patterned, but general suite of marks identifies it as Lapland: well-defined auriculars, rusty wing patch, long wings, and stout bill.* CALIFORNIA, NOVEMBER

Chestnut-collared Longspur
Calcarius ornatus　CODE 1

BREEDING MALE.
A small, short-winged, fairly short-billed longspur. All plumages show extensive white in tail. Breeding male black below; head black and white; throat buff; nape chestnut.
MONTANA, JUNE

BREEDING FEMALE.
Pale sandy gray with variable black flecking on underparts. Female McCown's similar, but has rufous bar on wing (median coverts), which is lacking on Chestnut-collared.
MONTANA, JUNE

IMMATURE FEMALE.
Similar to McCown's; a little darker overall, with more contrast to facial pattern; McCown's has more white in tail. Bill thinner and shorter than McCown's.
NEVADA, OCTOBER

L 6"　　ws 10.5"　　wt 1 oz (28 g)

▸ two adult molts per year; complex alternate strategy

▸ strong seasonal, age-related, sex-related differences

▸ black on breast sometimes has chestnut highlights

Prime breeding habitat requires frequent fire and grazing. Favors taller grass than McCown's; where both species nest, separation according to vegetation height is often strictly observed. Numbers declining due to habitat deterioration on breeding and wintering grounds.

Song, given in flight, a short series of rich, buzzy, downslurred elements. Flight calls include a mellow *kiddle* and a variable rattle, often interspersed.

McCown's Longspur
Calcarius mccownii　CODE 2

BREEDING MALE. *The grayest longspur; large-billed, short-winged, and short-tailed. Black below is not as extensive as Chestnut-collared. Rufous wing bar (median coverts) often prominent.*
MONTANA, JUNE

BREEDING FEMALE.
Drabber than male; breast usually flecked with black. Rufous wing bar usually evident. All plumages show large bill, short wings, and short tail with much white.
MONTANA, JUNE

IMMATURE FEMALE.
Pale and plain. Upperparts mottled buff-and-sandy; underparts pale gray-brown with little streaking. Black-streaked cap contrasts weakly with paler eyebrow.
CALIFORNIA, OCTOBER

L 6"　　ws 11"　　wt 0.8 oz (23 g)

▸ two adult molts per year; complex alternate strategy

▸ strong seasonal, age-related, sex-related differences

▸ amount of black on breast variable in all plumages

Shortgrass prairie specialist; breeds in arid rangeland with prostrate cactuses and abundant cow patties. Winters, often with Horned Larks, in similar habitat; also in ploughed fields and dry lakebeds. Breeding range has contracted significantly; still contracting.

◀ Song, given in flight, an endless warble with distinctive lack of cadence. Flight calls include a short rattle and an abrupt *weert*, often interspersed.

Rustic Bunting

Emberiza rustica   CODE 3

BREEDING MALE. *Head mainly black with thick white line behind eye; chin white. Upperparts bright rusty; white below with long rusty streaks on breast and flanks.* ASIA, MAY

BREEDING FEMALE. *Black on head replaced by reddish brown; immature drabber. All plumages flash white in outer tail feathers in flight; all have relatively long, pinkish bill.* EUROPE, JUNE

L 6"	ws 9.5"	wt 0.7 oz (20 g)

▸ two adult molts per year; complex alternate strategy

▸ strong sex-related and moderate age-related differences

▸ amount of rusty on breast of juvenile variable

Common Asian breeder; nests in shrubby areas, east to Kamchatka Peninsula. Regular migrant in spring to western Aleutians; rare in fall. Casual elsewhere in Alaska and along Pacific coast south to California. When agitated, flicks tail and raises crown into short crest.

Song, sometimes given by migrants in Alaska, is a soft bubbling warble, sung rapidly: *eeleery-weary-wee-cheery.* Call note a sharp *chiv.*

McKay's Bunting

Plectrophenax hyperboreus   CODE 2

BREEDING MALE. *All plumages are whiter than corresponding plumages of Snow Bunting. Breeding male has all-white back, limited black in wings, and nearly white tail.* ALASKA, JULY

BREEDING FEMALE. *Similar to female Snow, but head has little if any black smudging; wings and back have more white. Identification of hybrids and variants not well worked out.* ALASKA, JULY

NONBREEDING. *Has warm rusty highlights like Snow Bunting, but much reduced in extent. Juvenile is pale dusky overall.* ALASKA, MARCH

L 6.75"	ws 14"	wt 1.9 oz

▸ one adult molt per year; complex basic strategy

▸ strong seasonal and age-related differences

▸ amount of white in wings and tail variable

All aspects of biology poorly known. Breeds only on Hall and St. Matthew Islands in Bering Sea. Winters mainly along coast of western mainland Alaska; extent of dispersal elsewhere unknown. Apparently hybridizes with Snow Bunting, but details are unknown.

All vocalizations usually reported to be identical to Snow Bunting, raising doubts about full-species status of McKay's; more field observations needed.

Snow Bunting
Plectrophenax nivalis   CODE 1

NONBREEDING. *Most winterers in lower 48 are dirty white overall, with variable rusty highlights on face, breast, and upperparts.* MINNESOTA, NOVEMBER

BREEDING MALE. *Black-and-white breeding garb is acquired through wear; despite dramatic differences between summer and winter, Snow Bunting has only one plumage per year.* ALASKA, JUNE

IN FLIGHT. *Throughout year, both sexes appear largely white from below. In distant flight, black outer wing contrasts sharply with white on rest of wing.* YUKON TERRITORY, APRIL

BREEDING FEMALE. *Duskier than breeding male: dark back and wings finely streaked and spotted with white; head usually smudged with a little black. Juvenile much duskier.* ALASKA, JUNE

L 6.75"   WS 14"   WT 1.5 OZ

▸ one adult molt per year; complex basic strategy

▸ strong seasonal and age-related differences

▸ Bering Sea breeders larger; have more white on back

In core winter range, occurs in flocks in weedy fields in open country; also occurs along seacoasts and lakeshores. Flocks can be either single-species or mixed-species. Males depart early for arctic breeding grounds, where they compete for nest sites in rock crevices.

JUVENILE. *Sooty overall. At rest, shows relatively little white in wing; juvenile McKay's Bunting shows more white. Juveniles are not seen south of the Arctic.* NUNAVUT, AUGUST

Song a buzzy warble: *zeeet azeery jurry jersey zeer jeet.* Flight calls include a longspur-like rattle, a descending *suwuoo,* and a very harsh buzz.

Buntings & Allies

FAMILY CARDINALIDAE

13 species: 9 ABA CODE 1   1 CODE 2   3 CODE 4

These colorful birds are members of a fairly small New World assemblage (approximately 42 species) of uncertain taxonomic affinities. All taxonomic authorities place the Cardinalidae somewhere toward the rear of the avian checklist, but the family's exact placement within the nine-primaried oscines (including warblers, sparrows, and blackbirds) is not agreed upon. In particular, it turns out that several cardinalid species should probably be assigned to the tanager family; conversely, several tanagers may actually be cardinalids. In the years ahead, birders can expect to see a fair bit of checklist shuffling where the family Cardinalidae is concerned.

Most of our cardinalids exhibit strong sexual dimorphism, with males being much more brightly colored than females. Identification of breeding males is ordinarily a cinch, but identification of females sometimes requires careful study of bill color and structure, as well as plumage contrast and pattern. Because of their gaudy colors, males are popular with aviculturists. This means that sightings out of range may refer to escaped birds, not to actual vagrants; the problem is acute with Northern Cardinal and Painted Bunting, which are especially popular as cage birds. Another identification conundrum involves hybridization: Indigo and Lazuli Buntings frequently hybridize with each other in the western Great Plains, and so do Rose-breasted and Black-headed Grosbeaks.

Cardinalids are seed eaters, with conical bills that run the gamut from fairly small (buntings) to massive (grosbeaks and cardinals). Bill structure is a reflection of diet, not of actual evolutionary relationships. For example, the large-billed Blue Grosbeak and small-billed Indigo Bunting are quite closely related (a fact that eluded taxonomists until recently), despite obvious differences in bill structure.

Migratory strategies in the family Cardinalidae are literally all over the map. Several species (for example, the Rose-breasted Grosbeak and Indigo Bunting) migrate long distances across the Gulf of Mexico. In the West, several other species stop in late summer at "molt-migration" sites between breeding and wintering grounds; the Lazuli Bunting is a conspicuous example. Still other species (the Pyrrhuloxia and Northern Cardinal) are nonmigratory.

Conservation priorities for the Cardinalidae are focused mainly on the highly migratory species. Most notable in this regard is the Dickcissel, which winters in immense flocks in a small region of the llanos of central Venezuela, where the species faces heavy persecution. Also of note is the recent realization that molt-migration sites may be of critical importance to cardinalids and other species. For example, marshes in the Desert Southwest, where Lazuli Buntings gather to molt, may be just as important in the annual cycle of that species as the breeding and wintering grounds.

Northern Cardinal
Cardinalis cardinalis   CODE 1

ADULT MALE, EASTERN. *Sleek and long-tailed; crest prominent; bill large and red. Adult male bright red all over, with irregular black patch around base of bill.* TEXAS, APRIL.

ADULT MALE, ARIZONA. *Similar to Eastern, but crest slightly longer and bill slightly thicker; black around base of bill reduced.* ARIZONA, FEBRUARY

JUVENILE. *Paler than adult female, but most have obvious reddish hues to wings, tail, and crest. Bill is dull, not red, but structure and body shape are as adult.* TEXAS, JUNE

ADULT FEMALE. *Largely warm tan-buff but with red in wings, tail, and crest; bill large and red, like male, with diminished black at base.*
MINNESOTA, JANUARY

L 8.75"   WS 12"   WT 1.6 OZ

▸ one adult molt per year; complex basic strategy

▸ strong age-related and sex-related differences

▸ Arizona cardinals have larger bills, longer crests

Conspicuous and familiar wherever it occurs. Most strongly associated with shrubby woodland edges, but ranges into forest interior and urban districts. Range expanding north and west, thanks to bird feeders, habitat alteration, and perhaps climate change.

◀ Both sexes sing. Song clear and ringing: *chur birdy birdy birdy birdy birdy burr* and *whit dyeer dyeer dyeer dyeer...* Call note a light, smacking *tyit.*

Pyrrhuloxia
Cardinalis sinuatus   CODE 1

ADULT MALE. *Mainly gray body has rosy highlights on wings, tail, underparts, face, and crest. All adults have yellow bill with strongly decurved culmen (upper mandible).*
TEXAS, NOVEMBER

DULL ADULT FEMALE. *Grayer than female Northern Cardinal; yellow bill has rounded culmen. Juvenile duller; bill uncolored. Juvenile Northern Cardinal warmer overall; bill structure different.*
TEXAS, MAY

L 8.75"   WS 12"   WT 1.3 OZ

▸ one adult molt per year; complex basic strategy

▸ strong sex-related and moderate age-related differences

▸ western birds smaller-billed and longer-tailed

The desert cardinal; occurs in dry washes, even well out in creosote bush flats. Where range overlaps with Northern Cardinal, Pyrrhuloxia favors drier habitats. Has benefited from agriculture and other human influences on the landscape; range slowly expanding northward.

Song pattern and delivery nearly identical to Northern Cardinal; individual notes average slightly thinner, but complete overlap. Call note a lisping *spitth*.

Black-headed Grosbeak
Pheucticus melanocephalus   CODE 1

BREEDING MALE. *Extensive burnt-orange on underparts extends to nape and rump. Nonbreeding male has less black above; orange not as intense.*
CALIFORNIA, APRIL

BREEDING FEMALE. *More buff overall than female Rose-breasted, especially below; breast streaking reduced. In all plumages, wing and tail patterns similar to Rose-breasted.*
CALIFORNIA, APRIL

BRIGHT ADULT FEMALE. *Some show extensive orange below; streaking can be extensive, but streaks always thin. In all plumages, bill duskier than Rose-breasted, especially upper mandible.*
CALIFORNIA, MAY

L 8.25"   WS 12.5"   WT 1.6 OZ

▸ two adult molts per year; complex alternate strategy

▸ strong sex-related, moderate age-related differences

▸ interior males have larger bills, lack eye stripe

Western counterpart of Rose-breasted; despite striking differences between adult males, the two species are closely related. Nests in diverse habitat types: from lush broadleaf riparian corridors to dry pine woods. Migrants visit feeding stations, orchards, and oases.

◂ All vocalizations similar to Rose-breasted. Song a little less vigorous, sung more slowly. Call note a little weaker: *chint*; flight call averages hoarser.

Rose-breasted Grosbeak

Pheucticus ludovicianus   CODE 1

ADULT FEMALE. *Most have coarse streaking on pale breast; female Black-headed has little or no streaking on buff breast. Pale bill always a good distinction from Black-headed.* TEXAS, APRIL

BREEDING MALE. *Mainly black-and-white with red patch on breast that forms inverted triangle; underwing linings flash red in flight. Nonbreeding male similar but scalier, "messier."* TEXAS, APRIL

IMMATURE MALE, FIRST SPRING. *Male plumages variable, but all show at least a hint of an inverted red triangle on breast.* TEXAS, MAY

IMMATURE MALE, FIRST SPRING. *In all plumages, oversized bill is entirely pale; hybrids and Black-headed have variable dark on bill, especially on upper mandible.* TEXAS, APRIL

L 8"   WS 12.5"   WT 1.6 OZ

▸two adult molts per year; complex alternate strategy

▸strong sex-related, age-related, seasonal differences

▸hybridizes extensively with Black-headed Grosbeak

Breeds in broadleaf woods, well into forest interior, but also in smaller lots and along edges. Avoids drier, sparser pinewoods. On breeding grounds, feeds high in trees. Among most frequent "eastern" vagrants out of range. Migrants and vagrants often visit feeders.

IMMATURE FEMALE, FIRST SPRING. *Some have thinner or reduced streaking below, with faint buff wash on breast; variant "pure" Rose-breasted sometimes impossible to separate from hybrids.* TEXAS, APRIL

◀Song a bright, sweet, lengthy warble; richer than Black-headed Grosbeak. Distinctive call note a squeaky *chink*. Flight call a short, rich, whistled *peep*.

Lazuli Bunting

Passerina amoena CODE 1

BREEDING MALE.
Bright blue upperparts with two white wing bars. Red-orange breast separates blue hood and white belly. CALIFORNIA, MAY

NONBREEDING MALE. *Variable; most show warm buff below and splotchy blue above with white wing bars. First-spring male similar.*
ARIZONA, DECEMBER

ADULT FEMALE.
Differs from female Indigo in having two cinnamon wing bars and cinnamon wash across breast; throat usually grayer than female Indigo.
CALIFORNIA, APRIL

L 5.5"   WS 8.75"   WT 0.5 oz (14 g)

▸ two adult molts per year; complex alternate strategy

▸ strong age-related and sex-related differences

▸ width and color of male breast band variable

Breeds in brushy habitats; especially characteristic of foothills and riparian areas. Breeding range expanding, probably in response to human modifications of landscape. After breeding, concentrates in large flocks in Desert Southwest, where it completes fall molt.

◀ Song more disorganized and less leisurely than Indigo: a hurried *wee-wee-wee-wee whit chew-chew-chew wit.* Call note and flight call similar to Indigo.

Blue Grosbeak

Passerina caerulea CODE 1

ADULT MALE. *Deep blue all over, with black at base of bill and two cinnamon wing bars. Larger-billed and larger overall than other Passerina buntings.*
TEXAS, APRIL

ADULT FEMALE.
Warm buff-brown all over. Distinguished from other Passerina buntings and female Brown-headed Cowbird by bill structure and cinnamon wing bars. TEXAS, APRIL

IMMATURE MALE, FIRST SPRING.
Amount of blue in plumage variable; most have at least some blue on head. Large bill and cinnamon wing bars always present.
TEXAS, APRIL

L 6.75"   WS 11"   WT 1 oz (28 g)

▸ two adult molts per year; complex alternate strategy

▸ strong sex-related and age-related differences

▸ eastern males darker; western birds longer-winged

Closely related to buntings. Nests in shrubby habitats: eastern birds in dry thickets and woodland edges, western birds in riparian habitats in arid landscapes. Frequent overshoot north of range. Actions usually slow and methodical, but sometimes excitedly raises crest.

Song a herky-jerky warble, distinctively buzzy. Call note distinct, a metallic *shint;* flight call a buzzy *jzzzd,* harsher than Indigo Bunting.

Indigo Bunting
Passerina cyanea   CODE 1

BREEDING MALE. *Completely blue, darker on head than rest of body. Bill relatively pale; bill size and body size smaller than Blue Grosbeak.* TEXAS, APRIL

IMMATURE MALE, FIRST SPRING. *Plumage mottled blue and brown. Most birds departing in early fall have acquired nonbreeding plumage; a few returning in spring retain nonbreeding plumage.* TEXAS, APRIL

TYPICAL ADULT FEMALE. *Pale sandy brown all over. Female Lazuli has brighter buff wash on breast (Indigo colder, often with diffuse streaks) and cinnamon wing bars. Female Indigo has paler throat than Lazuli.* TEXAS, APRIL

VARIANT ADULT FEMALE. *Some females (apparently only older individuals) acquire patchy blue highlights; can resemble immature males.* MAINE, MAY

L 5.5"   WS 8"   WT 0.5 oz (14 g)

▸ two adult molts per year; complex alternate strategy

▸ strong age-related, sex-related, seasonal differences

▸ hybridizes extensively with Lazuli Bunting

Common along brushy edges, power line rights-of-way, abandoned farms. Avoids both forest interior and districts with extensive human presence. Conspicuous nocturnal migrant; passage often continues well into morning. Status at western edge of range varies annually.

The ranges of the Indigo and Lazuli Buntings are expanding, with the result that the two species increasingly come into contact with one another. Hybrids are fairly common in the zone of overlap—especially in the foothills of the eastern Rockies and out onto the plains along creeks and rivers. First-generation hybrid males are fairly simple to recognize, but females and backcrosses are easily overlooked.

◀ Song a mix of buzzy and sweet notes, often paired: *swit swit, chew chew, spittit out chew.* Call note a smacking *spit*; flight call a buzzy, scratchy *zzzzt.*

Varied Bunting
Passerina versicolor CODE 2

BREEDING MALE. *Rich purple, plum, and red all over; deep hues cause bird to look black at a distance or in poor light.* ARIZONA, JULY

NONBREEDING MALE. *Retains blue on tail, rump, and face, but plum and red areas become brown. Immature male similar. All plumages shorter-winged than other buntings.* ARIZONA, JULY

ADULT FEMALE. *Pale overall, with little plumage contrast above or below. Bill appears bulbous, due to decurved culmen (upper mandible).* ARIZONA, JULY

L 5.5" WS 7.75" WT 0.4 oz (11 g)

▸ two adult molts per year; complex alternate strategy

▸ strong age-related, sex-related, seasonal differences

▸ western females and subadults more reddish-brown

An uncommon bunting of our southern borderlands; favors arid woodlands, thorn scrub, and overgrown clearings. Avoids residential districts. Breeding depends on rainfall; nesting sometimes postponed until August. May be expanding north in western portion of range.

Song more warbled, more indeterminate than Lazuli and Indigo; bright and sweet, with one or two buzzy notes. Call note a dry *spist*; flight call like Indigo.

Painted Bunting
Passerina ciris CODE 1

ADULT MALE. *Red, green, and blue plumage unmistakable, but lighting and foliage can obscure color and pattern. Distinctive plumage held all year.* ARIZONA, JULY

ADULT FEMALE. *Body structure similar to Indigo, but low-contrast plumage is notably green-hued, especially on upperparts.* TEXAS, APRIL

IMMATURE MALE. *Similar to adult female. Some are fairly yellow below, as here; others are more grayish. All have extensive green above.* TEXAS, MAY

L 5.5" WS 8.75" WT 0.5 oz (14 g)

▸ two adult molts per year; complex alternate strategy

▸ strong sex-related and age-related differences

▸ eastern males darker red, females darker yellow-green

Habitat as other buntings: semi-open areas with brush and scattered trees. Western birds molt during fall migration, eastern birds prior to fall migration. Both populations declining: eastern because of development of swampy thickets, western for unclear reasons.

Song a cheery warble; higher than Lazuli, usually of shorter duration. Call note a hollow, resonant *peek*; flight call, *pzzzd*, more buzzy than Indigo.

Dickcissel

Spiza americana   CODE 1

BREEDING MALE.
Yellow eyebrows with gray-and-white head; yellow breast separated from white chin by thick black crescent. Wing coverts have extensive chestnut.
TEXAS, APRIL

ADULT FEMALE.
All have at least a dab of yellow in eyebrow and on breast, plus chestnut in wings; some are duller than shown here. Also note long, thick bill and thin black throat stripe.
TEXAS, MAY

IMMATURE FEMALE.
Drab overall with yellow-tinged eyebrow; suggests female House Sparrow, but longer-billed and longer-winged, usually with a little streaking below.
CALIFORNIA, SEPTEMBER

L 6.25"   WS 9.75"   WT 1 oz (28 g)

▸ two adult molts per year; complex alternate strategy
▸ moderate age-related, sex-related, seasonal differences
▸ black bib of male variably mottled with white

Requires remnant or restored grasslands for breeding. Common in core Midwestern range; at periphery of range, numbers vary considerably from year to year, even within a single season. Frequent vagrant to West Coast; regular migrant in small numbers along East Coast.

◀ Song usually with two distinct parts: *chip chip wisss wisss wisss* (whence "Dickcissel"). Flight call, given by nocturnal migrants, a harsh, flatulent buzz.

CONSERVATION GOALS

A major goal for conservation biologists is to reverse population declines and stabilize populations. Many states and provinces have devised "Bird Conservation Plans" that specify these goals: for example, 0.2 Dickcissels per acre by 2010 or 150 pairs of Dickcissels in a particular location by 2012. These types of conservation plans are powerful tools, because they prescribe objectives that require funding and the creation of jobs to implement them.

Another type of conservation strategy, perhaps the best of all, is called an "Adaptive Management Plan." This type of plan explicitly acknowledges that populations of certain species are naturally unstable in the short term. Natural instability is a defining feature of the native grasslands of the Great Plains, and populations of birds, such as Dickcissels, have always fluctuated in response to drought, fire, and availability of food. A problem is that on top of natural sources of population instability, human disturbances, such as agriculture, livestock grazing, and suburbanization add challenges to population stability and growth. It is the combination of both natural and "unnatural" disturbances that can push many populations to the brink of extinction. Adaptive Management Plans characteristically embrace a hands-on approach to conservation—actively intervening on behalf of native species with revegetation, crop rotation, and controlled burns.

Blackbirds & Allies

FAMILY ICTERIDAE

25 species: 17 ABA CODE 1   5 CODE 2   1 CODE 4   2 CODE 5

A long with several wholly or nearly black-colored birds, the family Icteridae includes the festively colored orioles and meadowlarks. The icterids are a New World group, numbering about 98 species, with the greatest diversity at middle and low latitudes in both hemispheres.

It is convenient to recognize three major groups in North America: the blackbirds (including grackles and cowbirds), the orioles, and the meadowlarks. The Bobolink and Yellow-headed Blackbird, despite their blackbird-like morphology and behavior, appear to be more closely related to the meadowlarks than to other blackbirds.

Icterids display a sophisticated array of behaviors, including elaborate nest construction, complex vocalizations, diverse mating systems, and extreme gregariousness during the nonbreeding season. The Red-winged Blackbird, for example, has been the subject of several classic monographs on animal behavior, and the species provides endless fascination for a human visitor to its marshland breeding grounds. Several species of orioles build conspicuous and easily recognized nests, offering observers the chance to study an aspect of bird biology—called nidification—that most other bird species are highly secretive about. And the Great-tailed Grackle, despite its reputation for obnoxiousness, produces one of the most remarkable vocal arrays of any North American bird species.

Body structure in the icterids is highly varied, including such examples as the warbler-like Orchard Oriole, the crow-sized Great-tailed Grackle, and the oddly starling-like meadowlarks. The most uniform character in the icterids is the bill—dark, long, and slender in most species. But even bill structure is somewhat variable in the Icteridae; the Bobolink and especially the Brown-headed Cowbird have stout, conical, rather sparrow-like bills.

Several blackbird species are famously adaptable, and are reckoned as pests by both agricultural and conservation interests. Common Grackles are frequent nest predators, and Brown-headed Cowbirds lay their eggs in the nests of other species—a behavior called brood parasitism. The immediate consequences of nest predation and parasitism are devastating, but conservation biologists disagree about the large-scale and long-term impacts of grackles and cowbirds on populations of small songbirds. Other successful icterids include the Brewer's Blackbird and the Hooded and Scott's Orioles, all of which are expanding their ranges into human-modified habitats. But some species have not proven as adaptable. Bobolinks are poisoned in immense numbers on their South American wintering grounds, Rusty Blackbirds are decreasing sharply on their boreal breeding grounds for reasons that are unclear, and the little-known Audubon's Oriole is on the decline in Texas.

Bobolink
Dolichonyx oryzivorus CODE 1

BREEDING MALE. *Darker below than above. Nape has extensive cream-colored patch; rump, uppertail coverts, and scapulars white.*
CONNECTICUT, MAY

BREEDING FEMALE. *Long-winged and short-tailed. Cold buff all over, with streaking above and on flanks. Dark line behind eye and dark cap separated by tan eyebrow.*
MINNESOTA, JUNE

NONBREEDING. *Both sexes molt into buff-yellow plumage in late summer; juvenile similar but a little more colorful. Suggests an oversized, long-winged Ammodramus sparrow.*
NEW JERSEY, SEPTEMBER

| L 7" | WS 11.5" | WT 1.5 OZ | ♂>♀ |

▸ two adult molts per year; complex alternate strategy

▸ strong age-related, sex-related, seasonal differences

▸ returning spring males variably tinged with buff

Breeds in tall grass or mixed grass prairie; accepts hayfields and restored grasslands, but sensitive to disturbance. Leaves breeding grounds early in fall; many linger in freshwater coastal marshes to complete fall molt. Migrates in flocks, usually by day.

◀ Song, often given in flight, a complex, bubbly warble; fast and rich, many notes twangy. Call note a dull *chup*; flight call a distinctive rising *wenk?*

Brewer's Blackbird
Euphagus cyanocephalus CODE 1

BREEDING MALE. *Glossy all over, often with dark blue hue on head and metallic green hue on body; duller in winter. Shorter-tailed and shorter-billed than Common Grackle.*
CALIFORNIA, JUNE

ADULT FEMALE. *Lacks coppery tones of Rusty. Eye dark; female Rusty has pale eye. Bill a little shorter and thicker than Rusty. Bill of Rusty slightly decurved; bill of Brewer's straight.* TEXAS, MARCH

| L 9" | WS 15.5" | WT 2.2 OZ | ♂>♀ |

▸ one adult molt per year; complex basic strategy

▸ strong sex-related and weak seasonal differences

▸ females vary considerably in overall ground color

Familiar visitor to parking lots and golf courses, but also flourishes in natural habitats such as bogs and stream edges. Range expanding; apparently competes with Common Grackle where ranges overlap. After breeding season, joins large mixed-species blackbird flocks.

Song, with gasping quality, dies off quickly; a weak, rising *k'wheee?* May be either pure or raspy. Flight call a snappy *check*, usually separable from Rusty.

Rusty Blackbird
Euphagus carolinus CODE 1

ADULT MALE. *Less glossy than Brewer's Blackbird and Common Grackle. Slightly decurved bill is longer and thinner than Brewer's; tail much shorter than Common Grackle.* NEW JERSEY, DECEMBER

BREEDING FEMALE. *Dull gray-brown all over with little gloss. Separated from female Brewer's by pale eye and thinner, longer, slightly decurved bill.* NEW YORK, MAY

NONBREEDING MALE. *Starts to acquire coppery highlights by midsummer. Overall extent of coppery coloration variable, but most males show less than female.* MINNESOTA, OCTOBER

NONBREEDING FEMALE. *By early winter, most are extensive coppery-brown. Black around eye and base of bill contrasts with yellow eye and coppery color on face.* MINNESOTA, OCTOBER

L 9"	ws 14"	wt 2.1 oz	♂ > ♀

▸ one adult molt per year; complex basic strategy

▸ strong seasonal and sex-related differences

▸ eastern Canadian breeders average darker

The wilder counterpart of Brewer's. Breeds in wet northern forests; migrants and winterers use swamps and damp woods. Nonbreeding aggregations generally smaller than other blackbirds; does not usually mix with other species. Migrates early in spring, late in fall.

Song a squeaky gurgle, distinct from Brewer's, often with terminal rising whistle: *k'sh'weea'shweea'sh'shweeeee.* Flight call a short, dull *chut.*

Populations of several blackbird species are enjoying sustained growth. Red-winged and Brewer's Blackbirds, for example, along with all three of our cowbirds and all three of our grackles, have been increasing in number for decades. The Rusty Blackbird, however, has recently undergone a sharp decline. This drop-off has caught biologists by surprise, as the species was long thought to be secure. The reasons for the population losses are unknown. Researchers suspect that habitat conversion on the Rusty's boreal breeding grounds and wintering grounds may be an influence.

Bronzed Cowbird
Molothrus aeneus   CODE 1

ADULT MALE. *Bulkier overall and thicker-billed than Brown-headed; all adults have red eyes. Both sexes bullnecked; male puffs out ruff in display.* TEXAS, MAY

ADULT FEMALE. *Duller than male, lacking glossy blue-black in wings and tail. West of Texas, females paler: sandy gray below; dusky brown above. Juvenile similar; diffusely streaked below.* TEXAS, APRIL

L 8.75"	WS 14"	WT 2.2 oz	♂ > ♀

▸ two adult molts per year; complex alternate strategy

▸ moderate age-related and sex-related differences

▸ eastern females dull black, western females gray-brown

Habits and habitats similar to more widespread Brown-headed Cowbird: obligate brood parasite that has benefited from human modifications to the landscape; prefers slightly more open habitat than Brown-headed. Timing and extent of migration poorly understood.

Song, given in shimmying-and-hovering display, a lengthy series of high, sputtering whistles. Flight call a long, high, descending whistle: *tzeeeee.*

Shiny Cowbird
Molothrus bonariensis   CODE 2

ADULT MALE. *Glossy blue-black all over. Bill longer and pointier than Brown-headed; tail a little longer.* FLORIDA, APRIL

ADULT FEMALE. *Similar to female Brown-headed; female Shiny usually shows a pale eyebrow, weak or lacking in Brown-headed. Juvenile similar, but shows diffuse streaks below.* FLORIDA, MAY

L 7.5"	WS 11.5"	WT 1.3 oz	♂ > ♀

▸ one adult molt per year; complex basic strategy

▸ strong age-related and sex-related differences

▸ male gloss variable, especially on younger birds

Obligate brood parasite. Range expanding in South America. Reached Florida as natural colonist in 1985, and showed signs of spreading rapidly; however, expansion has slowed greatly. Uncommon on lawns, along roads, and at bird feeders; mainly in southern Florida.

Song a whistled jumble; timbre like Brown-headed, but lacks stuttering effect. Two flight calls: guttural rattle like Brown-headed; short, descending *tzeep.*

Brown-headed Cowbird

Molothrus ater   CODE 1

ADULT MALE. *Head glossy brown; body glossy blue-black or green-black. Appears all black at a distance or in poor light.* CALIFORNIA, JUNE

ADULT FEMALE. *Sooty gray all over, a little paler below than above; underparts weakly streaked. Small size, short tail, and conical bill make the species seem somewhat sparrow-like.* TEXAS, APRIL

JUVENILE. *Similar to female but a little brighter, with bolder streaking below; upperparts scaly. Often seen begging noisily from a foster parent smaller than itself.* OREGON, AUGUST

L 7.5"   ws 12"   wt 1.5 oz   ♂ > ♀

▸ one adult molt per year; complex basic strategy
▸ strong age-related and sex-related differences
▸ western birds smaller, western females paler

Common but unobtrusive; favors forest edges, orchards, parks, and suburbs. Obligate brood parasite; lays eggs in other species' nests. Long-term and large-scale effects on other species unclear. Originally found mainly in mid-continent; range has expanded greatly.

◀ Song, often presented by multiple males to a single female, a rising, tinkling gurgle. Flight calls include a three-note whistle and a long, guttural rattle.

BROOD PARASITISM

The Brown-headed Cowbird is one of the most maligned bird species in North America. A "brood parasite," its young are raised only by "foster parents" of other species. Females lay their eggs in the nests of another species, the cowbird eggs hatch quickly, and the chicks develop rapidly. The cowbird hatchlings evict the host's young from the nest, and their surrogate parents—evolutionarily programmed to feed a clamorous nestling—devote all their energy to feed the "parasite."

It is perhaps distressing to observe this process, but several points need to be borne in mind. First, the Brown-headed Cowbird is a native species that has parasitized the nests of other species since well before the arrival of humans in North America. Second, brood parasitism is a complex and fascinating behavior that has long intrigued evolutionary biologists. Third, it comes as a surprise to many people that the long-term and large-scale impacts of cowbird parasitism may be relatively minor, with several studies having failed to show population-level effects of cowbirds on their hosts.

Cowbird populations do have serious local effects, especially on Threatened and Endangered species. Most researches agree, for example, that cowbirds have negatively affected populations of "Least" Bell's Vireos in California and Golden-cheeked Warblers and Black-capped Vireos in Texas. The influence of cowbirds on most other songbirds is not at all clear, however, and cowbird control may be unwarranted in many instances.

Common Grackle

Quiscalus quiscula   CODE 1

ADULT MALE, "BRONZED." (Q. q. versicolor) *All adults are sleek and glossy, males especially so. Long tail looks keel-shaped in flight. Widespread "Bronzed" Grackle has glossy blue head and coppery body.* ILLINOIS, APRIL

ADULT MALE, "PURPLE." (Q. q. stonei) *Replaces "Bronzed" in the coastal plain north to mid-Atlantic states. Bill thinner and plumage more purplish overall than "Bronzed."* FLORIDA, APRIL

ADULT FEMALE. *Similar to male, but plumage less glossy. Many are noticeably smaller than males, but body size overlaps broadly.* TEXAS, APRIL

JUVENILE. *Sooty brown all over; often ratty-looking. Juvenile has dark eyes; adults of all populations have yellow eyes.* MISSOURI, JUNE

L 12.5"	WS 17"	WT 4 OZ	♂ > ♀

- ▸ one adult molt per year; complex basic strategy
- ▸ moderate age-related and weak sex-related differences
- ▸ "Purple" Grackle occurs east of Appalachians

One of the earliest passerine migrants in spring. Conspicuous in many aspects of life cycle: during diurnal migration, in courtship, and while strutting about lawns and golf courses. Secretive, however, at loosely colonial nest sites in conifers. Range expanding west.

◀ Song grating, with forced quality: *k'sharzh*, given as male spreads wings. Call note a harsh *keerzh*; flight call, heard frequently, a nasal *chap*.

ADULT FEMALE, "BRONZED." (Q. q. versicolor) *The most migratory subspecies of Common Grackle; all vagrants to the West appear to be of the "Bronzed" subspecies.* TEXAS, APRIL

Great-tailed Grackle

Quiscalus mexicanus CODE 1

ADULT MALE. *Larger overall than Common Grackle, with a proportionately longer and larger tail; much individual variation in size within populations. Plumage glossed purplish all over.* TEXAS, NOVEMBER

ADULT FEMALE. *Smaller and shorter-tailed than male. Upperparts dark brown; underparts vary from warm buff to cold gray. Most have pale buff eyebrow. Eye pale; juvenile has dark eye.*
ARIZONA, JANUARY

L 15–18" ws 19–23" wt 3.7–7.0 oz ♂ > ♀

▸ one adult molt per year; complex basic strategy

▸ strong sex-related and moderate age-related differences

▸ western adults smaller, western females paler

Originally a tropical species that extended north only into south Texas. Range expanded rapidly in late 20th century; now established in much of western U.S. Occurs in small to large aggregations in wet meadows and around feedlots. Conspicuous at nest colonies.

Song consists of diverse and expressive elements: for example, *wheeea! cha-cha-cha k'chick-k'chick-k'chick schwee schwee.* Call note more muffled than Common.

Boat-tailed Grackle

Quiscalus major CODE 1

ADULT MALE. *Longer-tailed than Common. In zone of overlap with Great-tailed (western Gulf coast), most adults are dark-eyed; Great-tailed pale-eyed. Atlantic coast Boat-tailed is pale-eyed.* TEXAS, MAY

ADULT FEMALE. *Most are warmer brown than female Great-tailed, but variable; eyebrow usually not as prominent as Great-tailed. Geographic variation in eye color as male.*
TEXAS, MAY

L 14.5–16.5" ws 17.5–23" wt 4.2–8.0 oz ♂ > ♀

▸ one adult molt per year; complex basic strategy

▸ strong sex-related and moderate age-related differences

▸ hybridizes with Great-tailed where ranges overlap

Coastal specialist, limited mainly to tidewater marshes. In summer, females gather in large colonies in breeding system called harem polygyny; in winter, both sexes assemble in compact, secretive flocks. Range expanding north; slowly invading inland habitats, too.

Song repertoire less rich than Great-tailed, consists of harsh notes given in series: *k'jeer k'jeer k'jeer...* and *chig chig chig...* Call note a dull *guck.*

Red-winged Blackbird

Agelaius phoeniceus   CODE 1

ADULT MALE. Prominence of red-and-yellow wing coverts ("epaulettes") variable: striking in display; conspicuous in flight; often reduced to thin sliver on bird at rest. FLORIDA, FEBRUARY

IMMATURE MALE. *Body heavily scalloped with rusty; variable. Some adult males acquire rusty scalloping in winter, too, but it is not as extensive as on immature.* NEW MEXICO, JANUARY

ADULT MALE, "BICOLORED." *(A. p. californicus) Amount of yellow in wing coverts varies geographically and individually; "Bicolored" Blackbird of California lacks yellow altogether.* CALIFORNIA, APRIL

ADULT FEMALE. *Variable, but always heavily streaked below; most have prominent eyebrow. Most are dark brown, with warm tones; some have pinkish tinge on breast.* TEXAS, MAY

ADULT FEMALE, "BICOLORED." *(A. p. californicus) Darker than most other females; mainly black-and-white, lacking warm tones. Combination of extensive white on breast and nearly black belly distinctive.* CALIFORNIA, MAY

L 8.75"   WS 13"   WT 1.8 oz   ♂ > ♀

▸ one adult molt per year; complex basic strategy

▸ strong sex-related and moderate age-related differences

▸ "Bicolored" Blackbird occurs in California

Perhaps the most abundant bird in North America. Breeds in wetlands of all sorts; wanders into croplands to feed. In winter, forms immense roosts; smaller parties visit feeders. Arrives very early in spring; males establish marshland territories immediately.

◀ Variable song usually contains long, grating rattle: *unka je'e'e'e'e'e' chiss.* Many calls: piercing *syeeer*, flat *chack*, light *tik*, and others.

Tricolored Blackbird

Agelaius tricolor   CODE 2

ADULT MALE. *Wing coverts red and white. Bill thinner and pointier than Red-winged; wings a little pointier. Immature male has gray scalloping all over; warmer rusty on Red-winged.*
CALIFORNIA, MAY

ADULT FEMALE. *Darker, colder than female Red-winged; breast and chin have dark streaks on light gray background. Female "Bicolored" Red-winged similar, but averages warmer and browner.*
CALIFORNIA, MAY

L 8.75"   WS 14"   WT 2.1 OZ   ♂ > ♀

▸ one adult molt per year; complex basic strategy
▸ strong sex-related and moderate age-related differences
▸ a few females show pinkish or peach highlights

Formerly abundant in cattail and bulrush marshes of California's Central Valley; has declined sharply in core range, but is also establishing new colonies east and north. Forms larger, denser colonies than Red-winged; prefers larger marshes with thicker vegetation.

Song rougher and lower than Red-winged; usually lacks long, wavering *je'e'e'e'e'e'* note of Red-winged. Call note a low, muffled *chutt.*

Yellow-headed Blackbird

Xanthocephalus xanthocephalus   CODE 1

ADULT MALE. *The "banana head," with a completely yellow hood. Rest of body is black, with white upperwing coverts; white patch largely concealed at rest.*
MONTANA, JUNE

ADULT FEMALE. *Brown crown and auriculars; yellow or buff eyebrow. Upper breast yellow, lower breast streaked white; rest of body brown. Immature similar but paler, more diffusely patterned.* UTAH, JUNE

IN FLIGHT. *White upperwing coverts of male are prominent in flight; all dark on female. All adults appear large overall and well-built.*
COLORADO, MAY

L 9.5"   WS 15"   WT 2.3 OZ   ♂ > ♀

▸ one adult molt per year; complex basic strategy
▸ strong age-related and sex-related differences
▸ body size increases from southeast to northwest

Unmistakable inhabitant of deep-water marshes with cattails, where it breeds in highly social colonies. Returns later in spring, leaves earlier in fall than many other blackbirds. In winter, small groups join large mixed-species flocks at feedlots and stubble fields.

Song loud and exceedingly harsh: *ugga gzzzzzzh.* Call notes include a liquid *plip* and a drier *chippit.*

Eastern Meadowlark

Sturnella magna CODE 1

NONBREEDING ADULT. *More washed-out than breeding plumage. Dark stripes on head contrast more strongly than on Western. Juvenile similar, but paler and scalier.*

FLORIDA, FEBRUARY

ADULT MALE. *All meadowlarks are plump and bobtailed; all breeders are bright yellow below, with broad black crescent on breast. Eastern has white malar region; yellow on Western.*

TEXAS, FEBRUARY

L 9.5"　　ws 14"　　wt 3.2 oz　　♂ > ♀

▸ one adult molt per year; complex basic strategy

▸ moderate seasonal and age-related variation

▸ "Lilian's" (*S. m. lilianae*) may be distinct species

Icon of farm pastures in eastern North America; often seen strutting along the ground or singing from fence post. Where ranges overlap in Midwest, Western Meadowlark occurs in drier habitats; in Desert Southwest, though, Western in moister habitats than "Lilian's".

ADULT, "LILIAN'S." (*S. m. lilianae*) *Inhabits deserts and grasslands in Southwest. Paler than eastern population. Tail shows more white; in flight, flashes 3–4 outer tail feathers.*

TEXAS, SEPTEMBER

IN FLIGHT. *Widespread eastern population flashes white in 2–3 outer tail feathers. All meadowlarks look flat-backed in flight; fly with short glides alternating with choppy downbeats.*

NEW JERSEY, JUNE

◀ Song consists of 4–6 sweet, downslurred whistles: *teeyee eeteeyah.* Call notes a piercing *zeert* and a rattling buzz; flight call a long, rising *sweeet.*

Western Meadowlark

Sturnella neglecta CODE 1

L 9.5"　　ws 14.5"　　wt 3.4 oz　　♂ > ♀

▸ one adult molt per year; complex basic strategy

▸ moderate age-related and seasonal differences

▸ birds in northwestern part of range slightly darker

Characteristic of western grasslands; also occurs in shrublands, even those with little grass. Habits much as Eastern Meadowlark. In early spring, territorial males are often the most conspicuous avian inhabitant of lonely open country. Range expanding northeast.

ADULT MALE. *Paler than eastern population of Eastern, but a little darker than "Lilian's" Eastern (S. m. lilianae). Malar region yellow; white on Western. Nonbreeding adult and juvenile paler.*

CALIFORNIA, MARCH

◀ Very loud song more complex than Eastern; mix of low gurgles and high whistles. Call notes a short *terp* and slow rattle; flight call a rising, nasal *sween.*

Baltimore Oriole

Icterus galbula   CODE 1

ADULT MALE. *Head entirely black; underparts deep orange. Orange upperwing coverts usually show as crescent on bird at rest; orange on tail extends to tips of outer feathers.*
TEXAS, APRIL

ADULT FEMALE. *Some individuals are like disheveled versions of the adult male: black hood splotched with orange; usually shows two white wing bars.*
TEXAS, APRIL

ADULT FEMALE. *Some adult females and most immatures of both sexes have little black on face; just a dusky wash. Almost all (except adult male) have pale region between eye and bill (lores).*
TEXAS, MAY

L 8.75"   WS 11.5"   WT 1.2 OZ

‣ one adult molt per year; complex basic strategy
‣ strong age-related and sex-related differences
‣ hybridizes extensively with Bullock's Oriole

Nests widely in eastern broadleaf woods, both in extensive tracts and in small groves; not averse to urban parks and leafy suburban districts. Usually feeds in canopy. Migrates day and night; migrants tarry at feeders, especially those supplemented with oranges.

◀ Song a series of rich whistles, delivered slowly. Call note a rising rattle; flight call a short, rising *weert?* Calls often incorporated into song.

Bullock's Oriole

Icterus bullockii   CODE 1

ADULT MALE. *Differs from adult male Baltimore in having extensive orange-yellow on head and large white wing panel. Hybrids with Baltimore show intermediate characters.*
CALIFORNIA, MAY

IMMATURE MALE. *Has black throat patch and thin black eye patch; plumage held through end of first year. Intensity of orange or orange-yellow variable in all plumages.*
CALIFORNIA, MAY

ADULT FEMALE. *Like female Baltimore, variable. Almost all show fairly bright orange on face (duskier on Baltimore) and dark loral region (paler on Baltimore).*
ARIZONA, APRIL

L 9"   WS 12"   WT 1.3 OZ

‣ one adult molt per year; complex basic strategy
‣ strong age-related and sex-related differences
‣ smaller in coastal southern California than elsewhere

Characteristic of riparian habitats from low deserts to mountain meadows; prominent in shelterbelts and around campgrounds. Departs early from breeding grounds to molt in Desert Southwest. Some migrate at night, but small groups are also often on the move by day.

◀ Song like Baltimore, a whistled jumble that incorporates calls, but notes weaker, overall delivery more hurried. Call note and flight call similar to Baltimore.

Hooded Oriole
Icterus cucullatus   CODE 1

ADULT MALE. *Black "beard" contrasts with orange underparts and head. Two white wing bars are prominent, especially the upper one. Tail black.* ARIZONA, APRIL

IMMATURE MALE. *Ground color variable, but tends toward yellow; more yellow-orange in adult male. Wing bars diminished; black "beard" averages narrower.* CALIFORNIA, APRIL

ADULT FEMALE. *Distinguished from female Bullock's by yellow belly (white in Bullock's) and longer tail; bill is thinner, longer, and more decurved than Bullock's. Longer-tailed and duller overall than female Orchard Oriole.* CALIFORNIA, JUNE

L 8"   ws 10.5"   wt 0.85 oz (24 g)

‣ one adult molt per year; complex basic strategy
‣ strong age-related and sex-related differences
‣ Texas birds brighter, shorter-winged, longer-tailed

Range expanding in Desert Southwest; especially fond of native and exotic tree-palms. Has flourished in suburban and even busy urban districts with adequate plantings; also inhabits oases and riparian strips, where it co-occurs with closely related Bullock's.

Variable, bunting-like song is faster and more variable than Bullock's. Call note a clipped *chut*; flight call a rising, musical *feerk?*

Audubon's Oriole
Icterus graduacauda   CODE 2

ADULT MALE. *Suggests Scott's, but has less black. Back yellow and most of breast yellow (back and breast black on Scott's). Head of juvenile dusky yellow, becoming black in first fall.* TEXAS, APRIL

ADULT FEMALE. *Variable, but averages paler than male, especially on wings; some females darker than shown here. All are large and yellow-hued, with a long, tapered bill.* TEXAS, FEBRUARY

L 9.5"   ws 12"   wt 1.3 oz

‣ one adult molt per year; complex basic strategy
‣ strong age-related and weak sex-related differences
‣ overall brightness and contrast of female variable

Little-known species that enters our area in southern Texas, where it inhabits a broad array of woodlands: riparian forest, thorn scrub, and live oak. More secretive than other orioles; actions sluggish. Some withdrawal in winter; not well understood. Numbers declining.

Song, sung by both sexes, consists of rich whistles, varying in pitch, given slowly: *wheeer... yare... eeyear... wurt? weeep.* Call note also whistled.

Scott's Oriole
Icterus parisorum CODE 1

ADULT MALE. *Head, breast, and back all black; bright yellow below. Black wings have two prominent wing bars, the upper bar half-yellow.* CALIFORNIA, APRIL

L 9"	ws 12.5"	wt 1.3 oz

▸ one adult molt per year; complex basic strategy

▸ strong age-related and sex-related differences

▸ amount of black on adult female highly variable

Locally common in arid environments: sparse juniper woodlands in foothills, lower-elevation Joshua tree forests, even out into creosote bush flats. Wanders unpredictably out of range, out of habitat. Range has expanded north; apparently still expanding.

Song a defining element of the desert sunrise: cheery, disjointed whistles, similar to Western Meadowlark. Calls low and gruff: *shup* and *jerp.*

IMMATURE MALE. *Similar to adult male, but plumage duskier. Younger birds, with diminished black on head, not always differentiable from adult females.* NEW MEXICO, MAY

ADULT FEMALE. *Dusky overall, especially on head; most have limited black on chin and breast. All Scott's have long, thin, straight bills; bill of Hooded more decurved.* CALIFORNIA, APRIL

IMMATURE FEMALE. *Uniform dusky coloration is extensive, covering all of head and upperparts; extends well onto breast. Long, thin, straight bill usually prominent.* ARIZONA, MAY

Spot-breasted Oriole
Icterus pectoralis   CODE 2

ADULT. *The only "bearded" oriole likely to be seen in Florida. Face bright orange; orange on breast dotted with namesake spots. Sexes similar, but female averages slightly duller.* FLORIDA, APRIL

IMMATURE. *Colors duller and pattern more muted than adult. Spots on breast often absent; "beard" diminished. All plumages flash white in wings.* FLORIDA, JANUARY

L 9.5"	ws 13"	wt 1.5 oz

- one adult molt per year; complex basic strategy
- moderate age-related and weak sex-related differences
- individual variation in brightness of orange

Native to Central America. Introduced in Miami area in early 20th century; was breeding by mid-century. Has never been abundant; occurs in small numbers in suburban districts with lush plantings. Population currently in decline; probably fewer than 100 pairs.

Song similar to Baltimore, but richer: clear whistles and a few deeper piping notes, *deer deer deer jare deer deer jare.* Call note a descending *jip.*

Altamira Oriole
Icterus gularis   CODE 2

ADULT. *Like an oversized, thick-billed Hooded. Adult vibrant orange; upper wing bar orange (white on Hooded). Female averages duller than male, but sexes not always distinguishable.* TEXAS, DECEMBER

IMMATURE. *Duller than adult: black on breast less extensive; wings not as contrastingly marked. Usually appears obviously larger with thicker base of bill than Hooded and Bullock's.* MEXICO, SEPTEMBER

L 10"	ws 14"	wt 2 oz

- one adult molt per year; complex basic strategy
- moderate age-related, weak sex-related differences
- individual variation in brightness of orange

With Audubon's, one of the two southern Texas oriole specialties; Altamira the more conspicuous, and probably more numerous. Altamira inhabits groves, especially near water. Striking nest is an oversized, pendulous affair, hung from the tip of a high branch.

Song like Audubon's, but not as distinctive: notes not as rich and widely spaced; scratchy *clench* sometimes mixed in. Call note whistled and abrupt.

Orchard Oriole

Icterus spurius CODE 1

ADULT MALE.
Extensive dark chestnut below; mainly black above. Chestnut areas can appear black or dark brown in poor light. Smaller and slighter than other orioles. TEXAS, APRIL

IMMATURE MALE.
Yellowish overall with variable greenish hue. Black on chin extends a short distance onto breast. Usually shows two white wing bars on dark olive wings. ILLINOIS, MAY

ADULT FEMALE.
Olive-yellow above with two white wing bars; dusky yellow below. Plumage, structure, and actions suggest a large warbler. TEXAS, APRIL

L 7.25"	WS 9.5"	WT 0.7 oz (20 g)

▸ two adult molts per year; complex alternate strategy
▸ strong age-related and sex-related differences
▸ northern birds larger, not as darkly colored

Active, warbler-like oriole; gleans arthropods from foliage. Breeds along forest edges and wooded waterways. Numbers vary locally; common in some places, absent from other areas of seemingly suitable habitat. Departs early in fall. Range slowly expanding.

Song bright and warbled; intermediate between buntings and other orioles. Call note a scratchy chatter like other orioles; flight call a flat, nasal *yennk.*

DIVERSITY PARADOX

Various regions of North America harbor more bird species at the present time than at any point in recorded history. Southern Florida, for example, harbors established populations of Purple Swamphens, Black-hooded Parakeets, Red-whiskered Bulbuls, Common Mynas, and Spot-breasted Orioles—all recent additions to the state's avifauna. Similar scenarios have played out in southern California and even more spectacularly in Hawaii.

Local and regional bird faunas have been supplemented with native species, too. Cities in the Desert Southwest host increasing numbers of Anna's Hummingbirds and Great-tailed Grackles, and the East Coast megalopolis is increasingly attractive to Herring Gulls, Northern Mockingbirds, and even Peregrine Falcons. The phenomenon is not restricted to urban regions. Fragmented forests and agricultural districts support a greater diversity of birds than did the habitats they replaced.

The paradox is striking. Across the continent, dozens—probably a few hundred—species are in trouble, yet species diversity is seemingly everywhere on the increase. What is happening is that humans are transforming landscapes in such ways as to favor population expansion by adaptable, generalist species; in the process, less-adaptable, specialist species lose out. Conservation biologists refer to this phenomenon as homogenization. The diversity paradox has received relatively little formal study to date, but it is guaranteed to be one of the central biological questions in the years to come.

Finches

FAMILY FRINGILLIDAE

23 species: 11 ABA CODE 1   5 CODE 2   1 CODE 3   4 CODE 4   2 CODE 5

Famous for their nomadic wanderings, the finches are best represented in North America in mountainous and boreal habitats. Worldwide, the finches are better represented in the Northern Hemisphere than in the Southern. Like a number of other families toward the end of the avian checklist, the finches currently find themselves in a position of taxonomic limbo. In the past, they were placed near the New World sparrows in the nine-primaried oscines group, but molecular evidence has caused controversy. One point that has been clarified by molecular evidence is the close relationship between the finches and the morphologically diverse honey-creepers of Hawaii.

Nearly all North American finches are prone to wandering. Most legendary in this regard is the Red Crossbill, which is segregated into at least nine "types" that disperse widely in search of conifer seeds; intriguingly, these different "types" may correspond to distinct species that are as yet unnamed. Several finches undergo periodic irruptions into regions far beyond the breeding range; examples include the Pine Siskin and Common Redpoll. Finches are also notable for winter wanderings within smaller geographic regions; roving flocks of rosy-finches are notorious for being unreliable to find at winter feeding stations. And not only the boreal species are nomadic; the Lawrence's Goldfinch, which breeds mainly in California's arid foothills, disperses erratically eastward in the fall and winter.

Finch vocalizations are complex, and several species are prized by aviculturists for their vocal talents. The goldfinches are good mimics, with the Lesser Goldfinch often including a surprising number of other species' songs in its repertoire. Many finches sing indeterminate songs that can be hard for birders to master, but the distinctive flight calls of most species are significantly useful in the identification process. The House Finch has been the subject of several influential studies by behavioral ecologists who study song learning and variation in dialects.

Not surprisingly, given their nomadic tendencies, several finch species have ranges that are currently unstable. Following deliberate release in New York in 1940, the House Finch quickly spread through eastern North America, but it is now declining in some places. The winter range of the Lesser Goldfinch is rapidly expanding, for reasons that await discovery. And the range of the Evening Grosbeak has been unstable, especially in the East, for at least a century; the planting of exotic trees and provisioning of sunflower seeds at winter feeding stations may be factors in their changing distribution.

House Finch
Carpodacus mexicanus CODE 1

ADULT MALE. *Typical adult, shown here, has broad red eyebrow, red on chin and breast, and red on rump. Underparts have coarse brown streaking.* CALIFORNIA, JANUARY

ADULT MALE. *Some males show extensive red, related to diet and age; even the brightest birds, as here, show brown auriculars and gray-brown back.* NEW JERSEY, OCTOBER

ADULT MALE. *Color variants, especially orange and yellow, are frequently seen. Distribution of color on this individual is typical of the species.* ARIZONA, MAY

JUVENILE. *Similar to female and immature, but ground color of underparts and wing bars tends toward dirty buff; dirty white in female and immature.* CALIFORNIA, JUNE

| L 6" | WS 9.5" | WT 0.75 oz (21 g) |

▸ one adult molt per year; complex basic strategy

▸ strong age-related and sex-related differences

▸ much individual variation in color of adult males

Occurs in small to medium flocks, often at feeders. Native to West; widespread in East, where introduced in 1940. Western birds inhabit cities, farms, and deserts. Eastern birds, experiencing disease-related decline, are limited mainly to areas of human habitation.

ADULT FEMALE. *Dirty gray-brown above; coarsely streaked below. Diffuse head pattern shows little contrast; ridge of culmen (upper mandible) curved. Immature (both sexes) similar.* CALIFORNIA, OCTOBER

◀ Variable song a bright, loose warble composed of twangy notes, often with rising, metallic note at end. Flight call twangy and metallic: *seewin* or *sweeoot?*

Purple Finch
Carpodacus purpureus   CODE 1

ADULT MALE, EASTERN. *Diffuse raspberry all over; color on House Finch not nearly as extensive. Streaking below reduced or absent; House Finch usually heavily streaked below.* NEW HAMPSHIRE, OCTOBER

ADULT FEMALE, EASTERN. *Streaked brown and white like female House, but more boldly patterned. Face pattern suggests female Rose-breasted Grosbeak.* NEW HAMPSHIRE, OCTOBER

ADULT FEMALE, WESTERN. *Midway in appearance between Cassin's and Eastern Purple; pale stripes on face not as thick as Eastern Purple, but lacks thin eye ring of Cassin's.* CALIFORNIA, JANUARY

L 6"   WS 10"   WT 0.9 oz (25 g)

▸ one adult molt per year; complex basic strategy

▸ strong age-related and sex-related differences

▸ western birds shorter-winged, females blurrier

Associated with large or small woodlands; breeds mainly in association with conifers, but feeds on buds and flowers of broadleaf trees. Eastern birds more migratory than western; both populations susceptible to periodic irruptions. Eastern birds may be in decline.

Song a long, rambling warble; less herky-jerky than House Finch, lacking metallic notes. Flight call a short *tik*, musical and liquid; also a vireo-like *eeseeyoo*.

Cassin's Finch
Carpodacus cassinii   CODE 1

ADULT MALE. *Crown often raised in fine crest. Head and breast imbued with rose; upperparts brown-pink. Undertail coverts may show a few fine streaks, but underparts otherwise plain.* CALIFORNIA, MARCH

ADULT FEMALE. *Plumage distinctively finer-grained than House and Purple. Bill longer and straighter than House and Purple; wings long and pointed. Immature (both sexes) similar.* CALIFORNIA, MARCH

L 6.25"   WS 11.5"   WT 0.9 oz (25 g)

▸ one adult molt per year; complex basic strategy

▸ strong age-related and sex-related differences

▸ individual variation in amount of red in males

Replaces Purple Finch in the Inter-mountain West. Closely tied to coni-fer forests, from the ponderosa pine zone up to near tree line. Conspicu-ous and typically numerous where present, but populations variable; most undergo regular latitudinal and altitudinal migrations.

Song with buzzy elements like House Finch, but more rapidly delivered, often longer overall; often imitates other species. Flight call a slurred *chiddeeup*, suggesting Western Tanager.

Pine Grosbeak
Pinicola enucleator CODE 1

ADULT MALE. *Large and long-tailed, with a stout black bill. All plumages have dark wings with two white wing bars; adult males are gray and red.*
MINNESOTA, JANUARY

ADULT MALE. *Some show greenish yellow highlights, as here. Interior West populations show least red; Canadian breeders usually show extensive red. Immature like adult female.*
MINNESOTA, JANUARY

ADULT FEMALE. *Mainly gray plumage shows bright olive-green highlights on head and rump. Called "snow parrot" for its size, color, tameness, feeding actions, and bill structure.*
MINNESOTA, JANUARY

L 9"   WS 14.5"   WT 2 OZ

▸ one adult molt per year; complex basic strategy

▸ strong age-related and sex-related differences

▸ geographic variation in amount of red on male

Very tame; easily approached as it feeds on the ground or in berry-laden shrubs. Breeds in moist spruce-fir forests, favoring clearings or other disturbed areas. More sedentary than most finches; eastern populations irrupt infrequently, western populations almost never.

Variable song a gentle warble, more leisurely than Cassin's Finch; individual notes slurred, with little buzz. Flight call a musical *wheeda-pwee* or shorter *pwee*.

White-winged Crossbill
Loxia leucoptera CODE 2

ADULT MALE. *Like Red Crossbill, has crossed mandibles that are often hard to see. Plumage pink in winter; pinkish red in summer. Thick white wing bars always prominent.*
MAINE, FEBRUARY

ADULT FEMALE. *Patterned like adult male, but pink of male replaced by dull yellow-green on female. Most adult females show diffuse streaking below.*
MAINE, JANUARY

JUVENILE. *Wing bars thinner than adult. (Some juvenile Red Crossbills also show thin wing bars.) In all plumages, White-winged ordinarily looks smaller-headed than Red.*
ALASKA, JUNE

L 6.5"   WS 10.5"   WT 0.9 OZ (25 g)

▸ one adult molt per year; complex basic strategy

▸ strong age-related and sex-related differences

▸ streaking on underparts of female and juvenile variable

Habits similar to Red Crossbill: wanders extensively in search of food; extracts seeds from cones with crossed bill. Southern limit of breeding range unstable. Range overlaps broadly with Red, but White-winged more partial to tamarack and especially spruce.

Variable song averages faster, drier, and shriller than Red: *je je je je je je* and *jet jet jet jet jet.* Flight call a harsh, downslurred *chet.*

Red Crossbill
Loxia curvirostra   CODE 1

ADULT MALE. *Crossed mandibles highly distinctive, but surprisingly hard to see. Typical adult males have dull brick-red coloration all over; dark wings lack wing bars.* CONNECTICUT, JUNE

ADULT FEMALE. *Variable, but most have dull green or greenish yellow cast. Differences in bill size and structure among "types" are difficult to assess in the field.* OREGON, AUGUST

IMMATURE MALE. *Most have orangish cast, intermediate between adult male and female. Field assessment of age and sex difficult; many show the "wrong" characters.* OREGON, AUGUST

JUVENILE. *Plumage resembles Carpodacus finches. Bill structure, if seen well, diagnostic; head appears oversized compared to genus Carpodacus.* OREGON, AUGUST

L 6.25"	WS 11"	WT 1.3 OZ

▸ one adult molt per year; complex basic strategy

▸ strong age-related and sex-related differences

▸ complex variation in adult color and bill size

Opportunistic and highly nomadic; wanders widely in search of conifer seeds. Specialized feeding behavior reflected in different bill morphologies, perhaps representing eight or more species; these different "types" often occur within the same range and habitat.

◀ Song highly variable but usually ringing, pulsing: *jeery j'jeery j'jeery...* Flight call of all "types" low, rich, and fairly pure: rising *jeet*, falling *jip*, and others.

ADULT MALE. *Crossbills sometimes descend to the ground (roadsides especially favored) to pick up grit to aid in digestion; they can be surprisingly approachable when feeding on the ground.* NEW HAMPSHIRE, MAY

Gray-crowned Rosy-Finch

Leucosticte tephrocotis   CODE 1

NONBREEDING ADULT, "INTERIOR." (L. t. tephrocotis) Population breeding in northern Rockies has chocolate body; gray extends from eye to rear of crown. Female duller than male. Forehead and chin black.
CALIFORNIA, MARCH

NONBREEDING ADULT, "HEPBURN'S." (L. t. littoralis) Widespread Alaskan breeder, extending south along Pacific slope of Rockies. Gray on head covers most of face.
COLORADO, MARCH

IMMATURE. Duller than adult, but basic pattern the same: gray extends from eye to crown and nape; darker on forehead and chin. All rosy-finches have yellow bills in winter.
COLORADO, JANUARY

L 6.25"   WS 13"   WT 0.9 oz (25 g)

▸ one adult molt per year; complex basic strategy

▸ moderate seasonal, age-related, sex-related differences

▸ geographic variation in amount of gray and black

The most widespread rosy-finch, with diverse migratory strategies: some populations are latitudinal migrants, some altitudinal, some sedentary. Breeds amid rocks above tree line. In winter, large flocks visit feeders in mountains and foothills, then suddenly depart.

Most frequently heard vocalization is the flight call, a muffled *choof*. Song consists of series of flight calls and *toop* notes given in quick succession.

Brown-capped Rosy-Finch

Leucosticte australis   CODE 2

BREEDING MALE. On breeding grounds, most rosy-finches are separated by geographic range; all breeders have dark bills. Brown-capped has low-contrast plumage; lacks light gray on head. COLORADO, JULY

NONBREEDING ADULT. Dark brown overall, with low-contrast plumage. Other rosy-finches show light gray on crown in nonbreeding plumage.
COLORADO, JANUARY

IMMATURE. Paler than adult; often scaly above, as in other rosy-finch species. Difficult to distinguish from pale immature Black and Gray-crowned.
NEW MEXICO, DECEMBER

L 6.25"   WS 13"   WT 0.9 oz (25 g)

▸ one adult molt per year; complex basic strategy

▸ moderate seasonal, age-related, sex-related variation

▸ some individuals show diffuse gray in crown

Scarce even in heart of breeding range in southern Rockies; greatly outnumbered by Horned Lark and American Pipit. In winter, assembles with other rosy-finches in large, mobile flocks; rarely descends even as low as foothills, preferring feeders in the mountains.

Flight call similar to Gray-crowned, but more ringing, less muffled: *cheer* or *choor*. Song composed of flight calls and similar calls, given quickly.

Black Rosy-Finch
Leucosticte atrata   CODE 2

NONBREEDING ADULT. *Pattern much like Interior Gray-crowned, but ground color black (chocolate in Gray-crowned). All rosy-finches have variable pinkish on belly, rump, and wings.* NEW MEXICO, DECEMBER

NONBREEDING. *Dark gray all over, with reduced pink. Identification of dull subadult rosy-finches is little studied, and extent of hybridization unknown; many may be unidentifiable.* NEW MEXICO, DECEMBER

L 6.25"   WS 13"   WT 0.9 oz (25 g)

▸ one adult molt per year; complex basic strategy

▸ moderate seasonal, age-related, sex-related variation

▸ hybridizes with Gray-crowned and Brown-capped

Habits much as other rosy-finches: for example, males of all species defend "floating" territories that move with their mates; winter flocks mobile, unpredictable. Like other rosy-finches, roosts in abandoned outbuildings; mineshafts may be especially valuable to Black.

Flight call similar to other rosy-finches, but averages a little harder: *chip*, somewhat like Red Crossbill. Song simple; similar to other rosy-finches.

Common Redpoll
Carduelis flammea   CODE 1

ADULT MALE. *Both sexes have red forehead ("poll"); adult male has rosy suffusion across breast, stronger in summer than winter. Flanks streaked, sometimes weakly so, as here.* NEW HAMPSHIRE, JANUARY

ADULT FEMALE. *Bill yellow in all plumages; bill small, but averages larger than Hoary. Juvenile similar to adult female, but browner overall; red in forehead duskier.* MINNESOTA, JANUARY

L 5.25"   WS 9"   WT 0.5 oz (14 g)

▸ one adult molt per year; complex basic strategy

▸ moderate seasonal, age-related, sex-related differences

▸ overall paleness, streakiness, and pinkness variable

Winter dispersal from breeding grounds in spruce-birch woods variable: in some years, large numbers move far south; in other years, only a few do so. In flight years, large flocks feed in birches and weedy fields; lone birds tend to find their ways to feeders.

Three main calls: a dry buzz; a clipped *chew*, often repeated; and a whistled *s'wee?* like Pine Siskin. Song consists of combinations of calls given quickly.

Hoary Redpoll

Carduelis hornemanni CODE 2

ADULT MALE. *Very pale, with limited streaking below and little or no pink on breast; variant Common can be almost as pale. All redpolls are small-billed; bill of Hoary is tiny.* NEW HAMPSHIRE, JANUARY

ADULT FEMALE. *All plumages of Hoary paler than corresponding plumage of Common. Wing feathers tipped frosty white ("hoary"). Immature darker than adult; some cannot be separated from Common.* ALASKA, JUNE

L 5.5"	ws 9"	wt 0.45 oz (13 g)

- one adult molt per year; complex basic strategy
- moderate seasonal, age-related, sex-related differences
- geographical and individual variation in paleness

Habits similar to those of closely related Common Redpoll. On breeding grounds, ranges farther out into shrubby tundra than Common. Irruptions linked to those of Common, but Hoary does not usually wander as far south; even in flight years, only a few reach the lower 48 states.

Call notes analogous to Common, but with the following differences: buzz is faster; *chew* notes are softer, more clipped; whistled note is purer.

Pine Siskin

Carduelis pinus CODE 1

ADULT MALE. *Small and streaky; bill unusually thin and pointy for a finch. Male has yellow in wings and tail, visible as thin slivers on bird at rest. Tail short; deeply notched.* CALIFORNIA, JANUARY

ADULT FEMALE. *Streaky overall; shows little or no yellow in wings and tail. Bill and tail structure aid in identification. Immature similar.* MAINE, JANUARY

L 5"	ws 9"	wt 0.5 oz (14 g)

- one adult molt per year; complex basic strategy
- moderate sex-related and weak age-related differences
- amount of yellow variable, especially on adult male

Breeds mainly in coniferous forests of the boreal zone. Winters widely; regionally common in irruption years, scarce and local otherwise. Flocks devour thistle seeds at feeders. Some winterers stay behind and breed well out of normal range.

◄ Flight call, often given by birds high overhead, a downslurred *cleeyoo*; call note a harsh, rising buzz, *zhreee-eeet?* Song a goldfinch-like jumble.

American Goldfinch
Carduelis tristis   CODE 1

BREEDING MALE. *Bright yellow with black wings and black cap. Bill pinkish orange in breeding plumage (both sexes); undertail coverts white in all adult plumages.* NORTH DAKOTA, JUNE

BREEDING FEMALE. *Similar to male, but yellow not as bright; upperparts heavily washed with olive brown. Breeding female has pale wing bars, lacking in breeding male.* NORTH DAKOTA, JUNE

NONBREEDING MALE. *Bright yellow of breeding plumage replaced with pale gray-yellow. Wings remain black, usually showing stronger wing bars. Black cap absent in nonbreeding plumage.* CALIFORNIA, JANUARY

NONBREEDING FEMALE. *Quite pale; some show little yellow. Juvenile more brightly colored, not as pale as nonbreeding female; has warm buff cast overall, with bright buff wing bars.* CALIFORNIA, FEBRUARY

L 5"   WS 9"   WT 0.45 oz (13 g)

▸ two adult molts per year; complex alternate strategy

▸ strong seasonal, age-related, sex-related differences

▸ geographic variation in paleness and plumage contrast

Familiar winter visitor to thistle feeders; common in summer at gardens with sunflowers, cosmos, and other composites. Waits until midsummer to breed; onset of nesting apparently tied to availability of thistles. Migrates mainly by day in high bounding flight.

◀ Song bright and bubbly, whence its popular moniker of "wild canary." Call note a rising *suwee?* Flight call a distinct *t'chip chip chip* ("potato chip").

EUROPEAN GOLDFINCH (C. carduelis). *Native to Old World. Formerly established in New York; escapes from captivity seen continent-wide. Face red, white, and black; wings black and yellow.* AFRICA, MARCH

Lesser Goldfinch
Carduelis psaltria CODE 1

ADULT MALE, BLACK-BACKED. *Tiny; noticeably smaller than American Goldfinch. Males in eastern portion of range tend to be black above; black-backed males sometimes seen farther west.*
TEXAS, MAY

ADULT MALE, GREEN-BACKED. *In most of range, male has dusky green back. All males are bright yellow below; all have white at base of primaries, conspicuous in flight.*
CALIFORNIA, MARCH

ADULT FEMALE. *Dull yellow gray all over; white at base of primaries not as extensive as on male. Undertail coverts usually yellow; white on American. Juvenile similar.*
CALIFORNIA, MARCH

L 4.5"   WS 8" WT 0.35 oz (10 g)

▸ two adult molts per year; complex alternate strategy
▸ strong age-related and sex-related differences
▸ marked geographic variation in back color of male

Nests in open woodlands of diverse sorts: low-elevation oak woods in Pacific coast region, montane pine woods in Rockies. Small flocks feed inconspicuously in weedy pastures and flower-filled meadows; most readily detected in flight. Winter range rapidly expanding.

Warbled song is rambling and halting, often including imitations of other species. In flight, mixes distinctive, kitten-like *dyeer* with harsh chatter.

Lawrence's Goldfinch
Carduelis lawrencei CODE 2

BREEDING MALE. *Gray and yellow, with a black face. Breast bright yellow; wings show extensive yellow. Nonbreeding plumage similar, but gray upperparts replaced with olive-brown.*
CALIFORNIA, APRIL

BREEDING FEMALE. *Lacks black face of male, but most have yellow on breast and wings. Tail of both sexes has white subterminal band, visible from below and in flight.*
CALIFORNIA, APRIL

ADULT FEMALE. *Some females lack yellow on breast and show reduced yellow in wings. Nonbreeding female and juvenile browner; juvenile diffusely streaked below.*
CALIFORNIA, APRIL

L 4.75" WS 8.25" WT 0.4 oz (11 g)

▸ two adult molts per year; complex alternate strategy
▸ strong age-related and sex-related differences
▸ amount of yellow variable, especially on female

Breeds in arid woodlands, often near water, in foothills of Pacific slope of California; breeding status irregular from year to year. After breeding, many disperse east into Desert Southwest, some reaching western Texas; magnitude of dispersal varies from year to year.

Song faster, more clipped than other goldfinches; often with imitations of other species. Call note an emphatic *pity-tea!* Flight call a rolling *deelip*.

Evening Grosbeak
Coccothraustes vespertinus   CODE 1

ADULT MALE. *A large, bullheaded, huge-billed finch. Male has yellow eyebrow and variable yellow on belly, back, and undertail coverts. White secondaries prominent in flight and at rest.* MONTANA, JUNE

ADULT FEMALE. *Like male, but paler; yellow and chocolate of male replaced by fairly neutral gray-brown on female. Juvenile of both sexes resembles adult female.* MONTANA, JUNE

L 8"   WS 14"   WT 2.1 oz

▸ two adult molts per year; complex alternate strategy

▸ strong age-related and sex-related differences

▸ western adults longer-billed; western females darker

Conspicuous at feeders, where rowdy flocks feast on sunflower seeds; comparatively subdued on boreal nesting grounds. Winter wanderings irregular; locally common in some years, absent in others. Eastern population in flux, due perhaps to changing land use.

Variable flight calls include a clear *cheep* like House Sparrow and a trilled *brrrr*, the two calls often intermixed. Song, apparently a short warble, is rarely heard.

Brambling
Fringilla montifringilla   CODE 3

NONBREEDING MALE. *Black above; orange-buff below. Orange extends to scapulars and upperwing coverts. Breeding male more strikingly black-and-orange; bill becomes black in summer.* EUROPE, MARCH

ADULT FEMALE. *Black upperparts of male replaced by gray on female. Most show extensive buff-orange on breast and wings. All plumages appear long-winged and fairly long-tailed.* EUROPE, FEBRUARY

L 6.25"   WS 11"   WT 0.75 oz (21 g)

▸ one adult molt per year; complex basic strategy

▸ strong age-related and seasonal differences

▸ amount of black on back and head variable

Migratory Eurasian species; breeds in northern birch forests. Regularly reaches western Aleutians; singles or small flocks arrive mainly in late spring. Rare elsewhere in Bering Sea region. Vagrants widely noted across North America, usually at feeders in winter.

Song a harsh, clanging buzz, *dwa'a'a'ng*, or a lazier *bweeeen*. Flight calls include a harsh *brzeezt* and a nasal *jeg*, often repeated.

Weaver-Finches
FAMILY PASSERIDAE

---

2 species: 1 ABA CODE 1   1 CODE 2

Two species of weaver-finches have been introduced to North America, with different results. The House Sparrow quickly spread across the continent following its 19th-century release in New York, whereas the Eurasian Tree Sparrow has expanded only slowly in the general vicinity of its 19th-century release site in St. Louis. Despite being called "sparrows," neither species is closely related to the New World sparrows in the family Emberizidae. Instead, they form a complex with several Old World families that consist of the waxbills, mannikins, and bishops—several species of which frequently escape from captivity in North America, and two of which (the Orange Bishop and Nutmeg Mannikin) are established and probably expanding in the Los Angeles area.

House Sparrow
Passer domesticus   CODE 1

L 6.25"	WS 9.5"	WT 1 oz (28 g)

▸ one adult molt per year; complex basic strategy
▸ strong seasonal, age-related, sex-related differences
▸ geographic differences in size, shape, and paleness

Introduced from Europe in the 1800s; now established, but declining, across most of continent. Occurs in farm country and in large cities; absent from deep woods and remote deserts. Sociable flocks hide quietly in shrubbery, then flush on whirring wings.

◀ Song a simple *chirp*, given quickly or slowly, in seemingly endless succession. In agitation, gives a dry *ji'ji'ji'ji'ji*. Fight call a dull *jigga*.

NONBREEDING MALE. *Freshly molted birds in fall are mainly gray below and on head; feathers wear down during winter, gradually exposing sharp pattern seen in spring and summer.*
COLORADO, DECEMBER

BREEDING MALE. *Distinctive patterns and colors acquired by plumage wear. Throat and breast black; cheeks and crown gray. Chestnut extends from eye to nape. Bill black on breeders.* TEXAS, APRIL

ADULT FEMALE. *Drab; tan and brown. Most show a broad buff stripe behind eye and a white wing bar. Overall build is hefty; short-tailed and short-winged. Juvenile similar.*
TEXAS, APRIL

Eurasian Tree Sparrow
Passer montanus   CODE 2

ADULT. *Crown chocolate; chin black. White face has round black spot below ear. Sexes similar, unlike House Sparrow; plumage similar throughout year, also unlike House Sparrow.* ILLINOIS, NOVEMBER

JUVENILE. *Duller than adult, but shows basic pattern of rusty crown and white cheek with dark spot. Juvenile bill yellow; gradually blackens.* MISSOURI, AUGUST

L 6"	WS 8.75"	WT 0.8 oz (23 g)

- ▸ one adult molt per year; complex basic strategy
- ▸ moderate age-related and slight seasonal differences
- ▸ amount of yellow and black on adult bill variable

Like House Sparrow, was introduced from Europe in the 19th century, but range has not expanded greatly. Also like House Sparrow, forms tight flocks and is associated with humans, but more inclined to woodlots, farms, and urban parks. Sedentary and unobtrusive.

Vocalizations similar to House Sparrow, but sharper, lacking liquid quality. Song a series of sharp *chip* notes; flight call a metallic *jet* or *tek*.

EXOTIC WEAVERS AND FINCHES

Bishops, Manikens, and finches in the families Ploceidae and Estrildidae are popular cage birds that frequently escape from captivity. Some have established self-sustaining populations in the wild.

ORANGE BISHOP (Euplectes fransiscanus). *Native to Africa. Established in Los Angeles area. Orange-and-black male stunning; female, shown here, suggests Grasshopper Sparrow.* NEW YORK, SEPTEMBER

NUTMEG MANNIKIN (Lonchura punctulata). *Native to Asia. Like Orange Bishop, also established in Los Angeles area. Warm brown above; stippled brown and white below.* ASIA, OCTOBER

ZEBRA FINCH (Taeniopygia guttata). *Native to Australia. Popular cage bird; escapes from captivity possible anywhere, as are escapes of other exotic weaver-finches (family Estrildidae).* AUSTRALIA, JANUARY

Acknowledgments

Author's Acknowledgments

Learning how to identify birds is inevitably a collaborative effort. Through interactions with other birders, we get better at recognizing bird calls, field marks, and flight patterns. Learning and teaching, sharing and discovering—those things all get rolled together in the same package. That has been my own experience as a birder, and that was the genesis and unfolding of the *Smithsonian Guide*. The guide is a distillation of everything that everybody has shared with me about birds over the years.

I have had especially rewarding interactions with the technical consultants for the *Smithsonian Guide*. They are Jessie Barry, Donna Dittmann, Jon Dunn, Nathan Pieplow, Bill Pranty, Peter Pyle, Noah Strycker, Rick Wright, and the Smithsonian Institution reviewers. These experts read large chunks of the galleys of the *Smithsonian Guide*, and they made many helpful corrections and improvements. The detailed range maps were created by Paul Lehman, one of North America's top experts on avian status and distribution. The audio DVD of birdsongs and other vocalizations was produced beautifully by Lang Elliot with contributions by Kevin Colver and Ted Mack.

Technical accuracy is helpful in a field guide only inasmuch as it is conveyed effectively, and I am grateful to the following persons for improving the rhetoric and presentation of the *Smithsonian Guide*: Allan Burns, Macklin Smith, and Jay Withgott. Also, Noah Strycker and Rick Wright stepped outside their role as technical consultants, and supplied me with good ideas about readability and audience. Elisabeth Ammon, Ned Brinkley, Steve Carbol, Cameron Cox, Chris Elphick, Kimball Garrett, Steve Howell, Alvaro Jaramillo, Betsy Neely, Bryan Patrick, Brian Patteson, Brian Sullivan, and Jessica Young answered questions and provided advice—sometimes without even realizing they were doing so.

The bird images that fill the pages of the *Smithsonian Guide* have been supplied by a veritable "who's who" of North American bird photographers: Brian Small, Mike Danzenbaker, Brian Wheeler, Jim Zipp, Kevin Karlson, Bill Schmoker, Garth McElroy, Brian Patteson, Robert Royse, and many others. Many excellent images were supplied by VIREO (Visual Resources for Ornithology), and I thank director Doug Wechsler for procuring the images from VIREO's amazing library of photographs.

George Scott of Scott & Nix conceived the *Smithsonian Guide*, oversaw all aspects of its creation, and was always there to help. I have met very few people as deeply committed to quality as George is. Charles Nix applied extreme care and exquisite craftsmanship both to the "big picture" of the entire guide and to innumerable little details on every page. Editor Paul Hess read every word of every draft of the *Smithsonian Guide*, and made untold contributions in the areas of technical accuracy and readability. Through it all, Paul modeled not only intellectual rigor and the utmost of professional competence, but also the virtues of patience, encouragement, and equanimity.

The *Smithsonian Guide* is ultimately a verbal and pictorial statement of my own experiences and impressions in the field. In this regard, I have enjoyed the companionship of literally thousands of other birders, especially in recent years in connection with my work for the American Birding Association, but three field companions stand apart from the rest. My daughter Hannah was not old enough to read during the time that I wrote the *Smithsonian Guide*, but she was old enough to provide boundless energy, bewitching commentary, and precocious field skills in the course of both short jaunts to the park and long day trips. My son Andrew wasn't even old enough to talk during the writing of the *Smithsonian Guide*, but he was an obliging, amiable, and keenly attentive companion. Finally, my wife Kei—who has accompanied me on birding expeditions over the years throughout North America and beyond—did more than merely tolerate the writing of the *Smithsonian Guide*. It is not even sufficient to say that she was amazingly patient. Rather, she cheerfully encouraged and actively contributed to my work on the guide. She counseled me on writing, she nudged me toward important human resources, and she promoted and organized sanity-preserving family birding trips throughout the gestation of the *Smithsonian Guide*.

Scott & Nix, Inc.

George Scott and Charles Nix extend many thanks to Russell Galen of Scovil, Chichak, Galen, the entire team of professionals at HarperCollins, including Jane Friedman, Bruce Nichols, Karen Lumley, Jean Marie Kelly, Lisa Hacken, and the acquiring editor Phil Friedman, and editor Mathew Benjamin, both formerly of HarperCollins. Our gratitude goes to Linda Readerman and Ngoc Trice of Imago and Imago/Bright Arts for their consistently excellent work and good humor. We thank Christina Wiginton and Carolyn Gleason at Smithsonian Books in Washington, D.C. for their helpful advice and coordination.

We thank the many photographers who contributed to the project, especially Brian E. Small, whose beautiful bird images are at the heart of this guide.

Special thanks are due to Dmitriy Baronov, who painstakingly and accurately recorded many thousands of photographic images. He also spent many hours carefully interpreting the range information for the more than 700 maps in the book. Thanks are also due to Whitney Grant, who adroitly created layouts and revisions, and finally, many thanks to Erich Nagler, Gary Robbins, Katharina Siefert, Jemme Aldridge, May Jampathom, James Montalbano, Caleb Clauset, Stephen O'Brien, and Chandi Perera.

Scott & Nix, Inc. welcomes comments and questions about this book addressed via e-mail to *contact@scottandnix.com*.

Recommended References and Resources

ALDERFER, J. AND J.L. DUNN. *Birding Essentials*. National Geographic Society, Washington, 2007.

AMERICAN BIRDING ASSOCIATION. *ABA Checklist: Birds of the Continental United States and Canada*, sixth edition. American Birding Association, Colorado Springs, 2002.

AMERICAN ORNITHOLOGISTS' UNION. *The A.O.U. Check-list of North American Birds*, third edition. American Ornithologists' Union, Washington, 1998.

BAICICH, P.J., AND C.J.O. HARRISON. *A Guide to the Nests, Eggs, and Nestlings of North American Birds*. Academic Press, San Diego, 1997.

BARROW, M.V. *A Passion for Birds: American Ornithology After Audubon*. Princeton University Press, Princeton, 1998.

EHRLICH, P.R., D.S. DOBKIN, AND D. Wheye. *The Birder's Handbook*. Simon & Schuster, New York, 1988.

HOWELL, S.N.G. "All You Ever Wanted to Know About Molt But Were Afraid to Ask. Part II: Finding Order Amid the Chaos." *Birding* 35:640–650, 2003.

KASTNER, J. *A World of Watchers*. Sierra Club Books, San Francisco, 1986.

KROODSMA, D. *The Singing Life of Birds*. Houghton Mifflin, Boston, 2005.

LEAHY, C.W. *The Birdwatcher's Companion*. Princeton University Press, Princeton, 2004.

PYLE, P. *Identification Guide to North American Birds, Part I: Columbidae to Ploceidae*, Slate Creek Press, Bolinas, 1997.

———. *Identification Guide to North American Birds, Part II: Anatidae to Alcidae*. Slate Creek Press, Bolinas, 2008.

SIBLEY, D.A. *Sibley's Birding Basics*. Knopf, New York, 2002.

SIBLEY, D.A., C. ELPHICK, AND J.B. DUNNING, EDS. *The Sibley Guide to Bird Life and Behavior*. Knopf, New York, 2001.

WEIDENSAUL, S. *Of a Feather: A Brief History of American Birding*. Harcourt, Orlando, 2007.

WHITE, L., ED. *Good Birders Don't Wear White: 50 Tips from North America's Top Birders*. Houghton Mifflin, Boston, 2006.

Websites

American Bird Conservancy. www.abcbirds.org

American Birding Association. www.aba.org

American Ornithologists' Union Check-list of North American Birds. www.aou.org/checklist/index.php3

BirdChat. listserv.arizona.edu/archives/birdchat.html

Birding on the Net. birdingonthe.net/birdmail.html

Birds of North America Online. bna.birds.cornell.edu/bna

eBird. ebird.org

Frontiers of Bird Identification. listserv.arizona.edu/archives/birdwg01.html

Partners in Flight. www.partnersinflight.org

SurfBirds. surfbirds.com

Periodicals

Audubon. Magazine of the National Audubon Society, New York, New York.

The Auk. Quarterly Journal of the American Ornithologists' Union, McLean, Virginia.

Birder's World. Waukesha, Wisconsin

Birding. American Birding Association, Colorado Springs, Colorado.

Bird Watcher's Digest. Klambach, Marietta, Ohio.

North American Birds. American Birding Association, Colordao Springs, Colorado.

Western Birds. Quarterly Journal of Western Field Ornithologists, San Diego, California.

WildBird. Bowtie, Los Angeles, California.

Photo Credits

Page ii: American Kestrel (Jim Zipp). **Page viii: Great Egret** (Dr. Joseph Turner/VIREO). **Page 3: Greater Yellowlegs** (Brian E. Small); **Mountain Plover** (Brian E. Small); **Sharp-tailed Sandpiper** (Brian E. Small); **Eurasian Curlew** (Jari Peltomäki/VIREO). **Page 4: Reddish Egret** (Brian E. Small); **Orange-crowned Warbler** (Robert Behrstock/VIREO); **Yellow Warbler** (Brian E. Small/VIREO); **American Robin** (Jim Zipp); **Surfbird** (Brian E. Small). **Page 6: Laughing Gull** (*Adult*, Brian E. Small); (*First Winter*, Brian E. Small); **Fox Sparrow** ("*Thick-billed*", Brian E. Small); ("*Slate-colored*", Lee Trott/VIREO); ("*Sooty*", Jim Culbertson/VIREO); ("*Red*", James M. Wedge/VIREO); **White-throated Sparrow** (*White-striped Morph*, Brian E. Small); (*Tan-striped Morph*, Gerard Bailey/VIREO). **Page 8: Chestnut-sided Warbler** (Brian E. Small). **Page 9: Sprague's Pipit** (Brian E. Small). **Page 10: Sage Thrasher** (Brian E. Small). **Page 11: Brown-capped Rosy-Finch** (Brian E. Small). **Page 12: Common Moorhen** (Brian E. Small). **Page 13: Black Oystercatcher** (Brian E. Small). **Page 14: Wilson's Warbler** (Brian E. Small). **Page 15: Carolina Wren** (Brian E. Small). **Page 16: Common Tern** (Jim Zipp); **Red-winged Blackbird** (Brian E. Small). **Page 21: Painted Bunting** (*Adult Male*, Brian E. Small); (*Adult Female*, Brian E. Small). **Page 22: Herring Gull** (*Juvenile*, Kevin T. Karlson); (*Immature*, Bill Schmoker); (*Adult*, Brian E. Small). **Page 23: Yellow-rumped Warbler** (*Breeding*, Brian E. Small); (*Nonbreeding*, Brian E. Small); **Common Loon** (*Breeding*, Kevin T. Karlson); (*Nonbreeding*, Kevin T. Karlson). **Page 25: Northern Parula** (Brian E. Small); **Yellow Rail** (*In Flight*, Matt White/VIREO). **Page 26: Saltmarsh Sharp-tailed Sparrow** (Brian E. Small). **Page 30: Black-bellied Whistling-Duck** (*Adult*, Brian E. Small); (*Adult*, Brian E. Small). **Fulvous Whistling-Duck** (*Adult*, Brian E. Small); (*In Flight*, Bill Schmoker). **Page 31: Greater White-fronted Goose** (*Adult*, Robert Royse/VIREO); (*Juvenile*, Herbert Clarke/VIREO); (*In Flight*, Arthur Morris/VIREO). **Swan Goose** (Bill Schmoker). **Graylag Goose** (Jari Peltomäki/VIREO). **Barnacle Goose** (Hanne & Jens Eriksen/VIREO). **Page 32: Snow Goose** (*Adult "Lesser"*, Arthur Morris/VIREO); (*Juvenile*, Bill Schmoker); (*Adult Blue*, Brian E. Small); (*In Flight*, Bill Schmoker). **Page 33: Ross's Goose** (*Adult*, Brian Wheeler); (*Juvenile*, Tony Leukering/VIREO); (*Hybrid*, Bill Schmoker). **Emperor Goose** (*Immature*, Brian E. Small). **Page 34: Canada Goose** (*Immature*, Brian E. Small); (*Adult*, Bill Schmoker). **Cackling Goose** (*Adult*, Mike Danzenbaker); (*Adult*, Mike Danzenbaker); (*In Flight*, Bill Schmoker). **Page 35: Brant** (*Adult "Black"*, Brian E. Small); (*Adult "Atlantic"*, Bill Schmoker). **Tundra Swan** (*Adult*, Brian E. Small); (*Adults*, Bill Schmoker); (*Juvenile*, Steven Holt/VIREO). **Page 36: Trumpeter Swan** (*Adult*, Bill Schmoker); (*Adult*, Brian E. Small). **Whooper Swan** (*Adult*, Mike Danzenbaker); (*Adult*, Mike Danzenbaker). **Mute Swan** (*Adult*, Bill Schmoker); (*In Flight*, Bill Schmoker). **Page 37: Wood Duck** (*Breeding Male*, Brian E. Small); (*Adult Female*, Brian E. Small); (*In Flight*, Alan & Sandy Carey/VIREO). **Muscovy Duck** (*Feral Adult*, Brian E. Small). **Page 38: American Wigeon** (*Breeding Male*, Brian E. Small); (*Adult Female*, Brian E. Small); (*In Flight*, Jim Zipp). **Eurasian Wigeon** (*Immature Male*, Brian E. Small); (*Adult Female*, Arthur Morris/VIREO). **Page 39: American Black Duck** (*Adult Male*, Jukka Jantunen/VIREO); (*Adult Female*, Jim Culbertson/VIREO); (*Hybrid*, Mike Danzenbaker). **Mallard** (*Breeding Male*, Brian E. Small); (*Adult Female*, Brian E. Small); (*In Flight*, Bill Schmoker); (*Domestic*, Bill Schmoker). **Page 40: Mottled Duck** (*Adult Male*, Brian E. Small); (*In Flight*, Kevin T. Karlson). **Gadwall** (*Breeding Male*, Brian E. Small); (*Adult Female*, Brian E. Small); (*In Flight*, Jukka Jantunen/VIREO). **Page 41: Northern Shoveler** (*Breeding Male*, Brian E. Small); (*Immature Female*, Brian E. Small). **Cinnamon Teal** (*Breeding Male*, Brian E. Small); (*Adult Female*, Brian E. Small). **Page 42: Blue-winged Teal** (*Breeding Male*, Brian E. Small); (*Adult Female*, Brian E. Small); (*In Flight*, Doug Wechsler/VIREO). **Green-winged Teal** (*Breeding Male*, Brian E. Small); (*Adult Female*, Brian E. Small). **Page 43: Northern Pintail** (*Breeding Male*, Brian E. Small); (*Adult Male In Flight*, Bill Schmoker). **Garganey** (*Breeding Male*, Steve Young/VIREO); (*Adult Female*, Ray Tipper/VIREO). **Page 44: Redhead** (*Breeding Male*, Brian E. Small); (*Adult Female*, Brian E. Small). **Canvasback** (*Breeding Male*, Brian E. Small); (*Adult Female*, Brian E. Small). **Page 45: Common Pochard** (*Adult Male*, Mike Danzenbaker); (*Adult Female*, Mike Danzenbaker). **Tufted Duck** (*Adult Male*, Brian E. Small); (*Adult Female*, Brian E. Small). **Page 46: Ring-necked Duck** (*Breeding Male*, Jim Zipp); (*Adult Female*, Brian E. Small). **Greater Scaup** (*Breeding Male*, Brian E. Small); (*Adult Female*, Brian E. Small); (*In Flight*, Jukka

Jantunen/VIREO). **Page 47: Lesser Scaup** (*Breeding Male*, Brian E. Small); (*Adult Female*, Brian E. Small); (*In Flight*, Kevin T. Karlson). **Common Eider** (*Breeding Male*, Robert Royse/VIREO); (*Adult Female*, Mike Danzenbaker). **Page 48: King Eider** (*Adult Female*, Mike Danzenbaker); (*Immature Male*, Brian E. Small); (*In Flight*, Kevin T. Karlson); (*Breeding Male*, Kevin T. Karlson). **Spectacled Eider** (*Breeding Male*, Kevin T. Karlson). **Page 49: Steller's Eider** (*Breeding Male*, Jari Peltomäki/VIREO); (*Adult Female*, Jari Peltomäki/VIREO). **Harlequin Duck** (*Breeding Males*, Jim Zipp); (*Adult Female*, Brian E. Small); (*In Flight*, Kevin T. Karlson); (*Breeding Male*, Jim Zipp). **Page 50: Surf Scoter** (*Adult Male*, Brian E. Small); (*Adult Female*, Brian E. Small); (*In Flight*, Kevin T. Karlson). **Black Scoter** (*Pair*, Kevin T. Karlson); (*Female*, Brian E. Small); (*In Flight*, Kevin T. Karlson). **Page 51: White-winged Scoter** (*Adult Male*, Brian E. Small); (*Female*, Brian E. Small); (*Female*, Brian E. Small); (*In Flight*, Bill Schmoker). **Bufflehead** (*Breeding Male*, Brian E. Small); (*Adult Female*, Brian E. Small); (*In Flight*, Bill Schmoker). **Page 52: Long-tailed Duck** (*Breeding Male*, Brian E. Small); (*Breeding Female*, Brian E. Small); (*In Flight*, Kevin T. Karlson); (*Immature Male*, Brian E. Small). **Page 53: Common Goldeneye** (*Breeding Male*, Brian E. Small); (*Adult Female*, Brian E. Small); (*Immature Male*, Brian E. Small); (*In Flight*, Bill Schmoker). **Barrow's Goldeneye** (*Breeding Male*, Brian E. Small); (*Adult Female*, Brian E. Small); (*Immature Male*, Brian E. Small); (*In Flight*, Jukka Jantunen/VIREO). **Page 54: Common Merganser** (*Breeding Male*, Brian E. Small); (*Adult Female*, Brian E. Small). **Red-breasted Merganser** (*Breeding Male*, Brian E. Small); (*Adult Female*, Kevin T. Karlson); (*Immature*, Brian E. Small). **Page 55: Hooded Merganser** (*Breeding Male*, Brian E. Small); (*Adult Female*, Brian E. Small). **Smew** (*Breeding Male*, Brian E. Small); (*Adult Female*, Brian E. Small). **Page 56: Ruddy Duck** (*Breeding Male*, Brian E. Small); (*Adult Female*, Brian E. Small); (*Nonbreeding Male*, Brian E. Small). **Masked Duck** (*Breeding Male*, Brian E. Small); (*Adult Female*, Brian E. Small). **Page 58: Plain Chachalaca** (*Adult*, Brian E. Small). **Gray Partridge** (*Adult Male*, Bob de Lange/VIREO). **Chukar** (*Adult*, Brian E. Small). **Page 59: Himalayan Snowcock** (*Adult Male*, James D. Bland/VIREO). **Ring-necked Pheasant** (*Adult Male*, Mike Danzenbaker); (*Adult Female*, Brian E. Small). **Green Pheasant** (Mike Danzenbaker). **Indian Peafowl** (Adrian & Jane Binns/VIREO). **Page 60: Sharp-tailed Grouse** (*Adult Male*, Brian E. Small); (*Adult*, Brian E. Small). **Greater Prairie-Chicken** (*Adult Male*, Brian E. Small); (*Adult Female*, Brian E. Small). **Page 61: Lesser Prairie-Chicken** (*Adult Male*, Brian E. Small); (*Adult Female*, Brian E. Small). **Greater Sage-Grouse** (*Adult Male*, Brian E. Small); (*Adult Female*, Brian E. Small). **Page 62: Gunnison Sage-Grouse** (*Displaying Male*, Lance Beeny/VIREO); (*Adult Female*, Kevin T. Karlson). **Spruce Grouse** (*Adult Male*, Brian E. Small); (*Adult Female*, Kevin T. Karlson). **Page 63: Sooty Grouse** (*Adult Male*, Mike Danzenbaker); (*Adult Male*, Brian E. Small). **Dusky Grouse** (*Adult Male*, Brian E. Small); (*Adult Female*, Brian E. Small). **Page 64: Ruffed Grouse** (*Adult Gray Morph*, Brian E. Small); (*Adult Red Morph*, Jim Zipp). **Willow Ptarmigan** (*Breeding Male*, Brian E. Small); (*Breeding Female*, Brian E. Small). **Page 65: White-tailed Ptarmigan** (*Breeding Male*, Brian E. Small); (*Breeding Female*, Brian E. Small); (*Winter Adult*, Brian E. Small); (*Summer-Autumn Adult*, Brian E. Small); (*In Flight*, Bill Schmoker). **Page 66: Rock Ptarmigan** (*Breeding Male*, Kevin T. Karlson); (*Adult Male*, Brian E. Small); (*Breeding Female*, Kevin T. Karlson); (*Breeding Male*, Ronald M. Saldino/VIREO). **Page 67: Wild Turkey** (*Adult Male*, Brian E. Small); (*Adult Male*, Brian E. Small); (*Nonbreeding Male*, Brian E. Small). **Montezuma Quail** (*Adult Male*, Brian E. Small); (*Adult Female*, Brian E. Small). **Page 68: Northern Bobwhite** (*Adult Male*, Brian E. Small); (*Adult Male*, Brian E. Small); (*Adult Female*, Brian E. Small); (*Adult Female, "Masked"*, Rick and Nora Bowers/VIREO). **Page 69: Gambel's Quail** (*Adult Male*, Brian E. Small); (*Adult Female*, Brian E. Small). **California Quail** (*Adult Male*, Brian E. Small); (*Adult Female*, Brian E. Small). **Page 70: Mountain Quail** (*Adult Male*, Brian E. Small). **Scaled Quail** (*Adult*, Jim Zipp); (*Adult*, Greg Lasley/VIREO); (*Hybrid*, Brian E. Small). **Page 72: Common Loon** (*Breeding Adult*, Brian E. Small); (*Nonbreeding*, Kevin T. Karlson); (*Juvenile*, Brian E. Small); (*Nonbreeding*, Bill Schmoker); (*In Flight*, Martin Meyers/VIREO). **Page 73: Yellow-billed Loon** (*Breeding Adult*, Gerrit Vyn/VIREO); (*Juvenile*, Bill Schmoker). **Red-throated Loon** (*Breeding Adult*, Brian E. Small); (*Nonbreeding Adult*, Brian E. Small). **Page 74: Pacific Loon** (*Adult Breeding*, Robert Royse/VIREO); (*Juvenile*, Arthur Morris/VIREO). **Arctic Loon** (*Breeding Adult*, Jari Peltomäki/VIREO); (*Subadult*, David Tipling/VIREO). **Page 76: Pied-billed Grebe**

(*Breeding Adult*, Brian E. Small); (*Nonbreeding Adult*, Brian E. Small). **Horned Grebe** (*Breeding Adult*, Brian E. Small); (*Nonbreeding Adult*, Brian E. Small). **Page 77: Least Grebe** (*Breeding Adult*, Brian E. Small). **Eared Grebe** (*Nonbreeding Adult*, Brian E. Small); (*Breeding Adult*, Brian E. Small); (*In Flight*, Bill Schmoker). **Page 78: Red-necked Grebe** (*Breeding Adult*, Brian E. Small); (*Nonbreeding Adult*, Brian E. Small); (*Downy Young*, Brian E. Small). **Clark's Grebe** (*Breeding Adult*, Brian E. Small); (*Nonbreeding Adult*, Brian E. Small). **Page 79: Western Grebe** (*Breeding Adult*, Brian E. Small); (*Nonbreeding Adult*, Brian E. Small); (*Rushing Ceremony*, Gary Neuchterlein); (*Weed Ceremony*, Gary Neuchterlein). **Page 81: Black-footed Albatross** (*Above*, Brian E. Small); (*Below*, Peter LaTourrette/VIREO). **Laysan Albatross** (*Fresh Adult*, Brian E. Small); (*Below*, George L. Armistead/VIREO). **Page 82: Short-tailed Albatross** (*Adult*, Brian E. Small); (*Immature*, Jeff Poklen). **Northern Fulmar** (*Light Individual*, Brian E. Small); (*Dark Individual*, Bill Schmoker); (*Intermediate Individual*, Brian E. Small). **Page 83: Black-capped Petrel** (*Fresh Adult*, Brian Patteson/VIREO); (*Worn Adult*, Mike Danzenbaker); (*Below*, Brian Patteson). **Herald Petrel** (*Intermediate Morph*, Mike Danzenbaker); (*Above*, Brian Patteson). **Page 84: Fea's Petrel** (*Below*, Brian Patteson); (*Above*, Brian Patteson). **Murphy's Petrel** (*Adult*, Brian E. Small); (*Adult*, Brian E. Small). **Page 85: Mottled Petrel** (*Above*, Greg Lasley/VIREO); (*Below*, George L. Armistead/VIREO). **Cook's Petrel** (*Above*, Tony Palliser/VIREO); (*Below*, Tadao Shimba/VIREO). **Page 86: Cory's Shearwater** (*Below*, Mike Danzenbaker); (*Above*, Brian Patteson/VIREO); (*Adult*, Geoff Malosh/VIREO). **Greater Shearwater** (*Juvenile*, Bill Schmoker). **Page 87: Audubon's Shearwater** (*Adult*, Mike Danzenbaker). **Sooty Shearwater** (*Above*, Brian Patteson/VIREO); (*Below*, Glen Tepke/VIREO); (*Flock at Sea*, Brian E. Small). **Page 88: Short-tailed Shearwater** (*Adult*, Mike Danzenbaker). **Manx Shearwater** (*Adult*, Mike Danzenbaker). **Black-vented Shearwater** (*Typical Adult*, Mike Danzenbaker). **Page 89: Pink-footed Shearwater** (*Dark Adult*, Brian E. Small); (*Juvenile*, Brian E. Small). **Flesh-footed Shearwater** (*Typical Adult*, Brian E. Small). **Page 90: Buller's Shearwater** (*Above*, Mike Danzenbaker); (*Below*, Mike Danzenbaker). **Wilson's Storm-Petrel** (*Worn Adult*, Brian Patteson); (*Adult*, Glen Tepke/VIREO). **Page 91: Band-rumped Storm-Petrel** (*Adult*, Mike Danzenbaker). **White-faced Storm-Petrel** (Tony Palliser/VIREO). **Leach's Storm-Petrel** (*Above*, Mike Danzenbaker); (*Below*, Mike Danzenbaker). **Page 92: Black Storm-Petrel** (*Juvenile*, Mike Danzenbaker); (*Mixed-species Flock*, Ronald M. Saldino/VIREO). **Ashy Storm-Petrel** (*Worn Adult*, Mike Danzenbaker); (*Below*, Peter LaTourrette/VIREO). **Page 93: Least Storm-Petrel** (*Above*, Mike Danzenbaker); (*Below*, Mike Danzenbaker). **Fork-tailed Storm-Petrel** (*Adult*, Mike Danzenbaker); (*Adult*, Mike Danzenbaker). **Page 95: Red-billed Tropicbird** (*Juvenile*, Brian Patteson); (*Adult*, Adrian & Jane Binns/VIREO). **White-tailed Tropicbird** (*Adult*, Brian E. Small); (*Adult*, Brian E. Small). **Red-tailed Tropicbird** (Brian E. Small). **Page 96: Northern Gannet** (*Adult*, Brian Wheeler); (*Immature*, George L. Armistead/VIREO). **Masked Booby** (*Adult*, Mike Danzenbaker); (*Juvenile*, Hanne & Jens Eriksen/VIREO). **Page 97: Brown Booby** (*Adult Female*, Brian E. Small); (*Juvenile*, Adrian & Jane Binns/VIREO). **Blue-footed Booby** (Brian E. Small). **Red-footed Booby** (Brian E. Small). **Page 98: Brown Pelican** (*Nonbreeding Adult*, Brian E. Small); (*Breeding Adult*, Brian E. Small); (*In Flight*, Bill Schmoker); (*Breeding Adult*, Brian E. Small). **Page 99: American White Pelican** (*Breeding Adult*, Steven Holt/VIREO); (*Nonbreeding Adult*, Brian E. Small); (*In Flight*, Kevin T. Karlson). **Great Cormorant** (*Adults*, David Tipling/VIREO); (*Juvenile*, Richard Crossley/VIREO). **Page 100: Neotropic Cormorant** (*Breeding Adult*, Brian E. Small); (*Juvenile*, Kevin T. Karlson); (*In Flight*, Kevin T. Karlson). **Double-crested Cormorant** (*Breeding Adult*, Brian E. Small); (*Juvenile*, Brian E. Small); (*In Flight*, Brian E. Small). **Page 101: Brandt's Cormorant** (*Nonbreeding Adult*, Brian E. Small); (*Molting Adult*, Brian E. Small). **Pelagic Cormorant** (*Breeding Adult*, Brian E. Small); (*Nonbreeding Adult*, Brian E. Small). **Page 102: Red-faced Cormorant** (*Breeding Adult*, Arthur Morris/VIREO); (*In Flight*, Arthur Morris/VIREO). **Anhinga** (*In Flight*, Brian E. Small); (*Breeding Male*, Brian E. Small); (*Adult Female*, Brian E. Small). **Page 103: Magnificent Frigatebird** (*Adult Male*, Brian E. Small); (*Adult Female*, Brian E. Small); (*Juvenile*, Brian E. Small); (*Adult Female*, Arthur Morris/VIREO); (*Adult Male*, Arthur Morris/VIREO). **Page 105: Great Blue Heron** (*Nonbreeding Adult*, Brian E. Small); (*Breeding Adult*, Brian E. Small); (*Juvenile*, Brian E. Small); (*"Würdeman's"*, John Heidecker/VIREO); (*"Great White"*, Kevin T. Karlson). **Page 106: Great Egret** (*Adult*, Brian E. Small); (*Courting Adult*, Kevin T. Karlson). **Snowy Egret** (*Nonbreeding Adult*, Brian E. Small); (*Courting Adult*, Mike Danzenbaker). **Page 107: Cattle Egret** (*Breeding Adult*, Brian E. Small); (*Nonbreeding Adult*, Brian E. Small). **Little Blue Heron** (*Adult*, Brian E. Small); (*Juvenile*, Brian E. Small); (*Immature*, Brian E. Small). **Page 108: Tricolored Heron** (*Nonbreeding Adult*, Brian E. Small); (*Juvenile*, Brian E. Small); (*In Flight*, Kevin T. Karlson). **Reddish Egret** (*Dark Adult*, Brian E. Small); (*White Adult*, Brian E. Small); (*White Juvenile*, Brian E. Small). **Page 109: Yellow-crowned Night-Heron** (*Adult*, Brian E. Small); (*Juvenile*, Brian E. Small); (*In Flight*, Kevin T. Karlson). **Black-crowned Night-Heron** (*Adult*, Brian E. Small); (*Juvenile*, Brian E. Small). **Page 110: Green Heron** (*Adult*, Brian E. Small); (*Juvenile*, Brian E. Small). **American Bittern** (*Adult*, Brian E. Small); (*In Flight*, Bill Schmoker). **Page 111: Least Bittern** (*Adult Male*, Brian E. Small); (*Adult Female*, Kevin T. Karlson). **White Ibis** (*Adult*, Brian E. Small); (*Subadult*, Brian E. Small). **Page 112: White-faced Ibis** (*Breeding Adult*, Brian E. Small); (*Immature*, Brian E. Small). **Glossy Ibis** (*Breeding Adult*, Kevin T. Karlson); (*Juvenile*, Kevin T. Karlson); (*In Flight*, Kevin T. Karlson). **Page 113: Roseate Spoonbill** (*Adult*, Brian E. Small); (*Juvenile*, Brian E. Small). **Wood Stork** (*Adult*, Brian E. Small); (*Juvenile*, Kevin T. Karlson). **Page 115: Greater Flamingo** (*Adult*, Brian E. Small); (*In Flight*, Kevin T. Karlson); (*Flock*, Kevin T. Karlson). **Chilean Flamingo** (J.H. Dick/VIREO). **Page 117: Turkey Vulture** (*Adult*, Brian E. Small); (*In Flight*, Brian E. Small). **Black Vulture** (*In Flight*, Jim Zipp); (*Adult*, Brian E. Small). **California Condor** (Brian E. Small). **Page 118: Osprey** (*Adults*, Kevin T. Karlson); (*Juveniles*, Kevin T. Karlson); (*Adult Female*, Brian Wheeler). **Hook-billed Kite** (*Adult Male*, Brian Wheeler); (*Adult Male*, Brian Wheeler). **Page 119: White-tailed Kite** (*Adult*, Brian Wheeler); (*Adult*, Brian Wheeler); (*Juvenile*, Brian Wheeler). **Mississippi Kite** (*Adult Male*, Brian Wheeler); (*Juvenile*, Brian Wheeler); (*Adult Male*, Brian Wheeler). **Page 120: Swallow-tailed Kite** (*Adult*, Brian Wheeler); (*Adult Above*, Jim Zipp); (*Adult Below*, Brian Wheeler). **Snail Kite** (*Adult Male*, Jim Zipp); (*Adult Female*, Brian E. Small). **Page 121: Sharp-shinned Hawk** (*Adult*, Brian E. Small); (*Juvenile*, Jim Zipp); (*Juvenile*, Brian Wheeler). **Cooper's Hawk** (*Adult Male*, Brian Wheeler); (*Juvenile*, Brian Wheeler); (*Juvenile*, Brian Wheeler). **Page 122: Northern Goshawk** (*Adult Male*, Brian Wheeler); (*Juvenile*, Brian Wheeler). **Northern Harrier** (*Adult Male*, Jim Zipp); (*Adult Female*, Brian Wheeler); (*Juvenile*, Brian E. Small). **Page 123: Bald Eagle** (*Adult*, Jim Zipp); (*Immature, Third Year*, Jim Zipp); (*Adult*, Jim Zipp); (*Immature, Second Year*, Jim Zipp); (*Immature, Second Year*, Brian E. Small). **Page 124: Golden Eagle** (*Adult*, Mike Danzenbaker); (*Juvenile*, Brian Wheeler); (*Adult*, Brian Wheeler). **Page 125: Ferruginous Hawk** (*Light-morph Adult*, Jim Zipp); (*Dark-morph Adult*, Brian Wheeler); (*Light-morph Adult*, Brian Wheeler); (*Light-morph Juvenile*, Brian Wheeler); (*Light-morph Adult*, Brian Wheeler). **Page 126: Rough-legged Hawk** (*Light-morph Adult Male*, Brian Wheeler); (*Light-morph Adult Female*, Brian Wheeler); (*Light-morph Adult Female*, Brian Wheeler); (*Dark-morph Adult Male*, Brian Wheeler); (*Dark-morph Adult Juvenile*, Brian Wheeler); (*Dark-morph Adult Male*, Brian Wheeler). **Page 127: Swainson's Hawk** (*Light-morph Adult*, Brian Wheeler); (*Light-morph Juvenile*, Brian Wheeler); (*Intermediate-morph Adult*, Brian Wheeler); (*Dark-morph Adult*, Brian Wheeler); (*Light-morph Adult*, Brian E. Small). **Page 128: Red-shouldered Hawk** (*Eastern Adult*, Brian Wheeler); (*California Adult*, Brian Wheeler); (*Florida Adult*, Brian Wheeler). **Page Broad-winged Hawk** (*Light-morph Adult*, Brian Wheeler); (*Light-morph Adult*, Brian Wheeler); (*Light-morph Juvenile*, Brian Wheeler); (*Juvenile*, Brian Wheeler). **Page 129: Short-tailed Hawk** (*Dark-morph Juvenile*, Brian Wheeler); (*Light-morph Juvenile*, Brian Wheeler); (*Dark-morph Adult*, Brian Wheeler); (*Light-morph Adult*, Brian Wheeler). **White-tailed Hawk** (*Adult Male*, Brian Wheeler); (*Adult Female*, Brian Wheeler); (*Immature*, Brian Wheeler). **Page 130–131: Red-tailed Hawk** (*Adult, Eastern*, Jim Zipp); (*Light-morph Adult, Western*, Brian E. Small); (*Juvenile*, Brian Wheeler); (*Adult, "Harlan's"*, Brian Wheeler); (*Intermediate-morph Adult, Western*, Brian Wheeler); (*Adult, Eastern*, Jim Zipp); (*Adult, "Krider's"*, Brian Wheeler); (*Adult, "Harlan's"*, Brian Wheeler); (*Intermediate-morph Adult, Western*, Bill Schmoker); (*Dark-morph*, Jim Zipp). **Page 132: Harris's Hawk** (*Adult*, Brian Wheeler); (*Juvenile*, Brian E. Small); (*Juvenile*, Brian Wheeler). **Zone-tailed Hawk** (*Adult*, Brian Wheeler); (*Juvenile*, Brian Wheeler). **Page 133: Gray Hawk** (*Adult*, Brian Wheeler); (*Juvenile*, Jim Zipp); (*Juvenile*, Brian E. Small). **Common Black-Hawk** (*Adult*, Brian E. Small); (*Juvenile*, Brian Wheeler). **Page 134: American Kestrel** (*Adult Male*, Jim Zipp); (*Adult Male*, Brian Wheeler); (*Adult Female*, Brian Wheeler). **Aplomado Falcon** (Brian Wheeler). **Page 135: Merlin** (*Adult Male, Northern "Taiga"*, Brian Wheeler); (*In Flight*, Jim Zipp); (*Adult Male, "Prairie"*, Brian Wheeler); (*In Flight*, Brian Wheeler); (*Adult, "Black"*, Martin Meyers/VIREO). **Page 136: Prairie Falcon** (*Adult*, Brian Wheeler); (*Juvenile*, Brian Wheeler). **Peregrine Falcon** (*Adult*, Jim Zipp); (*Adult*, Brian Wheeler); (*Juvenile*, Brian Wheeler). **Page 137: Gyrfalcon** (*Gray-morph Adult*, Jim Zipp); (*Gray-morph Adult*, Jim Zipp); (*White-morph Adult*, Brian E. Small). **Crested Caracara** (*Adult*, Brian E. Small); (*Subadult*, Brian E. Small). **Page 139: Clapper Rail** (*Adult Atlantic Coast*, Brian E. Small); (*Adult Western*, Brian E. Small); (*Adult Gulf Coast*, Kevin T. Karlson). **King Rail** (*Adult*, Brian E. Small); (*Juvenile*, Earl H. Harrison/VIREO). **Page 140: Virginia Rail** (*Adult*, Brian E. Small); (*Older Juvenile*, Brian E. Small). **Sora** (*Breeding Male*, Brian E. Small); (*Juvenile*, Jukka Jantunen/VIREO). **Page 141: Black Rail** (*Adult*, Brian E. Small). **Yellow Rail** (*Adult*, Brian E. Small); (*In Flight*, Matt White/VIREO). **Page 142: Common Moorhen** (*Breeding Adult*, Brian E. Small); (*Juvenile*, Brian E. Small). **Purple Gallinule** (*Adult*, Brian E. Small); (*Juvenile*, Arthur Morris/VIREO). **Purple Swamphen** (Kevin T. Karlson). **Page 143: American Coot** (*Nonbreeding Adult*, Brian E. Small), (*Juvenile*, Hugh P. Smith, Jr. & Susan C. Smith/VIREO). **Whooping Crane** (*Adult*, Brian E. Small); (*In Flight*, Joanne Williams/VIREO). **Page 144: Sandhill Crane** (*Adult "Greater"*, Bill Schmoker); (*Adult "Lesser"*, Bill Schmoker); (*Juvenile*, Stephen J. Lang/VIREO). **Limpkin** (*Adult*, Brian E. Small). **Common Crane** (Jorge Sierra/VIREO). **Page 146: Killdeer** (*Adult*, Brian E. Small); (*In Flight*, Mike Danzenbaker); (*Chick*, Mike Danzenbaker); (*Defensive Display*, Kevin T. Karlson). **Northern Lapwing** (Mike Danzenbaker). **Page 147: Semipalmated Plover** (*Breeding Adult*, Brian E. Small); (*Adult*, Mike Danzenbaker); (*Juvenile*, Kevin T. Karlson). **Common Ringed Plover** (*Breeding Adult*, Mike Danzenbaker). **Page 148: Wilson's Plover** (*Breeding Adult*, Brian E. Small); (*Nonbreeding*, Brian E. Small). **Snowy Plover** (*Breeding Adult*, Kevin T. Karlson); (*Nonbreeding*, Kevin T. Karlson). **Page 149: Piping Plover** (*Breeding Adult*, Brian E. Small); (*Nonbreeding*, Brian E. Small); (*Juvenile*, Kevin T. Karlson). **Mountain Plover** (*Breeding Adult*, Brian E. Small); (*Juvenile*, Bill Schmoker); (*In Flight*, Bill Schmoker). **Page 150: Lesser Sand-Plover** (*Breeding Adult*, Mike Danzenbaker); (*Juvenile*, Mike

Danzenbaker). **Black-bellied Plover** (*Breeding Adult*, Brian E. Small); (*Nonbreeding Adult*, Brian E. Small); (*In Flight*, Mike Danzenbaker); (*Juvenile*, Mike Danzenbaker). **Page 151: American Golden-Plover** (*Breeding Adult*, Brian E. Small); (*Juvenile*, Brian E. Small); (*In Flight*, Richard Crossley/VIREO). **Pacific Golden-Plover** (*Breeding Adult*, Brian E. Small); (*Immature*, Brian E. Small). **European Golden-Plover** (Rolf Nussbaumer/VIREO). **Page 153: American Oystercatcher** (*Adult*, Brian E. Small); (*In Flight*, Doug Wechsler/VIREO). **Black Oystercatcher** (*Adult*, Brian E. Small); (*Worn Adult*, Mike Danzenbaker). **Page 154: American Avocet** (*Nonbreeding Male*, Brian E. Small); (*Breeding Female*, Brian E. Small); (*In Flight*, Brian E. Small). **Black-necked Stilt** (*Adult Male*, Mike Danzenbaker); (*In Flight*, Mike Danzenbaker). **Page 156: Spotted Sandpiper** (*Breeding Adult*, Brian E. Small); (*Immature*, Brian E. Small); (*In Flight*, Bill Schmoker). **Common Sandpiper** (*Breeding Adult*, Mike Danzenbaker). **Page 157: Solitary Sandpiper** (*Breeding Adult*, Mike Danzenbaker); (*Nonbreeding*, Mike Danzenbaker); (*In Flight*, Richard Crossley/VIREO). **Wood Sandpiper** (*Breeding Adult*, Mike Danzenbaker); (*Juvenile*, Mike Danzenbaker). **Page 158: Lesser Yellowlegs** (*Breeding Adult*, Kevin T. Karlson); (*Juvenile*, Kevin T. Karlson); (*In Flight*, Mike Danzenbaker). **Greater Yellowlegs** (*Breeding Adult*, Mike Danzenbaker); (*Nonbreeding*, Brian E. Small). **Page 159: Willet** (*Breeding Adult, Western*, Brian E. Small); (*Breeding Adult, Eastern*, Kevin T. Karlson); (*Nonbreeding*, Brian E. Small); (*In Flight*, Brian E. Small). **Common Greenshank** (*Breeding Adult*, Mike Danzenbaker). **Page 160: Wandering Tattler** (*Breeding Adult*, Brian E. Small); (*Nonbreeding*, Brian E. Small); (*In Flight*, Mike Danzenbaker). **Gray-tailed Tattler** (*Breeding Adult*, Mike Danzenbaker). **Page 161: Upland Sandpiper** (*Breeding Adult*, Brian E. Small); (*In Flight*, Richard Crossley/VIREO); (*Adult*, Kevin T. Karlson). **Terek Sandpiper** (*Nonbreeding*, Mike Danzenbaker). **Page 162: Whimbrel** (*Adult*, Brian E. Small); (*In Flight*, Mike Danzenbaker). **Long-billed Curlew** (*Adult*, Kevin T. Karlson); (*In Flight*, Kevin T. Karlson). **Page 163: Bristle-thighed Curlew** (*Immature*, Brian E. Small); (*In Flight*, Doug Wechsler/VIREO). **Marbled Godwit** (*Breeding Adult*, Brian E. Small); (*In Flight*, Kevin T. Karlson). **Page 164: Hudsonian Godwit** (*Breeding Male*, Kevin T. Karlson); (*Immature*, Brian E. Small); (*In Flight*, Mike Danzenbaker). **Bar-tailed Godwit** (*Breeding Male*, Brian E. Small); (*In Flight*, Mike Danzenbaker). **Page 165: Black-tailed Godwit** (*Adult Male*, Patricio Robles Gil/VIREO); (*Juvenile*, Steve Young/VIREO); (*In Flight*, Mike Danzenbaker). **Ruddy Turnstone** (*Breeding male*, Brian E. Small); (*Nonbreeding Adult*, Brian E. Small); (*In Flight*, Mike Danzenbaker). **Page 166: Black Turnstone** (*Breeding Adult*, Brian E. Small); (*Immature*, Brian E. Small); (*In Flight*, Mike Danzenbaker). **Surfbird** (*Breeding Adult*, Brian E. Small); (*Nonbreeding*, Brian E. Small); (*In Flight*, Mike Danzenbaker). **Page 167: Rock Sandpiper** (*Nonbreeding Adult*, Brian E. Small); (*Breeding Adult*, Brian E. Small); (*Breeding Adult*, Sam Fried/VIREO). **Purple Sandpiper** (*Nonbreeding Adult*, Kevin T. Karlson); (*Breeding Adult*, Mike Danzenbaker); (*In Flight*, Mike Danzenbaker). **Page 168: Sanderling** (*Nonbreeding Adult*, Brian E. Small); (*Breeding Adult*, Brian E. Small); (*Juvenile*, Brian E. Small); (*In Flight*, Mike Danzenbaker); (*In Flight*, Kevin T. Karlson). **Page 169: Red Knot** (*Breeding Adult*, Kevin T. Karlson); (*Adults*, Kevin T. Karlson); (*Juvenile*, Brian E. Small); (*In Flight*, Mike Danzenbaker); (*Nonbreeding Adult*, Arthur Morris/VIREO). **Page 170: Semipalmated Sandpiper** (*Adult Male*, Arthur Morris/VIREO); (*Juvenile Female*, Kevin T. Karlson); (*In Flight*, Julian R. Hough/VIREO). **Western Sandpiper** (*Breeding Adult*, Brian E. Small); (*Nonbreeding*, Brian E. Small); (*Juvenile*, Brian E. Small). **Page 171: Least Sandpiper** (*Breeding Adult*, Kevin T. Karlson); (*Juvenile*, Kevin T. Karlson); (*In Flight*, Mike Danzenbaker). **Long-toed Stint** (*Breeding Adult*, Martin Hale/VIREO). **Page 172: Red-necked Stint** (*Bright Adult*, Brian E. Small); (*Juvenile*, Christian Artuso/VIREO). **Temminck's Stint** (*Breeding Adult*, Mike Danzenbaker); (*Juvenile*, Mike Danzenbaker). **Little Stint** (H. Clarke/VIREO). **Page 173: Baird's Sandpiper** (*Breeding Adult*, Brian E. Small); (*Nonbreeding*, Bill Schmoker); (*Juvenile*, Brian E. Small); (*In Flight*, Richard Crossley/VIREO). **White-rumped Sandpiper** (*Breeding Adult*, Mike Danzenbaker); (*Nonbreeding*, Kevin T. Karlson); (*In Flight*, Mike Danzenbaker). **Page 174: Pectoral Sandpiper** (*Breeding Adult*, Mike Danzenbaker); (*Displaying Male*, Kevin T. Karlson); (*Juvenile*, Mike Danzenbaker); (*In Flight*, Mike Danzenbaker). **Sharp-tailed Sandpiper** (*Adult*, Brian E. Small); (*Juvenile*, Brian E. Small). **Page 175: Dunlin** (*Breeding Adult*, Brian E. Small); (*Nonbreeding*, Mike Danzenbaker); (*Juvenile*, Paul Lehman). **Curlew Sandpiper** (*Breeding Female*, Bill Schmoker); (*Nonbreeding*, Martin Hale/VIREO); (*Juvenile*, Mike Danzenbaker). **Page 176: Stilt Sandpiper** (*Breeding Adult*, Kevin T. Karlson); (*Nonbreeding Adult*, Kevin T. Karlson); (*Juvenile*, Kevin T. Karlson). **Buff-breasted Sandpiper** (*Breeding Adult*, Kevin T. Karlson); (*Juvenile*, Mike Danzenbaker). **Page 177: Short-billed Dowitcher** (*Adult Breeding, Western*, Brian E. Small); (*Adult Breeding, Interior*, Kevin T. Karlson); (*Adult Breeding, Eastern*, Doug Wechsler/VIREO); (*Juvenile*, Arthur Morris/VIREO); (*In Flight*, Kevin T. Karlson). **Page 178: Long-billed Dowitcher** (*Breeding Adult*, Brian E. Small); (*In Flight*, Jukka Jantunen/VIREO); (*Nonbreeding Adult*, Mike Danzenbaker); (*Juvenile*, Mike Danzenbaker); (*Breeding Adult*, Brian E. Small). **Page 179: Wilson's Snipe** (*In Flight*, Mike Danzenbaker); (*Adult*, Jim Zipp); (*Adult*, Jim Zipp). **Common Snipe** (*In Flight*, Mike Danzenbaker); (*Adult*, David Tipling/VIREO). **Page 180: American Woodcock** (*Adult*, Jim Zipp); (*In Flight*, Richard Crossley/VIREO). **Wilson's Phalarope** (*Breeding Female*, Brian E. Small); (*Juvenile*, Mike Danzenbaker); (*In Flight*, Mike Danzenbaker). **Page 181: Red-necked Phalarope** (*Breeding Female*, Kevin T. Karlson); (*Juvenile*, Mike Danzenbaker); (*In Flight*, Mike Danzenbaker). **Red Phalarope** (*Breeding Female*, Kevin T. Karlson); (*Nonbreeding Adult*, Mike Danzenbaker); (*In Flight*, Mike Danzenbaker). **Page 182: Ruff** (*Displaying Male*, David Tipling/VIREO); (*Adult Female*, Brian E. Small); (*Juvenile*, Mike Danzenbaker). **Page 184: Ring-billed Gull** (*Breeding Adult*, Brian E. Small); (*In Flight*, Bill Schmoker); (*Juvenile*, Arthur Morris/VIREO); (*Immature, First Cycle*, Arthur Morris/VIREO); (*Immature, Second Cycle*, Jeff Poklen/VIREO). **Page 185: Mew Gull** (*Nonbreeding Adult*, Brian E. Small); (*Immature, First Cycle*, Brian E. Small); (*Juvenile*, Arthur Morris/VIREO); (*In Flight*, Jukka Jantunen/VIREO); (*Chick*, Tom Vezo/VIREO). **Page 186: California Gull** (*Breeding Adult*, Brian E. Small); (*Nonbreeding Adult*, Brian E. Small); (*Immature, First Cycle*, Jeff Poklen/VIREO); (*Nonbreeding Adult*, Jukka Jantunen/VIREO). **Black-tailed Gull** (Dr. Yuri Artukhin/VIREO). **Page 187: Herring Gull** (*Breeding Adult*, Kevin T. Karlson); (*Juvenile*, Kevin T. Karlson); (*Immature, First Cycle*, Brian E. Small); (*Immature, Second Cycle*, Bill Schmoker); (*Immature, Third Cycle*, Bill Schmoker). **Yellow-legged Gull** (Georges Olioso/VIREO). **Page 188: Thayer's Gull** (*Breeding Adult*, Brian E. Small); (*Juvenile*, Jeff Poklen/VIREO); (*In Flight*, Jukka Jantunen/VIREO). **Iceland Gull** (*Nonbreeding Adult*, Jeff Poklen/VIREO); (*In Flight*, Garth McElroy/VIREO); (*Juvenile*, Jeff Poklen/VIREO). **Page 189: Glaucous Gull** (*Adult*, Brian E. Small); (*Juvenile*, Brian E. Small); (*In Flight*, Arthur Morris/VIREO). **Glaucous-winged Gull** (*Adult*, Jim Zipp); (*Hybrid Adult*, Jeff Poklen/VIREO); (*Hybrid Immature*, Jeff Poklen/VIREO). **Page 190: Western Gull** (*Immature, First Cycle*, Brian E. Small); (*Breeding Adult, Southern*, Brian E. Small); (*Immature, Third Cycle*, Bill Schmoker). **Yellow-footed Gull** (*Breeding Adult*, Brian E. Small). **Page 191: Slaty-backed Gull** (*Nonbreeding Adult*, Brian E. Small); (*Immature, First Cycle*, Y.Artukhin/VIREO). **Lesser Black-backed Gull** (*Nonbreeding Adult*, Brian E. Small); (*Immature, First Cycle*, Brian E. Small). **Page 192: Great Black-backed Gull** (*Subadult, Third Cycle*, Brian E. Small); (*Immature, First Cycle*, Brian E. Small). **Kelp Gull** (George L. Armistead/VIREO). **Heermann's Gull** (*Breeding Adult*, Brian E. Small); (*Variant Adult*, Bill Schmoker); (*Immature, First Cycle*, Brian E. Small). **Page 193: Laughing Gull** (*Breeding Adult*, Bill Schmoker); (*Nonbreeding Adult*, Brian E. Small); (*Immature, First Cycle*, Arthur Morris/VIREO). **Franklin's Gull** (*Breeding Adult*, Kevin T. Karlson); (*Nonbreeding Adult*, Martin Hale/VIREO); (*Immature, First Cycle*, Brian E. Small). **Page 194: Black-headed Gull** (*Breeding Adult*, Dr. Yuri Artukhin/VIREO); (*Nonbreeding Adult*, Brian E. Small); (*Immature, First Cycle*, Jeff Poklen/VIREO). **Bonaparte's Gull** (*Breeding Adult*, Brian E. Small); (*Nonbreeding Adult*, Jukka Jantunen/VIREO); (*Immature, First Cycle*, Bill Schmoker). **Page 195: Little Gull** (*Breeding Adult*, Jari Peltomäki/VIREO); (*Nonbreeding Adult*, Mike Danzenbaker); (*Immature, First Cycle*, Mike Danzenbaker). **Sabine's Gull** (*Breeding Adult*, Kevin T. Karlson); (*Nonbreeding Adult*, Richard Crossley/VIREO); (*Juvenile*, Bill Schmoker). **Page 196: Black-legged Kittiwake** (*Breeding Adults*, Kevin T. Karlson); (*Nonbreeding Adult*, Brian E. Small); (*Juvenile*, Jeff Poklen/VIREO). **Red-legged Kittiwake** (*Immature*, Arthur Morris/VIREO); (*Breeding Adult*, Mike Danzenbaker). **Page 197: Ross's Gull** (*Breeding Adult*, Eugene Potapov/VIREO); (*Nonbreeding Adult*, Steve Young/VIREO); (*Immature, First Cycle*, Joe Fuhrman/VIREO). **Ivory Gull** (*Adult*, Hanne & Jens Eriksen/VIREO); (*Juvenile*, Cindy Creighton/VIREO). **Page 198: Forster's Tern** (*Breeding Adult*, Rob Curtis/VIREO); (*Juvenile*, Kevin T. Karlson); (*In Flight*, Rick and Nora Bowers/VIREO). **Common Tern** (*Breeding Adult*, Brian E. Small); (*Immature*, Brian E. Small); (*In Flight*, Richard Crossley/VIREO). **Page 199: Arctic Tern** (*Breeding Adult*, Brian E. Small); (*Juvenile*, Barry Miller/VIREO); (*In Flight*, Kevin T. Karlson). **Roseate Tern** (*Breeding Adult*, Kevin T. Karlson); (*Juvenile*, Arthur Morris/VIREO); (*In Flight*, Arthur Morris/VIREO). **Page 200: Gull-billed Tern** (*Nonbreeding Adult*, Kevin T. Karlson); (*Breeding Adult*, Brian E. Small). **Sandwich Tern** (*Breeding Adult*, Brian E. Small); (*Nonbreeding Adult*, Brian E. Small); (*Adult In Flight*, Greg Lasley/VIREO). **Page 201: Elegant Tern** (*Breeding Adult*, Brian E. Small); (*Nonbreeding Adult*, Jeff Poklen/VIREO); (*In Flight*, Martin Meyers/VIREO). **Royal Tern** (*Breeding Adult*, Brian E. Small); (*Nonbreeding Adult*, Doug Wechsler/VIREO); (*Adult In Flight*, Brian E. Small). **Page 202: Caspian Tern** (*Adult In Flight*, Brian E. Small); (*Nonbreeding Adult*, Brian E. Small). **Least Tern** (*Immature*, Arthur Morris/VIREO); (*Breeding Adult*, Brian E. Small); (*In Flight*, Arthur Morris/VIREO). **Page 203: Black Tern** (*Breeding Adult*, Brian E. Small); (*Nonbreeding Adult*, Brian E. Small); (*In Flight*, Brian E. Small); (*Juvenile*, Richard Crossley/VIREO). **White-winged Tern** (Martin Hale/VIREO). **Page 204: Aleutian Tern** (*Breeding Adult*, Martin Hale/VIREO); (*In Flight*, Mike Danzenbaker). **Sooty Tern** (*Juvenile*, Robert L. Pitman/VIREO); (*Adult*, Peter G. Connors/VIREO); (*In Flight*, Brian E. Small). **Page 205: Bridled Tern** (*Juvenile*, Hanne & Jens Eriksen/VIREO); (*Adults*, Adrian & Jane Binns/VIREO); (*In Flight*, Brian Patteson/VIREO). **Brown Noddy** (*Adults*, Brian E. Small). **Page 206: Black Noddy** (*Immature*, Brian E. Small). **Black Skimmer** (*Juvenile*, Kevin T. Karlson); (*Breeding Adult*, Arthur Morris/VIREO); (*In Flight*, Kevin T. Karlson). **Page 208: South Polar Skua** (*Intermediate-morph Adult*, Brian Patteson); (*Dark-morph Adult*, Martin Meyers/VIREO). **Great Skua** (*Adult*, David Tipling/VIREO); (*Adult*, Adrian & Jane Binns/VIREO). **Page 209: Pomarine Jaeger** (*Light-morph Breeding Adult*, Jeff Poklen/VIREO); (*Light-morph Nonbreeding Adult*, George L. Armistead/VIREO); (*Immature*, Bill Schmoker); (*Dark-morph Breeding Adult*, Brian Patteson); (*Light-morph Breeding Adult*, Brian Patteson). **Page 210: Parasitic Jaeger** (*Light-morph Breeding Adult*, Bill Schmoker); (*Dark-morph Adult*, Mike Danzenbaker); (*Head Detail*, Kevin T. Karlson); (*Molting Immature*, Tony Palliser/VIREO); (*Light-morph Breeding Adult*, Kevin T. Karlson). **Page 211: Long-tailed Jaeger** (*Breeding Adult*, Kevin T. Karlson); (*Nonbreeding Adult*, Mike Danzenbaker); (*Light-morph Juvenile*, Mike Danzenbaker); (*Head Detail*, Arthur Morris/VIREO); (*Light-morph Breeding Adult*, Tom Vezo/VIREO). **Page 213:**

Common Murre (*Breeding Adult, Bridled Morph*, Steven Holt/VIREO); (*Nonbreeding Adult*, Brian E. Small). **Thick-billed Murre** (*Breeding Adult*, Brian E. Small); (*Nonbreeding Adult*, Brian E. Small). **Page 214: Razorbill** (*Breeding Adult*, Mike Danzenbaker); (*Subadult*, Scott Elowitz/VIREO). **Dovekie** (*Breeding Adult*, Hanne & Jens Eriksen/VIREO); (*Nonbreeding Adult*, Glen Tepke/VIREO). **Page 215: Black Guillemot** (*Breeding Adult*, Tom J. Ulrich/VIREO); (*Nonbreeding Adult*, Kevin T. Karlson); (*In Flight*, Jari Peltomäki/VIREO). **Pigeon Guillemot** (*Breeding Adults*, Brian E. Small); (*Immature*, Jeff Poklen); (*Juvenile*, Jim Zipp). **Page 216: Atlantic Puffin** (*Breeding Adult*, Richard & Susan Day/VIREO); (*Nonbreeding*, John Brian Patrick Patteson/VIREO); (*In Flight*, Tom Vezo/VIREO). **Horned Puffin** (*Breeding Adult*, Arthur Morris/VIREO); (*Nonbreeding Adult*, Peter LaTourrette/VIREO); (*In Flight*, Arthur Morris/VIREO). **Page 217: Tufted Puffin** (*Breeding Adults*, Brian E. Small); (*Subadult*, Rick and Nora Bowers/VIREO); (*In Flight*, Arthur Morris/VIREO). **Rhinoceros Auklet** (*Breeding Adult*, Jukka Jantunen/VIREO); (*Juvenile*, Jukka Jantunen/VIREO); (*In Flight*, Brian E. Small). **Page 218: Cassin's Auklet** (*Adult*, Robert L. Pitman/VIREO); (*In Flight*, Glen Tepke/VIREO). **Ancient Murrelet** (*Breeding Adult*, Mike Danzenbaker); (*Nonbreeding*, Mike Danzenbaker). **Xantus's Murrelet** (*Adult, Northern*, Jeff Poklen). **Page 219: Craveri's Murrelet** (*Adult*, Don Roberson/VIREO); (*In Flight*, Les Chibana). **Marbled Murrelet** (*Molting Adult*, Tom Middleton/VIREO); (*Nonbreeding Adult*, Jim Zipp); (*In Flight*, Brian E. Small/VIREO). **Long-billed Murrelet** (Brian E. Small). **Page 220: Kittlitz's Murrelet** (*Nonbreeding Adult*, Tim Zurowski/VIREO); (*In Flight*, Robert H. Day/VIREO). **Parakeet Auklet** (*Breeding Adult*, Brian E. Small); (*In Flight*, Brian Sullivan/VIREO). **Page 221: Crested Auklet** (*Breeding Adult*, Mike Danzenbaker); (*Immature*, Mike Danzenbaker). **Whiskered Auklet** (*Breeding Adult*, Martin Hale/VIREO). **Least Auklet** (*Breeding Adult*, Brian E. Small). **Page 223: Rock Pigeon** (*Variants*, Brian E. Small); (*Wild-type Adult*, Brian E. Small); (*In Flight*, Jukka Jantunen/VIREO). **White-crowned Pigeon** (*Adult*, Brian E. Small). **Page 224: Band-tailed Pigeon** (*Adult*, Mathew Tekulsky/VIREO); (*In Flight*, Jukka Jantunen/VIREO); (*Juvenile*, Brian E. Small). **Red-billed Pigeon** (*Adult*, Harold Stiver/VIREO). **Page 225: Spotted Dove** (*Adult*, Marvin R. Hyett, M.D./VIREO). **Eurasian Collared-Dove** (*Adult*, Bill Schmoker); (*Juveniles*, Jari Peltomäki/VIREO). **African Collared-Dove** (Hanne & Jens Eriksen/VIREO). **Page 226: White-winged Dove** (*Adult*, Brian E. Small); (*In Flight*, Tom Vezo/VIREO). **White-tipped Dove** (*Adult*, Brian E. Small); (*Adult*, Rick and Nora Bowers/VIREO). **Page 227: Mourning Dove** (*Adult*, Brian E. Small); (*Adult In Flight*, Tom Vezo/VIREO); (*Juvenile*, Rolf Nussbaumer/VIREO). **Ruddy Ground-Dove** (*Adult Male*, Brian E. Small). **Page 228: Inca Dove** (*Adult*, Brian E. Small); (*Adult*, Brian E. Small). **Common Ground-Dove** (*Adult Male*, Brian E. Small); (*Adult Female*, Brian E. Small); (*Adult*, Greg Lasley/VIREO). **Page 230: Monk Parakeet** (*Adults*, Brian E. Small). **Rose-ringed Parakeet** (Adrian & Jane Binns/VIREO). **Black-hooded Parakeet** (Brian E. Small). **Peach-faced Lovebird** (Peter Craig-Cooper/VIREO). **Page 231: Red-crowned Parrot** (*Adult*, Brian E. Small). **Green Parakeet** (*Adult*, Brian E. Small). **Thick-billed Parrot** (Rick and Nora Bowers/VIREO). **Lilac-crowned Parrot** (Rick and Nora Bowers/VIREO). **Page 232: White-winged Parakeet** (*Adult*, John H. Boyd III/VIREO). **Budgerigar** (*Wild-type Adult*, Jim Culbertson/VIREO). **Yellow-chevroned Parakeet** (Adrian & Jane Binns/VIREO). **Page 234: Yellow-billed Cuckoo** (*Adult*, Tom J. Ulrich/VIREO); (*Juvenile*, Magill Weber/VIREO). **Page 234: Black-billed Cuckoo** (*Adult*, Tom J. Ulrich/VIREO); (*Juvenile*, Richard Crossley/VIREO). **Page 235: Mangrove Cuckoo** (*Adult*, Brian E. Small). **Page 235: Common Cuckoo** (*Gray-morph Adult Male*, Hanne & Jens Eriksen/VIREO); (*Red-morph Female*, Jari Peltomäki/VIREO). **Page 235: Oriental Cuckoo** (Doug Wechsler/VIREO). **Page 236: Smooth-billed Ani** (*Adult*, John Dunning/VIREO); (*Juvenile*, Doug Wechsler/VIREO). **Page 236: Groove-billed Ani** (*Adult*, Rolf Nussbaumer/VIREO); (*Adult*, Brian E. Small). **Page 237: Greater Roadrunner** (*Adult*, Dustin Huntington/VIREO); (*Juvenile*, Arthur Morris/VIREO); (*Adult*, Tom Vezo/VIREO); (*Foraging*, Dale & Marian Zimmerman/VIREO). **Page 239: Barn Owl** (*Adult Male*, Rick and Nora Bowers/VIREO); (*Adult Female*, Martin Meyers/VIREO). **Page 239: Great Horned Owl** (*Adult*, Brian E. Small); (*Fledgling*, Jim Zipp). **Page 240: Eastern Screech-Owl** (*Gray-morph Adult*, Jim Zipp); (*Red-morph Adult*, Jim Zipp). **Page 240: Western Screech-Owl** (*Adult*, Brian E. Small); (*Adult*, Brian E. Small). **Page 241: Whiskered Screech-Owl** (*Adult*, Brian E. Small). **Page 241: Flammulated Owl** (*Adult*, Brian E. Small). **Page 242: Elf Owl** (*Adult*, Brian E. Small). **Page 242: Northern Pygmy-Owl** (*Adult*, Bill Schmoker). **Page 243: Ferruginous Pygmy-Owl** (*Adult*, Brian E. Small); (*Adult*, Bill Schmoker). **Page 243: Burrowing Owl** (*Adult*, Brian E. Small); (*Juvenile*, Brian E. Small). **Page 244: Short-eared Owl** (*Adult*, Jim Zipp); (*Adult*, Jim Zipp); (*In Flight*, Jim Zipp). **Page 244: Long-eared Owl** (*Adult*, Brian E. Small); (*In Flight*, Brian E. Small). **Page 245: Barred Owl** (*Adult*, Jim Zipp); (*In Flight*, Jim Zipp). **Page 245: Spotted Owl** (*Adult, "Northern"*, Kevin Schafer/VIREO); (*Immature, "Mexican"*, Jim Zipp). **Page 246: Northern Saw-whet Owl** (*Adult With Prey*, Brian E. Small); (*Juvenile*, Dr. Edgar T. Jones/VIREO). **Page 246: Boreal Owl** (*Adult*, Brian E. Small); (*Juvenile*, Dan Roby & Karen Brink/VIREO). **Page 247: Northern Hawk-Owl** (*Adult*, Jim Zipp); (*In Flight*, Jim Zipp). **Page 247: Great Gray Owl** (*Adult*, Brian E. Small). **Page 248: Snowy Owl** (*Adult male*, Warren Greene/VIREO); (*Immature*, Brian Henry/VIREO). **Page 248: Common Nighthawk** (*Adult*, Brian E. Small); (*In Flight*, Kevin T. Karlson). **Page 249: Lesser Nighthawk** (*At Rest*, Rolf Nussbaumer/VIREO); (*In Flight*, Brian E. Small). **Page 249: Antillean Nighthawk** (*Adult Female*, Joe Fuhrman/VIREO); (*In Flight*, Mike Danzenbaker). **Page 250: Whip-poor-will** (*Adult Female, "Eastern"*, Warren Greene/VIREO); (*Adult Female, "Arizona"*, Brian E. Small). **Page 250: Buff-collared Nightjar** (Rick and Nora Bowers/VIREO). **Page 250: Chuck-will's-widow** (*Adult*, Kevin T. Karlson); (*Adult*, Brian E. Small). **Page 251: Common Poorwill** (*Adult*, Brian E. Small); (*Adult Male*, Rick and Nora Bowers/VIREO). **Page 251: Common Pauraque** (*Adult Male*, Christian Artuso/VIREO); (*Adult Female*, Brian E. Small). **Page 253: Chimney Swift** (*Adult*, Mike Danzenbaker); (*Adult*, Greg Lasley/VIREO). **Page 253: Vaux's Swift** (*Adult*, Mike Danzenbaker); (*Adult*, Mike Danzenbaker). **Page 254: Black Swift** (*Adult*, Bill Schmoker); (*At Rest*, Mike Danzenbaker). **Page 254: White-throated Swift** (*Adult*, Brian E. Small); (*Adult*, Kevin Smith/VIREO). **Page 255: Ruby-throated Hummingbird** (*Adult Male*, Rolf Nussbaumer/VIREO); (*Adult Male*, Mike Danzenbaker); (*Immature Male*, Sid & Shirley Rucker/VIREO); (*Adult Female*, Glenn Bartley/VIREO). **Page 256: Allen's Hummingbird** (*Adult Male*, Dr. Joseph Turner/VIREO); (*Juvenile Male*, Mathew Tekulsky/VIREO). **Page 256: Rufous Hummingbird** (*Adult Male*, Brian E. Small); (*Female*, Brian E. Small). **Page 257: Black-chinned Hummingbird** (*Adult Male*, Brian E. Small); (*Juvenile Female*, Bill Schmoker). **Page 257: Broad-tailed Hummingbird** (*Adult Male*, Rolf Nussbaumer/VIREO); (*Adult Female*, Bill Schmoker). **Page 258: Calliope Hummingbird** (*Adult Male*, Brian E. Small); (*Adult Female*, Bill Schmoker). **Page 258: Anna's Hummingbird** (*Adult Male*, Brian E. Small); (*Adult Female*, Brian E. Small). **Page 259: Costa's Hummingbird** (*Adult Male*, Brian E. Small); (*Adult Female*, Brian E. Small). **Page 259: Magnificent Hummingbird** (*Adult Male*, Brian E. Small); (*Adult Female*, Brian E. Small). **Page 260: Blue-throated Hummingbird** (*Adult Male*, Brian E. Small); (*Adult Female*, Brian E. Small). **Page 260: Broad-billed Hummingbird** (*Adult Male*, Mike Danzenbaker); (*Female*, Adrian & Jane Binns/VIREO). **Page 261: White-eared Hummingbird** (*Adult Male*, Jim Zipp); (*Adult Female*, Bill Schmoker). **Page 261: Berylline Hummingbird** (*Adult*, Rick and Nora Bowers/VIREO); (*Immature*, Sid & Shirley Rucker/VIREO). **Page 262: Lucifer Hummingbird** (*Adult Male*, Mike Danzenbaker); (*Adult Female*, Bill Schmoker). **Page 262: Violet-crowned Hummingbird** (*Adult*, Dan True/VIREO). **Page 262: Buff-bellied Hummingbird** (*Adult*, Brian E. Small). **Page 262: Green Violet-ear** (Glenn Bartley/VIREO). **Page 264: Elegant Trogon** (*Adult Male*, Brian E. Small); (*Adult Female*, Brian E. Small); (*Juvenile*, Bill Schmoker). **Page 264: Eared Quetzal** (Rick and Nora Bowers/VIREO). **Page 265: Belted Kingfisher** (*Adult Male*, Brian E. Small); (*Adult Female*, Brian E. Small); (*In Flight*, Greg Lasley/VIREO); (*Hovering*, Jukka Jantunen/VIREO); (*Adult Male*, Jukka Jantunen/VIREO). **Page 266: Ringed Kingfisher** (*Adult Male*, Brian E. Small); (*Adult Female*, Brian E. Small). **Green Kingfisher** (*Adult Male*, Brian E. Small); (*Adult Female*, Brian E. Small). **Page 268: Downy Woodpecker** (*Adult Male, Western*, Brian E. Small); (*Adult Female, Western*, Brian E. Small); (*Juvenile*, Bill Schmoker); (*Adult Female, Eastern*, Brian E. Small). **Page 269: Hairy Woodpecker** (*Adult Male, Western*, Brian E. Small); (*Adult Female, Eastern*, Brian E. Small). **Red-cockaded Woodpecker** (*Adult Female*, Stephen G. Maka/VIREO); (*Adult Male*, Brian E. Small). **Page 270: Ladder-backed Woodpecker** (*Adult Male*, Brian E. Small); (*Adult Female*, Brian E. Small). **Nuttall's Woodpecker** (*Adult Male*, Brian E. Small); (*Adult Female*, Brian E. Small). **Page 271: Arizona Woodpecker** (*Adult Male*, Brian E. Small); (*Adult Female*, Brian E. Small). **White-headed Woodpecker** (*Adult Male*, Brian E. Small); (*Adult Female*, Brian E. Small). **Page 272: Red-headed Woodpecker** (*Adult*, Brian E. Small); (*Juvenile*, R. & N. Bowers/VIREO). **Black-backed Woodpecker** (*Adult Male*, Brian E. Small); (*Adult Female*, Brian E. Small). **Page 273: American Three-toed Woodpecker** (*Adult Male, Northwestern*, Jukka Jantunen/VIREO); (*Adult Male, Eastern*, Garth McElroy/VIREO); (*Adult Female, Intermountain West*, Brian E. Small). **Page 274: Red-bellied Woodpecker** (*Adult Male*, Brian E. Small); (*Adult Female*, Brian E. Small); (*Juvenile*, Brian E. Small). **Page 275: Gila Woodpecker** (*Adult Male*, Brian E. Small); (*Adult Female*, Brian E. Small). **Golden-fronted Woodpecker** (*Adult Male*, Brian E. Small); (*Adult Female*, Brian E. Small). **Page 276: Acorn Woodpecker** (*Adult Male*, Brian E. Small); (*Adult Female*, Brian E. Small). **Lewis's Woodpecker** (*Adult*, Brian E. Small); (*Juvenile*, Brian E. Small). **Page 277: Yellow-bellied Sapsucker** (*Adult Male*, Brian E. Small); (*Adult Female*, Brian E. Small); (*Juvenile*, Arthur Morris/VIREO). **Page 278: Red-naped Sapsucker** (*Adult Male*, Brian E. Small); (*Hybrid*, Brian E. Small); (*Adult Female*, Brian E. Small). **Red-breasted Sapsucker** (*Adult Male*, Paul Bannick/VIREO); (*Adult Female*, Mike Danzenbaker). **Page 279: Williamson's Sapsucker** (*Adult Male*, Brian E. Small); (*Adult Female*, Brian E. Small). **Gilded Flicker** (*Adult Male*, Brian E. Small); (*Adult Female*, Brian E. Small). **Page 280: Northern Flicker** (*Adult Male, "Yellow-shafted"*, Brian E. Small); (*Adult Male, "Red-shafted"*, Brian E. Small); (*Adult Female, "Yellow-Shafted"*, Fred Truslow/VIREO); (*Adult Female, "Red-shafted"*, Brian E. Small); (*Intergrade*, Bill Schmoker). **Page 281: Pileated Woodpecker** (*Adult Male*, Brian E. Small); (*Adult Female*, Brian E. Small); (*In Flight*, Brian E. Small). **Page 283: Northern Beardless-Tyrannulet** (*Adult*, Brian E. Small). **Acadian Flycatcher** (*Adult*, Greg Lasley/VIREO). **Yellow-bellied Flycatcher** (*Adult*, Arthur Morris/VIREO). **Page 284: Alder Flycatcher** (*Adult*, Rick and Nora Bowers/VIREO); (*Adult*, Fred Truslow/VIREO). **Willow Flycatcher** (*Adult*, Greg Lasley/VIREO); (*Juvenile*, Brian E. Small). **Page 285: Least Flycatcher** (*Adult*, Brian E. Small); (*Immature*, Brian E. Small). **Gray Flycatcher** (*Adult*, Brian E. Small); (*Nonbreeding*, Rick and Nora Bowers/VIREO). **Page 286: Dusky Flycatcher** (*Adult*, Brian E. Small); (*Adult*, Brian E. Small). **Hammond's Flycatcher** (*Adult*, Brian E. Small); (*Immature*, Brian E. Small). **Page 287: Pacific-slope Flycatcher** (*Adult*, Rick and

Nora Bowers/VIREO); (*Juvenile*, Rick and Nora Bowers/VIREO). **Cordilleran Flycatcher** (*Adult*, Brian E. Small); (*Adult*, Bill Schmoker). **Page 288: Buff-breasted Flycatcher** (*Adult*, Brian E. Small); (*Adult*, Brian E. Small). **Olive-sided Flycatcher** (*Adult*, Brian E. Small); (*Adult*, Glenn Bartley/VIREO). **Page 289: Eastern Wood-Pewee** (*Adult*, Brian E. Small); (*Adult*, Brian E. Small). **Western Wood-Pewee** (*Adult*, Joe Fuhrman/VIREO); (*Adult*, Garth McElroy/VIREO). **Page 290: Greater Pewee** (*Adult*, Rick and Nora Bowers/VIREO); (*Adult*, Brian E. Small). **Eastern Phoebe** (*Adult*, Brian E. Small); (*Juvenile*, Arthur Morris/VIREO). **Page 291: Say's Phoebe** (*Adult*, Jim Zipp); (*Juvenile*, Rick and Nora Bowers/VIREO). **Black Phoebe** (*Adult*, Brian E. Small); (*Juvenile*, Alan David Walther/VIREO). **Page 292: Vermilion Flycatcher** (*Adult Male*, Brian E. Small); (*Adult Female*, Brian E. Small); (*Juvenile*, Brian E. Small). **Great Crested Flycatcher** (*Adult*, Brian E. Small); (*Adult*, Brian E. Small). **Page 293: Brown-crested Flycatcher** (*Adult*, Brian E. Small); (*Adult*, Brian E. Small). **Ash-throated Flycatcher** (*Adult*, Brian E. Small). **La Sagra's Flycatcher** (Doug Wechsler/VIREO). **Page 294: Dusky-capped Flycatcher** (*Adult*, Rick and Nora Bowers/VIREO); (*Adult*, Rick and Nora Bowers/VIREO). **Gray Kingbird** (*Adult*, Marvin R. Hyett, M.D./VIREO); (*Adult*, Adrian & Jane Binns/VIREO). **Page 295: Eastern Kingbird** (*Adult*, Brian E. Small); (*Juvenile*, John Heidecker/VIREO); (*In Flight*, Jukka Jantunen/VIREO). **Page 296: Thick-billed Kingbird** (*Adult*, Brian E. Small); (*Immature*, Martin Meyers/VIREO). **Cassin's Kingbird** (*Adult*, Rick and Nora Bowers/VIREO); (*Adult*, Brian E. Small). **Page 297: Western Kingbird** (*Adult*, Brian E. Small); (*Immature*, Dr. Michael Stubblefield/VIREO); (*In Flight*, Brian E. Small); (*In Flight*, Bill Schmoker); (*Juvenile*, Bob Steele/VIREO). **Page 298: Tropical Kingbird** (*Adult*, Brian E. Small). **Couch's Kingbird** (*Adult*, Richard & Susan Day/VIREO). **Page 299: Scissor-tailed Flycatcher** (*Adult*, Adrian & Jane Binns/VIREO); (*Juvenile*, Rolf Nussbaumer/VIREO). **Sulphur-bellied Flycatcher** (*Adult*, Brian E. Small). **Fork-tailed Flycatcher** (Peter Alden/VIREO). **Page 300: Great Kiskadee** (*Adult*, Brian E. Small). **Rose-throated Becard** (*Adult Male*, Brian E. Small); (*Adult Female*, Greg Lasley/VIREO). **Page 302: Loggerhead Shrike** (*Adult*, Jim Zipp); (*Immature*, Brian E. Small). **Northern Shrike** (*Adult*, Arthur Morris/VIREO); (*In Flight*, Jari Peltomäki/VIREO). **Page 304: Red-eyed Vireo** (*Adult*, Brian E. Small); (*Adult*, Brian E. Small); (*Immature*, Rick and Nora Bowers/VIREO). **Yellow-green Vireo** (*Adult*, Brian E. Small). **Page 305: Black-whiskered Vireo** (*Adult*, Brian E. Small). **Philadelphia Vireo** (*Adult*, Brian E. Small); (*Adult*, Brian E. Small); (*In Flight*, Doug Wechsler/VIREO). **Page 306: Gray Vireo** (*Adult*, Brian E. Small). **Warbling Vireo** (*Adult*, *Western*, Brian E. Small); (*Adult*, *Eastern*, Brian E. Small). **Page 307: Hutton's Vireo** (*Adult*, *Pacific-slope*, Brian E. Small); (*Adult*, *Interior*, Brian E. Small). **Bell's Vireo** (*Adult*, *Midwest*, Robert Royse/VIREO); (*Adult*, *Southwest*, Brian E. Small). **Page 308: White-eyed Vireo** (*Adult*, Brian E. Small); (*Bill Detail*, Doug Wechsler/VIREO). **Yellow-throated Vireo** (*Adult*, Rick and Nora Bowers/VIREO); (*Adult*, Jim Zipp). **Page 309: Plumbeous Vireo** (*Adult*, Brian E. Small). **Blue-headed Vireo** (*Adult*, Brian E. Small). **Black-capped Vireo** (*Adult Male*, Brian E. Small). **Page 310: Cassin's Vireo** (*Adult*, Brian E. Small); (*Adult*, Brian E. Small). **Page 312: Blue Jay** (*In Flight*, Jim Zipp); (*Adult*, Jim Zipp). **Steller's Jay** (*Adult*, *Interior*, Bill Schmoker); (*Adult*, *Pacific Slope*, Brian E. Small). **Page 313: Western Scrub-Jay** (*Adult*, *"California"*, Brian E. Small); (*Adult*, *"Woodhouse's"*, Bill Schmoker). **Island Scrub-Jay** (*Adult*, Brian E. Small); (*Adult*, Peter LaTourrette/VIREO). **Page 314: Mexican Jay** (*Adult*, Jim Zipp); (*Juvenile*, Brian E. Small). **Pinyon Jay** (*Adult*, Brian E. Small); (*Juvenile*, Brian E. Small). **Page 315: Florida Scrub-Jay** (*Adult*, Brian E. Small). **Green Jay** (*Adult*, Brian E. Small). **Brown Jay** (*Adult*, Brian E. Small). **Page 316: Gray Jay** (*Adult*, *Eastern*, Jim Zipp); (*Adult*, *Pacific-slope*, Brian E. Small); (*Adult*, *Interior*, Bill Schmoker); (*Juvenile*, Brian E. Small). **Page 317: Black-billed Magpie** (*Adult*, Brian E. Small); (*Juvenile*, Brian E. Small); (*In Flight*, Martin Meyers/VIREO); (*Adult*, Rolf Nussbaumer/VIREO); (*Adult*, Arthur Morris/VIREO). **Page 318: Yellow-billed Magpie** (*Adult*, Brian E. Small). **Fish Crow** (*Adult*, Brian E. Small). **Tamaulipas Crow** (*Adult*, Jim Culbertson/VIREO). **Page 319: Clark's Nutcracker** (*Adult*, Brian E. Small); (*In Flight*, Martin Meyers/VIREO). **American Crow** (*Adult*, Jim Zipp); (*In Flight*, Bill Schmoker). **Page 320: Northwestern Crow** (*Adult*, Brian E. Small); (*In Flight*, Jim Zipp). **Chihuahuan Raven** (*Adult*, Brian E. Small). **Page 321: Common Raven** (*Adult*, Brian E. Small); (*In Flight*, Brian E. Small); (*In Flight*, Jim Zipp); (*In Flight*, Mike Danzenbaker); (*Adult*, Alvaro Jamarillo). **Page 323: Horned Lark** (*Adult Male*, Bill Schmoker); (*Adult Female*, Garth McElroy/VIREO); (*Juvenile*, Jukka Jantunen/VIREO); (*In Flight*, Bill Schmoker). **Sky Lark** (*Adult*, Mike Danzenbaker). **Page 325: Barn Swallow** (*Adult Male*, Brian E. Small); (*Adult Female*, Brian E. Small); (*In Flight*, Rob Curtis/VIREO); (*Juvenile*, Rob & Ann Simpson/VIREO); (*Adult and Juveniles*, Greg Lasley/VIREO). **Page 326: Cliff Swallow** (*Adult*, Brian E. Small); (*Juvenile*, Brian E. Small); (*In Flight*, Jim Zipp); (*In Flight*, Greg Lasley/VIREO); (*Nest Colony*, Arthur Morris/VIREO). **Page 327: Cave Swallow** (*Adult*, Brian E. Small); (*In Flight*, Brian Sullivan/VIREO). **Northern Rough-winged Swallow** (*In Flight*, Robert Shantz/VIREO); (*Adult*, Brian E. Small); (*Juvenile*, Doug Wechsler/VIREO). **Page 328: Bank Swallow** (*Adult*, Brian E. Small). **Tree Swallow** (*Juvenile*, Bill Schmoker); (*Adult Male*, Brian E. Small); (*Adult Female*, Brian E. Small). **Page 329: Violet-green Swallow** (*Adult Male*, Brian E. Small); (*Adult Female*, Brian E. Small); (*In Flight*, Jukka Jantunen/VIREO). **Purple Martin** (*Adult Male*, Brian E. Small); (*Adult Female*, Brian E. Small). **Page 331: Black-capped Chickadee** (*Adult*, Brian E. Small); (*Adult*, Brian E. Small). **Carolina Chickadee** (*Adult*, Brian E. Small). **Page 332: Mexican Chickadee** (*Adult*, Jim Zipp); (*Adult*, Martin Meyers/VIREO). **Mountain Chickadee** (*Fresh Adult*, *Intermountain West*, Brian E. Small); (*Worn Adult*, *Pacific Slope*, Brian E. Small). **Page 333: Boreal Chickadee** (*Fresh Adult*, Jim Zipp); (*Worn Adult*, Brian E. Small). **Chestnut-backed Chickadee** (*Adult*, *Northern*, Jim Zipp); (*Adult*, *Southern*, Brian E. Small). **Page 334: Gray-headed Chickadee** (*Adult*, Aaron Lang/VIREO). **Tufted Titmouse** (*Adult*, Brian E. Small); (*Adult*, Richard Crossley/VIREO); (*Juvenile*, Rick and Nora Bowers/VIREO). **Page 335: Bridled Titmouse** (*Adult*, Jim Zipp). **Black-crested Titmouse** (*Adult*, Brian E. Small). **Oak Titmouse** (*Adult*, Brian E. Small). **Page 336: Juniper Titmouse** (*Adult*, Brian E. Small). **Verdin** (*Adult Male*, Brian E. Small); (*Adult Female*, Brian E. Small); (*Juvenile*, Bob Steele/VIREO). **Page 337: Bushtit** (*Male*, *Interior*, Jim Zipp); (*Female*, *Interior*, Bill Schmoker); (*Male*, *Pacific Slope*, Brian E. Small); (*Adult at Nest*, Rob Curtis/VIREO); (*Female*, *Pacific Slope*, Brian E. Small). **Page 338: Red-breasted Nuthatch** (*Male*, Brian E. Small); (*Female*, Brian E. Small); (*In Flight*, Kevin Smith/VIREO). **Pygmy Nuthatch** (*Adult*, Jim Zipp); (*Juvenile*, Brian E. Small). **Page 339: White-breasted Nuthatch** (*Adult Male*, *Eastern*, Jim Zipp); (*Adult Female*, *Eastern*, Garth McElroy/VIREO); (*Adult Male*, *Interior West*, Garth McElroy/VIREO); (*Adult Male*, *Pacific Slope*, Brian E. Small). **Page 340: Brown-headed Nuthatch** (*Adult*, Brian E. Small). **Brown Creeper** (*Adult*, Brian E. Small); (*Adult*, Jim Zipp). **Page 342: House Wren** (*Adult*, Arthur Morris/VIREO); (*Juvenile*, James M. Wedge/VIREO); (*"Brown-throated" Wren*, Christian Artuso/VIREO). **Winter Wren** (*Worn Adult*, *Eastern*, Brian E. Small); (*Fresh Adult*, *Western*, Tom Vezo/VIREO); (*Adult*, *Bering Sea*, Brian E. Small). **Page 343: Bewick's Wren** (*Adult*, Brian E. Small); (*Adult*, Brian E. Small). **Carolina Wren** (*Adult*, Brian E. Small); (*Adult*, Rob Curtis/VIREO). **Page 344: Marsh Wren** (*Adult*, *Western*, Brian E. Small); (*Adult*, *Eastern*, Steve Greer/VIREO). **Sedge Wren** (*Adult*, Brian E. Small); (*Adult*, Brian E. Small). **Page 345: Rock Wren** (*Adult*, Brian E. Small); (*Adult*, Brian E. Small). **Canyon Wren** (*Adult*, Brian E. Small); (*Adult*, Rick and Nora Bowers/VIREO). **Page 346: Cactus Wren** (*Adult*, Brian E. Small); (*Adult*, Brian E. Small). **Page 346: American Dipper** (*Adult*, Brian E. Small); (*Adult*, Bill Schmoker). **Page 347: Red-whiskered Bulbul** (*Adult*, Martin Hale/VIREO). **Page 349: Ruby-crowned Kinglet** (*Adult Male*, Brian E. Small); (*Adult Female*, Brian E. Small). **Golden-crowned Kinglet** (*Adult Male*, Brian E. Small); (*Adult Female*, Brian E. Small). **Page 350: Blue-gray Gnatcatcher** (*Breeding Male*, Greg Lasley/VIREO); (*Adult Female*, Brian E. Small). **Black-tailed Gnatcatcher** (*Adult Male*, Brian E. Small); (*Adult Female*, Rick and Nora Bowers/VIREO). **Black-capped Gnatcatcher** (Christian Artuso/VIREO). **Page 351: California Gnatcatcher** (*Breeding Male*, Brian E. Small); (*Adult Female*, Brian E. Small/VIREO). **Arctic Warbler** (*Adult*, Brian E. Small); (*Adult*, Brian E. Small). **Page 353: Northern Wheatear** (*Breeding Male*, *Western*, Brian E. Small); (*Nonbreeding Male*, *Eastern*, Steve Young/VIREO); (*Breeding Female*, Brian E. Small). **Bluethroat** (*Adult Male*, Brian E. Small). **Page 354: Siberian Rubythroat** (*Adult Male*, Mike Danzenbaker). **Eastern Bluebird** (*Adult Male*, Jim Zipp); (*Adult Female*, Brian E. Small); (*Juvenile*, Jim Zipp). **Page 355: Western Bluebird** (*Adult Male*, Brian E. Small); (*Adult Female*, Brian E. Small); (*Juvenile*, Brian E. Small); (*Adult Male and Nestlings*, Betty Randall/VIREO). **Page 356: Mountain Bluebird** (*Adult Male*, Bill Schmoker); (*Adult Female*, Brian E. Small); (*Juvenile*, Brian E. Small). **Townsend's Solitaire** (*Adult*, Brian E. Small); (*Juvenile*, Tony Leukering/VIREO). **Page 357: Varied Thrush** (*Adult Male*, Brian E. Small); (*Adult Female*, Jukka Jantunen/VIREO). **Rufous-backed Robin** (*Adult*, Rick and Nora Bowers/VIREO). **Aztec Thrush** (Dale & Marian Zimmerman/VIREO). **Page 358: American Robin** (*Adult Male*, Robert Royse); (*In Flight*, Jukka Jantunen/VIREO); (*Adult Female*, Brian E. Small); (*Juvenile*, Brian E. Small). **Fieldfare** (Rolf Nussbaumer/VIREO). **Page 359: Clay-colored Robin** (*Adult*, Brian E. Small). **Eyebrowed Thrush** (*Adult Male*, Doug Wechsler/VIREO). **Gray-cheeked Thrush** (*Adult*, Brian E. Small). **Page 360: Hermit Thrush** (*Adult*, *Eastern*, Gerard Bailey/VIREO); (*Adult*, *Intermountain West*, Bill Schmoker); (*Adult*, *Pacific Slope*, Brian E. Small). **Bicknell's Thrush** (*Adult*, Garth McElroy/VIREO). **Page 361: Swainson's Thrush** (*Adult*, *Eastern*, Brian E. Small); (*Adult*, *Eastern*, Brian E. Small); (*Adult*, *Pacific Slope*, Brian E. Small). **Page 362: Veery** (*Adult*, *Eastern*, Jim Zipp). **Wood Thrush** (*Adult*, Gerard Bailey/VIREO). **Page 363: Wrentit** (*Adult*, Brian E. Small). **Page 365: Northern Mockingbird** (*Adult*, Brian E. Small); (*Juvenile*, Brian E. Small); (*In Flight*, Martin Meyers/VIREO); (*Adult*, Brian E. Small). **Bahama Mockingbird** (Kathy Adams Clark/VIREO). **Page 366: Sage Thrasher** (*Fresh Adult*, Jim Zipp); (*Worn Adult*, Brian E. Small); (*Juvenile*, Brian E. Small). **Bendire's Thrasher** (*Adult*, Jim Zipp). **Page 367: Curve-billed Thrasher** (*Adult*, *Arizona*, Brian E. Small); (*Adult*, *Texas*, Brian E. Small); (*Juvenile*, Rick and Nora Bowers/VIREO). **California Thrasher** (*Adult*, Brian E. Small). **Page 368: Long-billed Thrasher** (*Adult*, Brian E. Small). **Brown Thrasher** (*Adult*, Brian E. Small); (*Adult*, Bill Schmoker). **Page 369: Crissal Thrasher** (*Adult*, Rick and Nora Bowers/VIREO). **Le Conte's Thrasher** (*Adult*, Brian E. Small). **Gray Catbird** (*Adult*, Brian E. Small). **Page 370: European Starling** (*Breeding Male*, Bill Schmoker); (*Nonbreeding Adult*, Bill Schmoker); (*In Flight*, Marvin R. Hyett, M.D./VIREO); (*Juvenile*, Brian E. Small). **Common Myna** (Brian E. Small). **Page 372: White Wagtail** (*Breeding Male*, *"Black-backed"*, Mike Danzenbaker); (*Breeding Male*, *"Gray-backed"*, Doug Wechsler/VIREO); (*Adult Female*, Mike Danzenbaker). **Eastern Yellow Wagtail** (*Breeding Male*, Brian E. Small); (*Adult Female*, Brian E. Small). **Page 373: Sprague's Pipit** (*Juvenile*, Brian E. Small); (*Adult*, Brian E. Small); (*Adult*, Rick and Nora Bowers/VIREO). **Olive-backed Pipit** (*Adult*, Mike Danzenbaker). **Page 374: American Pipit** (*Breeding Adult*, *Southern Rockies*, Brian E. Small); (*Nonbreeding Adult*, Rick and Nora Bowers/VIREO); (*Breeding Adult*, *Northern*, Garth McElroy).

Red-throated Pipit (*Adult Male*, Brian E. Small). **Page 376: Cedar Waxwing** (*Adult*, Brian E. Small); (*Juvenile*, Rolf Nussbaumer/VIREO); (*Adult and Nestlings*, Warren Greene/VIREO); (*Immature*, Garth McElroy/VIREO); (*Flock*, Richard Crossley/VIREO). **Page 377: Bohemian Waxwing** (*Adult*, Brian E. Small); (*Immature*, Garth McElroy/VIREO). **Phainopepla** (*Adult Male*, Brian E. Small); (*Female*, Brian E. Small). **Page 379: Olive Warbler** (*Adult Male*, Jim Zipp); (*Adult Female*, Brian E. Small). **Tennessee Warbler** (*Adult Male*, Brian E. Small); (*Adult Female*, Brian E. Small/VIREO). **Page 380: Orange-crowned Warbler** (*Adult Male*, Brian E. Small/VIREO); (*Immature, Eastern*, Arthur Morris/VIREO); (*Adult, Pacific Slope*, Brian E. Small); (*Adult, Intermountain West*, Brian E. Small/VIREO); (*Fall Adult, Intermountain West*, Manuel Grosselet/VIREO). **Page 381: Nashville Warbler** (*Adult Male, Eastern*, Jim Zipp); (*Adult Male, Western*, Brian E. Small); (*Immature*, Brian E. Small). **Virginia's Warbler** (*Adult Male*, Brian E. Small); (*Immature*, Bill Schmoker). **Page 382: Colima Warbler** (*Adult*, Rick and Nora Bowers/VIREO); (*Adult*, Mike Danzenbaker). **Lucy's Warbler** (*Adult Male*, Rick and Nora Bowers/VIREO); (*Adult Female*, Rick and Nora Bowers/VIREO). **Page 383: Blue-winged Warbler** (*Adult Male*, Brian E. Small); (*Immature Female*, Gerard Bailey/VIREO); (*Adult Male Hybrid*, Jim Zipp); (*"Brewster's" Warbler*, Jim Zipp); (*"Lawrence's" Warbler*, Jim Zipp). **Page 384: Golden-winged Warbler** (*Adult Male*, Jim Zipp); (*Adult Female*, Jim Zipp). **Chestnut-sided Warbler** (*Breeding Male*, Brian E. Small); (*Immature Female*, Jim Zipp); (*Immature*, Brian E. Small). **Page 385: Northern Parula** (*Breeding Male*, Brian E. Small); (*Adult Female*, Jim Zipp); (*Immature Female*, John Heidecker/VIREO). **Tropical Parula** (*Adult Male*, Brian E. Small). **Page 386: Yellow Warbler** (*Breeding Male*, Brian E. Small); (*Immature male*, Brian E. Small); (*Breeding Male*, Joe Fuhrman/VIREO); (*Immature Female*, Brian E. Small/VIREO); (*"Mangrove" Warbler*, Greg Lasley/VIREO). **Page 387: Prairie Warbler** (*Adult Male*, Brian E. Small); (*Immature Female*, James M. Wedge/VIREO). **Yellow-throated Warbler** (*Adult Male, Inland*, Greg Lasley/VIREO); (*Adult Male, Coastal Plain*, Scott Elowitz/VIREO). **Page 388: Palm Warbler** (*Breeding Adult, "Yellow"*, Brian E Small); (*Nonbreeding, "Yellow"*, Dr Michael Stubblefield/VIREO); (*Breeding Adult, "Western"*, Rob Curtis/VIREO), (*Nonbreeding, "Western"*, Brian E. Small); (*Juvenile, "Western"*, Dr. Edgar T. Jones/VIREO). **Page 389: Cape May Warbler** (*Breeding Male*, Brian E. Small); (*Immature Male*, Rob Curtis/VIREO); (*Breeding Female*, Johann Schumacher/VIREO); (*Immature Male*, Richard Crossley/VIREO); (*Immature Female*, Rob Curtis/VIREO). **Page 390: Grace's Warbler** (*Adult Male*, Mike Danzenbaker); (*Adult Female*, Rick and Nora Bowers/VIREO). **Magnolia Warbler** (*Breeding Male*, Brian E. Small); (*Immature*, Brian E. Small). **Page 391: Yellow-rumped Warbler** (*Adult Male, "Audubon's"*, Brian E. Small); (*Nonbreeding, "Audubon's"*, Brian E. Small); (*Adult Male, "Myrtle"*, Jim Zipp); (*Nonbreeding, "Myrtle"*, Johann Schumacher/VIREO); (*Juvenile*, Jukka Jantunen/VIREO). **Page 392: Kirtland's Warbler** (*Breeding Male*, Robert Royse/VIREO); (*Adult Female*, Ron Austing/VIREO). **Black-throated Green Warbler** (*Breeding Male*, Jim Zipp); (*Immature Female*, James M. Wedge/VIREO). **Page 393: Townsend's Warbler** (*Adult Male*, Brian E. Small); (*Immature Female*, Brian E. Small). **Hermit Warbler** (*Adult Male*, Brian Sullivan/VIREO); (*Immature Female*, Brian E. Small); (*Hybrid*, Brian E. Small). **Page 394: Golden-cheeked Warbler** (*Breeding Male*, Brian E. Small); (*Adult Female*, Sid & Shirley Rucker/VIREO). **Black-throated Gray Warbler** (*Adult Male*, Brian E. Small); (*Immature Female*, Brian E. Small). **Page 395: Blackburnian Warbler** (*Breeding Male*, Jim Zipp); (*Immature*, Lee Trott/VIREO). **Cerulean Warbler** (*Breeding Male*, Jim Zipp); (*Breeding Female*, Brian E. Small). **Page 396: Pine Warbler** (*Adult Male*, Brian E. Small); (*Immature Female*, James M. Wedge/VIREO). **Bay-breasted Warbler** (*Breeding Male*, Brian E. Small); (*Breeding Female*, Brian E. Small). **Page 397: Blackpoll Warbler** (*Breeding Male*, Brian E. Small); (*Nonbreeding*, Bob Steele/VIREO); (*Breeding Female*, Rob Curtis/VIREO). **Page 398: Black-throated Blue Warbler** (*Adult Male*, Brian E. Small); (*Female*, Brian E. Small). **Black-and-white Warbler** (*Adult Male*, Brian E. Small); (*Immature Female*, Brian E. Small). **Page 399: American Redstart** (*Adult Male*, Brian E. Small); (*Immature Male*, Brian E. Small); (*Adult Female*, Brian E. Small). **Worm-eating Warbler** (*Adult*, Brian E. Small). **Page 400: Prothonotary Warbler** (*Adult Male*, Brian E. Small); (*Adult Female*, Brian E. Small). **Swainson's Warbler** (*Adult*, Doug Wechsler/VIREO); (*Adult*, Brian E. Small). **Page 401: Northern Waterthrush** (*Adult*, Rick and Nora Bowers/VIREO); (*Adult*, Jim Zipp). **Louisiana Waterthrush** (*Adult*, Jim Zipp); (*Adult*, Johann Schumacher/VIREO). **Page 402: Common Yellowthroat** (*Adult Male*, Brian E. Small); (*Adult Female*, Brian E. Small); (*Immature Male*, Brian E. Small); (*Adult Male*, Doug Wechsler/VIREO). **Gray-crowned Yellowthroat** (Jim Culbertson/VIREO). **Page 403: Ovenbird** (*Adult*, Rob Curtis/VIREO); (*Immature*, Lee Trott/VIREO). **MacGillivray's Warbler** (*Adult Male*, Brian E. Small); (*Immature Female*, Brian E. Small). **Page 404: Mourning Warbler** (*Adult Male*, Brian E. Small); (*Adult Female*, Rick and Nora Bowers/VIREO). **Connecticut Warbler** (*Adult Male*, Mike Danzenbaker); (*Immature Female*, Scott Elowitz/VIREO). **Kentucky Warbler** (*Adult Male*, Gerard Bailey/VIREO); (*Adult Female*, Brian E. Small). **Page 405: Wilson's Warbler** (*Adult Male*, Brian E. Small); (*Immature Female*, Barth Schorre/VIREO); (*Adult*, Brian E. Small). **Page 406: Hooded Warbler** (*Adult Male*, Robert Royse); (*Adult Female*, Adrian & Jane Binns/VIREO); (*Immature Female*, Brian E. Small). **Red-faced Warbler** (*Adult Male*, Brian E. Small). **Page 407: Canada Warbler** (*Breeding Male*, Jim Zipp); (*Adult Female*, Rob Curtis/VIREO); (*Immature Female*, Gerard Bailey/VIREO). **Painted Redstart** (*Adult*, Brian E. Small). **Page 408: Yellow-breasted Chat** (*Adult Male, Eastern*, Johann Schumacher/VIREO); (*Adult Female, Western*, Rick and Nora Bowers/VIREO). **Golden-crowned Warbler** (Doug Wechsler/VIREO). **Rufous-capped Warbler** (Rick and Nora Bowers/VIREO). **Slate-throated Redstart** (Rick and Nora Bowers/VIREO). **Page 410: Scarlet Tanager** (*Breeding Male*, Brian E. Small); (*Immature Male, First Spring*, Brian E. Small); (*Immature Male, First Fall*, Nathan Barnes/VIREO); (*Adult Female*, Brian E. Small); (*Adult Female*, Rolf Nussbaumer/VIREO). **Page 411: Western Tanager** (*Breeding Male*, Brian E. Small); (*Adult Female*, Brian E. Small); (*Immature Male*, Joe Fuhrman/VIREO); (*Adult Female*, Brian E. Small). **Flame-colored Tanager** (Brian E. Small). **Page 412: Summer Tanager** (*Adult Male, Eastern*, Brian E. Small); (*Adult Female*, Brian E. Small); (*Immature Male, First Spring*, Brian E. Small). **Hepatic Tanager** (*Adult Male*, Brian E. Small); (*Adult Female*, Rick and Nora Bowers/VIREO). **Page 414: White-collared Seedeater** (*Adult Male*, Adrian & Jane Binns/VIREO); (*Adult Female*, John Dunning/VIREO). **Olive Sparrow** (*Adult*, Greg Lasley/VIREO); (*Adult*, Brian E. Small). **Page 415: Eastern Towhee** (*Adult Male*, Robert Royse); (*Adult Female*, James M. Wedge/VIREO); (*Adult Male*, Brian E. Small); (*Immature*, Jim Culbertson/VIREO); (*Juvenile*, Johann Schumacher/VIREO). **Page 416: Spotted Towhee** (*Adult Male*, Brian E. Small); (*Adult Male*, Robert Royse); (*Adult Female*, Brian E. Small); (*Immature*, Glenn Bartley/VIREO); (*Juvenile*, Peter LaTourrette/VIREO). **Page 417: California Towhee** (*Adult*, Brian E. Small). **Canyon Towhee** (*Adult*, Brian E. Small). **Abert's Towhee** (*Adult*, Brian E. Small). **Page 418: Green-tailed Towhee** (*Adult*, Brian E. Small); (*Juvenile*, Brian E. Small). **Cassin's Sparrow** (*Adult*, Brian E. Small); (*Adult*, Brian E. Small); (*Adult*, Brian E. Small). **Page 419: Botteri's Sparrow** (*Adult*, Brian E. Small); (*Adult*, Brian E. Small). **Rufous-crowned Sparrow** (*Adult*, Brian E. Small); (*Adult*, Brian E. Small). **Page 420: Bachman's Sparrow** (*Adult*, Brian E. Small); (*Adult*, Brian E. Small). **Rufous-winged Sparrow** (*Adult*, Jim Zipp). **Page 421: Five-striped Sparrow** (*Adult*, Brian E. Small); (*Adult*, Brian E. Small). **Chipping Sparrow** (*Breeding Adult*, Brian E. Small); (*Nonbreeding Adult*, Brian E. Small); (*Juvenile*, Brian E. Small). **Page 422: Clay-colored Sparrow** (*Breeding Adult*, Brian E. Small); (*Nonbreeding Adult*, Johann Schumacher/VIREO). **Brewer's Sparrow** (*Adult*, Brian E. Small); (*Adult*, Brian E. Small); (*Immature*, Brian E. Small). **Page 423: Field Sparrow** (*Adult*, Brian E. Small), (*Adult*, Jim Zipp). **American Tree Sparrow** (*Adult*, Johann Schumacher/VIREO); (*Adult*, Rob Curtis/VIREO). **Page 424: Black-chinned Sparrow** (*Adult Male*, Brian E. Small); (*Juvenile*, Marvin R. Hyett, M.D./VIREO). **Black-throated Sparrow** (*Adult*, Jim Zipp); (*Juvenile*, Brian E. Small). **Page 425: Sage Sparrow** (*Adult, Interior*, Bill Schmoker); (*Adult, "Bell's"*, Brian E. Small). **Lark Bunting** (*Breeding Male*, Brian E. Small); (*Nonbreeding Male*, Brian E. Small); (*Adult Female*, Brian E. Small). **Page 426: Lark Sparrow** (*Immature*, Brian E. Small); (*Adult*, Brian E. Small); (*Adult*, Brian E. Small). **Vesper Sparrow** (*Adult*, Richard Crossley/VIREO); (*Adult*, Brian E. Small). **Page 427: Savannah Sparrow** (*Typical Adult*, Jim Zipp); (*Typical Adult*, Brian E. Small); (*Adult, "Ipswich"*, Dr. Michael Stubblefield/VIREO); (*Adult, "Large-billed"*, Brian E. Small); (*Adult, "Belding's"*, Brian E. Small). **Page 428: Grasshopper Sparrow** (*Adult*, Brian E. Small); (*Adult*, Brian E. Small); (*Adult*, Brian E. Small) **Henslow's Sparrow** (*Adult*, Robert Royse/VIREO); (*Adult*, Rick and Nora Bowers/VIREO). **Page 429: Baird's Sparrow** (*Adult*, Greg Lasley/VIREO); (*Adult*, Brian E. Small). **Le Conte's Sparrow** (*Adult*, Brian E. Small); (*Adult*, Rick and Nora Bowers/VIREO). **Page 430: Saltmarsh Sharp-tailed Sparrow** (*Adult*, Richard Crossley/VIREO); (*Adult, Hybrid*, Garth McElroy/VIREO). **Nelson's Sharp-tailed Sparrow** (*Adult, Coastal*, Brian E. Small); (*Adult, Coastal*, Brian E. Small); (*Adult, Prairie*, Brian E. Small). **Page 431: Seaside Sparrow** (*Adult, Gulf coast*, Brian E. Small); (*Adult, Atlantic Coast*, Doug Wechsler/VIREO); (*Adult, "Cape Sable"*, Ronald M. Saldino/VIREO); (*Juvenile*, George L. Armistead/VIREO). **Page 432: Fox Sparrow** (*Adult, "Red"*, James M. Wedge/VIREO); (*Adult, "Slate-Colored"*, Bill Schmoker); (*Adult, "Thick-billed"*, Brian E. Small); (*Adult, "Sooty"*, Jim Culbertson/VIREO); (*Adult, Intermediate*, Brian E. Small). **Page 433: Song Sparrow** (*Adult*, Brian E. Small); (*Adult*, Bill Schmoker); (*Adult, Aleutians*, Brian E. Small); (*Juvenile*, Garth McElroy/VIREO); (*Adult*, Rick and Nora Bowers/VIREO). **Page 434: Swamp Sparrow** (*Breeding Adult*, Garth McElroy/VIREO); (*Nonbreeding Adult*, Garth McElroy/VIREO); (*Juvenile*, Rick and Nora Bowers/VIREO); (*Breeding Adult, Mid-Atlantic Coast*, Doug Wechsler/VIREO). **Page 435: Lincoln's Sparrow** (*Adult*, Brian E. Small); (*Adult*, Brian E. Small). **Golden-crowned Sparrow** (*Breeding male*, Arthur Morris/VIREO); (*Nonbreeding Adult*, Brian E. Small); (*Immature*, Brian E. Small). **Page 436: White-throated Sparrow** (*Adult, White-striped Morph*, Brian E. Small); (*Adult, Tan-striped Morph*, Rick and Nora Bowers/VIREO); (*Immature*, Brian E. Small); (*Juvenile*, Jukka Jantunen/VIREO). **Page 437: White-crowned Sparrow** (*Black-lored Adult*, Brian E. Small); (*Adult, "Gambel's"*, Jim Zipp); (*Adult, "Nuttall's"*, Hugh P. Smith, Jr. & Susan C. Smith/VIREO); (*Immature*, Brian E. Small); (*Leucistic Adult, "Nuttall's"*, Bob Steele/VIREO). **Page 438: Harris's Sparrow** (*Breeding Adult*, Robert Royse/VIREO); (*Nonbreeding Adult*, Brian E. Small); (*Immature*, Brian E. Small). **Page 438–439: Dark-eyed Junco** (*Adult Male, "Slate-colored"*, Brian E. Small); (*Adult, "Oregon"*, Brian E. Small); (*Adult, "White-winged"*, Kayleen A. Niyo/VIREO); (*Adult, "Red-backed"*, Robert Royse); (*Adult, "Pink-sided"*, Bill Schmoker); (*Adult, "Cassiar"*, Brian E. Small); (*Adult, "Gray-headed"*, Brian E. Small); (*Juvenile*, Joe Fuhrman/VIREO). **Page 440: Yellow-eyed Junco** (*Adult*, Brian E. Small); (*Adult*, Rick and Nora Bowers/VIREO). **Smith's Longspur** (*Breeding Male*, Arthur Morris/VIREO); (*Breeding Female*, Tom Vezo/VIREO); (*Immature*, Brian E. Small). **Page 441: Lapland Longspur** (*Breeding Male*, Brian E. Small); (*Breeding Female*, Brian E. Small); (*Nonbreeding Male*, Bill Schmoker); (*Immature Male*, Brian E. Small).

Page 442: Chestnut-collared Longspur (*Breeding Male*, Brian E. Small); (*Breeding Female*, Brian E. Small); (*Immature Female*, Martin Meyers/VIREO). **McCown's Longspur** (*Breeding Male*, Brian E. Small); (*Breeding Female*, Brian E. Small); (*Immature Female*, Mike Danzenbaker). **Page 443: Rustic Bunting** (*Breeding Male*, Tadao Shimba/VIREO); (*Breeding Female*, Jari Peltomäki/VIREO). **McKay's Bunting** (*Breeding Male*, Dr. Erica M. Brendel/VIREO); (*Breeding Female*, Dr. Erica M. Brendel/VIREO); (*Nonbreeding*, Dan Logen/VIREO). **Page 444: Snow Bunting** (*Nonbreeding*, Brian E. Small); (*Breeding Male*, Brian E. Small); (*Breeding Female*, Brian E. Small); (*In Flight*, Jukka Jantunen/VIREO); (*Juvenile*, Martin Hale/VIREO). **Page 446: Northern Cardinal** (*Adult Male, Eastern*, Brian E. Small); (*Adult Male, Arizona*, Jim Zipp); (*Juvenile*, Brian E. Small); (*Adult Female*, Brian E. Small). **Page 447: Pyrrhuloxia** (*Adult Male*, Brian E. Small); (*Dull Adult Female*, Brian E. Small). **Black-headed Grosbeak** (*Breeding Male*, Brian E. Small); (*Breeding Female*, Brian E. Small); (*Immature Male, First Spring*, Brian E. Small). **Page 448: Rose-breasted Grosbeak** (*Breeding male*, Brian E. Small); (*Adult Female*, Brian E. Small); (*Immature Male, First Spring*, Brian E. Small); (*Immature Male, First Spring*, Brian E. Small); (*Immature Female, First Spring*, Brian E. Small). **Page 449: Lazuli Bunting** (*Breeding Male*, Brian E. Small); (*Nonbreeding Male*, Marvin R. Hyett, M.D./VIREO); (*Adult Female*, Brian E. Small). **Blue Grosbeak** (*Adult Male*, Brian E. Small); (*Adult Female*, Brian E. Small); (*Immature Male, First Spring*, Brian E. Small). **Page 450: Indigo Bunting** (*Breeding Male*, Brian E. Small); (*Immature Male, First Spring*, Brian E. Small); (*Variant Adult Female*, Garth McElroy/VIREO); (*Typical Adult Female*, Brian E. Small). **Page 451: Varied Bunting** (*Breeding Male*, Brian E. Small); (*Nonbreeding Male*, Tom Vezo/VIREO); (*Adult Female*, Brian E. Small). **Painted Bunting** (*Adult Male*, Brian E. Small); (*Adult Female*, Brian E. Small); (*Immature Male*, Greg Lasley/VIREO). **Page 452: Dickcissel** (*Breeding Male*, Brian E. Small); (*Adult Female*, Rick and Nora Bowers/VIREO); (*Immature Female*, Brian E. Small). **Page 454: Bobolink** (*Breeding Male*, Jim Zipp); (*Breeding Female*, Rick and Nora Bowers/VIREO); (*Nonbreeding*, Kevin T. Karlson). **Brewer's Blackbird** (*Breeding Male*, Brian E. Small); (*Adult Female*, Brian E. Small). **Page 455: Rusty Blackbird** (*Adult Male*, Richard Crossley/VIREO); (*Breeding Female*, Dr. Michael Stubblefield/VIREO); (*Nonbreeding Male*, Rick and Nora Bowers/VIREO); (*Nonbreeding Female*, Rick and Nora Bowers/VIREO). **Page 456: Bronzed Cowbird** (*Adult Male*, Bill Schmoker); (*Adult Female*, Brian E. Small). **Shiny Cowbird** (*Adult Male*, Brian E. Small); (*Adult Female*, Brian E. Small). **Page 457: Brown-headed Cowbird** (*Adult Male*, Brian E. Small); (*Adult Female*, Brian E. Small); (*Juvenile*, Brian E. Small). **Page 458: Common Grackle** (*Adult Male, "Bronzed"*, Rob Curtis/VIREO); (*Adult Male, "Purple"*, Arthur Morris/VIREO); (*Adult Female*, Brian E. Small); (*Juvenile*, Rick and Nora Bowers/VIREO); (*Adult Male, "Bronzed"*, Brian E. Small). **Page 459: Great-tailed Grackle** (*Adult Male*, Brian E. Small); (*Adult Female*, Brian E. Small). **Boat-tailed Grackle** (*Adult Male*, Brian E. Small); (*Adult Female*, Brian E. Small). **Page 460: Red-winged Blackbird** (*Adult Male*, Greg Lasley/VIREO); (*Adult Male, bicolored*, Peter LaTourrette/VIREO); (*Immature Male*, Dustin Huntington/VIREO); (*Adult Female*, Brian E. Small); (*Adult Female, "Bicolored"*, Peter LaTourrette/VIREO). **Page 461: Tricolored Blackbird** (*Adult Male*, Brian E. Small); (*Adult Female*, Brian E. Small). **Yellow-headed Blackbird** (*Adult Male*, Brian E. Small); (*In Flight*, Bill Schmoker). **Page 462: Eastern Meadowlark** (*Nonbreeding Adult*, Brian E. Small); (*Adult Male*, Brian E. Small); (*Adult, "Lillian's"*, Brian E. Small); (*In Flight*, Kevin T. Karlson). **Western Meadowlark** (*Adult Male*, Brian E. Small). **Page 463: Baltimore Oriole** (*Adult Male*, Brian E. Small); (*Adult Female*, Brian E. Small); (*Adult Female*, Brian E. Small). **Bullock's Oriole** (*Adult Male*, Brian E. Small); (*Immature Male*, Brian E. Small); (*Adult Female*, Brian E. Small). **Page 464: Hooded Oriole** (*Adult Male*, Brian E. Small); (*Immature Male*, Bob Steele/VIREO); (*Adult Female*, Brian E. Small). **Audubon's Oriole** (*Adult Male*, Arthur Morris/VIREO); (*Adult Female*, Greg Lasley/VIREO). **Page 465: Scott's Oriole** (*Adult Male*, Brian E. Small); (*Immature Male*, Brian E. Small); (*Adult Female*, Brian E. Small); (*Immature Female*, Rick and Nora Bowers/VIREO). **Page 466: Spot-breasted Oriole** (*Adult*, Adrian & Jane Binns/VIREO); (*Immature*, Jim Culbertson/VIREO). **Altamira Oriole** (*Adult*, Brian E. Small); (*Immature*, Adrian & Jane Binns/VIREO). **Page 467: Orchard Oriole** (*Adult Male*, Brian E. Small); (*Immature Male*, Rob Curtis/VIREO); (*Adult Female*, Brian E. Small). **Page 469: House Finch** (*Adult Male*, Brian E. Small); (*Adult Male*, Brian E. Small); (*Juvenile*, Hugh P. Smith, Jr. & Susan C. Smith/VIREO); (*Adult Female*, Brian E. Small); (*Adult Female*, Brian E. Small). **Page 470: Purple Finch** (*Adult Male, Eastern*, Garth McElroy); (*Adult Male, Eastern*, Garth McElroy); (*Adult Female, Western*, Brian E. Small). **Cassin's Finch** (*Adult Male*, Brian E. Small); (*Adult Female*, Brian E. Small). **Page 471: Pine Grosbeak** (*Adult Male*, Brian E. Small); (*Adult Male*, Brian E. Small; Jim Zipp); (*Adult Female*, Jim Zipp); (*Juvenile*, Rob Curtis/VIREO). **Page 472: Red Crossbill** (*Adult Male*, Jim Zipp); (*Adult Female*, Brian E. Small); (*Immature Male*, Brian E. Small); (*Juvenile*, Brian E. Small; *Adult Male*, Garth McElroy/VIREO). **Page 473: Gray-crowned Rosy-Finch** (*Nonbreeding Adult, Interior*, Brian E. Small); (*Nonbreeding Adult, "Hepburn's"*, Bill Schmoker); (*Immature*, Bill Schmoker). **Brown-capped Rosy-Finch** (*Breeding Male*, Bill Schmoker); (*Nonbreeding Adult*, Bill Schmoker); (*Immature*, Brian E. Small). **Page 474: Black Rosy-Finch** (*Nonbreeding Adult*, Brian E. Small); (*Immature*, Brian E. Small). **Common Redpoll** (*Adult Male*, Jim Zipp); (*Adult Female*, Brian E. Small). **Page 475: Hoary Redpoll** (*Adult Male*, Jim Zipp); (*Adult Female*, Brian E. Small). **Pine Siskin** (*Adult Male*, Brian E. Small); (*Adult Female*, Jim Zipp). **Page 476: American Goldfinch** (*Breeding Male*, Brian E. Small); (*Breeding Female*, Brian E. Small); (*Nonbreeding Male*, Brian E. Small); (*Nonbreeding Female*, Brian E. Small). **European Goldfinch** (Adrian & Jane Binns/VIREO). **Page 477: Lesser Goldfinch** (*Adult Male, black-backed*, Brian E. Small); (*Adult Male, Green-backed*, Brian E. Small); (*Adult Female*, Brian E. Small). **Lawrence's Goldfinch** (*Breeding Male*, Brian E. Small); (*Breeding Female*, Brian E. Small); (*Adult Female*, Brian E. Small). **Page 478: Evening Grosbeak** (*Adult Male*, Brian E. Small); (*Adult Female*, Brian E. Small). **Brambling** (*Nonbreeding Male*, David Tipling/VIREO); (*Adult Female*, Robin Chittenden/VIREO). **Page 479: House Sparrow** (*Breeding Male*, Brian E. Small); (*Adult Female*, Brian E. Small); (*Nonbreeding Male*, Bill Schmoker). **Page 480: Eurasian Tree Sparrow** (*Adult*, Rob Curtis/VIREO); (*Juvenile*, Martin Meyers/VIREO). **Orange Bishop** (Johann Schumacher/VIREO). **Nutmeg Mannikin** (Martin Hale/VIREO). **Zebra Finch** (Roger Brown/VIREO).

DVD Booklet

Mallard, American Wigeon, Northern Bobwhite, Gambel's Quail, Common Loon, Black-crowned Night-Heron, Virginia Rail, Sandhill Crane, Black-bellied Plover, Greater Yellowlegs, Least Sandpiper, Short-billed Dowitcher, Wilson's Snipe, Forster's Tern, Common Nighthawk, Wild Turkey, Wood Duck, American Coot, Solitary Sandpiper, Spotted Sandpiper, Swainson's Hawk, Eastern Screech-Owl, Great Horned Owl, Mourning Dove, Ring-billed Gull, Western Screech-Owl, Yellow-billed Cuckoo, Pileated Woodpecker, Northern Flicker, Anna's Hummingbird, Belted Kingfisher, Hairy Woodpecker, Red-bellied Woodpecker, Red-headed Woodpecker, Rufous Hummingbird, Western Grebe, Western Wood-Pewee, Least Flycatcher, Willow Flycatcher, Common Raven, Ash-throated Flycatcher, Bell's Vireo, Blue Jay, Blue-headed Vireo, Cassin's Vireo, Eastern Kingbird, Great Crested Flycatcher, Plumbeous Vireo, Red-eyed Vireo, Warbling Vireo, Tree Swallow, Cliff Swallow, Townsend's Solitaire, Wrentit, American Pipit, Barn Swallow, Bewick's Wren, Blue-gray Gnatcatcher, European Starling, Golden-crowned Kinglet, Horned Lark, Marsh Wren, Northern Mockingbird, Ruby-crowned Kinglet, Swainson's Thrush, Verdin, Black-capped Chickadee, Brown Creeper, Carolina Wren, Cedar Waxwing, Gray Catbird, Northern Rough-winged Swallow, White-breasted Nuthatch, Wood Thrush, American Redstart, Common Yellowthroat, Orange-crowned Warbler, Pine Warbler, Yellow Warbler, Yellow-breasted Chat, Yellow-rumped Warbler, Eastern Wood-Pewee, Chipping Sparrow, Dark-eyed Junco, Lincoln's Sparrow, Savannah Sparrow, Scarlet Tanager, Song Sparrow, Western Tanager, Vesper Sparrow, White-crowned Sparrow, Clay-colored Sparrow, Lark Bunting, Lapland Longspur (Brian E. Small); Chimney Swift (Mike Danzenbaker); Brown-headed Cowbird, American Goldfinch, Baltimore Oriole, Black-headed Grosbeak, Bullock's Oriole, Common Grackle, Dickcissel, Eastern Meadowlark, House Sparrow, Indigo Bunting, Lazuli Bunting, Northern Cardinal, Pine Siskin, Red Crossbill, Western Meadowlark, Rose-breasted Grosbeak (Brian E. Small); Bobolink, Bushtit, American Crow, Eastern Bluebird, Northern Waterthrush, Sora, American Kestrel, Broad-winged Hawk, Osprey, Red-tailed Hawk (Jim Zipp); Cooper's Hawk, Pied-billed Grebe, Ring-necked Pheasant, Killdeer, House Finch (Kevin T. Karlson); American Woodcock (Richard Crossley/VIREO); Scissor-tailed Flycatcher (R. Nussbaumer/VIREO); Tufted Titmouse (Jim Zipp); Barn Owl (Rick and Nora Bowers/VIREO); House Wren (Arthur Morris/VIREO); Red-winged Blackbird (Greg Lasley/VIREO); American Tree Sparrow (Rob Curtis/VIREO); Fox Sparrow (James M. Wedge/VIREO); Eastern Towhee, Spotted Towhee, American Robin (Robert Royse); McCown's Longspur (Greg Lasley/VIREO).

Contributor Websites

Mike Danzenbaker (www.avesphoto.com)
Kevin T. Karlson (www.kevinkarlsonphotography.com)
Garth McElroy (www.featheredfotos.com)
Brian Patteson (www.patteson.com)
Jeff Poklen (www.pbase.com/jpkln)
Robert Royse (www.roysephotos.com)
Bill Schmoker (www.schmoker.org)
Brian E. Small (www.briansmallphoto.com)
VIREO (Visual Resources for Ornithology),
　　The Academy of Natural Sciences (www.vireo.acnatsci.org)
Jim Zipp (www.jimzipp.com)

Smithsonian Field Guide to the Birds of North America Birdsong DVD

The *Smithsonian Guide* includes a DVD with 587 MP3 sound files of birdsongs and vocalizations for 138 bird species found in North America. Created by Lang Elliot of NatureSound Studio with Kevin Colver and Ted Mack, the files represent years of expert fieldwork and sound engineering to provide the finest quality digitally mastered birdsong recordings available to the public. The *Smithsonian Guide* is the first field guide to combine comprehensive photographic identification with a DVD of birdsongs and vocalizations. The combination of text, photographs, and maps in the printed guide along with the birdsongs and vocalizations on the DVD will give users a powerful suite of tools to assist in learning to identify birds by both sight and sound.

The selection of species in this collection of vocalizations is based on coverage of a variety of species, the complexity of songs, calls, notes, and other vocalizations, and the beauty of some of the songs. This DVD is by no means a comprehensive collection of songs and other vocalizations for North American birds. However, it provides an excellent representative variety of vocalizations for each selected species.

Users of the *Smithsonian Guide* may easily transfer the audio files to a computer with a DVD drive and copy them to a portable MP3 player to carry wherever they go birding. The popular iTunes® and Windows Media Player® programs (each available for both Macintosh and PC computers) are excellent tools for organizing and playing MP3 sound files. Many other software programs that play MP3 files are available to download free from the Internet for both Macintosh computers and PCs. A search on the Internet will reveal many choices that are suitable for a particular operating system and compatible computers and portable MP3 players.

The *Smithsonian Field Guide to the Birds of North America* Birdsong disk is formatted as a data/audio DVD. It contains over five and one-half hours of playing time and nearly a gigabyte of data. It will not work in an audio CD player or a CD-ROM drive that is common in older personal computers.

Associated with each MP3 file on the DVD is a digital color image of the species of bird as "album art." With the sound files on a computer or on a portable MP3 player, such as the new generation of iPods®, iPod Nano®, and iPhone®, users will have both digitally mastered sound and a crisp color photographic image of the birds. Consult the information that accompanies software programs and hardware devices mentioned above for instructions on how to display album art.

The MP3 audio files on the DVD are organized in separate folders by the name of the bird and a standard six-letter abbreviation. Each MP3 recording on the DVD is named with the text abbreviation of the bird's name, a number to indicate the species "track," a description of the vocalization, and in many cases, the state or province where it was recorded. For example, the Chipping Sparrow folder on the DVD contains the following MP3 files:

CHISPA_1. fast song NY
CHISPA_2. slow song NY
CHISPA_3. chattery song ND
CHISPA_4. dawn song (stereo) CO
CHISPA_5. dawn song (stereo) NY
CHISPA_6. alarm calls MI
CHISPA_7. interaction calls NY

The vocalizations of many species vary greatly among geographic regions and even among individual birds within the species. Some of the MP3s on the disk include songs from different regions of North America to enable identification of special vocal "dialects" in various areas of the continent.

Some files on the *Smithsonian* Birdsong DVD are labeled stereo, monaural, or binaural. These special types of recordings deliver an expanded dimension to the experience of listening to birdsongs. With good speakers, and especially with quality headphones, all the birdsongs and vocalizations included on the DVD give users the opportunity to focus on learning and enjoying some of the most beautiful sounds in nature.

In the printed field guide, a symbol ◀ preceding the voice description text in a species account indicates that a selection of vocalizations for that species is available on the DVD.

The following alphabetical index lists the corresponding text page for each species, including the abbreviation of the name included on the DVD. The folder insert in the DVD pouch in the back of this book includes thumbnail images of the bird species on the DVD, abbreviations of the species name, track listings, and page references for the vocalization tracks arranged by their order of appearance in the *Smithsonian Guide*.

SPECIES	ABBREVIATION	PAGE
American Coot	AMECOO	143
American Crow	AMECRO	319
American Goldfinch	AMEGOL	476
American Kestrel	AMEKES	134
American Pipit	AMEPIP	374
American Redstart	AMERED	399
American Robin	AMEROB	358
American Tree Sparrow	AMETRSP	423
American Wigeon	AMEWIG	38
American Woodcock	AMEWOO	180
Anna's Hummingbird	ANNHUM	258
Ash-throated Flycatcher	ASTHFL	293
Baltimore Oriole	BALORI	463
Barn Owl	BARNOW	239
Barn Swallow	BARSWA	325
Bell's Vireo	BELVIR	307
Belted Kingfisher	BELKIN	265
Bewick's Wren	BEWWRE	343
Black-bellied Plover	BLBEPL	150
Black-capped Chickadee	BLCACH	331
Black-crowned Night-Heron	BCNIHE	109
Black-headed Grosbeak	BLHEGR	447
Blue Jay	BLUJAY	312
Blue-gray Gnatcatcher	BLGRGN	350
Blue-headed Vireo	BLHEVI	309
Bobolink	BOBOLI	454
Broad-winged Hawk	BRWIHA	128
Brown Creeper	BROCRE	340
Brown-headed Cowbird	BRHECO	457
Bullock's Oriole	BULORI	463

SPECIES	ABBREVIATION	PAGE	SPECIES	ABBREVIATION	PAGE
Bushtit	BUSHTI	337	Least Sandpiper	LEASAN	171
Carolina Wren	CARWRE	343	Lincoln's Sparrow	LINSPA	435
Cassin's Vireo	CASVIR	310	Mallard	MALLAR	39
Cedar Waxwing	CEDWAX	376	Marsh Wren	MARWRE	344
Chimney Swift	CHISWI	253	McCown's Longspur	MCCLON	442
Chipping Sparrow	CHISPA	421	Mourning Dove	MOUDOV	227
Clay-colored Sparrow	CLCOSP	422	Northern Bobwhite	NORBOB	68
Cliff Swallow	CLISWA	326	Northern Cardinal	NORCAR	446
Common Grackle	COMGRA	458	Northern Flicker	NORFLI	280
Common Loon	COMLOO	72	Northern Mockingbird	NORMOC	365
Common Nighthawk	COMNIG	248	Northern Waterthrush	NORWAT	401
Common Raven	COMRAV	321	Northern Rough-winged Swallow	NRWISW	327
Common Yellowthroat	COMYEL	402	Orange-crowned Warbler	ORCRWA	380
Cooper's Hawk	COOHAW	121	Osprey	OSPREY	118
Dark-eyed Junco	DAEYJU	438	Pied-billed Grebe	PIBIGR	76
Dickcissel	DICKCI	452	Pileated Woodpecker	PILWOO	281
Eastern Bluebird	EASBLU	354	Pine Siskin	PINSIS	475
Eastern Kingbird	EASKIN	295	Pine Warbler	PINWAR	396
Eastern Meadowlark	EASMEA	462	Plumbeous Vireo	PLUVIR	309
Eastern Screech-Owl	EASCOW	240	Red Crossbill	REBEWO	472
Eastern Towhee	EASTOW	415	Red-bellied Woodpecker	REDCRO	274
Eastern Wood-Pewee	EAWOPE	289	Red-eyed Vireo	REEYVI	304
European Starling	EURSTA	370	Red-headed Woodpecker	REHEWO	272
Forster's Tern	FORTER	198	Red-tailed Hawk	RETAHA	130
Fox Sparrow	FOXSPA	432	Red-winged Blackbird	REWIBL	460
Gambel's Quail	GMQUA	69	Ring-billed Gull	RIBIGU	184
Golden-crowned Kinglet	GOCRKI	349	Ring-necked Pheasant	RINEPH	59
Gray Catbird	GRACAT	369	Rose-breasted Grosbeak	ROBRGR	448
Great Crested Flycatcher	GRCRFL	292	Ruby-crowned Kinglet	RUCRKI	349
Great Horned Owl	GRHOOW	239	Rufous Hummingbird	RUFHUM	256
Greater Yellowlegs	GREYEL	158	Sandhill Crane	SANCRA	144
Hairy Woodpecker	HAIWOO	269	Savannah Sparrow	SAVSPA	427
Horned Lark	HORLAR	323	Scarlet Tanager	SCATAN	410
House Finch	HOUFIN	469	Scissor-tailed Flycatcher	SCTAFL	299
House Sparrow	HOUSPA	479	Short-billed Dowitcher	SHBIDO	177
House Wren	HOUWRE	342	Solitary Sandpiper	SOLSAN	157
Indigo Bunting	INDBUN	450	Song Sparrow	SONSPA	433
Killdeer	KILLDE	146	Sora	SORA	140
Lapland Longspur	LAPLON	441	Spotted Sandpiper	SPOSAN	156
Lark Bunting	LARBUN	425	Spotted Towhee	SPOTOW	416
Lazuli Bunting	LAZBUN	449	Swainson's Hawk	SWAHAW	127
Least Flycatcher	LEAFLY	285	Swainson's Thrush	SWATHR	361

SPECIES	ABBREVIATION	PAGE
Townsend's Solitaire	TOWSOL	356
Tree Swallow	TRESWA	328
Tufted Titmouse	TUFTIT	334
Verdin	VERDIN	336
Vesper Sparrow	VESSPA	426
Virginia Rail	VIRRAI	140
Warbling Vireo	WARVIR	306
Western Grebe	WESGRE	79
Western Meadowlark	WESMEA	462
Western Screech-Owl	WESCOW	240
Western Tanager	WESTAN	411
Western Wood-Pewee	WEWOPE	289
White-breasted Nuthatch	WHBRNU	339
White-crowned Sparrow	WHCRSP	437
Wild Turkey	WILTUR	67
Willow Flycatcher	WILFLY	284
Wilson's Snipe	WILSNI	179
Wood Duck	WOODUC	37
Wood Thrush	WOOTHR	362
Wrentit	WRENTI	363
Yellow Warbler	YELWAR	386
Yellow-billed Cuckoo	YEBICU	234
Yellow-breasted Chat	YEBRCH	408
Yellow-rumped Warbler	YERUWA	391

The track names of the sound files on the DVD include state and/or province and territory name abbreviations for the continental U.S. and Canada that indicate where the recording was made. Here is the standard list of these abbreviations for both countries.

CONTINENTAL UNITED STATES OF AMERICA

AL	Alabama	NE	Nebraska
AK	Alaska	NV	Nevada
AZ	Arizona	NH	New Hampshire
AR	Arkansas	NJ	New Jersey
CA	California	NM	New Mexico
CO	Colorado	NY	New York
CT	Connecticut	NC	North Carolina
DE	Delaware	ND	North Dakota
FL	Florida	OH	Ohio
GA	Georgia	OK	Oklahoma
ID	Idaho	OR	Oregon
IL	Illinois	PA	Pennsylvania
IN	Indiana	RI	Rhode Island
IA	Iowa	SC	South Carolina
KS	Kansas	SD	South Dakota
KY	Kentucky	TN	Tennessee
LA	Louisiana	TX	Texas
ME	Maine	UT	Utah
MD	Maryland	VT	Vermont
MA	Massachusetts	VA	Virginia
MI	Michigan	WA	Washington
MN	Minnesota	WV	West Virginia
MS	Mississippi	WI	Wisconsin
MO	Missouri	WY	Wyoming
MT	Montana		

CANADIAN PROVINCES AND TERRITORIES

AB	Alberta	NU	Nunavut
BC	British Columbia	ON	Ontario
MB	Manitoba	PE	Prince Edward Island
NB	New Brunswick	QC	Quebec
NL	Newfoundland & Labrador	SK	Saskatchewan
NT	Northwest Territories	YT	Yukon
NS	Nova Scotia		

Glossary of Terms

This glossary includes brief explanations of a selection of terms that are used in the *Smithsonian Guide*. Each entry is followed by a species name or section and page number where the term appears in the text. Many of the terms here are used throughout the guide.

Accidental. Occurring so rarely away from the normal range that there is no clear pattern of geographic distribution. LESSER SAND-PLOVER, P. 150

Adult. Final plumage cycle; sometimes called "definitive" plumage in technical literature. BLACK-THROATED SPARROW, P. 424

Alternate Plumage. Plumage other than basic plumage ("alternates" with basic plumage). Usually, but not always, corresponds to breeding plumage. BLACK OYSTERCATCHER, P. 153

Auriculars. Ear covert feathers or "ear patch," a region on the side of the face that often stands out from surrounding feathers. STILT SANDPIPER, P. 176

Austral. Pertaining to the Southern Hemisphere; used especially for seabirds in reference to breeding locations. FLESH-FOOTED SHEARWATER, P. 89

Axillaries. Underwing coverts closest to the body; informally called "wingpits" or "armpits." KING EIDER, P. 48

Backcross. Second-generation hybrid with one hybrid parent and one purebred parent. INDIGO BUNTING, P. 450

Bare Parts. Unfeathered parts, including the eyes, bill, and feet. KILLDEER, P. 146

Basic Plumage. A bird's baseline plumage, acquired by way of a complete molt of all feathers, usually in late summer or early fall; all birds have a basic plumage, which may or may not alternate annually with an alternate plumage. MEXICAN CHICKADEE, P. 322

Boreal Forest (Boreal Zone). Subarctic ecological zone (south of the tundra) characterized by harsh winters and consisting primarily of conifer forests. LINCOLN'S SPARROW, P. 435

Belly. Underparts below the breast and in front of the vent. BROWN JAY, P. 315

Breast. Underparts below the throat and above the belly. MONK PARAKEET, P. 230

Brood Parasite. A bird that lays its eggs in the nest of another bird. SHINY COWBIRD, P. 456

Call Note. Usually simple vocalization such as a *chip* or *tseep*; often different from the flight call of a species. KENTUCKY WARBLER, P. 405

Canopy. Upper levels of tree foliage, where many caterpillars and other insects feed. BALTIMORE ORIOLE, P. 463

Carpal Bar. Patch on the upperwing, generally appearing as a long stripe, created by contrast between greater coverts and the rest of the wing. LEACH'S STORM-PETREL, P. 91

Casual. Having an established, but not annual, pattern of occurrence. AZTEC THRUSH, P. 357

Cere. Fleshy area at the base of the upper mandible. ZONE-TAILED HAWK, P. 132

Chick. Downy young. GAMBEL'S QUAIL, P. 69

Chihuahuan Desert. Easternmost North American Desert, characterized by summer rains and expanses of creosote bush. SCALED QUAIL, P. 70

Clinal. Showing gradual variation across a region; clinal variation often mirrors an environmental gradient such as aridity or elevation. SONG SPARROW, P. 433

Colony (Colonial). Assemblage of nests often placed extremely close to one another. CLIFF SWALLOW, P. 326

Complex Alternate Strategy. Having two molts per plumage cycle (per year), plus an additional molt in the first plumage cycle (first year of life). CHESTNUT-SIDED WARBLER, P. 384

Complex Basic Strategy. Having one molt per plumage cycle (per year), plus an additional molt in the first plumage cycle (first year of life). YELLOW-BILLED CUCKOO, P. 234

Coverts. Groups of feathers that overlap the primaries and secondaries on the wings and the main feathers of the tail. SOOTY TERN, P. 204

Primary Coverts. Upperwing tract of feathers at the base of the primaries; prominent in soaring flight, but largely concealed at rest. COMMON GROUND-DOVE, P. 228

Secondary Coverts. Array of several upperwing feather tracts at the base of the secondaries; coverts' arrangement and contrast with the secondaries often create wing bars. DUSKY-CAPPED FLYCATCHER, P. 294

Undertail Coverts. Tract of feathers, sometimes called the "crissum," between the vent and the base of the flight feathers of the tail. PACIFIC GOLDEN-PLOVER, P. 151

Underwing Coverts. Array of several feather tracts on the underside of the wings; invisible at rest, but sometimes prominent in flight. AMERICAN ROBIN, P. 358

Uppertail Coverts. Tract of feathers lying over the base of the flight feathers of the tail. BLACK-FOOTED ALBATROSS, P. 81

Upperwing Coverts. The primary coverts plus the secondary coverts. TRICOLORED HERON, P. 108

Crissum. Area extending from the tail to the vent, including the undertail coverts. RED-WHISKERED BULBUL, P. 347

Culmen. Ridge of the upper mandible (bill). PYRRHULOXIA, P. 447

Cycle. Period of molt; used particularly in descriptions of molt strategies in gulls. RING-BILLED GULL, P. 184

Decurved. Curving downward. LUCIFER HUMMINGBIRD, P. 262

Dimorphic. Having two distinct morphs. REDDISH EGRET, P. 108

Display. Ritualized behavior that communicates information; often but not always associated with courtship, territorial defense, and protection of young. BRISTLE-THIGHED CURLEW, P. 163

Diurnal. Active during daytime. BURROWING OWL, P. 243

Eastern Deciduous Forest. Diverse woodland that dominates New England and the mid-Atlantic, plus parts of the Midwest and Southeast; recovering from deforestation. WOOD THRUSH, P. 362

Eclipse. "Female-like" plumage of male ducks, briefly held during the summer months; although drab, this plumage is actually a duck's alternate plumage. CINNAMON TEAL, P. 41

Endangered. Legal designation of a species or subspecies as being in immediate risk of extinction or extirpation. RED-COCKADED WOODPECKER, P. 269

Escape (Escapee). Individual bird that originated as a captive. WHOOPER SWAN, P. 36

Established. Population of an exotic (non-native) bird species that is breeding and self-sustaining in the wild. BLACK-HOODED PARAKEET, P. 230

Eyebrow. Slightly arched line of feathers lying directly above the eye; often contrasts with adjacent regions; sometimes called "supercilium." PALM WARBLER, P. 388

Eye Ring. Thin, contrasting circle around the eye. Eye rings are feathered, but orbital rings are bare. LEAST FLYCATCHER, P. 285

Eye Stripe. Thin line of feathers extending both in front of and behind the eye; sometimes called "transoccipital line." CHIPPING SPARROW, P. 421

Exotic. Non-native; occurring by deliberate release or accidental escape from captivity. CHILEAN FLAMINGO, P. 115

Extirpated. Regionally but not globally extinct; sometimes termed "regionally extinct." SHARP-TAILED GROUSE, P. 60

Family. In taxonomy, a group of genera that are believed to share a common evolutionary ancestor. ARCTIC WARBLER, P. 351

Feral. Wild bird or population of birds that originated in captivity. MUSCOVY DUCK, P. 37

Flank. Narrow area between the belly and the wing, visible on a perched bird below the folded wing. TUFTED TITMOUSE, P. 334

Fledge (Fledgling). To acquire the first "true" set of feathers enabling flight; also, to leave the nest as a fledgling after acquiring these feathers. INTRODUCTION P. 21

Flight Call. Vocalization typically given in flight; often different from a bird's call note. Many nocturnal migrants can be identified by their flight calls. GRAY-CHEEKED THRUSH, P. 359

Flight Feathers. Large feathers of the wings and tail, technically known as the remiges (wings) and rectrices (tail), that propel a bird in flight. SWAINSON'S HAWK, P. 127

Gadfly Petrel. A seabird in the genus *Pterodroma.* HERALD PETREL, P. 83

Gape. Area where the upper bill (maxilla) and lower bill (mandible) meet. SAY'S PHOEBE, P. 291

Gonydeal Angle. Pointed protrusion along the lower mandible; can be valuable for identification of gulls and jaegers. PARASITIC JAEGER, P. 210

Gorget. Feathers on the throat and chin of a hummingbird; typically dazzling and iridescent. BLACK-CHINNED HUMMINGBIRD, P. 257

Great Basin. Very arid region of mountains and high valleys; bounded on the west by the Sierra Nevada and on the east by the southern Rockies. FLAMMULATED OWL, P. 241

Great Plains. Broad north–south band in mid-continent bounded on the west by the Rockies and on the east by the Eastern deciduous forest; dominated by diverse grassland ecosystems. LAPLAND LONGSPUR, P. 441

Ground Color. Overall color of a bird, which often contrasts with streaking, spotting, splotching, and other patterns. SPRUCE GROUSE, P. 62

Group. In taxonomy, an assemblage of closely related subspecies. YELLOW-RUMPED WARBLER, P. 391

Hybrid. Offspring of a mixed-species pair. BLUE-WINGED WARBLER, P. 383

Immature. Age-class between juvenile and adult; distinctive in some species, but not evident or difficult to recognize in many other species. GOLDEN-CROWNED SPARROW, P. 435

Insectivorous. Eating insects and other arthropods, such as spiders. TOWNSEND'S WARBLER, P. 393

Intermountain West. Huge, roughly square-shaped region extending west to east from the Sierra Nevada to the Rockies, north to south from Canadian border to Mexican border. ORANGE-CROWNED WARBLER, P. 380

Introgression. Mixing of the genes of two populations (including separate species) where their ranges meet. AMERICAN BLACK DUCK, P. 39

Irruption. Greater-than-average dispersal, usually away from breeding grounds; often triggered by a regional reduction in food supply. PURPLE FINCH, P. 470

Juvenile. Bird that has fledged into its first coat (plumage) of "true" feathers. NORTHERN MOCKINGBIRD, P. 365

Kleptoparasitism. Stealing of food by one bird from another. BROWN BOOBY, P. 97

Lek. Site where males gather and perform courtship displays to attract females; usually refers to grouse, although other birds show lek-type behavior. SHARP-TAILED GROUSE, P. 60

Leucistic. Having abnormally white or pale feathers; sometimes called "partially albino" or "albinistic." WHITE-CROWNED SPARROW, P. 437

Lores (Loral). Stripe or patch extending from a bird's eye to the base of its bill; may be either feathered or bare. GREAT EGRET, P. 106

Malar. Region where the lower portion of face abuts the throat; if conspicuous, appears as a stripe or patch extending diagonally downward from the base of the bill. CASSIN'S KINGBIRD, P. 296

Mandible. One half of the bill. SMOOTH-BILLED ANI, P. 236

Lower Mandible. Lower half of the bill. NORTHERN PARULA, P. 385

Upper Mandible. Upper half of the bill, sometimes called "maxilla." COMMON LOON, P. 72

Mantle. The back, scapulars, and upper-wing. RING-BILLED GULL, P. 184

Microhabitat. Fine-scale differences in plant species composition, water conditions, and type of soil that can influence the occurrence and distribution of birds. ROCK WREN, P. 345

Midstory. Middle sections of a tree or forest layer; area between the canopy and the understory. BLACK-THROATED BLUE WARBLER, P. 398

Mob (Mobbing). Aggressive behavior, usually in flight, of a smaller species of bird toward a larger bird or other agressor. EASTERN KINGBIRD, P. 295

Mojave Desert. Hot, arid desert region occupying parts of Nevada, California, and Arizona; most precipitation falls in the winter. COSTA'S HUMMINGBIRD, P. 259

Molt. Regular replacement of one set of feathers by another; distinct from feather wear and bleaching. Birds molt from one plumage to another. HOUSE SPARROW, P. 479

Molt Strategy. Regular temporally- or seasonally-based pattern of molts and plumages. MASKED DUCK, P. 56

Complex Alternate Strategy. Having two molts per plumage cycle (per year), plus an additional molt in the first plumage cycle (first year of life). CHESTNUT-SIDED WARBLER, P. 384

Complex Basic Strategy. Having one molt per plumage cycle (per year), plus an additional molt in the first plumage cycle (first year of life). YELLOW-BILLED CUCKOO, P. 234

Simple Alternate Strategy. Having two molts per plumage cycle (per year). ANHINGA, P. 102

Simple Basic Strategy. Having one molt per plumage cycle (per year). CORY'S SHEARWATER, P. 86

Morph. Regularly occurring genetic variant, usually described in birds as a particular color form. SNOW GOOSE, P. 32

Nape. Region above the back and below the crown; informally, the back of the neck. GOLDEN EAGLE, P. 124

Nine-primaried oscines. Subgroup of presumably related passerines that have nine primary feathers on each wing.

Nuchal. Referring to the nape. NORTHERN FLICKER, P. 280

Obligate. Necessary; required for a species to survive. TURKEY VULTURE, P. 117

Orbital Ring. Thin circle of bare skin around the eye. BLACK-BILLED CUCKOO, P. 234

Order. In taxonomy, a group of families that are believed to share a common evolutionary ancestor. ELEGANT TROGON, P. 264

Pacific Slope. Westernmost sliver of North America, extending from the Alaskan Panhandle southward, and extending inland to the crests of the Sierra Nevada, Cascades, and Canadian Rockies. YELLOW-BILLED MAGPIE, P. 318

Passerine. A perching bird belonging to the order Passeriformes: in North America, all species from the tyrant-flycatchers through the weaver-finches. INTRODUCTION, P. 11

Peep(s). Informal term used generally for smaller shorebirds in the genus *Calidris*. LEAST SANDPIPER, P. 171

Pelagic. Offshore ocean waters. ARCTIC TERN, P. 199

Polymorphic. Having two or more morphs; when only two morphs, usually called "dimorphic." BEWICK'S WREN, P. 343

Position Note. In *Empidonax* flycatchers, a simple vocalization used to indicate territory (but sometimes heard on migration). CORDILLERAN FLYCATCHER, P. 287

Plumage. Coat of feathers, usually molted (replaced) one or two times per year after a bird's first year of life. YELLOW WARBLER, P. 386

Alternate Plumage. Plumage other than basic plumage ("alternates" with basic plumage). Usually, but not always, corresponds to breeding plumage. BLACK OYSTERCATCHER, P. 153

Basic Plumage. Baseline plumage; all birds have a basic plumage, which may or may not alternate annually with an alternate plumage. MEXICAN CHICKADEE, P. 332

Plumage Cycle. Annual sequence of molts and plumages from one prebasic molt (usually in the fall) to the next prebasic molt. BAY-BREASTED WARBLER, P. 396

Prairie. Extensive expanse of grass; usually used in reference to native grasslands. CLAY-COLORED SPARROW, P. 422

Midgrass Prairie. Grassland of intermediate height; occurs at the interface between shortgrass and tallgrass prairie. SPRAGUE'S PIPIT, P. 373

Shortgrass Prairie. Imperiled ecosystem of southwestern Great Plains; dominant vegetation is low grasses. LESSER PRAIRIE-CHICKEN, P. 61

Tallgrass Prairie. Extensive ecosystem covering much of the northern Great Plains and extending east to the Great Lakes; dominant vegetation is high grasses. GREATER PRAIRIE-CHICKEN, P. 60

Prebasic Molt. Complete molt of all feathers, typically in late summer or early fall, which results in basic plumage. DUSKY FLYCATCHER, P. 286

Precocial. Having young that are active and semi-independent almost immediately after hatching. RED-NECKED GREBE, P. 78

Primaries. Outer flight feathers of the wings, typically 9–11; often the longest and most prominent feathers. FERRUGINOUS HAWK, P. 125

Primary Coverts. Upperwing tract of feathers at the base of the primaries; prominent in soaring flight, but largely concealed at rest. COMMON GROUND-DOVE, P. 228

Primary Projection. On a bird with its wings folded, the extension of the primaries beyond the tips of the tertials; birds with long primary projections often migrate longer distances than birds with short primary projections. YELLOW-THROATED VIREO, P. 308

Primary Tips. Outer extent of primary feathers, often darker than the rest of the primaries.

BLACK-LEGGED KITTIWAKE, P. 196

Riparian. Terrestrial habitat along the edges of seeps, streams, and rivers. VERDIN, P. 336

Rufescent. Tinged with reddish-brown. KING RAIL, P. 139

Rump. Portion of upperparts between the back and uppertail coverts. CAVE SWALLOW, P. 327

Scapulars. Feather tracts along the edges of the upper back; often prominent on birds at rest. WESTERN BLUEBIRD, P. 355

Secondaries. Inner flight feathers of the wings. LESSER SCAUP, P. 47

Secondary Coverts. Array of several upperwing feather tracts at the base of the secondaries. DUSKY-CAPPED FLYCATCHER, P. 294

Simple Alternate Strategy. Having two molts per plumage cycle (per year). ANHINGA, P. 102

Simple Basic Strategy. Having one molt per plumage cycle (per year). CORY'S SHEARWATER, P. 86

Sky islands. Isolated mountain ranges of the Desert Southwest. ARIZONA WOODPECKER, P. 271

Song. Often-elaborate vocalization used to attract mates or establish territories; typically given by males, but some females sing. OLIVE SPARROW, P. 414

Sonoran Desert. Hot region of southern Arizona and northwestern Mexico, characterized by cacti and other succulents; receives rain in both summer and winter. RUFOUS-WINGED SPARROW, P. 420

Speculum. Imprecise term to denote a colorful patch of feathers on the wings; most often refers to colorful secondaries on waterfowl. MALLARD, P. 39

Subadult. Catchall term for plumages—in most instances, juvenile and immature—that precede adult plumage. BALD EAGLE, P. 123

Subspecies. Geographically discrete breeding population that is distinguishable from other populations on the basis of plumage or other characters, but that is not reproductively isolated. FOX SPARROW, P. 432

Superspecies. Complex of species that have recently radiated from a common ancestor; they share many features and often hybridize. MOTTLED DUCK, P. 40

Supraloral. Region lying directly above the lores; usually extends from just above the eye to the base of the upper mandible. VIRGINIA RAIL, P. 140

Taiga. Widely used as a synonym for boreal spruce-fir forest below the tundra; more specifically used for the region where the forest merges with the tundra in an open, swampy area of scattered conifers and tundra vegetation. WHIMBREL, P. 162

Taxonomy (Taxonomic). The science of naming and classifying birds according to standardized rules based on Linnaeus' system of nomenclature. SOUTH POLAR SKUA, P. 208

Tertials. Innermost flight feathers, sometimes prominent on birds at rest. SHORT-BILLED DOWITCHER, P. 177

Tundra. Treeless habitat with year-round ice beneath the soil; there are three major classes of tundra. GREATER WHITE-FRONTED GOOSE, P. 31

Alpine Tundra. Treeless habitat atop high mountains. WHITE-TAILED PTARMIGAN, P. 65

Dry Tundra. Treeless arctic habitat, usually rocky. HOARY REDPOLL, P. 475

Wet Tundra. Treeless arctic habitat, usually with extensive ponds and lakes in low-lying regions. YELLOW-BILLED LOON, P. 73

Undertail Coverts. Tract of feathers, sometimes called the "crissum," between the vent and the base of the flight feathers of the tail. PACIFIC GOLDEN-PLOVER, P. 151

Underwing Coverts. Array of several feather tracts on the underside of the wings; invisible at rest, but sometimes prominent in flight. AMERICAN ROBIN, P. 358

Uppertail Coverts. Tract of feathers lying over the base of the flight feathers of the tail. BLACK-FOOTED ALBATROSS, P. 81

Upperwing Coverts. The primary coverts plus the secondary coverts. TRICOLORED HERON, P. 108

Vagrant (Vagrancy). Bird that has wandered from its primary range; the occurrence of many vagrants is somewhat predictable and repeatable. ASH-THROATED FLYCATCHER, P. 293

Vent. Portion of a bird's underparts below its belly and above its undertail coverts. RED-BELLIED WOODPECKER, P. 274

Wear (Worn). Feathers of birds become abraded, bleached by sunlight, and break down due to age and other environmental causes; these feathers may appear different when compared to "fresh" newly-grown feathers. BLACK-FOOTED ALBATROSS, P. 81

Wing Bar. Stripe on upperwing created by contrasting tips of secondary coverts. Wing bars often occur in pairs. HUTTON'S VIREO, P. 307

Wing Length. Measurement of a single outstretched wing from tip of the wing to where the wing joins the body of the bird at its shoulder.

Wingspan. Measurement from wing tip to wing tip of the outstretched wings of a bird in flight.

Metric Conversions

U.S. measurements for average body length (bill tip to tail tip), wingspan, and body weight are included in the main species accounts in the *Smithsonian Guide*. Gram weight equivalents are also included for birds that weigh one ounce or less. Here is a guide for converting U.S. to metric measurements.

1 ounce = 28.3 grams

1 pound = 0.45 kilograms

1 inch = 25.4 millimeters

1 inch = 2.5 centimeters

1 foot (12 inches) = 0.3 meters

American Birding Association Checklist

The American Birding Association (ABA) Checklist includes native North American breeding species, regular visitors, casuals and accidentals from other regions that are believed to have strayed here without direct human aid, and well-established introduced species that are now part of our avifauna.

Revisions are published annually in the ABA's *Birding* Magazine and updated regularly at the ABA website (www.aba.org/checklist).

Each species on the Checklist is assigned a code number that is included here after each entry. The status of each species is summarized in the ABA Checklist as follows:

Code 1. Occurs widely and regularly in North America.

Code 2. Regular within a restricted region of North America.

Code 3. Rare but regular; occurs annually, usually in small numbers.

Code 4. Casual; well-defined pattern of occurrence, but not annual.

Code 5. Accidental; one or a few records with no defined pattern.

Code 6. Extinct or otherwise impossible to observe in the wild.

Ducks, Geese, and Swans

- [ ] Black-bellied Whistling-Duck 1
- [ ] Fulvous Whistling-Duck 1
- [ ] Bean Goose 3
- [ ] Pink-footed Goose 4
- [ ] Greater White-fronted Goose 1
- [ ] Lesser White-fronted Goose 5
- [ ] Emperor Goose 2
- [ ] Snow Goose 1
- [ ] Ross's Goose 1
- [ ] Brant 1
- [ ] Barnacle Goose 5
- [ ] Cackling Goose 1
- [ ] Canada Goose 1
- [ ] Mute Swan 1
- [ ] Trumpeter Swan 1
- [ ] Tundra Swan 1
- [ ] Whooper Swan 3
- [ ] Muscovy Duck 3
- [ ] Wood Duck 1
- [ ] Gadwall 1
- [ ] Falcated Duck 4
- [ ] Eurasian Wigeon 3
- [ ] American Wigeon 1
- [ ] American Black Duck 1
- [ ] Mallard 1
- [ ] Mottled Duck 1
- [ ] Spot-billed Duck 5
- [ ] Blue-winged Teal 1
- [ ] Cinnamon Teal 1
- [ ] Northern Shoveler 1
- [ ] White-cheeked Pintail 4
- [ ] Northern Pintail 1
- [ ] Garganey 3
- [ ] Baikal Teal 4
- [ ] Green-winged Teal 1
- [ ] Canvasback 1
- [ ] Redhead 1
- [ ] Common Pochard 3
- [ ] Ring-necked Duck 1
- [ ] Tufted Duck 3
- [ ] Greater Scaup 1
- [ ] Lesser Scaup 1
- [ ] Steller's Eider 2
- [ ] Spectacled Eider 2
- [ ] King Eider 1
- [ ] Common Eider 1
- [ ] Harlequin Duck 1
- [ ] Labrador Duck 6
- [ ] Surf Scoter 1
- [ ] White-winged Scoter 1
- [ ] Black Scoter 1
- [ ] Long-tailed Duck 1
- [ ] Bufflehead 1
- [ ] Common Goldeneye 1
- [ ] Barrow's Goldeneye 1
- [ ] Smew 3
- [ ] Hooded Merganser 1
- [ ] Common Merganser 1
- [ ] Red-breasted Merganser 1
- [ ] Masked Duck 3
- [ ] Ruddy Duck 1

Curassows and Guans

- [ ] Plain Chachalaca 2

Partridges, Grouse, Turkeys, and Old World Quail

- [ ] Chukar 2
- [ ] Himalayan Snowcock 2
- [ ] Gray Partridge 2
- [ ] Ring-necked Pheasant 1
- [ ] Ruffed Grouse 1
- [ ] Greater Sage-Grouse 1
- [ ] Gunnison Sage-Grouse 2
- [ ] Spruce Grouse 1
- [ ] Willow Ptarmigan 1
- [ ] Rock Ptarmigan 1
- [ ] White-tailed Ptarmigan 2
- [ ] Dusky Grouse 2
- [ ] Sooty Grouse 2
- [ ] Sharp-tailed Grouse 2

☐ Greater Prairie-Chicken 2
☐ Lesser Prairie-Chicken 2
☐ Wild Turkey 1

New World Quail

☐ Mountain Quail 1
☐ Scaled Quail 1
☐ California Quail 1
☐ Gambel's Quail 1
☐ Northern Bobwhite 1
☐ Montezuma Quail 2

Loons

☐ Red-throated Loon 1
☐ Arctic Loon 2
☐ Pacific Loon 1
☐ Common Loon 1
☐ Yellow-billed Loon 2

Grebes

☐ Least Grebe 2
☐ Pied-billed Grebe 1
☐ Horned Grebe 1
☐ Red-necked Grebe 1
☐ Eared Grebe 1
☐ Western Grebe 1
☐ Clark's Grebe 1

Albatrosses

☐ Yellow-nosed Albatross 4
☐ Shy Albatross 4
☐ Black-browed Albatross 5
☐ Wandering Albatross 5
☐ Laysan Albatross 2
☐ Black-footed Albatross 1
☐ Short-tailed Albatross 3

Shearwaters and Petrels

☐ Northern Fulmar 1
☐ Great-winged Petrel 5
☐ Herald Petrel 3
☐ Murphy's Petrel 3
☐ Mottled Petrel 3
☐ Bermuda Petrel 4
☐ Black-capped Petrel 2
☐ Galapagos/Hawaiian ("Dark-rumped") Petrel* 5
☐ Fea's Petrel 3
☐ Cook's Petrel 2
☐ Stejneger's Petrel 4
☐ Bulwer's Petrel 5
☐ Streaked Shearwater 4
☐ Cory's Shearwater 1
☐ Cape Verde Shearwater 5
☐ Pink-footed Shearwater 1

☐ Flesh-footed Shearwater 3
☐ Greater Shearwater 1
☐ Wedge-tailed Shearwater 4
☐ Buller's Shearwater 2
☐ Sooty Shearwater 1
☐ Short-tailed Shearwater 2
☐ Manx Shearwater 2
☐ Black-vented Shearwater 2
☐ Audubon's Shearwater 1
☐ Little Shearwater 5

Storm-Petrels

☐ Wilson's Storm-Petrel 1
☐ White-faced Storm-Petrel 4
☐ European Storm-Petrel 5
☐ Black-bellied Storm-Petrel 5
☐ Fork-tailed Storm-Petrel 2
☐ Ringed Storm-Petrel 5
☐ Leach's Storm-Petrel 1
☐ Ashy Storm-Petrel 2
☐ Band-rumped Storm-Petrel 2
☐ Wedge-rumped Storm-Petrel 5
☐ Black Storm-Petrel 2
☐ Least Storm-Petrel 2

Tropicbirds

☐ White-tailed Tropicbird 3
☐ Red-billed Tropicbird 3
☐ Red-tailed Tropicbird 3

Boobies and Gannets

☐ Masked Booby 3
☐ Blue-footed Booby 4
☐ Brown Booby 3
☐ Red-footed Booby 4
☐ Northern Gannet 1

Pelicans

☐ American White Pelican 1
☐ Brown Pelican 1

Cormorants

☐ Brandt's Cormorant 1
☐ Neotropic Cormorant 1
☐ Double-crested Cormorant 1
☐ Great Cormorant 1
☐ Red-faced Cormorant 2
☐ Pelagic Cormorant 1

Darters

☐ Anhinga 1

Frigatebirds

☐ Magnificent Frigatebird 1
☐ Great Frigatebird 4
☐ Lesser Frigatebird 5

Bitterns, Herons, and Allies

☐ American Bittern 1
☐ Yellow Bittern 5
☐ Least Bittern 1
☐ Great Blue Heron 1
☐ Great Egret 1
☐ Chinese Egret 5
☐ Little Egret 4
☐ Western Reef-Heron 5
☐ Snowy Egret 1
☐ Little Blue Heron 1
☐ Tricolored Heron 1
☐ Reddish Egret 1
☐ Cattle Egret 1
☐ Chinese Pond-Heron 5
☐ Green Heron 1
☐ Black-crowned Night-Heron 1
☐ Yellow-crowned Night-Heron 1

Ibises and Spoonbills

☐ White Ibis 1
☐ Scarlet Ibis 5
☐ Glossy Ibis 1
☐ White-faced Ibis 1
☐ Roseate Spoonbill 1

Storks

☐ Jabiru 5
☐ Wood Stork 1

New World Vultures

☐ Black Vulture 1
☐ Turkey Vulture 1
☐ California Condor 6

Flamingos

☐ Greater Flamingo 3

Hawks, Kites, Eagles, and Allies

☐ Osprey 1
☐ Hook-billed Kite 3
☐ Swallow-tailed Kite 1

☐ White-tailed Kite 1
☐ Snail Kite 2
☐ Mississippi Kite 1
☐ Bald Eagle 1
☐ White-tailed Eagle 4
☐ Steller's Sea-Eagle 4
☐ Northern Harrier 1
☐ Sharp-shinned Hawk 1
☐ Cooper's Hawk 1
☐ Northern Goshawk 1
☐ Crane Hawk 5
☐ Common Black-Hawk 2
☐ Harris's Hawk 1
☐ Roadside Hawk 5
☐ Red-shouldered Hawk 1
☐ Broad-winged Hawk 1
☐ Gray Hawk 2
☐ Short-tailed Hawk 2
☐ Swainson's Hawk 1
☐ White-tailed Hawk 2
☐ Zone-tailed Hawk 2
☐ Red-tailed Hawk 1
☐ Ferruginous Hawk 1
☐ Rough-legged Hawk 1
☐ Golden Eagle 1

Caracaras and Falcons

☐ Collared Forest-Falcon 5
☐ Crested Caracara 1
☐ Eurasian Kestrel 4
☐ American Kestrel 1
☐ Merlin 1
☐ Eurasian Hobby 4
☐ Red-footed Falcon 5
☐ Aplomado Falcon 4
☐ Gyrfalcon 2
☐ Peregrine Falcon 1
☐ Prairie Falcon 1

Rails, Gallinules, and Coots

☐ Yellow Rail 2
☐ Black Rail 2
☐ Corn Crake 5
☐ Clapper Rail 1
☐ King Rail 1
☐ Virginia Rail 1
☐ Sora 1
☐ Paint-billed Crake 5
☐ Spotted Rail 5
☐ Purple Gallinule 1
☐ Common Moorhen 1
☐ Eurasian Coot 5
☐ American Coot 1

Limpkins

☐ Limpkin 2

Cranes

☐ Sandhill Crane 1
☐ Common Crane 4
☐ Whooping Crane 2

Thick-knees

☐ Double-striped Thick-knee 5

Lapwings and Plovers

☐ Northern Lapwing 4
☐ Black-bellied Plover 1
☐ European Golden-Plover 4
☐ American Golden-Plover 1
☐ Pacific Golden-Plover 2
☐ Lesser Sand-Plover 3
☐ Greater Sand-Plover 5
☐ Collared Plover 5
☐ Snowy Plover 1
☐ Wilson's Plover 1
☐ Common Ringed Plover 2
☐ Semipalmated Plover 1
☐ Piping Plover 2
☐ Little Ringed Plover 5
☐ Killdeer 1
☐ Mountain Plover 2
☐ Eurasian Dotterel 4

Oystercatchers

☐ Eurasian Oystercatcher 5
☐ American Oystercatcher 1
☐ Black Oystercatcher 1

Stilts and Avocets

☐ Black-winged Stilt 5
☐ Black-necked Stilt 1
☐ American Avocet 1

Jacanas

☐ Northern Jacana 4

Sandpipers, Phalaropes, and Allies

☐ Terek Sandpiper 3
☐ Common Sandpiper 3
☐ Spotted Sandpiper 1
☐ Green Sandpiper 4
☐ Solitary Sandpiper 1
☐ Gray-tailed Tattler 3
☐ Wandering Tattler 1
☐ Spotted Redshank 4
☐ Greater Yellowlegs 1

☐ Common Greenshank 3
☐ Willet 1
☐ Lesser Yellowlegs 1
☐ Marsh Sandpiper 5
☐ Wood Sandpiper 2
☐ Common Redshank 5
☐ Upland Sandpiper 2
☐ Little Curlew 4
☐ Eskimo Curlew 6
☐ Whimbrel 1
☐ Bristle-thighed Curlew 2
☐ Far Eastern Curlew 4
☐ Slender-billed Curlew 6
☐ Eurasian Curlew 4
☐ Long-billed Curlew 1
☐ Black-tailed Godwit 3
☐ Hudsonian Godwit 1
☐ Bar-tailed Godwit 2
☐ Marbled Godwit 1
☐ Ruddy Turnstone 1
☐ Black Turnstone 1
☐ Surfbird 1
☐ Great Knot 4
☐ Red Knot 1
☐ Sanderling 1
☐ Semipalmated Sandpiper 1
☐ Western Sandpiper 1
☐ Red-necked Stint 3
☐ Little Stint 4
☐ Temminck's Stint 3
☐ Long-toed Stint 3
☐ Least Sandpiper 1
☐ White-rumped Sandpiper 1
☐ Baird's Sandpiper 1
☐ Pectoral Sandpiper 1
☐ Sharp-tailed Sandpiper 3
☐ Purple Sandpiper 1
☐ Rock Sandpiper 2
☐ Dunlin 1
☐ Curlew Sandpiper 3
☐ Stilt Sandpiper 1
☐ Spoon-billed Sandpiper 4
☐ Broad-billed Sandpiper 4
☐ Buff-breasted Sandpiper 1
☐ Ruff 3
☐ Short-billed Dowitcher 1
☐ Long-billed Dowitcher 1
☐ Jack Snipe 5
☐ Wilson's Snipe 1
☐ Common Snipe 3
☐ Pin-tailed Snipe 5
☐ Eurasian Woodcock 5
☐ American Woodcock 1
☐ Wilson's Phalarope 1

☐ Red-necked Phalarope 1
☐ Red Phalarope 1

Pratincoles

☐ Oriental Pratincole 5

Gulls, Terns, and Skimmers

☐ Laughing Gull 1
☐ Franklin's Gull 1
☐ Little Gull 3
☐ Black-headed Gull 3
☐ Bonaparte's Gull 1
☐ Heermann's Gull 1
☐ Gray-hooded Gull 5
☐ Belcher's Gull 5
☐ Black-tailed Gull 4
☐ Mew Gull 1
☐ Ring-billed Gull 1
☐ California Gull 1
☐ Herring Gull 1
☐ Yellow-legged Gull 4
☐ Thayer's Gull 2
☐ Iceland Gull 2
☐ Lesser Black-backed Gull 3
☐ Slaty-backed Gull 3
☐ Yellow-footed Gull 2
☐ Western Gull 1
☐ Glaucous-winged Gull 1
☐ Glaucous Gull 1
☐ Great Black-backed Gull 1
☐ Kelp Gull 4
☐ Sabine's Gull 1
☐ Black-legged Kittiwake 1
☐ Red-legged Kittiwake 2
☐ Ross's Gull 3
☐ Ivory Gull 3
☐ Brown Noddy 2
☐ Black Noddy 3
☐ Sooty Tern 2
☐ Bridled Tern 2
☐ Aleutian Tern 2
☐ Least Tern 1
☐ Large-billed Tern 5
☐ Gull-billed Tern 1
☐ Caspian Tern 1
☐ Black Tern 1
☐ White-winged Tern 4
☐ Whiskered Tern 5
☐ Roseate Tern 2
☐ Common Tern 1
☐ Arctic Tern 1
☐ Forster's Tern 1
☐ Royal Tern 1
☐ Sandwich Tern 1

☐ Elegant Tern 1
☐ Black Skimmer 1

Skuas and Jaegers

☐ Great Skua 3
☐ South Polar Skua 2
☐ Pomarine Jaeger 1
☐ Parasitic Jaeger 1
☐ Long-tailed Jaeger 1

Auks, Murres, and Puffins

☐ Dovekie 2
☐ Common Murre 1
☐ Thick-billed Murre 1
☐ Razorbill 1
☐ Great Auk 6
☐ Black Guillemot 1
☐ Pigeon Guillemot 1
☐ Long-billed Murrelet 4
☐ Marbled Murrelet 1
☐ Kittlitz's Murrelet 2
☐ Xantus's Murrelet 2
☐ Craveri's Murrelet 3
☐ Ancient Murrelet 2
☐ Cassin's Auklet 1
☐ Parakeet Auklet 2
☐ Least Auklet 2
☐ Whiskered Auklet 2
☐ Crested Auklet 2
☐ Rhinoceros Auklet 1
☐ Atlantic Puffin 1
☐ Horned Puffin 1
☐ Tufted Puffin 1

Pigeons and Doves

☐ Rock Pigeon 1
☐ Scaly-naped Pigeon 5
☐ White-crowned Pigeon 1
☐ Red-billed Pigeon 3
☐ Band-tailed Pigeon 1
☐ Oriental Turtle-Dove 4
☐ Eurasian Collared-Dove 1
☐ Spotted Dove 1
☐ White-winged Dove 1
☐ Zenaida Dove 5
☐ Mourning Dove 1
☐ Passenger Pigeon 6
☐ Inca Dove 1
☐ Common Ground-Dove 1
☐ Ruddy Ground-Dove 3
☐ White-tipped Dove 2
☐ Key West Quail-Dove 4
☐ Ruddy Quail-Dove 5

Lories, Parakeets, Macaws, and Parrots

- ☐ Budgerigar   3
- ☐ Monk Parakeet   2
- ☐ Carolina Parakeet   6
- ☐ Green Parakeet   2
- ☐ Thick-billed Parrot   6
- ☐ White-winged Parakeet   2
- ☐ Red-crowned Parrot   2

Cuckoos, Roadrunners, and Anis

- ☐ Common Cuckoo   3
- ☐ Oriental Cuckoo   4
- ☐ Yellow-billed Cuckoo   1
- ☐ Mangrove Cuckoo   2
- ☐ Black-billed Cuckoo   1
- ☐ Greater Roadrunner   1
- ☐ Smooth-billed Ani   3
- ☐ Groove-billed Ani   2

Barn Owls

- ☐ Barn Owl   1

Typical Owls

- ☐ Flammulated Owl   2
- ☐ Oriental Scops-Owl   5
- ☐ Western Screech-Owl   1
- ☐ Eastern Screech-Owl   1
- ☐ Whiskered Screech-Owl   2
- ☐ Great Horned Owl   1
- ☐ Snowy Owl   2
- ☐ Northern Hawk Owl   2
- ☐ Northern Pygmy-Owl   2
- ☐ Ferruginous Pygmy-Owl   3
- ☐ Elf Owl   2
- ☐ Burrowing Owl   1
- ☐ Mottled Owl   5
- ☐ Spotted Owl   2
- ☐ Barred Owl   1
- ☐ Great Gray Owl   2
- ☐ Long-eared Owl   2
- ☐ Stygian Owl   5
- ☐ Short-eared Owl   2
- ☐ Boreal Owl   2
- ☐ Northern Saw-whet Owl   2

Goatsuckers

- ☐ Lesser Nighthawk   1
- ☐ Common Nighthawk   1
- ☐ Antillean Nighthawk   3
- ☐ Common Pauraque   2
- ☐ Common Poorwill   1
- ☐ Chuck-will's-widow   1

- ☐ Buff-collared Nightjar   4
- ☐ Whip-poor-will   1
- ☐ Gray Nightjar   5

Swifts

- ☐ Black Swift   2
- ☐ White-collared Swift   4
- ☐ Chimney Swift   1
- ☐ Vaux's Swift   1
- ☐ White-throated Needletail   5
- ☐ Common Swift   5
- ☐ Fork-tailed Swift   4
- ☐ White-throated Swift   1
- ☐ Antillean Palm-Swift   5

Hummingbirds

- ☐ Green Violet-ear   4
- ☐ Green-breasted Mango   4
- ☐ Broad-billed Hummingbird   2
- ☐ White-eared Hummingbird   3
- ☐ Xantus's Hummingbird   5
- ☐ Berylline Hummingbird   3
- ☐ Buff-bellied Hummingbird   2
- ☐ Cinnamon Hummingbird   5
- ☐ Violet-crowned Hummingbird   2
- ☐ Blue-throated Hummingbird   2
- ☐ Magnificent Hummingbird   2
- ☐ Plain-capped Starthroat   4
- ☐ Bahama Woodstar   5
- ☐ Lucifer Hummingbird   2
- ☐ Ruby-throated Hummingbird   1
- ☐ Black-chinned Hummingbird   1
- ☐ Anna's Hummingbird   1
- ☐ Costa's Hummingbird   1
- ☐ Calliope Hummingbird   1
- ☐ Bumblebee Hummingbird   5
- ☐ Broad-tailed Hummingbird   1
- ☐ Rufous Hummingbird   1
- ☐ Allen's Hummingbird   1

Trogons

- ☐ Elegant Trogon   2
- ☐ Eared Quetzal   4
- ☐ Eurasian Hoopoe   5

Kingfishers

- ☐ Ringed Kingfisher   2
- ☐ Belted Kingfisher   1
- ☐ Green Kingfisher   2

Woodpeckers and Allies

- ☐ Eurasian Wryneck   5
- ☐ Lewis's Woodpecker   1
- ☐ Red-headed Woodpecker   1
- ☐ Acorn Woodpecker   1
- ☐ Gila Woodpecker   1
- ☐ Golden-fronted Woodpecker   1
- ☐ Red-bellied Woodpecker   1
- ☐ Williamson's Sapsucker   1
- ☐ Yellow-bellied Sapsucker   1
- ☐ Red-naped Sapsucker   1
- ☐ Red-breasted Sapsucker   1
- ☐ Great Spotted Woodpecker   4
- ☐ Ladder-backed Woodpecker   1
- ☐ Nuttall's Woodpecker   1
- ☐ Downy Woodpecker   1
- ☐ Hairy Woodpecker   1
- ☐ Arizona Woodpecker   2
- ☐ Red-cockaded Woodpecker   2
- ☐ White-headed Woodpecker   2
- ☐ American Three-toed Woodpecker   2
- ☐ Black-backed Woodpecker   2
- ☐ Northern Flicker   1
- ☐ Gilded Flicker   2
- ☐ Pileated Woodpecker   1
- ☐ Ivory-billed Woodpecker   6

Tyrant-Flycatchers

- ☐ Northern Beardless-Tyrannulet   2
- ☐ Greenish Elaenia   5
- ☐ Caribbean Elaenia   5
- ☐ Tufted Flycatcher   5
- ☐ Olive-sided Flycatcher   1
- ☐ Greater Pewee   2
- ☐ Western Wood-Pewee   1
- ☐ Eastern Wood-Pewee   1
- ☐ Cuban Pewee   5
- ☐ Yellow-bellied Flycatcher   1
- ☐ Acadian Flycatcher   1
- ☐ Alder Flycatcher   1
- ☐ Willow Flycatcher   1

- ☐ Least Flycatcher   1
- ☐ Hammond's Flycatcher   1
- ☐ Gray Flycatcher   1
- ☐ Dusky Flycatcher   1
- ☐ Pacific-slope Flycatcher   1
- ☐ Cordilleran Flycatcher   1
- ☐ Buff-breasted Flycatcher   2
- ☐ Black Phoebe   1
- ☐ Eastern Phoebe   1
- ☐ Say's Phoebe   1
- ☐ Vermilion Flycatcher   1
- ☐ Dusky-capped Flycatcher   2
- ☐ Ash-throated Flycatcher   1
- ☐ Nutting's Flycatcher   5
- ☐ Great Crested Flycatcher   1
- ☐ Brown-crested Flycatcher   1
- ☐ La Sagra's Flycatcher   4
- ☐ Great Kiskadee   2
- ☐ Social Flycatcher   5
- ☐ Sulphur-bellied Flycatcher   2
- ☐ Piratic Flycatcher   5
- ☐ Variegated Flycatcher   5
- ☐ Tropical Kingbird   2
- ☐ Couch's Kingbird   2
- ☐ Cassin's Kingbird   2
- ☐ Thick-billed Kingbird   2
- ☐ Western Kingbird   1
- ☐ Eastern Kingbird   1
- ☐ Gray Kingbird   2
- ☐ Scissor-tailed Flycatcher   1
- ☐ Fork-tailed Flycatcher   4
- ☐ Rose-throated Becard   3
- ☐ Masked Tityra   5

Shrikes

- ☐ Brown Shrike   4
- ☐ Loggerhead Shrike   1
- ☐ Northern Shrike   1

Vireos

- ☐ White-eyed Vireo   1
- ☐ Thick-billed Vireo   5
- ☐ Bell's Vireo   1
- ☐ Black-capped Vireo   2
- ☐ Gray Vireo   2
- ☐ Yellow-throated Vireo   1
- ☐ Plumbeous Vireo   1
- ☐ Cassin's Vireo   1
- ☐ Blue-headed Vireo   1
- ☐ Hutton's Vireo   1
- ☐ Warbling Vireo   1
- ☐ Philadelphia Vireo   1
- ☐ Red-eyed Vireo   1

Yellow-green Vireo 3
Black-whiskered Vireo 2
Yucatan Vireo 5

Jays and Crows

Gray Jay 1
Steller's Jay 1
Blue Jay 1
Green Jay 2
Brown Jay 3
Florida Scrub-Jay 2
Island Scrub-Jay 2
Western Scrub-Jay 1
Mexican Jay 2
Pinyon Jay 1
Clark's Nutcracker 1
Black-billed Magpie 1
Yellow-billed Magpie 2
Eurasian Jackdaw 4
American Crow 1
Northwestern Crow 1
Tamaulipas Crow 3
Fish Crow 1
Chihuahuan Raven 1
Common Raven 1

Larks

Sky Lark 2
Horned Lark 1

Swallows

Purple Martin 1
Cuban Martin 5
Gray-breasted Martin 5
Southern Martin 5
Brown-chested Martin 5
Tree Swallow 1
Mangrove Swallow 5
Violet-green Swallow 1
Bahama Swallow 4
Northern Rough-winged Swallow 1
Bank Swallow 1
Cliff Swallow 1
Cave Swallow 1
Barn Swallow 1
Common House-Martin 4

Chickadees and Titmice

Carolina Chickadee 1
Black-capped Chickadee 1
Mountain Chickadee 1
Mexican Chickadee 2
Chestnut-backed Chickadee 1

Boreal Chickadee 1
Gray-headed Chickadee 3
Bridled Titmouse 2
Oak Titmouse 1
Juniper Titmouse 1
Tufted Titmouse 1
Black-crested Titmouse 2

Verdin

Verdin 1

Bushtits

Bushtit 1

Nuthatches

Red-breasted Nuthatch 1
White-breasted Nuthatch 1
Pygmy Nuthatch 1
Brown-headed Nuthatch 1

Creepers

Brown Creeper 1

Wrens

Cactus Wren 1
Rock Wren 1
Canyon Wren 1
Carolina Wren 1
Bewick's Wren 1
House Wren 1
Winter Wren 1
Sedge Wren 1
Marsh Wren 1

Dippers

American Dipper 1

Bulbuls

Red-whiskered Bulbul 2

Kinglets

Golden-crowned Kinglet 1
Ruby-crowned Kinglet 1

Old World Warblers and Gnatcatchers

Middendorff's Grasshopper-Warbler 4
Lanceolated Warbler 5
Willow Warbler 5
Wood Warbler 5
Dusky Warbler 4
Yellow-browed Warbler 5
Arctic Warbler 2
Lesser Whitethroat 5

Blue-gray Gnatcatcher 1
California Gnatcatcher 2
Black-tailed Gnatcatcher 1
Black-capped Gnatcatcher 4

Old World Flycatchers

Narcissus Flycatcher 5
Mugimaki Flycatcher 5
Taiga Flycatcher 4
Dark-sided Flycatcher 4
Gray-streaked Flycatcher 4
Asian Brown Flycatcher 5
Spotted Flycatcher 5

Thrushes

Siberian Rubythroat 3
Bluethroat 2
Siberian Blue Robin 5
Red-flanked Bluetail 4
Northern Wheatear 2
Stonechat 4
Eastern Bluebird 1
Western Bluebird 1
Mountain Bluebird 1
Townsend's Solitaire 1
Orange-billed Nightingale-Thrush 5
Black-headed Nightingale-Thrush 5
Veery 1
Gray-cheeked Thrush 1
Bicknell's Thrush 2
Swainson's Thrush 1
Hermit Thrush 1
Wood Thrush 1
Eurasian Blackbird 5
Eyebrowed Thrush 3
Dusky Thrush 4
Fieldfare 4
Redwing 4
Clay-colored Robin 3
White-throated Robin 5
Rufous-backed Robin 3
American Robin 1
Varied Thrush 1
Aztec Thrush 4

Babblers

Wrentit 1

Mockingbirds and Thrashers

Gray Catbird 1
Northern Mockingbird 1
Bahama Mockingbird 4
Sage Thrasher 1
Brown Thrasher 1
Long-billed Thrasher 2
Bendire's Thrasher 2
Curve-billed Thrasher 1
California Thrasher 1
Crissal Thrasher 2
Le Conte's Thrasher 2
Blue Mockingbird 5

Starlings

European Starling 1

Accentors

Siberian Accentor 4

Wagtails and Pipits

Eastern Yellow Wagtail 1
Citrine Wagtail 5
Gray Wagtail 4
White Wagtail 2
Tree Pipit 5
Olive-backed Pipit 3
Pechora Pipit 4
Red-throated Pipit 3
American Pipit 1
Sprague's Pipit 2

Waxwings

Bohemian Waxwing 1
Cedar Waxwing 1

Silky-Flycatchers

Gray Silky-flycatcher 5
Phainopepla 1

Olive Warbler

Olive Warbler 2

Wood-Warblers

Bachman's Warbler 6
Blue-winged Warbler 1
Golden-winged Warbler 2
Tennessee Warbler 1
Orange-crowned Warbler 1
Nashville Warbler 1
Virginia's Warbler 1
Colima Warbler 2

- ☐ Lucy's Warbler 1
- ☐ Crescent-chested Warbler 5
- ☐ Northern Parula 1
- ☐ Tropical Parula 3
- ☐ Yellow Warbler 1
- ☐ Chestnut-sided Warbler 1
- ☐ Magnolia Warbler 1
- ☐ Cape May Warbler 1
- ☐ Black-throated Blue Warbler 1
- ☐ Yellow-rumped Warbler 1
- ☐ Black-throated Gray Warbler 1
- ☐ Golden-cheeked Warbler 2
- ☐ Black-throated Green Warbler 1
- ☐ Townsend's Warbler 1
- ☐ Hermit Warbler 1
- ☐ Blackburnian Warbler 1
- ☐ Yellow-throated Warbler 1
- ☐ Grace's Warbler 1
- ☐ Pine Warbler 1
- ☐ Kirtland's Warbler 3
- ☐ Prairie Warbler 1
- ☐ Palm Warbler 1
- ☐ Bay-breasted Warbler 1
- ☐ Blackpoll Warbler 1
- ☐ Cerulean Warbler 1
- ☐ Black-and-white Warbler 1
- ☐ American Redstart 1
- ☐ Prothonotary Warbler 1
- ☐ Worm-eating Warbler 1
- ☐ Swainson's Warbler 2
- ☐ Ovenbird 1
- ☐ Northern Waterthrush 1
- ☐ Louisiana Waterthrush 1
- ☐ Kentucky Warbler 1
- ☐ Connecticut Warbler 2
- ☐ Mourning Warbler 1
- ☐ MacGillivray's Warbler 1
- ☐ Common Yellowthroat 1
- ☐ Gray-crowned Yellowthroat 4
- ☐ Hooded Warbler 1
- ☐ Wilson's Warbler 1
- ☐ Canada Warbler 1
- ☐ Red-faced Warbler 2
- ☐ Painted Redstart 2
- ☐ Slate-throated Redstart 4
- ☐ Fan-tailed Warbler 4
- ☐ Golden-crowned Warbler 4
- ☐ Rufous-capped Warbler 4
- ☐ Yellow-breasted Chat 1

Bananaquits

- ☐ Bananaquit 4

Tanagers

- ☐ Hepatic Tanager 2
- ☐ Summer Tanager 1
- ☐ Scarlet Tanager 1
- ☐ Western Tanager 1
- ☐ Flame-colored Tanager 4
- ☐ Western Spindalis 4

Emberizids

- ☐ White-collared Seedeater 3
- ☐ Yellow-faced Grassquit 4
- ☐ Black-faced Grassquit 5
- ☐ Olive Sparrow 2
- ☐ Green-tailed Towhee 1
- ☐ Spotted Towhee 1
- ☐ Eastern Towhee 1
- ☐ Canyon Towhee 1
- ☐ California Towhee 1
- ☐ Abert's Towhee 1
- ☐ Rufous-winged Sparrow 2
- ☐ Cassin's Sparrow 1
- ☐ Bachman's Sparrow 2
- ☐ Botteri's Sparrow 2
- ☐ Rufous-crowned Sparrow 1
- ☐ Five-striped Sparrow 3
- ☐ American Tree Sparrow 1
- ☐ Chipping Sparrow 1
- ☐ Clay-colored Sparrow 1
- ☐ Brewer's Sparrow 1
- ☐ Field Sparrow 1
- ☐ Worthen's Sparrow 5
- ☐ Black-chinned Sparrow 1
- ☐ Vesper Sparrow 1
- ☐ Lark Sparrow 1
- ☐ Black-throated Sparrow 1
- ☐ Sage Sparrow 1
- ☐ Lark Bunting 1
- ☐ Savannah Sparrow 1
- ☐ Grasshopper Sparrow 1
- ☐ Baird's Sparrow 2
- ☐ Henslow's Sparrow 2
- ☐ Le Conte's Sparrow 1
- ☐ Nelson's Sharp-tailed Sparrow 1
- ☐ Saltmarsh Sharp-tailed Sparrow 1
- ☐ Seaside Sparrow 1
- ☐ Fox Sparrow 1
- ☐ Song Sparrow 1
- ☐ Lincoln's Sparrow 1

- ☐ Swamp Sparrow 1
- ☐ White-throated Sparrow 1
- ☐ Harris's Sparrow 1
- ☐ White-crowned Sparrow 1
- ☐ Golden-crowned Sparrow 1
- ☐ Dark-eyed Junco 1
- ☐ Yellow-eyed Junco 2
- ☐ McCown's Longspur 2
- ☐ Lapland Longspur 1
- ☐ Smith's Longspur 2
- ☐ Chestnut-collared Longspur 1
- ☐ Pine Bunting 5
- ☐ Little Bunting 4
- ☐ Rustic Bunting 3
- ☐ Yellow-throated Bunting 5
- ☐ Yellow-breasted Bunting 5
- ☐ Gray Bunting 5
- ☐ Pallas's Bunting 5
- ☐ Reed Bunting 4
- ☐ Snow Bunting 1
- ☐ McKay's Bunting 2

Cardinals, Saltators, and Allies

- ☐ Crimson-collared Grosbeak 4
- ☐ Northern Cardinal 1
- ☐ Pyrrhuloxia 1
- ☐ Yellow Grosbeak 4
- ☐ Rose-breasted Grosbeak 1
- ☐ Black-headed Grosbeak 1
- ☐ Blue Bunting 4
- ☐ Blue Grosbeak 1
- ☐ Lazuli Bunting 1
- ☐ Indigo Bunting 1
- ☐ Varied Bunting 2
- ☐ Painted Bunting 1
- ☐ Dickcissel 1

Blackbirds

- ☐ Bobolink 1
- ☐ Red-winged Blackbird 1
- ☐ Tricolored Blackbird 1
- ☐ Tawny-shouldered Blackbird 5
- ☐ Eastern Meadowlark 1
- ☐ Western Meadowlark 1
- ☐ Yellow-headed Blackbird 1
- ☐ Rusty Blackbird 1
- ☐ Brewer's Blackbird 1
- ☐ Common Grackle 1
- ☐ Boat-tailed Grackle 1
- ☐ Great-tailed Grackle 1

- ☐ Shiny Cowbird 2
- ☐ Bronzed Cowbird 1
- ☐ Brown-headed Cowbird 1
- ☐ Black-vented Oriole 5
- ☐ Orchard Oriole 1
- ☐ Hooded Oriole 1
- ☐ Streak-backed Oriole 4
- ☐ Bullock's Oriole 1
- ☐ Spot-breasted Oriole 2
- ☐ Altamira Oriole 2
- ☐ Audubon's Oriole 2
- ☐ Baltimore Oriole 1
- ☐ Scott's Oriole 1

Fringilline and Cardueline Finches and Allies

- ☐ Common Chaffinch 5
- ☐ Brambling 3
- ☐ Gray-crowned Rosy-Finch 1
- ☐ Black Rosy-Finch 2
- ☐ Brown-capped Rosy-Finch 2
- ☐ Pine Grosbeak 1
- ☐ Common Rosefinch 4
- ☐ Purple Finch 1
- ☐ Cassin's Finch 1
- ☐ House Finch 1
- ☐ Red Crossbill 1
- ☐ White-winged Crossbill 2
- ☐ Common Redpoll 1
- ☐ Hoary Redpoll 2
- ☐ Eurasian Siskin 5
- ☐ Pine Siskin 1
- ☐ Lesser Goldfinch 1
- ☐ Lawrence's Goldfinch 2
- ☐ American Goldfinch 1
- ☐ Oriental Greenfinch 4
- ☐ Eurasian Bullfinch 4
- ☐ Evening Grosbeak 1
- ☐ Hawfinch 4

Old World Sparrows

- ☐ House Sparrow 1
- ☐ Eurasian Tree Sparrow 2

* Formerly classified as a single species, Dark-rumped Storm-Petrel; now separated as two species, but which cannot be distinguished from each other in the field.

Index

Quick Index

See the Species Index for a complete listing of all birds in the *Smithsonian Guide*
by standard English and scientific names.